Endocrine Interactions of Insect Parasites and Pathogens

EXPERIMENTAL BIOLOGY REVIEWS

Series advisors:

D.W. Lawlor
AFRC Institute of Arable Crops Research, Rothamsted Experimental Station, Harpenden, Hertfordshire AL5 2JQ, UK

M. Thorndyke
School of Biological Sciences, Royal Holloway, University of London, Egham, Surrey TW20 0EX, UK

Environmental Stress and Gene Regulation
Sex Determination in Plants
Plant Carbohydrate Biochemistry
Programmed Cell Death in Animals and Plants – SEB Symposium Series Vol. 52
Biomechanics in Animal Behaviour
Cell and Molecular Biology of Wood Formation
Molecular Mechanisms of Metabolic Arrest
Environment and Animal Development: genes, life histories and plasticity
Brain Stem Cells – SEB Symposium Series Vol. 53
Endocrine Interactions of Insect Parasites and Pathogens

Forthcoming titles include:
Vertebrate Biomechanics and Evolution
Osmoregulation and Drinking in Vertebrates

Endocrine Interactions of Insect Parasites and Pathogens

J.P. EDWARDS
Invertebrate Physiology & Novel Pest Management Strategies, Central Science Laboratory, Sand Hutton, York, UK

R.J. WEAVER
Invertebrate Physiology & Novel Pest Management Strategies, Central Science Laboratory, Sand Hutton, York, UK

BIOS

First published in 2001

A CIP catalogue record for this book is available from the British Library.

ISBN 1 85996 217 3

BIOS Scientific Publishers Ltd
9 Newtec Place, Magdalen Road, Oxford OX4 1RE, UK.
Tel. +44 (0)1865 726286. Fax. +44 (0)1865 246823
World Wide Web home page: http://www.bios.co.uk/

Distributed exclusively in the United States, its dependent territories, Canada, Mexico, Central and South America, and the Caribbean by Springer-Verlag New York Inc, 175 Fifth Avenue, New York, USA, by arrangement with BIOS Scientific Publishers Ltd, 9 Newtec Place, Magdalen Road, Oxford OX4 1RE, UK.

Production Editor: Paul Barlass.
Typeset by Saxon Graphics Ltd, Derby, UK.
Printed by The Cromwell Press, Trowbridge, UK.

Front cover inset: Photomicrograph of the embryo of the braconid parasitoid *Meteorus gyrator* showing teratocytes in the serosal membrane (courtesy of Howard A Bell, CSL).

Contents

Contributors

Nancy E. Beckage, Departments of Entomology & Cell Biology and Neuroscience, 5429 Boyce Hall, University of California-Riverside, Riverside, CA 92521, USA.

Catherine Béliveau, Laurentian Forestry Centre, Canadian Forest Service, Natural Resources Canada, PO Box 3800, Sainte-Foy, Quebec, G1V 4C7, Canada and Département de biochimie, Pavillon Charles-Eugène Marchand, Université Laval, Sainte-Foy, Quebec, G1K 7P4, Canada.

Howard A. Bell, Central Science Laboratory, Department for Environment, Food and Rural Affairs, Sand Hutton, York YO41 1LZ, UK.

Guy Bellemare, Département de biochimie, Pavillon Charles-Eugène Marchand, Université Laval, Sainte-Foy, Quebec, G1K 7P4, Canada.

Jared Bramlett, Graduate Center for Toxicology, Chandler Medical Center, University of Kentucky, Lexington, KY 40536-0305, USA.

Sandra L. Brandt, USDA, ARS, Biological Control of Insects Research Laboratory, 1503 S. Providence Rd., Columbia, MO 65203, USA.

Jian-Hua Chen, School of Biological Sciences, University of Liverpool, Crown Street, Liverpool, L69 3GS, UK.

Petra Christiansen-Weniger, Aphid Biology Group, Department of Biology, Imperial College at Silwood Park, Ascot SL5 7PY, UK.

Fernando L. Consoli, Texas A&M University, College Station, TX, USA.

Jenny S. Cory, NERC Centre for Ecology and Hydrology – Oxford, Mansfield Road, Oxford OX1 3SR, UK.

Thomas A. Coudron, USDA, ARS, Biological Control of Insects Research Laboratory, 1503 S. Providence Rd., Columbia, MO 65203, USA.

Michel Cusson, Laurentian Forestry Centre, Canadian Forest Service, Natural Resources Canada, PO Box 3800, Sainte-Foy, Quebec, G1V 4C7, Canada.

Douglas L. Dahlman, Department of Entomology, University of Kentucky, Lexington, KY 40546-0091, USA.

John P. Edwards, Central Science Laboratory, Department for Environment, Food and Rural Affairs, Sand Hutton, York YO41 1LZ, UK.

Dale B. Gelman, Insect Biocontrol Laboratory, USDA/ARS, Bldg 306, Room 323, BARCE, 10300 Baltimore Avenue, Beltsville, Maryland 20705, USA.

Rosemary S. Hails, NERC Centre for Ecology and Hydrology – Oxford, Mansfield Road, Oxford OX1 3SR, UK.

Jim Hardie, Aphid Biology Group, Department of Biology, Imperial College at Silwood Park, Ascot SL5 7PY, UK.

Hilary Hurd, Centre for Applied Entomology and Parasitology, School of Life Sciences, Keele University, Keele, Staffordshire ST5 5BG, UK

Davy Jones, Graduate Center for Toxicology, Chandler Medical Center, University of Kentucky, Lexington, KY 40536-0305, USA.

Maureen Knop-Wright, USDA, ARS, Biological Control of Insects Research Laboratory, 1503 S. Providence Rd., Columbia, MO 65203, USA.

Marlène Laforge, Laurentian Forestry Centre, Canadian Forest Service, Natural Resources Canada, PO Box 3800, Sainte-Foy, Quebec, G1V 4C7, Canada and Département de biochimie, Pavillon Charles-Eugène Marchand, Université Laval, Sainte-Foy, Quebec, G1K 7P4, Canada.

Beatrice Lanzrein, Institute of Cell Biology, University of Berne, Baltzerstrasse 4, CH-3012 Bern, Switzerland.

Anic Levasseur, Laurentian Forestry Centre, Canadian Forest Service, Natural Resources Canada, PO Box 3800, Sainte-Foy, Quebec, G1V 4C7, Canada and Département de biochimie, Pavillon Charles-Eugène Marchand, Université Laval, Sainte-Foy, Quebec, G1K 7P4, Canada.

Carla Malva, Istituto Internazionale di Genetica e Biofisica, CNR, via Marconi 10, 80121 Napoli, Italy.

Gay C. Marris, Central Science Laboratory, Department for Environment, Food and Rural Affairs, Sand Hutton, York YO41 1LZ, UK.

David R. O'Reilly, Department of Biology, Imperial College of Science, Technology and Medicine, Imperial College Road, London SW7 2AZ, UK.

Francesco Pennacchio, Dipartimento di Biologia, Difesa e Biotecnologie Agro-Forestali, Università della Basilicata, Macchia Romana, 85100 Potenza, Italy and Istituto Internazionale di Genetica e Biofisica, CNR, via Marconi 10, 80121 Napoli, Italy.

Rita Pfister-Wilhelm, Institute of Cell Biology, University of Berne, Baltzerstrasse 4, CH-3012 Bern, Switzerland.

Rico L. Rana, Department of Entomology, University of Kentucky, Lexington, KY 40546-0091, USA.

Huw H. Rees, School of Biological Sciences, University of Liverpool, Crown Street, Liverpool L69 7ZB, UK.

Steven M. Sait, School of Biological Sciences, University of Liverpool, Brownlow Street, Liverpool L69 3GS, UK.

Don Stoltz, Department of Microbiology and Immunology, Dalhousie University, Halifax, Nova Scotia, B3H 4H7, Canada.

Hajime Takeuchi, School of Biological Sciences, University of Liverpool, Crown Street, Liverpool L69 7ZB, UK.

Philip C. Turner, School of Biological Sciences, University of Liverpool, Crown Street, Liverpool L69 7ZB, UK.

S. Bradleigh Vinson, Department of Entomology, Texas A&M University, College Station, TX 77843, USA.

Felix von Niederhäusern, Institute of Cell Biology, University of Berne, Baltzerstrasse 4, CH-3012 Bern, Switzerland.

Susanne Wache, Graduate Center for Toxicology, Chandler Medical Center, University of Kentucky, Lexington, KY 40536-0305, USA.

Robert J. Weaver, Central Science Laboratory, Department for Environment, Food and Rural Affairs, Sand Hutton, York YO41 1LZ, UK.

Bruce A. Webb, Department of Entomology, University of Kentucky, Lexington, KY 40546-0091, USA.

Katharine R. Wilson, NERC Centre for Ecology and Hydrology - Oxford, Mansfield Road, Oxford OX1 3SR, UK.

Abbreviations

CA	corpora allata
CC	corpora cardiaca
CCAP	crustacean cardioactive peptide
CGRP	calcitonin gene reactive protein
CsPDV	*C. sonorensis* polydnavirus
CSR	critical size ratio
2-D PAGE	two-dimensional polyacrylamide gel electrophoresis
20E	20-hydroxyecdysone
EcR	ecdysteroid receptor
EcRE	ecdysone response element
EGT	ecdysone glucosyltransferase
EH	eclosion hormone
ETH	ecdysis-triggering hormone
GBP	growth-blocking peptide
GC/MS	gas chromatography/mass spectrometry
HPLC	high-performance liquid chromatography
HTIF	haemocyte translation inhibiting factor
JH	juvenile hormone
JHE	juvenile hormone esterase
JHSP	juvenile hormone storage proteins
MRSP	methionine-rich storage proteins
NBAD	*N*-β-alanyldopamine
OB	occlusion body
ORF	open reading frame
PDV	polydnavirus
PETH	pre-ecdysis-triggering hormone
PG	prothoracic gland
PPO	prophenoloxidase
PTTH	prothoracicotropic hormone
RG	reniform gland
RIA	radioimmunoassay
RP-HPLC	reversed-phase high-performance liquid chromatography
SDS-PAGE	sodium dodecyl sulfate polyacrylamide gel electrophoresis
TSP	teratocyte secreted protein
UDPGT	UDP-glucuronyltransferase
UG	U-shaped gland
USP	ultraspiracle
Vg	vitellogenin
VHDL	very high density lipoproteins
VLP	virus-like particle

Preface

When parasites and pathogens of insects attack their hosts, they entrust their development and survival, and that of their progeny, to the environment of the host's body. In doing so, these organisms are subjected to the ongoing physiological processes that are consequence of the natural biochemical phenomena that are associated with normal growth, development and behaviour occurring in the host insect. These physiological processes include the natural "immune" responses of hosts to infection and, in particular, the critical changes in levels of various hormones that determine the onset and result of growth, moulting and metamorphosis. The invading organisms have either, to conform to the patterns determined by the host's own developmental programme(s) or, to take-over the physiology of the host and manipulate it to suit the needs of the parasite or pathogen. Thus, during the process of co-evolution, the struggle for existence between the parasites and pathogens and their hosts has led to the emergence of sophisticated systems used by the invading organisms to take control of the internal physiology of the host. Most spectacular among these are the strategies designed to manipulate the endocrine system of the host in order to regulate optimally the host's internal milieu to ensure the survival of the parasite or pathogen. The study of the mechanisms whereby these creatures modify the life processes of their hosts, not only provides insight into the co-evolutionary process, but also provides important leads for the discovery of new and specific ways of controlling insect pests.

In August 2000, the Society for Experimental Biology (SEB) held a meeting at Queen's College, Cambridge University, part of which was a symposium exploring the endocrine interactions of parasites and pathogens with their insect hosts. This symposium was attended by many of the leading researchers in the field, and this volume is a compendium of the reviews presented at the meeting. Taken together, the contents of this volume represent a summary of the current state of scientific knowledge concerning the endocrine interactions of insect pathogens and parasites with their hosts. We hope that these reviews will stimulate further research interest in this field, and lead to a greater and more detailed understanding of the mechanisms by which host endocrinology is modulated.

1

Endocrine changes in lepidopteran larvae: potential challenges to parasitoid development and survival

John P. Edwards, Robert J. Weaver and Gay C. Marris

1. Introduction

The parasitic Hymenoptera are a group of insects, containing many thousands of species, that parasitize insects and other arthropods and are collectively known as parasitoid wasps (see review by Godfray, 1994). Parasitoids attack a wide range of hemimetabolous and holometabolous insects, and the adult female wasps lay their eggs either in or on the host insect. From that point onwards, the parasitoids are constrained to complete their larval development in or on a single host – often in or on a single host developmental stage. Parasitoids of hemimetabolous insects are frequently less stage-specific than those that parasitize holometabolous groups. This may be because the nymphal and adult stages of hemimetabolous insects frequently share identical habitat, food material, and general behaviour. By contrast, parasitoids that attack holometabolous insects are often stage-specific (i.e. they attack and develop entirely in one host developmental stage, e.g. eggs, larvae, or pupae), although parasitoid development spanning more than one host developmental stage (e.g. egg-larval) is not unknown. Moreover, when parasitoids attack a particular host stage, they frequently show a distinct preference for hosts of a particular age within the selected developmental stage (e.g. a particular larval stadium).

The range of holometabolous hosts utilized by hymenopteran parasitoids extends to several insect orders including Coleoptera, Diptera, and even other Hymenoptera, but the casual observer might be forgiven for believing that parasitoids display 'an

Endocrine Interactions of Insect Parasites and Pathogens, edited by J.P. Edwards and R.J. Weaver.

inordinate fondness' for the larval stages of the Lepidoptera (butterflies and moths). In this chapter, we examine the cycles of changing endocrine events associated with growth, moulting and metamorphosis of lepidopteran insects, and consider these in relation to the challenges that they might pose to the development and survival of endo- and ectoparasitoids. We explore some of the strategies that might be employed by parasitoids to deal with the problems posed by host endocrine systems, and highlight some selected examples of the ways in which parasitoids manipulate the hormones of their lepidopteran hosts.

After an adult female parasitoid wasp has accomplished the various tasks associated with host location, host selection, and acceptance (Van Alphen and Vet, 1986; Vinson, 1976), she will deposit her eggs in or on the host. From this point onwards, the developing parasitoid eggs and larvae will face a range of physiological challenges that could, potentially, reduce their chances of survival. These challenges include ensuring adequate supplies of food and oxygen for the duration of the development of the parasitoid, as well as avoiding death due to disease, predation, or some form of secondary parasitism. However, it seems most likely that the two most dangerous potential threats to parasitoid survival are present in the forms of the various host 'immune' defence mechanisms, and the endogenous variations in host endocrinology that are crucial for regulating host development. The potential hazard from the host immune defence system is outside the scope of this chapter, which will concentrate specifically on the endocrine interactions of parasitic Hymenoptera with the developmental stages (i.e. eggs, larvae and pupae) of their lepidopteran hosts.

2. The endocrine system regulating development, moulting and metamorphosis in insects

2.1 *Introduction to insect hormone systems*

Hormones are internal, blood-borne, chemical signals that control changes in physiology, biochemistry and behaviour. They may modulate single, relatively rapid, highly-specific processes, or longer-term, general changes in internal physiology. Hormones usually act on target organs and tissues via second messenger systems such as cyclic AMP, and the production of the hormones themselves is often modulated by neurohormonal signals from the central nervous system (especially the brain). This complex interaction of environmental stimuli that initiates brain and nerve signals, leading to hormone production, receptor activation, and physiological effects on target organs, comprises the endocrine system of all living organisms.

Thus, like most other life forms, insects have a complex and extensive endocrine system. Almost every aspect of insect life, from embryonic development, through larval growth and metamorphosis, to adulthood, reproduction, and eventual death, is in some way controlled by hormone-regulated systems. More than 20 different hormones have been identified in insects during the past 50 years, and yet more hormones and hormone regulators are being discovered each year. The majority of insect hormones are produced either by neurosecretory cells in the insect central nervous system, especially the brain, or by specialized endocrine organs (corpora allata (CA), corpora cardiaca (CC), prothoracic glands (PG), etc.). The latter are invariably under the control of regulating hormones produced in the brain (e.g. allatostatins, prothoracicotropic hormone (PTTH)). The brain is the interface co-ordinating the internal physiology of the insect with changes in the outside environment (photoperiod, food supply, etc.) and with internal changes coincident with growth, reproduction, or other developmental changes (moulting, metamorphosis, reproduction, etc.).

The primary focus of this chapter is the endocrine interactions that may arise when the developing stages of parasitoid wasps are necessarily obliged to spend the greater part of their pre-adult life in close proximity to the tissues (and therefore the hormones) of the developing stages of their hosts. For this reason, our discussions will be concerned predominantly with those endocrine factors that are most crucially relevant to the successful growth, moulting and metamorphosis of developing insect stages. It is our view that the two most important insect hormone groups present (or absent) during development are the juvenile hormones (JH) and the ecdysteroid moulting hormones. Certainly, it seems probable, for reasons outlined below, that these two hormone groups are the most likely to pose difficulties for the development and survival of parasitoid eggs and larvae.

Parasitoids are insects that live in or on the bodies of other insects and, in doing so, place themselves in an environment (the host) that is a constantly changing hormonal milieu. Since the processes of growth, moulting and metamorphosis in parasitoids are also regulated by JH, ecdysteroids and other components of the insect endocrine system, the presence of these materials in the host tissues presents a continually changing, and potentially challenging environment for normal developmental programming of parasitoid eggs, larvae and pupae. Given this potential challenge, parasitoid larvae have evolved several life-styles to enable them to cope with the potentially 'unfriendly' endocrine environment presented by the host (Beckage, 1985; Lawrence, 1986a). Thus, parasitoids may complete their development in synchrony with changes in host endocrinology, or they may manipulate aspects of host endocrinology in order to defend themselves from the potential hazard, and to optimize their own survival (i.e. they conform, or they regulate; Lawrence, 1986a; Vinson and Iwantsch, 1980a).

2.2 *Insect development and the insect endocrine system*

Insect growth and development. An understanding of the general workings of the insect endocrine system is made easier by a consideration of the main developmental strategies employed by different insect groups. Basically, insects develop in one of two ways: they undergo incomplete metamorphosis (hemimetabolous development), or complete metamorphosis (holometabolous development). Since the organisms discussed in this chapter (moths and parasitoid wasps) are both holometabolous insects, we shall concentrate on the endocrine regulation of development in this sub-group.

After the eggs have hatched, holometabolous insects grow and develop through several larval stages (stadia), each stadium beginning and ending with a moult (ecdysis). During each larval stadium the growing insect must reach a critical size (Nijhout, 1975; Nijhout and Williams, 1974a, b), and its tissues must become responsive to the various developmental hormones (Nijhout and Wheeler, 1982) before the moult can take place. For the majority of larval development, the moulting hormones (Section 2.5) initiate moulting in the presence of JH (Section 2.3), and this combination ensures that the outcome of the moult will lead to another larval stage. However, in the last larval stadium, JH is largely absent, and this is the signal that, in response to moulting hormone, insect development is directed towards pupation and metamorphosis. Thus, at the end of the final larval stadium, the insect moults again to form an 'intermediate' stage (the pupa). During the pupal stage, the adult organs and tissues are formed by the differentiation of small groups of cells (imaginal discs) that have remained largely undifferentiated during larval development. Finally, at the end of the pupal stage, the last moult takes place, and the adult insect emerges from within the pupal cuticle. This complete metamorphosis enables optimal physiological division of labour between the various stages – the larva feeding and growing, the pupa facilitating the tremendous change from larva to adult, and the adult optimized for reproduction and dispersal. This carefully controlled sequence of growth, moulting and metamorphosis is regulated by the endocrine system in all insects, including moths and wasps.

The growth of lepidopteran larvae begins rather slowly. The lepidopteran egg contains sufficient nutritive resources, in the form of yolk proteins (vitellins), to sustain embryonic development. However, by the time the lepidopteran larva hatches these resources are exhausted, and the 1st-instar larva usually consumes some, or all, of the eggshell before starting to feed on the appropriate food material (e.g. the leaves of the foodplant). Larval growth, in the early larval stadia is relatively slow, but increases substantially (as does food consumption and utilization) as the larva progresses through the larval development. Thus, for example, the newly-hatched larvae of the noctuid, *Lacanobia oleracea,* weigh only about 0.1 mg. However, during the antepenultimate (4th), penultimate (5th) and final (6th) larval stadia, the mean larval weight at each stadium increases to 52 mg, 161 mg and 431 mg, respectively, representing an overall 4000-fold increase in larval weight between the first and final stadia. Importantly, most of this increase (88%) occurs during the last two larval stadia (Corbitt *et al.*, 1996). Furthermore, during the final larval stadium of lepidopteran insects, there is a substantial release of storage proteins from the fat body into the haemolymph (see below). Thus, the earlier larval stages of lepidopteran hosts represent rather meagre fare for parasitoids, whereas the last two larval stadia are, by comparison, a veritable banquet. From the point of view of the developing parasitoid, these later stages of host development represent the ideal environment in terms of potential available nutritional resource (see Chapter 11). However, the latter stages of host development also present the most serious potential challenges to developing parasitoids in terms of changing host endocrinology (and possibly increased host immune competence; Gupta, 1991). Host hormones change dramatically during pre-adult development (see *Figures 6* and *8*), particularly between the penultimate and final larval stadia. These endocrine changes in host larvae intrinsically constitute a potential hazard to parasitoid development and survival, because the host hormones may impinge on the endocrine processes affecting the parasitoid larva (see Section 6). Furthermore, there are additional consequences associated with parasitoid development in the latter portions of host larval development. The most obvious of these is the normal change in host behaviour that is associated with the preparation for pupation.

In the final-instar larva, food consumption and growth ceases about two-thirds of the way into the final larval stadium. At this time, the behaviour of the larvae changes, they begin to wander away from the food source and seek pupation sites. Often, the wandering behaviour is linked with burrowing (e.g. in soil) and with the formation of a subterranean 'cell' in which a silken lining is produced, and in which the larva (now rather shrunken in appearance) eventually pupates. Larvae exhibiting this 'pre-pupal' behaviour (and morphology) are often referred to as pre-pupae. Obviously, both endo- and ecto-parasitoids of larval hosts might derive significant benefit from preventing this host pupation, and the associated behaviours. Moreover, an additional significant physiological consequence associated with host pupation, is the massive release of storage proteins into the haemolymph of final-instar larvae (Haunerland, 1996; Telfer and Kunkel, 1991). This release of storage proteins is thought to be regulated by the endocrine system of the host. These storage proteins, which are a natural consequence of the host being allowed to survive to the last larval stadium, represent an enormous nutritional bonus for any developing parasitoid. Thus, parasitoids have two developmental options: either to permit the host to develop as far as the last larval stage (and therefore to benefit from additional nutritional resources), or to avoid the potentially disastrous consequences of the endocrinological changes and altered behaviour that are associated with host pupation.

The insect endocrine system. The brain constitutes the main insect endocrine centre. The major organs of the insect endocrine system, and some of their respective hormones, are shown in *Figure 1*. Neuropeptides produced in the brain are directed towards target organs (including endocrine organs) and tissues either via neurohormonal connections or via the haemolymph. Non-peptide hormones like JH and ecdysteroids are synthesized in specific endocrine organs, and appear mainly to be secreted

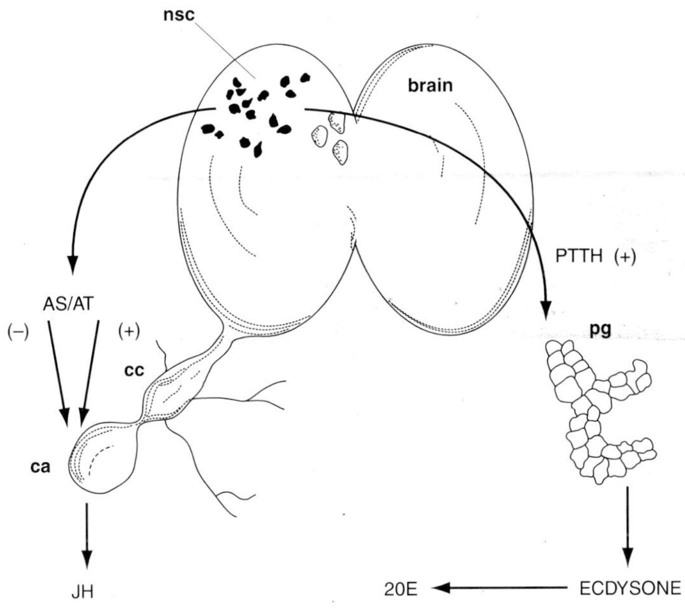

Figure 1. *Schematic diagram of the main components of the insect endocrine system. The brain and its associated neurosecretory cells (nsc) produce regulatory neuropeptides that control the activity of the endocrine organs. These neurohormones may pass to the target endocrine organs via neurohormonal connections like those passing through the corpora cardiaca (cc), or may be secreted directly into the haemolymph. Allatostatins (AS) and allatotropins (AT) respectively inhibit or stimulate the corpora allata (ca) to produce juvenile hormones (JH). Prothoracicotropic hormone (PTTH) acts on the prothoracic glands (pg) to stimulate the production of 3-dehydroecdysone and/or ecdysone, which is subsequently converted to the active moulting hormone, 20-hydroxyecdysone (20E).*

into the haemolymph. The major organs of the insect endocrine system are the neurosecretory cells of the brain, the CC, the CA, the PG, the sub-oesophageal ganglion, and various other ganglia in the ventral nerve cord.

2.3 *Insect juvenile hormones*

Insect juvenile hormones (JH) are terpenoid molecules containing an epoxide ring and a methyl ester, produced by paired retrocerebral endocrine glands, the corpora allata (CA). They are secreted into the haemolymph through which they are transported to target tissues bound to a protective carrier protein. The majority of insect orders (including the Hymenoptera) appear to use only one hormone (JH III; Trautmann *et al.*, 1974), while Lepidoptera have at least five closely related JH homologues (JH 0, JH I, JH II, 4-methyl JH I, and JH III). It is likely that the first three of these homologues occur only in the Lepidoptera (Schooley *et al.*, 1984). The flies (Diptera) seem to utilize yet another variant (JH III bis-epoxide, Richards *et al.*, 1989) as well as JH III and, as yet, the exact chemical nature of the JH in the Hemiptera (true bugs) remains unknown. *Figure 2* shows the chemical structure of the known, naturally occurring insect juvenile hormones.

The CA are regulated by the brain through neurohormones that stimulate or inhibit JH production (Section 2.4). Juvenile hormones have two major, yet distinct, roles in the life history of insects (although they also impact on many other physiological processes such as diapause, polymorphism and behaviour). During the growth and development of the larvae, the JH serve to maintain the developing insect in the larval

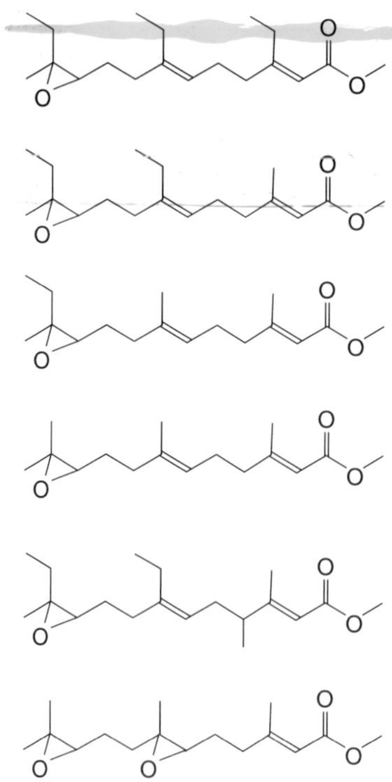

JH-0

JH-I

JH-II

JH-III

4-methyl JH-I

JH-III bis epoxide

Figure 2. Chemical structures of the known insect juvenile hormones.

state at each successive larval moult. Thus, the hormones are present for the majority of larval development, and their presence inhibits the initiation of metamorphosis. During the final larval stadium, the levels of JH in the blood are reduced to almost zero, and this absence of JH is the signal for the process of metamorphosis to proceed. Often, a small peak of JH is present at the end of the final larval stadium, and this may be associated with initiating behaviours (e.g. wandering) appropriate for the forthcoming pupation (see *Figure 6*). During metamorphosis, JHs remain absent for the majority of pupal development, but reappear in the adult insect, where they assume their other role as gonadotrophic hormones. The changing levels of JH are a result of changes in the rate of hormone biosynthesis by the CA, balanced by the rate of metabolism of the hormones by degradative enzymes (De Kort and Granger, 1996; Hammock, 1985). One of the major JH degradation products (JH acid) can, under some circumstances, be reconverted to the active ester form of the hormone in specific tissues (e.g. imaginal discs; Sparagana *et al.*, 1985). Although the production of JH acid is generally thought to be an inactivation step, recent evidence suggests that this metabolite may have some hormonal activity in certain tissues (Ismail *et al.*, 1998, 2000).

Despite the fact that the existence and structures of JH, and their primary roles in insect physiology, have been known for many years, we still do not understand exactly how and where they work. No receptor for JH has yet been fully identified, and several theories have been postulated for the mechanism(s) of action of JH (see Riddiford, 1985; Riddiford and Truman, 1993). However, the elucidation of the exact mechanism of action of JH, and the characterization of the JH receptor(s), remain two of the most intriguing challenges facing insect endocrinologists.

2.4 *Allatostatins and allatotropins: endogenous regulators of insect juvenile hormones*

Insect brains contain neurosecretory peptides that modulate the production of JH by the CA. These peptides (some of which have been characterized) are termed, respectively, allatotropins and allatostatins, according to whether they stimulate or inhibit the production of JH. The first (and so far only) insect allatotropin was identified in the moth *Manduca sexta* (Kataoka *et al.*, 1989). This molecule (or very similar molecules) has also been found in other lepidopterans including *L. oleracea* (Audsley *et al.*, 2000; Edwards *et al.*, 2001) and *Spodoptera frugiperda* (Oeh *et al.*, 2000). A gene encoding this peptide has also been found in *Pseudaletia unipuncta* (Truesdell *et al.*, 2000). It seems likely, therefore, that this neuropeptide is the functional allatotropin in most, if not all lepidopterans, and its structure is shown in *Figure 3*. However, there may be other brain factors that also stimulate CA activity in some lepidopterans (e.g. *Galleria mellonella*; Bogus and Scheller, 1996).

There is no information on the precise nature of an allatotropic neurohormone from any hymenopteran insects. However, the *M. sexta* allatotropin does stimulate JH production by CA of the honey bee, *Apis mellifera* (Rachinsky and Feldlaufer, 2000; Rachinsky *et al.*, 2000), and it is possible that lepidopterans and hymenopterans share the same allatotropic neurohormone.

Following the first identification of allatostatic neurohormones in the cockroach *Diploptera punctata* (Pratt *et al.*, 1989; Woodhead *et al.*, 1989), several closely related peptides were identified in other cockroaches, as well as in several other insect types, including lepidopterans and hymenopterans (Weaver *et al.*, 1998). These peptides share a common C-terminal sequence (Xaa-Phe-Gly-Leu/Ile-NH_2), but not all appear to function solely (or even at all) as inhibitors of JH production in all of the insect groups (except cockroaches) in which they are found. Thus, there is continuing speculation about the true role of these cockroach-type allatostatins, especially in lepidopterans. In the Lepidoptera, a neuropeptide allatostatin (*Figure 3*) with a completely different structure to the cockroach-type allatostatins was identified in the moth *M. sexta*, and shown to inhibit JH biosynthesis in that species and in *Heliothis virescens* (Kramer *et al.*, 1991). This peptide (or its precursor gene) has also been found to occur in several other lepidopterans, including *P. unipuncta* (Jansons *et al.*, 1996), *Spodoptera littoralis* and *L. oleracea* (Audsley *et al.*, 1998). In the latter species, this peptide is biologically active in inhibiting JH biosynthesis by larval and adult CA (Audsley *et al.*, 1999, 2000), and it seems likely that the *M. sexta* allatostatin is probably the major functional allatostatic neurohormone in the Lepidoptera. Again, however, it is possible that other brain-derived factors may also inhibit JH biosynthesis. In fact, in *M. sexta*, another allatoregulatory factor (allatinhibin), that is chemically unrelated to the *M. sexta* allatostatin, inhibits JH production in final-instar larvae (Bhaskaran *et al.*, 1990; Unni *et al.*, 1993).

Within the Hymenoptera, evidence for the existence of an allatostatic neuropeptide comes only from studies of the honeybee, *A. mellifera* (Kaatz, 1994) in which it is claimed that inhibition of JH production by the CA in this species is modulated by allatostatic peptides related to the cockroach-type allatostatins. Moreover, Stay *et al.* (1994) reported cockroach allatostatin-like immunoreactivity in honeybee brains. However, other studies in this insect failed to produce evidence that the CA of *A. mellifera* were sensitive to inhibition by cockroach-type allatostatins (Rachinsky and Feldlaufer, 2000). At present, therefore, the nature of the allatostatin(s) in parasitic hymenopterans remains unknown.

2.5 *Endocrine regulation of insect moulting – ecdysteroids and PTTH*

Insect moulting is generally a response to the principal moulting hormone, 20-hydroxyecdysone (20E), which is produced by enzymatic conversion of the 'pro-hormones' ecdysone, and/or 3-dehydroecdysone, which are produced by the PG. Although 20E is regarded as the primary moulting hormone in insects, other ecdysteroids (including some found naturally in plants) may also have moulting hormone activity (Böcking and Hiruma, 1998). *Figure 4* shows the structures of some insect ecdysteroids. The PG themselves are regulated by a brain neuropeptide hormone, prothoracicotropic hormone (PTTH) and, possibly, by other neuropeptides like ecdysiotropin (Gelman and Beckage, 1995) and prothoracicostatin (Tanaka *et al.*, 1999). In addition, the production of PTTH may be regulated by JH in lepidopteran larvae (Gilbert and Schneiderman, 1959; Hiruma *et al.*, 1978; Lonard *et al.*, 1996; Rountree and Bollenbacher, 1986). The chemical nature of PTTH has been investigated in two lepidopterans (*M. sexta* and *Bombyx mori*). In *M. sexta* PTTH appears to constitute a

(a) H-Gly-Phe-Lys-Asn-Val-Glu-Met-Met-Thr-Ala-Arg-Gly-Phe-NH_2

(b) pGlu-Val-Arg-Phe-Arg-Gln-Cys-Tyr-Phe-Asn-Pro-Ile-Ser-Cys-Phe-OH

Figure 3. *Structure (amino acid sequence) of* Manduca sexta *allatotropin (a) and allatostatin (b). These peptides probably act as allatostatins and allatotropins in all lepidopterans.*

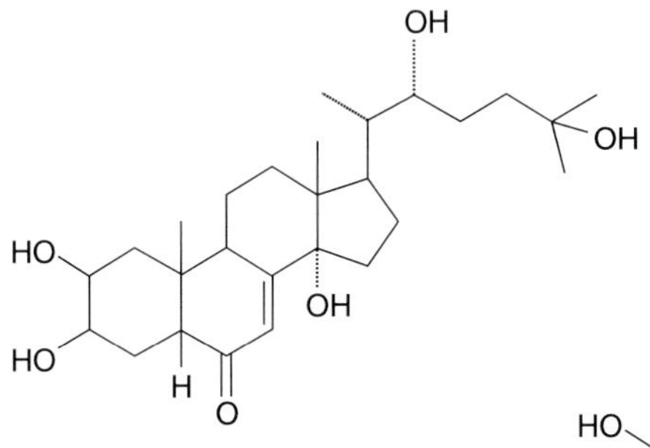

Ecdysone

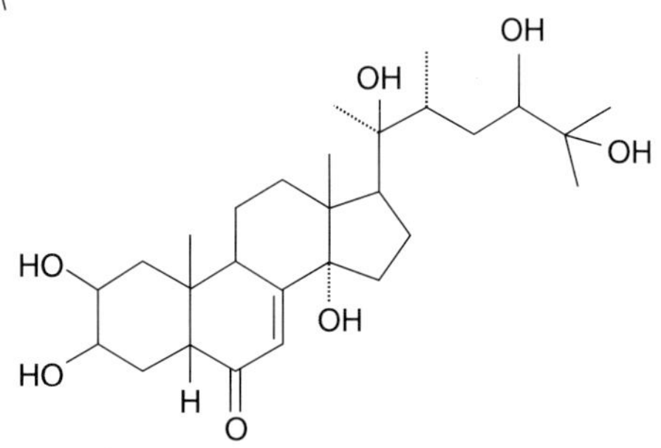

Makisterone A

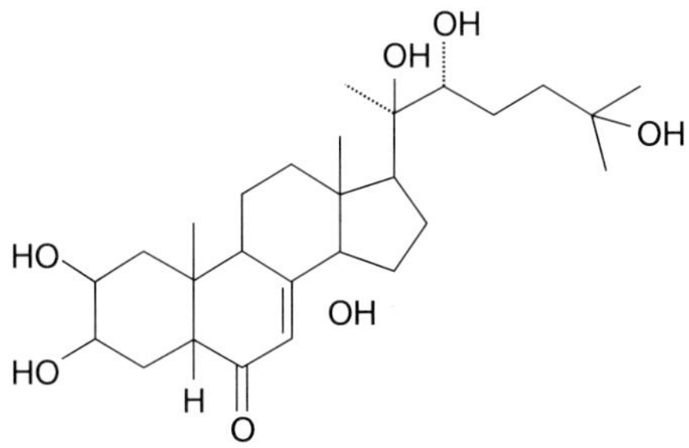

20-Hydroxyecdysone

Figure 4. *Chemical structure of some insect ecdysteroids.*

family of polypeptides of two main size classes (see Bollenbacher and Granger, 1985). In the silkworm, *B. mori*, PTTH is a homodimeric protein of two 109-amino acid subunits (Kataoka *et al.*, 1991; Noguti *et al.*, 1995). There is, as yet, no information about the structure or nature of the hymenopteran PTTH. However, it seems probable that in both moths and wasps, some form of PTTH will stimulate the PG to produce the ecdysteroid precursors of the moulting hormone (20E).

Moulting hormone initiates the multitude of complex biochemical processes associated with the phenomenon of moulting. Since moulting is a process that allows size increase in developing insects, the prothoracic glands usually degenerate as the insect approaches adulthood (usually during the pupal stage; Locke, 1985). However, ecdysteroids do have reproductive functions in adults of some insects and, in these cases, the main source of adult ecdysteroids appears to be the ovaries or testes. Like other steroid hormones, ecdysteroids are thought to act directly on the gene via classical nuclear receptors. At the molecular level, the ecdysteroid receptors (EcR) bind with the ultraspiracle protein (a vertebrate RXR homologue) and this complex binds with 20E. This hormone–receptor combination binds to an ecdysone response element (EcRE) to activate gene transcription (Yao *et al.*, 1993).

Ecdysteroids act in concert with other hormones (mainly the ecdysis- triggering hormone, eclosion hormone and bursicon, the 'tanning hormone'), to co-ordinate the complex process of moulting from partial digestion of the old cuticle, through apolysis (the separation of the old cuticle from the new) and ecdysis (the act of moulting), to the secretion and hardening (sclerotization) of the newly-formed cuticle. However, relatively little is known about these hormones in insects generally, although much of what is known stems from studies in the moth *M. sexta*. In this moth, whilst ecdysteroids initiate and regulate many of the physiological processes leading up to the moult (e.g. apolysis), the timing of each ecdysis is regulated by the eclosion hormone produced by neurosecretory cells in the brain (Truman, 1985; Truman and Riddiford, 1970; Truman *et al.*, 1981; Ewer *et al.*, 1997).

There is no unequivocal information on the corresponding hormones in the parasitic Hymenoptera, but it seems safe to assume that the mechanisms operating in these insects are largely similar to those in lepidopterans. Thus, it is probable that ecdysis-triggering hormone and eclosion hormone control the timing of ecdysis, while the physiological and biochemical events necessary for moulting are initiated by 20E. After ecdysis, bursicon initiates the tanning and hardening (sclerotization) of the newly formed cuticle. The action of moulting hormones in stimulating the moulting processes is also intimately linked to the effects of the juvenile hormones, as the latter are crucial in regulating metamorphosis. In simple terms, ecdysteroids, eclosion hormone and bursicon determine that a moult will occur and that the process proceeds normally, while JH determines the nature or outcome of that particular moult (i.e. larval to larval, larval to pupal, etc.). It is likely that lepidopterans and hymenopterans share the same biologically active moulting hormone (i.e. 20E). Thus, the untimely presence of this molecule in the host may constitute a potential hazard to developing parasitoids. It is also clear that the secretion of host ecdysteroids, and the consequent shedding of the host's cuticle will be of particular importance to ectoparasitoids. However, at least in some endoparasitoids, host ecdysteroids may actually be important for the growth of the parasitoid larva (Nakahara *et al.*, 2000).

2.6 *Other insect hormones*

Figure 5 shows the chemical nature (amino acid sequences) of a few of the other known insect hormones that have been found in lepidopterans. Most of these other insect hormones are peptides or proteins, and most are produced in the insect brain and/or central nervous system. By comparison with JH and the ecdysteroids, very little is known about the levels or activity of the majority of these other hormones, although their manipulation by parasitoid wasps cannot be excluded. Indeed, there is some evidence that parasitism leads to an accumulation of several neuropeptides in the cerebral neurosecretory system of host larvae (Zitnan *et al.*, 1995a). Some of these hormones are probably more significant regulators of physiology in adult insects, but others (e.g. diuretic hormones and diapause hormones) could impinge greatly on the developing stages of host insects, and thus constitute a potential threat to the development and survival of parasitoids. Further information about JH, ecdysteroids, PTTH, and other insect hormones is available in the excellent book by H.F. Nijhout (Nijhout, 1994).

(a) Diuretic hormone from the Tobacco hornworm moth (*Manduca sexta*).

H-Arg-Met-Pro-Ser-Leu-Ser-Ile-Asp-Leu-Pro-Met-Ser-Val-Leu-Arg-Gln-Lys-Leu-Ser-Leu-Glu-Lys-Glu-Arg-Lys-Val-Lys-Ala-Leu-Arg-Ala-Ala-Ala-Asn-Arg-Asn-Phe-Leu-Asn-Asp-Ile-NH_2

(b) Diapause hormone from the Silkworm moth (*Bombyx mori*).

Thr-Asp-Met-Lys-Asp-Glu-Ser-Asp-Arg-Gly-Ala-His-Ser-Glu-Arg-Gly-Ala-Leu-Cys-Phe-Gly-Pro-Arg-Leu-NH_2

(c) Pheromone biosynthesis activating neuropeptide (PBAN) from the Corn earworm moth (*Helicoverpa (Heliothis) zea*).

Leu-Ser-Asp-Asp-Met-Pro-Ala-Thr-Pro-Ala-Asp-Gln-Glu-Met-Tyr-Arg-Gln-Asp-Pro-Glu-Glen-Ile-Asp-Asp-Ser-Arg-Thr-Lys-Tyr-Phe-Ser-Pro-Arg-Leu-NH_2

(d) Adipokinetic hormone from the Tobacco hornworm moth (*Manduca sexta*).

pGlu-Leu-Thr-Phe-Thr-Ser-Ser-Trp-Gly-NH_2

Figure 5. *Amino acid sequences of some other lepidopteran hormones.*

3. Changes in juvenile hormones during lepidopteran development

3.1 *Measurement of juvenile hormones in insects*

Since the discovery of the chemical structure of a naturally occurring insect juvenile hormone (Röller *et al.*, 1967) there has been great interest in identifying and measuring these hormones in various insect taxa. Fortunately, the Lepidoptera have been particularly well studied in this respect. However, because of the diverse methodologies employed by different researchers (see Edwards *et al.*, 1995), the accuracy and validity of published data are variable. In general, measurements of JH using the gas chromatography/mass spectrometry (GC/MS) method developed by Bergot *et al.* (1981a) or variations thereof (e.g. Rembold and Lackner, 1985; Teal *et al.*, 2000) are regarded as being the most accurate and, in addition, permit precise identification of the various JH homologues. Radioimmunoassay (RIA) methods (Baehr *et al.*, 1976; Goodman *et al.*, 1990; Granger and Goodman, 1983; Strambi *et al.*, 1981) are also fairly reliable, but do not always distinguish unequivocally between homologues, while other (bioassay) techniques (e.g. the *Galleria* wax test) are much less reliable, although such methods do permit measurement of biological activity. *In vitro* techniques (Pratt and Tobe, 1974) opened the way for assessing JH production by isolated CA, but the very low rates of JH biosynthesis by larval CA have, until recently (e.g. Audsley, *et al.*, 2000; Bogus and Scheller, 1996; Jones and Yin, 1989; Pfister-Wilhelm and Lanzrein, 1996) largely confined such *in vitro* studies to the CA of adults. In the following sections, we present a summary of what is known about the nature and levels of JH in eggs, larvae and pupae of lepidopteran insects.

3.2 *Juvenile hormones in lepidopteran eggs*

There is ample evidence that the eggs and/or embryos of lepidopterans contain juvenile hormones. The eggs of *M. sexta* contain JH I, as well as two rather unusual homologues (JH 0 and 4-methyl JH I) that appear to be confined to this species and other lepidopterans such as *Hyalophora cecropia* and *H. virescens* (Bergot *et al.*, 1980, 1981a, b). In addition, JH 0 was also detected in egg-laying females (but not in virgins) of *L. oleracea* (Edwards *et al.*, 1995) suggesting that this homologue may also be present in eggs of this species. Data on the presence of JH in other lepidopteran eggs/embryos is rather sparse, although Grossniklaus-Bürgin and Lanzrein (1990a) reported JH activity in extracts of eggs of *Tricoplusia ni*, and Gharib *et al.* (1983) and Gharib and De Reggi (1983) reported relatively high levels of JH I, JH II and JH III in pre- and post-diapause, developing embryos of *B. mori.*

Thus, it seems likely that JHs are present in all lepidopteran eggs. Moreover, the analyses of JH in this developmental stage have produced indications that the hormones are not present at the earliest stage of embryonic development (i.e. soon after the eggs are laid), but that JH levels increase later in embryonic development, before declining immediately prior to hatching (Bergot *et al.*, 1980, 1981b; Gharib and De Reggi, 1983; Grossniklaus-Bürgin and Lanzrein, 1990a). Furthermore, the apparent absence of JH in newly laid eggs suggests that these hormones are not of maternal origin. However, this assumption should be tempered by the fact that adult lepidopterans are renowned for having very high JH levels (Baker *et al.*, 1985; Gilbert and Schneiderman, 1961).

Despite their relatively small size, the eggs of lepidopterans contain remarkably high levels of JH. In two-day-old *M. sexta* eggs, total JH levels were about 5.3 ng g^{-1} (Bergot *et al.*, 1981b) – considerably more JH than was detected in larvae of the same species (see below). By contrast, in one-day-old *H. virescens* eggs, the same authors reported total JH 0, I and II levels of ~ 0.16 ng g^{-1} – a level that is not markedly different from that found in several larval noctuids (see below). In eggs of *S. littoralis*, Steiner *et al.* (1999) reported levels of JH I (0.0–8.6 ng g^{-1}) and JH II (0.6–21.1 ng g^{-1}) depending on the age of the eggs. These authors also reported low levels of JH III (0.03–0.7 ng g^{-1}) in *S. littoralis* eggs. The results of Steiner *et al.* (1999) also indicate that JH levels are relatively low in early embryonic development, and increase in older embryos before declining prior to hatching. These embryonic JH levels are substantially higher than those detected in the haemolymph of newly moulted last-instar *S. littoralis* larvae (~1.8 pmol ml^{-1}) by Zimowska *et al.* (1989). Very high levels of JH (~26 ng g^{-1} JH II equivalents) were reported in embryos of *T. ni* (Grossniklaus-Bürgin and Lanzrein, 1990a), and Gharib and De Reggi (1983) reported combined levels of JH I, II and III, which were more than two orders of magnitude higher (~2000 ng g^{-1}) in eggs of *B. mori*, than those found in *M. sexta* (Bergot *et al.*, 1981b). Notwithstanding these discrepancies, it is clear that lepidopteran eggs/embryos (particularly during the later phases of embryogenesis) do contain significant levels of JH, and that these could pose a potential hazard to any developing egg or egg-larval parasitoid.

3.3 *Juvenile hormones in lepidopteran larvae*

There is much more information regarding the nature and levels of JH in larval lepidopterans than in any other insect group or developmental stage. Juvenile hormones in only a few larval lepidopterans have been studied using the GC/MS technique of Bergot *et al.* (1981a). These include *M. sexta* (Baker *et al.*, 1987), *T. ni* (Jones *et al.*, 1990b), and *L. oleracea* (Edwards *et al.*, 1995). In *L. oleracea* larvae JH II was the predominant homologue with lesser quantities of JH I also being present. Juvenile hormone III was detected inconsistently (and at very low levels) in some samples of

4th, 5th and 6th-instar larvae. Total JH levels were generally higher in 4th-instar larvae than in 5th instars (mean total JH I + JH II = 1.07 ng g^{-1} and 0.63 ng g^{-1} respectively). However, despite the quantitative differences in JH levels in these two stadia, the changing pattern of JH levels was similar in both. Thus, during the first four days of each stadium, combined JH I and II levels remained relatively constant, before increasing in five-day-old (clear head capsule) larvae just prior to the moult to the next stadium (Edwards *et al.*, 1995). In 6th (final)-instar larvae JH were essentially absent for the first half of the larval stadium. During the second half of the 6th larval stadium (on days 6, 7 and 8) there was consistent evidence of the presence of low levels of JH II. Finally, in larvae at the end of the wandering phase (day 9) and in day 10 pre-pupae (i.e. ~24 h prior to pupation), an increase in JH II levels (to ~0.04 ng g^{-1}) was detected, as were pro-rata increases in JH I (Edwards *et al.*, 1995). Unfortunately, in *M. sexta* JH levels were only measured in the final larval stadium (Baker *et al.*, 1987), and in *T. ni* JH levels were measured throughout the final stadium and in larvae that were at the very end of the penultimate larval stadium (Jones *et al.*, 1990b). For this reason, there is little information about detailed fluctuations of the JH levels in antepenultimate and penultimate-instar larvae in these lepidopteran species. In *L. oleracea*, Edwards *et al.* (1995) found that there was a noticeable increase in total JH levels in clear head capsule 4th and 5th-instar larvae (i.e. those approaching the moult to the subsequent stadium). Interestingly, Jones *et al.* (1990b) also found a relatively high titre (approx. 0.42 ng g^{-1}) of JH II just prior to the moult to the final stadium in *T. ni*. By contrast, in *M. sexta* relatively high levels of JH I and II (0.3–1.5 ng g^{-1}) are found immediately following the moult to the penultimate (4th) and final (5th) larval stadia (Baker *et al.*, 1987; Edwards, 1987; Edwards *et al.*, 1983). In *T. ni* the pattern of changing JH (II) levels (Jones *et al.*, 1990b) is remarkably similar to that found in *L. oleracea*. Thus, following relatively high JH levels prior to the moult to the last stadium, JH fell to very low levels during the majority of the final stadium before increasing again approximately 24 h before pupation. A similar pattern, involving both JH II and JH I, also occurs in final-instar *M. sexta* larvae (Baker *et al.*, 1987).

Other lepidopteran species in which larval JH levels have been investigated include: *G. mellonella* (Hsiao and Hsiao, 1977; Plantevin *et al.*, 1984; Rembold and Sehnal, 1987); *S. littoralis* (Steiner *et al.*, 1999; Zimowska *et al.*, 1989); *Mamestra brassicae* and *Mamestra (Lacanobia) oleracea* (Vargas *et al.*, 1976, 1992); *Pieris brassicae* (Mauchamp *et al.*, 1979; Vargas *et al.*, 1976); *B. mori* (Niimi and Sakurai, 1997; Plantevin *et al.*, 1987); *Lymantria dispar* (Jones and Yin, 1989); *Cerura vinula* and *Araschnia levana* (Koch, 1996).

Understandably there are some contradictions in these results (see Section 3.1) but several consensus points arrive from these studies. First, JH I and JH II are the predominant (possibly the only) homologues found in lepidopteran larvae. Second, JH are essentially absent during the greater part of the final larval stadium, and in pupae. Third, levels of JH in lepidopteran larvae are relatively low by comparison with levels found in adult stages and, probably, with levels found in eggs.

Most of the above studies have concentrated on last-instar larvae, and a few have included the penultimate larval stadium. Data on JH levels in earlier lepidopteran larval stages is extremely sparse. A comprehensive study (Steiner *et al.*, 1999) showed that 1st stadium larvae of *S. littoralis* contained more than 12 ng g^{-1} JH II and about 1.5 ng g^{-1} JH I. In 2nd stadium larvae levels of both homologues were lower at ~3.8 ng g^{-1} JH

II and zero JH I. In 3rd stadium larvae JH I was again essentially absent, and JH II levels were recorded at ~1.5 ng g^{-1}. In 4th, 5th and 6th-stadium *S. littoralis* larvae Steiner *et al.* (1999) measured JH levels in haemolymph (as opposed to whole bodies) and so these results were expressed as ngJH ml^{-1} haemolymph. These values are not easy to equate directly with the whole body titres measured in earlier larvae, but show roughly similar levels (2–4 ng ml^{-1} JH II and zero JH I) in 4th and 5th-stadium larvae. By contrast, JH were essentially absent from the haemolymph of 6th (final)-stadium larvae, until about the time of the formation of the pre-pupal stage (day 12) whereupon a small peak (approx. 0.5 ng ml^{-1}) of JH II was detected. Interestingly, Steiner *et al.* (1999) also reported the presence of JH III in eggs and in all larval stadia of *S. littoralis*. This homologue has often (but inconsistently) been found in other larval lepidopterans, but its apparent presence may be artefactual (see Edwards *et al.*, 1995). The changing pattern of JH I and II levels is remarkably consistent within the noctuid species *T. ni* (Jones *et al.*, 1990b), *L. oleracea* (Edwards *et al.*, 1995) and *S. littoralis* (Steiner *et al.*, 1999), especially in relation to penultimate and final larval stadia. Similarities in the overall patterns of JH in antepenultimate stadium larvae are also apparent between *L. oleracea* (Edwards *et al.*, 1995) and *S. littoralis* (Steiner *et al.*, 1999). From all of these studies, a general pattern of the relative levels and fluctuations of JH throughout larval development of lepidopteran insects begins to emerge. A diagrammatic representation of these generalized changes in JH levels during the course of development from the egg to the pupal stage in a 'typical' lepidopteran is shown in *Figure 6*.

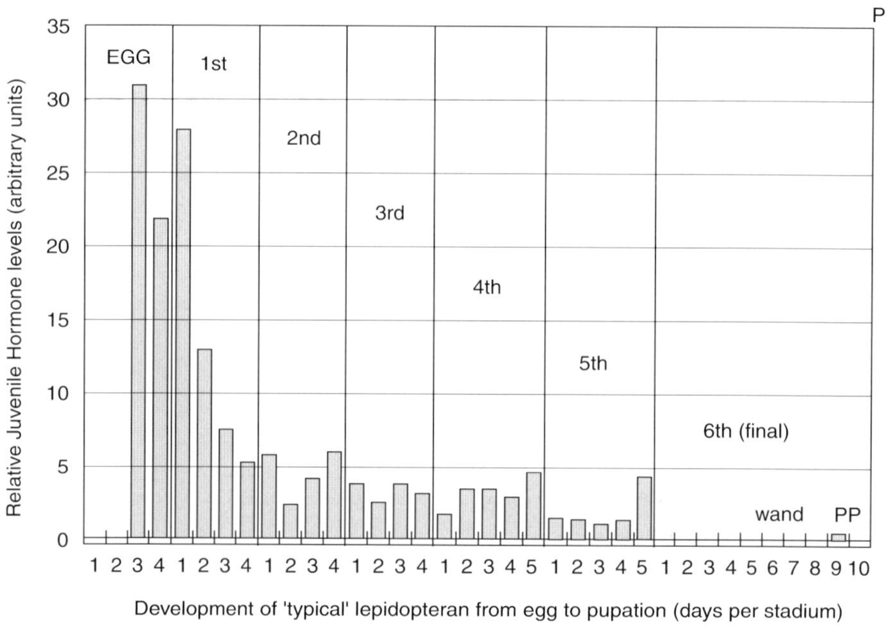

Figure 6. *Schematic representation of changes in juvenile hormones during lepidopteran development from egg to pupation. From about two thirds of the way through the final larval stadium, larvae stop feeding and begin pre-metamorphosis behaviours like wandering (wand) and burrowing. This behaviour usually culminates with the formation of a pre-pupa (PP) which rapidly undergoes metamorphosis to the pupal stage (P).*

In essence, JHs are present throughout the majority of larval development. JH levels may be highest in late embryos and in 1st-stadium larvae. In all larval stages (except the final stadium) JH levels often peak just prior to, or just after, each ecdysis. In the final larval stadium JH are essentially absent, but small peaks of JH may appear towards the end of this developmental stage. The presence of low levels of JH at this time may be associated with the change from feeding behaviour to the behaviours (wandering, burrowing, spinning, etc.) that are necessary precursors to pupation.

3.4 *Juvenile hormones in lepidopteran pupae*

Again, compared with larval stages, relatively little is known about changing JH levels in lepidopteran pupae. Koch (1996) did not detect any JH for the first few days of pupal development in *C. vinula* and *A. levana*. In *G. mellonella*, Rembold and Sehnal (1987) found that JHs (predominantly JH II) were essentially absent during the majority of pupal development, but increased at the pharate adult stage. Hsiao and Hsaio (1977) obtained similar results in pupae of *G. mellonella*. Shu *et al.* (1997) were also unable to detect JH during the majority of pupal development of *Diatrea grandiosella*, but did record substantial quantities (~12.5–15.0 ng ml^{-1}) of JH II and JH III respectively, in haemolymph of late pupae. Edwards *et al.* (1995) measured JH levels in pupae of *L. oleracea* at selected times during pupal development, and found no JH present until the end of pupal development, when insects had reached the pharate adult stage (25 day old 'pupae'). In these pharate adults, males contained ~1.6 ng g^{-1}, and females ~2.5 ng g^{-1}, combined JH I and II respectively. Baker *et al.* (1987) detected no JH in newly formed (two-hour-old) pupae of *M. sexta*, but did not study older pupae. According to generally accepted theories on the role of JH in controlling metamorphosis in holometabolous insects, JH should be largely absent during pupal development (to permit full development of adult tissues). In fact, this presumption has been generally borne out by the above studies. It therefore seems reasonable to assume that JH are indeed absent for the major part of pupal development in most, if not all, lepidopterans. However, it is equally clear that pharate adults probably contain significant quantities of JH.

4. Changes in ecdysteroids during lepidopteran development

4.1 *Measurement of ecdysteroids in insects*

Levels of ecdysteroids (moulting hormones) in insects are usually measured using the RIA methods developed by Borst and O'Connor (1972) and Chang and O'Connor (1979), or subsequent variations thereof (see later). Several antibodies with differing specificities for different ecdysteroids have been employed. However, most suffer from a lack of absolute specificity for 20E – the ecdysteroid that is the actual moulting hormone in most insects (but see below). Greater precision in determining the chemical identity of the ecdysteroids can be achieved by chromatographic separation of different molecules (e.g. by high-performance liquid chromatography, HPLC) before RIA methods are used. The use of HPLC also permits the differentiation between biologically active ecdysteroids and some metabolites. In the following sections we review the current state of knowledge regarding the nature and changing levels of ecdysteroids during the development of lepidopterans.

4.2 *Ecdysteroids in lepidopteran eggs*

The eggs of lepidopterans certainly contain ecdysteroids and their conjugates (see review by Hoffmann and Lagueux, 1985). In a detailed and comprehensive study of *T. ni*, Grossniklaus-Bürgin and Lanzrein (1990a) found that 20E and other, more polar, ecdysteroids increased during embryonic development, before declining towards the time of egg hatch. In this study, 20E levels reached a maximum of ~90 ng g^{-1} in 36–40-hour-old embryos. A similar pattern of ecdysteroids in lepidopteran embryos has been reported for *P. brassicae* (Beydon *et al.*, 1989), *M. sexta* (Thompson *et al.*, 1988), and *B. mori* (Gharib and De Reggi, 1983). In general, levels of ecdysteroids in lepidopteran eggs appear to be lower than those found in later developmental stages (see below). The role of these embryonic ecdysteroids is probably associated with co-ordinating the embryonic moult that often occurs just prior to hatching.

4.3 *Ecdysteroids in lepidopteran larvae*

As with juvenile hormones, considerably more information about the nature and levels of ecdysteroids is available from studies with lepidopteran larvae than is available from measurements in eggs and embryos. Again, however, the majority of these studies in larvae have tended to concentrate only on the final larval stadium. Lepidopteran species, in which ecdysteroids have been investigated in larvae, include: *B. mori* (Calvez *et al.*, 1976); *Choristoneura fumiferana* (Lageaux *et al.*, 1976); *P. brassicae* (Lafont *et al.*, 1977); *G. mellonella* (Bollenbacher *et al.*, 1978; Hsiao and Hsiao, 1977); *Ephestia kuehniella* (Kallenborn and Mosbacher, 1983); *Ostrinia nubilalis* (Gelman and Woods, 1983) and *M. sexta* (Baker *et al.*, 1987; Bollenbacher *et al.*, 1975). Detailed profiles of ecdysteroids covering a substantial proportion of lepidopteran larval development are scarce. Fortunately, there are a number of exceptions to this, and these are now discussed in some detail. In an extensive study of *M. sexta*, Bollenbacher *et al.* (1981) recorded a peak of ecdysteroid activity (~2 μg ml^{-1} 20E equivalents) towards the end of the 4th (penultimate) larval stadium, a small peak (the pupal commitment peak) between days 4 and 5 of the 5th (final) larval stadium, and another substantial peak on day 7 that coincided with the formation of the pre-pupal stage. Interestingly, these authors stated that 20E was the predominant ecdysteroid present in the peaks preceding the larval and pupal ecdyses. Plantevin *et al.* (1984) found a similar pattern of ecdysteroids in penultimate and last-stadium larvae of *G. mellonella*, as did Koch (1996) in *C. vinula* and *A. levana*. The latter author also recorded a peak of ecdysteroid activity in haemolymph of antepenultimate (3rd)-stadium *C. vinula* larvae at about 2 days prior to the moult to the 4th stadium. In probably the most detailed analysis of ecdysteroids during the larval development of a moth, Grossniklaus-Bürgin and Lanzrein (1990a) measured ecdysteroids throughout larval development in *T. ni*. These authors reported relatively low levels of ecdysteroids (mostly ecdysone and 20E) during the feeding phases of each larval stadium, increased levels towards the cessation of feeding, and highest levels (800–1200 ng ml^{-1}) in penultimate and final-stadium larvae just (12–15 h) prior to the moult. Finally, in our own studies with *L. oleracea*, we have used the RIA methods described by Weaver *et al.* (1997) to measure daily fluctuations in the levels of total haemolymph ecdysteroids in 2nd, 3rd, 4th, 5th and 6th-stadium larvae, such that these data can be compared with the respective JH titres measured by Edwards *et al.* (1995). The results of these experiments are presented in *Figure 7*.

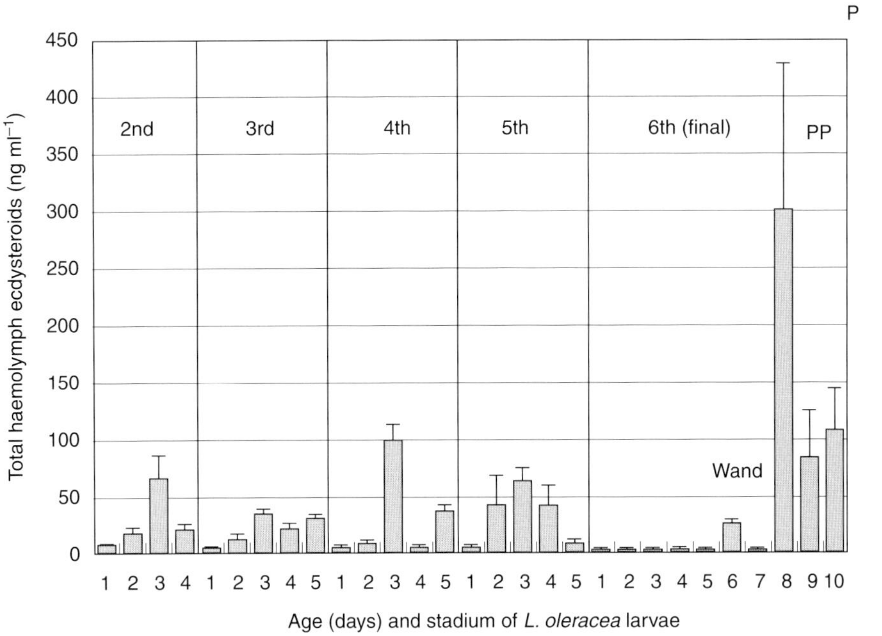

Figure 7. Daily changes in total haemolymph ecdysteroids during 2nd, 3rd, 4th, 5th and 6th larval stadia of Lacanobia oleracea *(mean values ± SEM, n = 8).*

We have examined all of the above reports, and have again produced a generalized picture of the likely changes in the relative levels of ecdysteroids during the egg (embryonic) and larval developmental stages of a 'typical' lepidopteran. In constructing this generalized scheme, we have placed particular emphasis on reported levels of 20E, rather than on published values for total ecdysteroids which may be 'skewed' by the presence of ecdysteroid metabolites and conjugates. This consensus representation of changing ecdysteroid levels during the pre-pupal development of a 'typical' lepidopteran is presented in *Figure 8*. In essence, peaks of ecdysteroid occur prior to the moult in each larval stadium. Overall levels of ecdysteroids appear to increase as larval development proceeds, peaks reaching maximum levels at the end of the final larval stadium. In the final stadium, there is consistent evidence for the appearance of a small ecdysteroid peak, prior to the main (moult-inducing) peak at the end of the stadium. The small peak is the pupal commitment peak (Truman *et al.*, 1974), and seems to be essential for the induction of the formation of the pupal stage.

4.4 *Ecdysteroids in lepidopteran pupae*

As with JH, extensive information about ecdysteroid levels in lepidopteran pupae is relatively sparse. However, it is clear that ecdysteroids are present throughout much of lepidopteran pupal development. The general pattern is that ecdysteroids are present at relatively low levels at the very beginning of pupal development, then increase towards the middle to end of pupal development, and decline during the pharate adult stage, towards the point of adult eclosion. This general pattern appears to hold true for several lepidopterans including *H. zea* (Holman and Meola, 1978), *A. levana* (Koch, 1996),

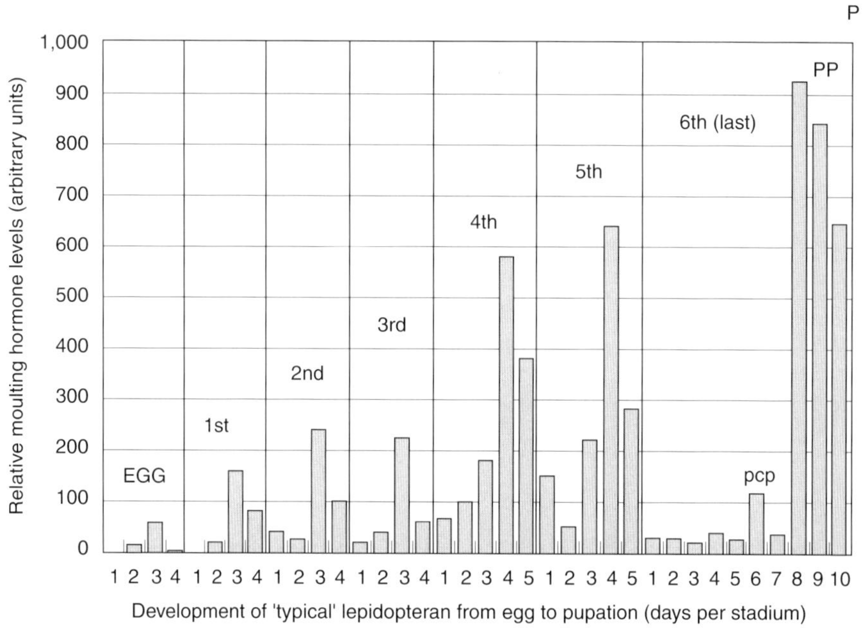

Figure 8. *Schematic representation of changes in ecdysteroids during lepidopteran development from egg to pupation. Note the small pupal commitment peak (pcp) of ecdysteroids during the final larval stadium, preceding the larger peak that initiates the metamorphic moult to the pupal stage. PP, pre-pupal stage; P, pupation.*

G. mellonella (Bollenbacher *et al.*, 1978; Hsiao and Hsiao, 1977; Plantevin *et al.*, 1984), *O. nubilalis* (Gelman and Woods, 1983) and *M. sexta* (Bollenbacher *et al.*, 1981). Interestingly, several of these studies clearly indicate sexual differences in ecdysteroid levels – female insects generally having higher ecdysteroid levels than males, possibly due to synthesis of ecdysteroids by developing ovarian tissues. Moreover, several authors provide evidence that, when present, pupal ecdysteroid levels can be substantially higher than those found during larval development.

5. Juvenile hormones and ecdysteroids in parasitic Hymenoptera

As indicated previously, hymenopterans are generally believed to produce and utilize only JH III. Much of this belief is based on studies with social hymenopterans, especially the honeybee, *A. mellifera,* which have identified or implicated this homologue as being the predominant (or only) JH in this species (Hagenguth and Rembold, 1978; Robinson *et al.*, 1991; Trautmann *et al.*, 1974). In addition, JH III has been identified as the primary (probably the only) JH in the bumblebee, *Bombus terrestris* (Bloch *et al.*, 2000a). Thus, it seems that JH III is the only JH produced by parasitoids, but we are aware of only very few parasitoid species in which the nature of the JH has been investigated. There is good evidence that the braconid egg-larval parasitoid, *Chelonus inanitus,* produces only JH III (Grossniklaus-Bürgin and Lanzrein, 1990b; Steiner *et al.,* 1999). The latter authors measured developmental changes in JH III titre in

C. inanitus, and these data indicate that, in general, changes in JH levels during parasitoid development are essentially the same as those seen in other insects. Interestingly, JH levels in this parasitoid fluctuate independently of those in the host (Steiner *et al.*, 1999). It is reasonable to suppose that the roles of JH in regulating development and metamorphosis in parasitoid larvae follow the general insect scheme described above. Lawrence *et al.* (1990) found traces of JH III in first-instar larvae of the braconid parasitoid, *Biosteres longicaudatus*, and Höller *et al.* (1994) reported high levels (about 4 ng g^{-1}) of JH III in the ovaries of adult females of the wasp *Dendrocerus carpenteri* (in this species, JH is apparently used as an oviposition marking pheromone). Interestingly, Joiner *et al.* (1973) reported that extracts of the teratocytes of *Cardiochiles nigriceps*, contained JH activity, but this fascinating observation has not yet been reported for teratocytes from any other parasitoid species.

The exact nature and developmental variations in ecdysteroids in developing parasitoids has, with a few notable exceptions, received very little attention from insect endocrinologists. Evidence from other hymenopterans (especially the honeybee, *A. mellifera*) suggests that, in addition to ecdysone and 20E, *A. mellifera* produces significant amounts of another ecdysteroid (makisterone A); at least during some developmental stages (Feldlaufer *et al.*, 1985, 1986). It has also been suggested that makisterone A may be the predominant ecdysteroid in *B. terrestris* (Bloch *et al.*, 2000b). However, there is as yet no convincing evidence that this ecdysteroid is a major component of any parasitoid ecdysteroid complex. In fact, in those few studies in which parasitoid ecdysteroids have been identified, the evidence points strongly to the presence mainly of ecdysone and 20E, together with various, more polar, ecdysteroid metabolites. This is true in the eulophid ectoparasitoid, *Euplectrus plathypenae* (Kelly and Coudron, 1990), the ichneumonid ectoparasitoid, *Diapetimorpha introita* (Gelman *et al.*, 2000), and in the braconid endoparasitoids, *Ascogaster reticulatus* (Brown *et al.*, 1993) and *Cotesia congregata* (Gelman *et al.*, 1999).

Thus, while parasitoids (with JH III) and their lepidopteran hosts (with JH I and II) appear to use different JH homologues to regulate embryonic development, pupation and metamorphosis, these two insect groups appear to have the same type of moulting hormone (i.e. 20E).

6. Parasitoid development in the hormonal environment of host lepidopterans

In the previous sections, we have described the general patterns of JH and moulting hormones that are likely to be encountered by the developing stages of parasitoid wasps as they develop in close proximity to host tissues. In this section we explore how, and whether, the presence of such host hormones could present a challenge to the successful completion of parasitoid development and reproduction. In considering the potential adverse effects of host hormones (assuming that they are present) on parasitoid biology, we must begin by assuming that several critical conditions are met. First, the developing parasitoid must be at a stage of development that is susceptible to the physiological action of exogenous hormones. Second, the host's hormones must be able to 'penetrate' the parasitoid's system in a way (and in a chemical configuration) that will result in some biological activity in the developing parasitoid. Third, the chemical nature of the host's hormone must afford sufficient potency to impinge on the parasitoid system. Fourth, the presence of the host hormone(s) must be out of synchrony with those of the developing parasitoid. Lastly, the parasitoid must not be able to compensate for an altered hormonal regime. If (and only if) these conditions are met, will the host's endocrine milieu constitute a serious potential hazard to the development and survival of the parasitoid. In circumstances where these conditions prevail, the developing parasitoid will be under considerable pressure to modify the endocrine physiology of the host (i.e. to manipulate hormonal control). Thus, there are a number of factors that are critically important in assessing the potential hazards (to parasitoids) associated with the presence of host hormones, and these can be summarized as follows:

- the presence (or absence) of the host hormone;
- the sensitivity of the particular parasitoid stage to any host hormone present;
- the likelihood of the intact host hormone entering the parasitoid system;
- the relative potency of the host hormone homologue in the parasitoid;
- the ability of the parasitoid to negate the effects of the host hormone.

There are marked fluctuations in the patterns of both JH and ecdysteroids during the course of lepidopteran development. Moreover, there are periods when either one or both hormone types are effectively absent in the host. Obviously, these particular 'hormone-free windows' do not represent a potential endocrinological challenge to parasitoid development (unless the parasitoid is, in some way, dependent on the host's hormones for its own development). However, despite the fact that parasitoids invariably complete their pre-adult development more rapidly than their hosts, it seems unlikely that any parasitoid would be able to complete its development and metamorphosis entirely without some contact with host hormones at some time or other. Possible exceptions to this might include egg parasitoids that were somehow able to incapacitate the host embryo before any embryonic production of hormones could occur, or pupal parasitoids that could destroy the host's ability to produce hormones.

Given that most parasitoids will be exposed to host hormones at some time, we shall now consider the other factors that contribute to the potential danger presented to developing parasitoids by host hormones. At this point, it is important to make a clear distinction between the direct physiological effects of host hormones on the development of parasitoids, and any indirect effects that host hormones might have (for example, the potentially catastrophic effects of host moulting on the survival of ectoparasitoids). Additionally, we must not preclude the possibility that parasitoids might actually require the presence of host hormones, such that, at certain times, they could sequester these molecules and utilize them to regulate their own development, including moulting and metamorphosis. Lastly, because the release of host storage proteins in the last larval stadium constitutes a nutritional bonus for the final stages of the developing parasitoid and, as this release is under endocrine control, there may be advantages from host endocrine manipulation that are not directly related to any possible direct adverse effects of host hormones on the development of parasitoid larvae.

From the multitude of studies on insect endocrinology, it now seems likely that all JH homologues will have *some* activity in *all* insects. The same is probably true with regard to those 'moulting hormones' that are close homologues of 20E (e.g. makisterone A). However, whilst these various homologues might all exhibit some biological activity, it is equally likely that there will be great differences in their relative potency depending on the insect species exposed (see Staal, 1975; 1977a). As we have seen, hymenopterans probably produce and utilize only JH III, whereas the

Lepidoptera use predominantly JH I and JH II as their JHs. Thus, when considering the likely impact of host hormones on developing parasitoids, we should be mindful of the fact that the biological activity of a particular hormone may be high in the host, but much reduced in the parasitoid and *vice versa*.

If a particular hormone is present in the host at any particular time, it must be able to enter the parasitoid system in order to exert any potential deleterious activity. The major contact between developing ectoparasitoids and the hormones of their hosts will derive from the fact that ectoparasitoids *ingest* host tissues and haemolymph and, therefore, consume any hormonally active materials present in the host. By contrast, endoparasitoids not only ingest host hormones but are also *bathed* in what is, essentially, an aqueous solution of these molecules (the host haemolymph). Thus, ectoparasitoids are potentially more exposed to any host hormones that are active following ingestion. By contrast, endoparasitoids have to contend with ingested hormones, as well as those present in the host haemolymph or tissues that may be able to penetrate the cuticle of the developing parasitoid. Of course, we can not preclude the possibility that host hormones could be absorbed through the cuticle of ectoparasitoids too (especially through those parts, e.g. mouthparts and pedicels, in intimate contact with the host).

There is some debate about the activity of ecdysteroids (particularly 20E) following ingestion by insects. Neither the noctuid *L. oleracea* nor the sphingid *Acherontia atropos* were affected by very high levels (~400 ppm) of 20E in the diet (Blackford and Dinan, 1997c), whereas other species seem susceptible to ingested 20E (Blackford and Dinan, 1997b; Kubo *et al.*, 1983). In addition, there is contradictory evidence as to whether ecdysteroids are able easily to penetrate the insect cuticle (Hasegawa and Ata, 1972 and references therein). However, in general, it seems that ecdysteroids are more likely to be active following ingestion, while JHs are probably more easily taken up by absorption through the cuticle. Irrespective of the mode of entry of host hormones into the parasitoid system, even an efficient penetration of the host hormone would not necessarily guarantee that the molecule would be active in the recipient parasitoid system. This is because all insects have efficient enzyme systems capable of deactivating hormones like 20E and JH, and there is no reason to suppose that these enzymes would not be effective in metabolizing host hormones. Nevertheless, it is reasonable to assume that, at certain times, developing parasitoids will be exposed to levels of host ecdysteroids and JH that could challenge their normal development and survival. By contrast with ecdysteroids and JH, it is extremely unlikely that any of the host's neuropeptide hormones (e.g. allatostatin) would be able either to pass through the parasitoid cuticle, or to remain active following passage through the parasitoid digestive system.

The different developmental stages of insects differ markedly in their sensitivity to the effects of insect hormones. In general, developing insects are potentially 'susceptible' to exposure to exogenous JH, only at three critical times during their pre-adult development. These times (the early egg stage, the final larval stadium, and the early phases of pupal development) coincide with periods when the insect's own JH levels are necessarily low.

The physiological effects of naturally occurring JH are mimicked by the effects of synthetic JH analogues, and much of what we know about the physiological and morphological effects of excess JH derives from studies with these synthetic molecules (see Edwards and Menn, 1981; Hammock and Quistad, 1981; Retnakaran *et al.*, 1985; Staal, 1975). Normally, JH is present in insects at relatively high levels from about halfway through embryonic development until the beginning of the final larval stadium

(*Figure 6*). It is therefore not surprising that insects exposed to exogenous JH, during all but the early embryonic and final larval stages, usually exhibit little apparent effect – either on the developing insect, or on the subsequent adult. However, exposure of early embryonic stages (i.e. newly deposited eggs), frequently results in disruption of embryonic development, and prevents hatching. Embryos at later stages of development (i.e. older eggs), like most larval stages, generally remain unaffected by unnatural exposure to JH. By marked contrast, exposure of final-instar larvae to JH almost invariably produces a range of morphogenetic effects before, during, and after, the process of metamorphosis. These effects are often dose-dependent – higher levels (or more potent homologues) tend to be more effective in maintaining the juvenile (larval) state, such that supernumerary larval stages may be produced, and pupation may be delayed or even prevented. If final-instar larvae of holometabolous insects are exposed to lower levels of JH (or to less potent homologues) they usually complete the process of pupation in an apparently normal manner. However, during the subsequent course of pupal development the consequence of prior exposure to JH during the final larval stadium is revealed as a severe (and often fatal) developmental disruption. In such insects, adult emergence is severely inhibited and this is frequently associated with the formation of (non-viable) insects that appear intermediate between pupal and adult forms. In some insects, similar effects result when newly formed pupae are exposed to JH (see Edwards and Menn, 1981 and references therein).

Thus, from the point of view of parasitoid development in the host's hormonal milieu, there are three primary critical times at which host JH could pose a substantial threat to the survival of the parasite. These are (i) immediately after the parasitoid eggs have been laid by the adult female wasp, (ii) when the developing parasitoids enter their final larval stadium, and (iii) soon after the parasitoid has completed pupation. However, it is possible that the potential threat posed by the untimely presence of host JH might be more serious for endoparasitoids than for ectoparasitoids.

Whilst the inappropriate or untimely presence of host JH might pose clear problems for developing parasitoids, it is less easy to see how the absence of host JH could directly and adversely impinge on parasitoid development. This is mainly because parasitoids will, presumably, be producing their own JH as and when necessary. However, the natural absence of JH during the host's final larval stadium is the endocrine cue that initiates the formation of the host pupa. Moreover, this endocrine change in last-stadium host larvae will also initiate behavioural changes (e.g. cessation of feeding, wandering behaviour, burrowing and cocoon formation) during the latter portion of this stadium. These behavioural changes in host larvae, coupled with the physical formation of the host pupa, are potentially disastrous for endoparasitoids and ectoparasitoids alike. In such circumstances, the indirect effects of reduced (or absent) host JH might have led to the evolution of mechanisms to counteract reduced host JH during this period. For example, stimulation of the production of host JH (or a reduction of metabolic deactivation) would tend to result in deferred host pupation, and the absence of any associated behaviours in the host. Conversely, should there be advantage to the parasitoid in prematurely encouraging pre-pupal behaviour (but not pupation), this might be accomplished by prematurely reducing host JH production or its action. Such hormonal manipulation in earlier host instars may also result in the beneficial induction of the (premature) release of storage proteins into the haemolymph of host larvae. Reduction of JH levels in pre-adult insects usually results in premature initiation of metamorphosis and in the production of (non-viable) premature pupae. Such phenomena are apparent when insects are treated with a variety of natural and synthetic molecules that exhibit anti juvenile hormone activity (Schooley and Edwards, 1996; Staal, 1986). Similar effects could be expected to result if parasitoids were able to induce the production of, or secrete, host allatostatins.

For reasons discussed above, the presence of host ecdysteroids might be less of a hazard for developing endoparasitoids than the presence of host JH. In general, the presence of ecdysteroids (specifically 20E) will induce moulting in all larval stages for much of the time (providing the insect tissues have developed sufficiently to be responsive to this hormone). If host ecdysteroids are present, and are out of phase with endoparasitoid moulting, the parasitoid could be induced to moult at an inappropriate time. Such untimely moulting of larval insects (which can also be induced experimentally by injection of ecdysteroids) is almost always accompanied by difficulties in completing the moult successfully. Often, these difficulties are manifest as the inability of the larval insect to completely shed the old cuticle, (especially at the head and appendages), rupture of the cuticle, cessation of feeding, and the eventual death of the affected larvae. Clearly, there is potential here for host ecdysteroids to constitute a hazard to endoparasitoid survival because of the possibility that there may be induction of parasitoid ecdysis at developmentally inappropriate times. Again, this will depend crucially on whether or not host ecdysteroids are able to enter the endoparasitoid in physiologically relevant quantities and configurations.

By contrast, host ecdysteroids present a more or less continual, and potentially catastrophic, danger for ectoparasitoids – simply because these hormones initiate the shedding of the host cuticle on which these parasitoids must reside. Thus, it is not surprising that ectoparasitoids have evolved mechanisms for manipulating host ecdysteroids, not because of their direct impact on the physiology of moulting cycles in the parasite, but because it is essential for survival that the parasitoid larvae do not become detached and cast from the host together with the cuticular exuvium.

7. Manipulating hormonal control – strategies for parasitoid survival

Parasitoids have evolved many mechanisms to counteract the potentially hostile environment of the host. In relation to dealing with host endocrinology, the different groups of parasitoids seem to have exploited many of the available options that, potentially, enable them either to increase, or decrease, host hormone levels in response to any potential hazard that these potent biochemical agents might pose. If host hormones do indeed constitute a potential hazard to the parasitoid, then, in the majority of circumstances, the response will be to eliminate or reduce the presence or activity of the threatening molecule. Parasitoids that have evolved systems designed to counteract the impact of host hormones in this way are the archetypal 'regulators' envisaged by some of the pioneers in this field (e.g. Lawrence, 1986a; Vinson and Iwantsch, 1980a). However, in some circumstances, parasitoids may actually derive advantage by stimulating the (untimely) presence or activity of host hormones. An example of this strategy might be the stimulation of host JH levels during the host's final larval stadium in order to defer or prevent host pupation. Thus, having evolved effective mechanisms for manipulating host hormones in order to avoid their potential deleterious effects, parasitoids may have further evolved to exploit these mechanisms for their own purposes (i.e. to optimally regulate the host's endocrine physiology for the exclusive benefit of the parasitoid). Originally, such mechanisms may have evolved to protect developing parasitoids from the direct effects of the hormones of their hosts, but the acquisition of this manipulative ability may have provided even greater adaptive advantage for parasitoids in their co-evolutionary battle for survival against their hosts. Thus, there may have been an evolutionary development of parasitoids from 'conformers' to 'regulators' and ultimately, perhaps, to 'exploiters'.

8. Manipulating hormonal control – choosing the optimal time

In Section 2.2 we have seen that, as a result of changes in the rate of growth (and therefore size) of lepidopteran larvae, there is a clear nutritional advantage for parasitoids to utilize host larvae when these have reached the penultimate or ultimate larval stadia (i.e. when the host body weight is close to maximum). There is also an additional nutritional benefit in utilizing final-instar host larvae because, during this developmental stage, there is a substantial release and availability of highly nutritious storage proteins from the host fat body into the haemolymph. In fact, even when parasitoids attack the earlier developmental stages of hosts (e.g. eggs and younger larvae), they often delay their own development in order that the maximum growth rate of the parasitoid larvae coincides with the later stages of host development. However, there are some dangers inherent in this strategy. This is particularly true when endoparasitoids attempt to exploit the final larval stadium of host insects. Thus, although the nutritional resource provided by the host may be maximal at this time, the final-instar host larvae are destined to undergo metamorphosis. The act of host pupation, not only presents the developing parasitoid with the prospect of having to emerge through a thick and sclerotized pupal cuticle, but also exposes the parasitoid to the commensurate pre-pupal behaviours of the host (burrowing, cocoon-spinning, etc.). Faced with these problems, parasitoids may either abandon the host before the onset of these events (as many do), or delay or prevent the act of pupation and its associated behaviours. Such host manipulation may be accomplished by increasing host JH levels in the final stadium, and/or by modifying the pattern of host ecdysteroids in this stage. Interestingly, when host ecdysteroids are manipulated, this is often done at a specific time in the development of the final-instar host larva, such that although pupation is prevented, the release of storage proteins still occurs.

An alternative strategy for developing parasitoids might be to exploit the host when it has reached the penultimate stage of larval development. Here, the nutritional resource of the host would not be as great as it would be in the final larval stadium, but the prospect of host pupation is negated. In fact, endoparasitoids that complete their development in the penultimate stadium of the host often deliberately manipulate the host JH and ecdysteroids to mimic the profiles normally seen in final stadium host larvae. Presumably, one reason for this is to induce (prematurely) the release of storage proteins. However, where this occurs, the host larva is also prevented from pupating, and many such insects remain developmentally arrested as precocious pre-pupae.

9. Manipulating hormonal control – ways and means

Before beginning to discuss the various ways in which parasitoids might manipulate the hormonal systems of their hosts, it is worth remembering that many parasitoids obviate the necessity for such action simply by avoiding the potential problems posed by the host's endocrine system. Perhaps the most obvious examples of this strategy can be seen in the 'conformers' (Lawrence, 1986a; Vinson and Iwantsch, 1980a) – parasitoids that synchronize their own endocrine systems with those of the host insect and thus avoid any 'endocrinological conflict of interests'. However, there are other cases in which parasitoids may reduce the potentially damaging impact of host hormones, by attacking host developmental stages at times when the host endocrine system is inoperative or quiescent. For example, some egg parasitoids (e.g. *Trichogramma* spp.) show a marked preference for the newly deposited eggs of their hosts (Pak, 1986; Reznik and Umarova, 1990; Ruberson and Kring, 1993), and this, among other factors (Strand, 1986), may be linked with the fact that the (potentially ovicidal) JH content of freshly deposited eggs is likely to be much lower than that in more mature eggs (Section 3.2). Similarly, parasitoids of lepidopteran pupae are unlikely to come into contact with host JH, since these hormones are essentially absent during pupal development (Section 3.4).

Nonetheless, it is clear that many parasitoids will be exposed to host hormones, the effects of which could be potentially harmful to the survival of the parasitoid, or be indirectly at risk from the natural consequences of host endocrine secretions (e.g. the effects of host moulting on the survival of ectoparasitoids). In many (if not all) of these cases, parasitoids would need to protect themselves against these threats to survival by manipulating the endocrine system of their hosts. Similarly, by exploiting host endocrinology (e.g. to prematurely induce, or postpone, certain physiological events in host larvae), parasitoids may also significantly increase their chances of survival. In the following sections, we discuss the many mechanisms that parasitoids could employ to effect a change in host endocrinology.

9.1 *Potential mechanisms for down-regulating host juvenile hormones*

At first sight, it might appear that parasitoids, confronted with the potentially harmful effects of host JH, would need to reduce the level and/or activity of the hormone concerned. In fact, as we shall see below, this is a relatively unusual phenomenon. Perhaps this can be explained by the fact that, with the exception of the very earliest stages of embryonic development, the last stages of larval development, and for a short time at the beginning of the pupal stage, developing insects are generally unaffected by exogenous JH.

The reduction of host JH (where necessary or appropriate for parasitoid survival) might be achieved in a number of ways. Parasitoids could reduce JH biosynthesis by host CA, either by direct interference with the enzymes involved in the biosynthetic pathway, or indirectly by down regulating CA activity using allatostatic neuropeptides. Alternatively, parasitoids could reduce host JH by increasing the rate of JH degradation (i.e. by increasing the activity of JH metabolizing enzymes such as JH esterases). In relation to the first option, it is noteworthy that there are substantial differences in the enzymic sequence and substrate utilization during JH biosynthesis between JH III producers (like parasitoids) and the Lepidoptera (which produce predominantly JH I and JH II). These differences occur both in early and in late steps in JH biosynthesis (Schooley and Baker, 1985). Lepidopterans derive the basic building blocks of JH I and II from propionate and homomevalonate, whilst cockroaches, locusts, (and probably all JH III producers) utilize mainly acetate and mevalonate. In addition, during the terminal stages of JH biosynthesis, epoxidation of the JH molecule in lepidopterans occurs before methylation to the ester, whereas in JH III producers, it appears that esterification precedes epoxidation (Schooley and Baker, 1985). These small, but significant differences could provide parasitoids with mechanisms for reducing the rate of JH I and II biosynthesis in host CA, without compromising their own production of JH III.

In relation to the possible deployment of allatostatic neuropeptides in order to reduce host JH production, we have already seen that there is some evidence pointing to the possibility that hymenopterans utilize cockroach-type allatostatins, whereas it is becoming increasingly clear that the *M. sexta* peptide (Kramer *et al.*, 1991) is probably the functional allatostatin in all lepidopterans (Section 2.4). Because these peptides are very different in structure (Section 2.4), and because cockroach-type allatostatins do not appear to affect lepidopteran CA and *vice versa* (Audsley *et al.*, 1999; Kramer *et al.*, 1991), there is again ample opportunity for selective inhibition of host CA, independent of the activity of the parasitoid's own endocrine glands. Of course, this assumes that parasitoids are able to produce and secrete the appropriate peptides into the haemolymph of the host (the *M. sexta* allatostatin is biologically active when injected into the haemocoel of *L. oleracea*, but is inactive if given orally; N. Audsley, personal communication). In addition, secretion of allatostatins would need to be more or less constant since, for example, the half-life of the *M. sexta* peptide in the haemolymph of larvae of *L. oleracea* is estimated to be between 2 and 5 minutes (N. Audsley, personal communication). Currently however, no reported case in which parasitoids manipulate host allatostatins exists, although parasitoids are certainly capable of inducing the accumulation of neuropeptides in the host central nervous system (Zitnan *et al.*, 1995a). Thus, given the possibility that parasitoids use allatostatic peptides that differ markedly from those of their lepidopteran hosts (Section 2.4), and that parasitoids display remarkable resourcefulness when manipulating host endocrine systems, the manipulation of host allatostatins as a means to reduce host JH remains a tantalizing possibility.

A further possibility for down-regulating JH comes from the opportunity to exploit JH metabolizing enzymes. Because the finite quantities of JH present in insect haemolymph will be a balance between the amount synthesized by the CA and the rate at which the hormones are metabolized and excreted, insects possess highly efficient enzyme systems capable of degrading JH. These mainly comprise the JH-specific esterases present in the haemolymph, and the epoxide hydrolases which may be present in other tissues (see review by Hammock, 1985).

It is perhaps surprising, given the above, that there is to date no unequivocal evidence that parasitoids use either interference with JH biosynthesis directly, or via allatostatins, to down-regulate host JH. In fact, examples of parasitoid-induced reductions in host JH levels are rather few and far between. Moreover, it appears that in most cases where parasitoids reduce host JH, they do this not because the host hormones represent a threat to the normal developmental programme of the parasitoid, but in order to induce the premature onset of metamorphosis (or the behaviours associated with pupation) in the host. Perhaps the best understood example of this is the interaction between the braconid parasitoid, *C. inanitus,* and its host, *S. littoralis*. In this relationship, the host is induced to undergo premature metamorphosis in the penultimate larval stadium. This is accomplished by the parasitoid larva up-regulating JH esterases (Steiner *et al.*, 1999) and/or prematurely inactivating host CA (Pfister-Wilhelm and Lanzrein, 1996). In this case, the parasitoid larva plays an important role in the manipulation of host JH levels (Steiner *et al.*, 1999). In the relationship between the parasitoid, *Copidosoma floridanum*, and its host, *T. nis*, (Strand *et al.*, 1990, 1991) there is evidence that host JH levels are increased in the 3rd and 4th larval stadia (see below), but decreased in the 5th stadium – again by increasing the metabolism of JH by JH esterases. Given that several other parasitoids are known to induce premature onset of metamorphosis behaviour in their hosts, it is likely that more examples of parasitoids reducing host JH will be discovered.

In summary, our current state of knowledge suggests that down-regulation of host JH by parasitoids is a relatively unusual event. Where such depletion of host JH does occur, it seems that this is predominantly accomplished by up-regulation of JH metabolisms (usually via stimulation of JH esterases). Such down-regulation of JH normally occurs in the penultimate larval stadium of the host's development. Moreover, the down-regulation of host JH appears to occur not as a defensive response by the developing parasitoid, but in order to induce premature pre-metamorphosis behaviour in the host and, particularly, to stimulate the premature release of host storage proteins. To date, there is little evidence that reduction in host JH is induced by direct action on the host CA, by indirect action (e.g. via allatostatins), by interference with the feedback regulation of JH levels, or by interaction with host JH receptors.

9.2 *Potential mechanisms for up-regulating host juvenile hormones*

By contrast with the apparent lack of examples in which parasitoids are known to reduce host JH, there are many cases in which parasitoids have been shown to increase the levels of JH in their lepidopteran hosts. Such hormonal manipulation is, self-evidently, not a response to any threat posed by host JH, but rather a deliberate manipulation of host development in order to optimize the survival of the parasitoid. As with down-regulation of host JH, there are several ways in which such manipulation might be accomplished. These include (i) secretion of JH (II) directly into the host haemolymph, (ii) stimulation of host CA (either directly or by production of neuropeptide alloproteins, (iii) decreasing metabolism of JH (e.g. by down-regulating JH esterase), (iv) re-activating host hormone metabolites (e.g. converting JH acids to the biologically active esters), or (v) stimulating host JH receptors.

Whilst some adult female parasitoids are known to produce relatively high levels of JH in their ovaries (Höller *et al.*, 1994), and to use this molecule as an oviposition marking pheromone, there is no evidence that female wasps inject JH into the host in an attempt to modify host endocrinology. Conversely, Joiner *et al.* (1973) suggested that the teratocytes of the parasitoid, *C. nigriceps,* might produce and secrete JH III into the body of host larvae. However, in almost every case where parasitoids have been shown to increase the JH content of host larvae, this is accomplished by down-regulation of JH metabolism (predominantly JH esterase). In *H. virescens* larvae, injection with teratocytes from the parasitoid, *Microplitis croceipes,* markedly reduced JH esterase activity (Zang *et al.*, 1992), and a protein secreted by the teratocytes inhibits protein synthesis and JH esterase activity in host fat body cultures maintained *in vitro* (Schepers, *et al.*, 1998). The reduction in JH metabolism (which presumably results in abnormally high JH at a critical time) has been linked to incomplete larval-pupal moulting and delayed pupation (Zang and Dahlman, 1989). Interestingly, the mechanisms responsible for the down-regulation of JH esterase in this relationship are thought to result from suppressed transcription of JH esterase genes and/or post transcriptional or translational modification of the protein (Dong *et al.*, 1996).

In *H. virescens* larvae parasitized by a different braconid (*Microplitis demolitor*), a similar reduction in JH esterase activity appears to be achieved by polydnaviruses (PDV), present in the calyx fluid, and injected by the adult female wasp at the time of oviposition (Dover *et al.*, 1995b). Again, the reduction in JH esterase activity in parasitized larvae is associated with developmental arrest and failure to pupate (Strand *et al.*, 1988a; Strand and Dover, 1991). The effects of the PDV are enhanced by venom from the female wasp (Strand *et al.*, 1994) and, although *M. demolitor* teratocytes can also produce developmental disruption in host larvae, these effects are augmented by the activity of the PDV (Strand and Wong, 1991). When larvae of the gypsy moth (*L. dispar*) are parasitized by the braconid *Glyptapanteles liparidis*, JH II and III levels in late 4th-stadium larvae are substantially higher than those in unparasitized larvae (Schopf and Rembold, 1993; Schopf *et al.*, 1996). These increases in JH levels were associated with depleted JH esterase activity in parasitized larvae (Schopf *et al.*, 1996). This is yet one more example in which increased JH levels are associated (probably causitively) with reduced JH esterase activity. In this case, however, there is also evidence that the parasitoid larva is secreting JH III into the host haemolymph – JH III titres were 10 – 100 times higher in parasitized 4th-instar larvae than in unparasitized 4th instars (Schopf and Rembold, 1993). However, other mechanisms may also lead to increased JH levels. For example, in 3rd and 4th-stadium larvae of *T. ni* parasitized by *C. floridanum*, titres of JH were higher than in unparasitized larvae (Strand *et al.*, 1991) although the activity of JH esterase did not differ between parasitized and unparasitized larvae (Strand *et al.*, 1990).

In summary, our current state of knowledge suggests that up-regulation of host JH by parasitoids is a relatively common event. Where such increases in host JH levels occur, it seems that this is predominantly accomplished by down-regulation of JH esterases. There is little evidence that parasitoid larvae secrete JH (III) into host haemolymph – in any case, this would be a dangerous strategy in view of the fact that parasitoid larvae are usually in their final stadium and approaching pupation, at the time when up-regulation of host JH is most frequently observed. Increased JH in parasitized hosts usually occurs in final-instar host larvae, where it is probably associated with postponing host wandering, burrowing and/or spinning behaviour and delaying or preventing pupation. To date, there is little evidence that parasitoid-induced increased host JH is achieved by direct action on the host CA, by indirect action (e.g. via allatotropins), by interference with the feedback regulation of JH levels, or by interaction with host JH receptors. Similarly, we are unaware of any incidence of parasitoids inducing the conversion of primary metabolites host JH (e.g. JH acids) back to the active ester forms of the hormones. Finally, the majority of instances in which parasitoids are known to manipulate host JH levels involve endoparasitoids. This may be because these parasitoids are potentially at greater risk from the action of host JH than are ectoparasitoids. However, we have seen that most cases of manipulation of host JH by parasitoids are not, at first sight, defensive measures to counteract the potential physiological effects of host JH on the developing parasitoid, but rather are designed to modify host development and/or behaviour. Moreover, recent studies indicate that some ectoparasitoids (e.g. *Eulophus pennicornis*) also produce marked changes in the activity of host JH esterases –particularly during the host's final larval stadium (G.C. Marris, unpublished results).

9.3 *Potential mechanisms for down-regulating host ecdysteroids*

In Section 2.5, we have described how ecdysteroids (ecdysone and/or its precursors) are produced by the prothoracic glands, and converted to the active moulting hormone, 20E. The PGs are themselves under the stimulatory control of PTTH produced in the insect brain. Following initiation of the biochemical and physiological events associated with moulting, 20E is metabolized to inactive, more polar products. Thus, parasitoids could reduce the levels of host ecdysteroids in a number of ways. First, by inhibiting the production of moulting hormone precursors by the PGs either directly (by causing degeneration of the glands, or in some other way by suppressing ecdysone biosynthesis), or indirectly, by interfering with the production or activity of PTTH. Second, parasitoids could block the normal conversion of ecdysone to 20E. Third, parasitoids might induce more rapid degradation of 20E in a way that is similar to the production of inactive metabolic conjugates by the action of the ecdysteroid glucosyl transferase machinery employed by some insect baculoviruses (see Chapter 14). Finally, parasitoids could interfere with host ecdysteroid function by inactivating or blocking host ecdysteroid receptors.

9.4 *Potential mechanisms for up-regulating host ecdysteroids*

Up-regulation of host ecdysteroids might be achieved by stimulation of PG activity (either directly, or via PTTH), by stimulating the conversion of ecdysteroid precursors to 20E, by inhibiting metabolic degradation of 20E, by secretion of ecdysteroids into the host haemocoel, or by stimulating host ecdysteroid receptors.

In the following section we shall examine some instances in which the modulation of lepidopteran host ecdysteroids by parasitoids have been studied in detail, in order to identify which, if any, of these potential mechanisms for hormonal manipulation are employed.

In relation to the up-regulation of host ecdysteroids by parasitoids, while there is some evidence that this does actually occur (see below), there is little information about how it is achieved. Obviously, the stimulation of ecdysteroid biosynthesis by host PGs, or the acceleration of the enzymatic formation of 20E, would be prime candidates. However, the larva of at least one parasitoid (*A. reticulatus*) has been shown to release ecdysteroids *in vitro* (Brown *et al.*, 1993), and therefore has the potential to increase host ecdysteroids in this manner.

Beckage and Riddiford (1978) began to study the interaction between the parasitoid *C. congregata* (= *Apanteles congregatus*) and its host, *M. sexta*. In this system, parasitism is associated with developmental arrest and inhibition of moulting linked with effects on the endocrine system during the later stages of host development (Beckage and Riddiford, 1982a). In parasitized larvae the altered ecdysteroid profile, associated with the inhibition of host moulting, was initially thought to be due to the inhibition of the conversion of ecdysone to 20E by ecdysone 20-monooxygenases (Beckage and Templeton, 1986). Certainly, there was no evidence for degeneration of the PG in parasitized insects, or in insects injected with *Cotesia* PDV (Beckage and de Buron, 1993). Gelman *et al.* (1998) showed that manipulation of *M. sexta* ecdysteroid levels by *C. congregata* was achieved in two ways. First, parasitization neutralizes the ability of the host to maintain normal levels of ecdysteroids and second, induces an increase in host ecdysteroids at a time that coincides with the emergence of the parasitoids from the host. It appears that the PG from parasitized hosts are less responsive to stimulation by PTTH, but remain active *in vitro* for longer than those removed from unparasitized hosts (Kelly *et al.*, 1998). These authors suggested that parasitism circumvents the normal PTTH control of ecdysteroid synthesis and, recently, Gelman *et al.* (1999) demonstrated that the larvae of *C. congregata* were capable of synthesizing and releasing ecdysteroids, including 20E. A more detailed analysis of the interaction between this parasitoid and its host *M. sexta* can be found in Chapter 3 of this volume.

When larvae of the noctuid, *H. virescens*, are parasitized by the braconid, *C. nigriceps*, they exhibit prolonged larval development and fail to pupate (Tanaka and Vinson, 1991a). The parasitoid suppresses the function of the host PGs and their ability to respond to PTTH (Tanaka and Vinson, 1991a), resulting in a marked reduction in ecdysteroid titres and changes in ecdysteroid profiles during the 5th larval stadium (Pennacchio *et al.*, 1992a, 1994a). Prothoracic glands from parasitized larvae are inactivated and produce less ecdysteroid than their unparasitized counterparts – a phenomenon ascribed to increased degradative metabolism of ecdysteroids (Pennacchio *et al.*, 1994b), reduced protein synthesis/transduction in the PG (Pennacchio *et al.*, 1997), and an inhibition of the PTTH signal transduction pathway (Pennacchio *et al.*, 1998b). Furthermore, these hormonal manipulations have been closely linked with the presence of PDV injected by the parasitoid at the time of oviposition (Pennacchio *et al.*, 1998a). A more detailed analysis of the endocrine interaction between *H. virescens* and *C. nigriceps* can be found in Chapter 7.

The development of the solitary endoparasitoid, *Microplitis croceipes*, is also closely related to that of its host, *H. virescens*, and involves developmental arrest mediated by hormonal manipulation (Webb and Dahlman, 1985, 1986). In this case, reduced ecdysteroid levels in parasitized larvae is again believed to be due to some modulation of the PTTH-PG axis (e.g. failure to release PG stimulatory factor), an insufficiency of sterol precursors, and/or the inhibition of release of PTTH or the failure of the PG to respond to this hormone (Dahlman *et al.*, 1990a). Further discussion of this system is presented in Chapter 4.

Several research groups have studied the endocrine interaction between the solitary, egg-larval endoparasitoids of the genus *Chelonus*, and their noctuid lepidopteran hosts. Typically, these parasitoids induce the premature onset of metamorphosis in the penultimate larval stadium of their hosts and then arrest further development in the precocious pre-pupal state. Jones *et al.* (1992b) reported substantial reductions in host ecdysteroids in precocious pre-pupae of *T. ni* parasitized by *Chelonus* near *curvimaculatus*. They concluded that the parasitoid larva inhibits the normal ability of the host to maintain its own changing pattern of ecdysteroid production, and initiates precocious pupal development by mimicking the pattern of ecdysteroids normally present in the host final larval stadium (Brown and Reed-Larsen, 1991; Jones *et al.*, 1992b; Jones, 1996). Such premature replication of the hormonal events, normally associated with the final larval stadium of unparasitized lepidopteran larvae, appears to be the common means utilized by chelonine wasps to induce premature metamorphosis in penultimate instar hosts. Such manipulation may involve parasitoid regulation of both ecdysteroids and JH (Grossniklaus-Bürgin *et al.*, 1989; Grossniklaus-Bürgin and Lanzrein, 1990b). Similarly, the parasitoid, *C. inanitus*, induces the premature onset of metamorphosis in penultimate instars of *S. littoralis*, again followed by developmental arrest of the host as a (premature) pre-pupa (Grossniklaus-Bürgin *et al.*, 1994). Once again, both JH and ecdysteroids in hosts are modified, and inhibition of host PG is associated with the presence of PDV and/or venom from the parasitoid (Grossniklaus-Bürgin *et al.*, 1998). A more detailed discussion of this system can be found in Chapter 5.

Manipulation of host ecdysteroids by parasitoids may also occur in final-instar host larvae. For example, ecdysteroids in 5th-stadium larvae of *T. ni* parasitized by *C. floridanum* peak about 24 h later than in unparasitized larvae (Strand *et al.*, 1990). This delay is associated with the fact that parasitized larvae cease feeding and initiate wandering one day later than unparasitized larvae. Yet again, this host manipulation stems from modifications to both host ecdysteroids and JH.

The majority of the above examples relate to endoparasitoids, but we have speculated that the control of hosts ecdysteroids (and therefore host moulting) may be more pertinent for ectoparasitoids. In the two ectoparasitoid systems that have been investigated, convincing evidence has been obtained that host ecdysteroids are down-regulated at critical times in order to prevent moulting. In both *T. ni* parasitized by *E. plathypenae* (Coudron *et al.*, 1990; Kelly and Coudron, 1990; and Chapter 9 of this volume) and *L. oleracea* parasitized by *Eulophus pennicornis* (Weaver *et al.*, 1997; and Chapter 8, this volume) inhibition of host moulting is associated with suppression of ecdysteroid production by the host PG. In the case of *Eulophus pennicornis*, effects on the PG are not direct, but may result from inhibition of the production or action of host PTTH (Marris *et al.*, 2001, and Chapter 8, this volume).

In summary, many parasitoids do modify the ecdysteroid endocrinology of their hosts. Ectoparasitoids appear to do this primarily to prevent host moulting. Endoparasitoids manipulate host ecdysteroids (often in conjunction with host JH) in order to induce premature initiation of metamorphosis (often followed by developmental arrest of the host as a precocious pre-pupa), or to delay the normal onset of metamorphosis, pupation, and associated behaviours. Manipulation of host ecdysteroids is often achieved by the inactivation of the host's own ecdysteroid producing system (usually by interfering with the action of PTTH on the PG) and by the production of ecdysteroids at inappropriate (for the host) times. There is some evidence that parasitoids may alter host ecdysteroid profiles by manipulating ecdysteroid metabolism, and some investigations point to the possibility that host ecdysteroid receptors (particularly in the epidermis) may also be a target (Coudron *et al.*, 1990).

10. Manipulating hormonal control – tools of the trade

It is clear that parasitoids utilize a variety of mechanisms to modify the hormone systems of lepidopteran hosts. The precise biochemical nature of these mechanisms is currently the subject of intense investigation by researchers. However, it has become apparent that parasitoids deploy a diverse armoury that includes a range of venom components (see Chapter 2), polydnaviruses, virus-like particles, and other maternal oviposition factors from adult female wasps, as well as teratocytes, enzymes, parasitism-specific proteins, and other secretions (including hormones) derived from the parasitoid larvae, in order to manipulate and exploit the host endocrine system for the ultimate survival of the parasitoid. Often, parasitoids modify both of the two developmentally important hormones of their lepidopteran hosts (i.e. JH and ecdysteroids), suggesting that this manipulation is both complex and precise – and, perhaps, indicating that during the course of this 'endocrinological co-evolution', parasitoid wasps have evolved a better 'understanding' of the subtleties and regulation of the endocrine systems of their insect hosts than is at present understood by insect physiologists. In the following chapters of this book, the reader will find detailed accounts of specific examples of the endocrine interactions of parasitoids, parasites and pathogens with their insect hosts. We hope that the information presented in this book will stimulate further exploration of this fascinating and important area of biological research.

Acknowledgements

The authors thank the Pesticides Safety Directorate of the Department for Environment, Food and Rural Affairs, UK, for financial support. We also thank Mrs J.E. Short for conducting ecdysteroid RIA in *L. oleracea*, and Mrs Anne Smith for the original artwork for Figure 1.

2

Identity and mode of action of the host endocrine disrupters from the venom of parasitoid wasps

Robert J. Weaver, Gay C. Marris, Howard A. Bell and John P. Edwards

1. Introduction

Parasitoid wasps have evolved a variety of means for regulating the physiology of their hosts for the benefit of their developing parasitoid offspring. These processes probably evolved at different times and independently in different groups (Gauld, 1988; Whitfield, 1992a, b; 1998). The endocrine targets for modulation are essentially the same in different hosts (see Chapter 1), but the mechanisms of control by different parasitoids may differ considerably. This chapter reviews the nature and identity of the venoms and associated secretions of the parasitic Hymenoptera. Most emphasis has been placed on factors of maternal origin that have been implicated in interfering, whether directly or indirectly, with the host endocrine system, although no venom components have been deliberately overlooked.

The principle groups of the parasitoid Hymenoptera together with species in the order with which they are dealt with in the text are shown in *Table 1*. Some parasitoid groups have been excluded, mainly because of a lack of relevant information concerning their venoms. The omitted groups include two superfamilies of the Symphyta (Orussidea and Stephanoidea) that are ectoparasitic on wood-boring beetles and siricid woodwasps, and may have paralysing venoms (Gauld and Bolton, 1988; Gauld, 1995; Whitfield, 1998). Also omitted are the Proctotrupoidea (except for Scelonidae), Trigonalyoidea, Megalyroidea, Ceraphronoidea, Evanioidea, and Cnipoidea. One family of the Cnipoidea, the Figitidae, includes species (e.g. *Leptopilina heterotoma*) that are believed to use viruses or virus-like particles to suppress the immune responses of their dipteran hosts (Rizki and Rizki, 1990). Whether these wasps also have venomous components that fall within the scope of this review is presently unknown.

Endocrine Interactions of Insect Parasites and Pathogens, edited by J.P. Edwards and R.J. Weaver.

Table 1. Phylogenetic relationship of some of the common parasitoid genera

Superfamily	Family/sub-family	Genera discussed in this chapter
ORUSSIDEA		
STEPHANOIDEA		
ICHNEUMONOIDEA		
	Ichneumonidae	*Pimpla, Venturia, Campoletis, Tranosema, Rhyssa*
	Braconidae	
	Braconinae	*Bracon (Habrobracon, Microbracon)*
	Cheloninae	*Chelonus*
	Microgastrinae	*Cotesia, Microplitis, Cardiochiles*
	Euphorinae	*Meteorus*
	Aphidinae	*Aphidius*
CHRYSIDOIDEA		
	Bethylidae	*Goniozus*
VESPOIDEA		
	Pompilidae	
APOIDEA		
	Sphecidae	
CYNIPOIDEA		
	Figitidae	*Leptopilina*
PROCTOTRUPOIDEA		
	Scelonidae	*Telonomus*
CHALCIDOIDEA		
	Agaonidae	(Fig wasps)
	Aphelinidae	
	Chalcididae	
	Elasmidae	
	Encyrtidae	
	Eucharitidae	
	Eulophidae	*Hyssopus, Euplectrus, Eulophus, Colpoclypeus*
	Eupelmidae	*Eupelmus*
	Eurytomidae	
	Leucospidae	
	Mymaridae	
	Ormyridae	
	Perilampidae	
	Pteromalidae	*Nasonia, Catolaccus, Asaphes*
	Signiphoridae	
	Tetracampidae	
	Torymidae	
	Trichogrammatidae	
PLATYGASTROIDEA		
TRIGONALYOIDEA		
MEGALYROIDEA		
CERAPHRONOIDEA		
	Megaspilidae	*Dendrocerus*
EVANIOIDEA		

2. Morphology of the venom apparatus in the parasitoid Hymenoptera

The comparative morphology of the venom apparatus in adult female parasitic wasps has been described for several Ichneumonidae and Braconidae (Edson and Vinson, 1979; Edson *et al.*, 1982; Moreno and Jimenez, 1993; Robertson, 1968; Togashi, 1963; van Marle, 1977; van Marle and Piek, 1986). In the ichneumons and braconids, the venom apparatus consists of a venom-secreting gland (the acid gland), which can be simple or branched, a storage reservoir (together with venom duct), and an associated Dufour's (or alkaline) gland. The duct of the venom reservoir, and that of the Dufour's gland, extend to a common oviduct or vagina, in which the secretions are combined with those of specialized areas of the lateral oviducts (e.g. the calyx tissue), and are ejected through a chitinous sting (or terebra), which also serves as the ovipositor. The contents of the venom reservoir, as well as other materials derived from epithelial cells of the lateral oviducts, which may include viruses, are injected into the host, either before, during, or after egg laying, and have effects on host immune defences and other aspects of host physiology, including development and behaviour (Coudron, 1991; Vinson and Iwantsch, 1980a). Venoms of many parasitoids cause host paralysis, which can be partial or complete, and transient or permanent, depending on the parasitoid species, and sometimes on that of the host (Piek, 1985).

Edson and Vinson (1979) examined the venom apparatus of 160 representative species of parasitoid wasps from 26 genera of Braconidae, and found two basic types of gland. The type 1 venom gland occurs in the more ancestral subfamilies (e.g. Braconinae), and consists of many glandular filaments and a cone-shaped muscular reservoir with an unevenly thickened cuticular lining. The type 2 venom apparatus is found in subfamilies having more advanced character traits (e.g. Cheloninae and Microgastrinae), and consists of only two glandular filaments with a thin-walled reservoir surrounded by relatively few muscles. The type 1 venom gland is found in wasps that are either ectoparasitoid or endoparasitoid, but in which the larvae of the wasp pupate inside the remains of the host. The type 2 gland is found predominantly in endoparasitoids that pupate outside the remains of the host (Edson and Vinson, 1979). The morphology of the braconid type 2 venom apparatus resembles the type of gland that is found in the Ichneumonidae (Togashi, 1963), and is indicative of an evolutionary continuity (Edson and Vinson, 1979). The structure of the venom apparatus of the Chalcidoidea has been less widely studied, although Ratcliffe and King (1967) provide details for *Nasonia vitripennis* (Pteromalidae), and Uematso and Sakanoshita (1987), studying the eulophid, *Euplectrus kuwanae*, describe a structure that is similar to that of the braconid type 2 morphology, except that it is composed of two reservoir pouches, and has a single tubular venom gland that originates from the apex of the main reservoir. The venom apparatus of the ichneumon wasp, *Pimpla hypochondriaca*, is shown *Figure 1*.

3. The paralysing and host-conditioning factors of the Ichneumonoidea

The superfamily, Ichneumonoidea, comprises two very large and biologically diverse parasitoid families (Ichneumonidae and Braconidae), that parasitize a wide variety of invertebrates. All species of the family Ichneumonidae are either internal or external

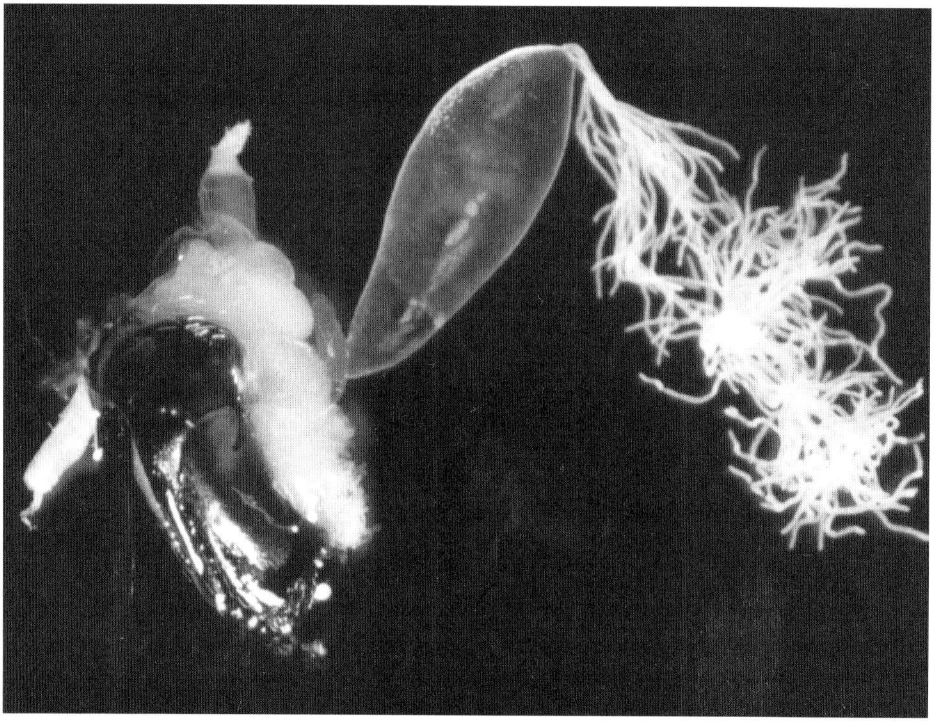

Figure 1. The venom apparatus of ichneumonid wasp, Pimpla hypochondriaca, *consisting of a filamentous venom gland (right) and thin-walled venom reservoir (centre).*

parasitoids of holometabolous insects, with preference for lepidopteran larvae. The basal lineages are ectoparasitic on wood-boring beetles and wasps, and other endophytic insects, the attacked larvae of which are paralysed (usually temporarily or incompletely) or sometimes killed during oviposition (Gauld, 1988; Piek, 1985; Piek and Spanjer, 1986; Whitfield, 1998). The venoms of the more derived groups of ichneumons are able to invoke a variety of developmental and immunological effects in host larvae, as well as altering the host's value as a nutritional resource (Coudron, 1991; see also Chapter 11).

3.1 *Venoms of the Ichneumonidae*

The solitary ichneumonid endoparasitoid, *Venturia canescens,* allows its host larva (*Ephestia kuhniella*) to feed and grow, and to finish larval development, even spinning a cocoon, before killing and consuming it. There is little to suggest a direct influence of *V. canescens* maternal factors on the host endocrine system. Host paralysis by *V. canescens* is either short-lived or absent, and may depend upon the age of the female wasp (Piek and Simon Thomas, 1969). A major component of the wasp's ovipositional secretion is a non-venom, virus-like particle (VLP) material, which is devoid of nucleic acid and is secreted onto the surface of the wasp egg as it passes through the calyx (Feddersen *et al.*, 1986). This coating protects the parasitoid egg from the immune response of the host. The VLP protein possesses antigenic determinants similar to host protein, hence the coated egg does not appear as a foreign body (Fedderson *et al.*, 1986; Hellers *et al.*, 1996; Theopold *et al.*, 1994). Ovarian calyx fluid from which the VLP has been removed contains a putative serine protease inhibitor (serpin) that may provide short-term or complementary inhibition of host immune defences, such as haemolymph phenoloxidase activation and haemocyte spreading (Beck *et al.*, 2000).

The Pimplinae are a large and cosmopolitan subfamily of ichneumons with a broad range of parasitic habits. Some (eg. *Rhyssa* sp.) are ectoparasitoid idiobionts that permanently immobilize the host prior to parasitization, whilst others are koinobionts, either ectoparasitoids of spiders or endoparasitoids of a variety of arthropods. Little is known of the nature or endocrine modulating effects (if any) of the paralysing venoms of the ectoparasitic species. The tribe Pimplini (e.g. *Pimpla* spp.) comprises predominantly endoparasitoids of the pupae or pre-pupae of lepidopteran hosts. Pimpline wasps inject various secretions into the host haemocoel during oviposition in order to prevent haemocyte-based encapsulation of the egg(s) or growing parasitoid larvae (Kilincer, 1975; Osman and Führer, 1979; Parkinson and Weaver, 1999). Parasitoids of pupae/pre-pupae are perhaps unlikely candidates to have a requirement to paralyse their host, or to manipulate the host endocrine system, provided that is, they can complete their larval growth, and emerge or kill the host before the end of the host pupal stage. Loughton (1965) observed that pupae of *Galleria mellonella* that were parasitized early in the pupal stage by *Pimpla turionellae* did not appear moribund, or in any way affected, by the event of parasitoid oviposition, and yet failed to emerge as adults. The early-attacked parasitized pupae still wriggled if warmed in the hand. Those pupae parasitized quite late in pupal development (i.e. within 24 h of eclosion) also failed to emerge as adults, but since this was prior to the hatch of the parasitoid egg, it seemed that the wasp must impart an eclosion-preventing substance at oviposition. Injections of *P. turionellae* venom, comprising one-fifth of venom gland equivalents, failed to elicit a similar response, but implantation of mature parasitoid eggs showed that the factor preventing emergence was derived from the developing parasitoid egg (Loughton, 1965). The accessory reproductive gland secretion of *P. turionellae*, which is co-injected into the host during oviposition, did not seem to block the cellular defence system of pupae of *G. mellonella*, but the venom secretion caused paralysis of host abdominal muscles, melanization of pupal organs, and finally resulted in death (Kilincer, 1975). Histochemical analysis of *P. turionellae* showed that the uterus gland secretion contains hyaluronic acid and lipoprotein, the alkaline (Dufour's) gland secretion contains lecithin and a cholesterol ester, and the secretion of venom (or poison) gland stains positively for neutral mucoprotein (Führer, 1973; Osman and Führer, 1979).

The venom of the related ichneumonid, *Pimpla hypochondriaca*, a pupal endoparasitoid of the tomato moth, *Lacanobia oleracea*, causes a flaccid and lethal paralysis when injected in sufficient quantity into the haemocoel of both larval and adult, as well as pupal stages of the host (Parkinson and Weaver, 1999). A heat stable component of the venom was found to be toxic to adults of the housefly, *Musca domestica*, and cockroach, *Blattella germanica*. An unexpected finding was the presence of strong phenoloxidase activity in samples of diluted venom. Phenoloxidase activity had not previously been identified in any hymenopteran venom or in venom from other sources (Parkinson and Weaver, 1999). When venom from *P. hypochondriaca* was heat-treated to remove all phenoloxidase activity it inhibited further phenoloxidase activity in untreated venom at a potency likely to preclude such enzyme activity in the wasp's own venom reservoir. The components of *Pimpla* venom have been fractionated according to size by high-resolution gel filtration chromatography (Parkinson and Weaver, 1999). Several classes of constituents of venom were separated, including phenoloxidase activity (~390 kDa), a paralytic component (27 kDa), a low molecular weight factor (<13 kDa) cytotoxic to *Spodoptera frugiperda* (Sf21) cells, and several other factors. Venom that was heat-treated yielded two fractions (12 kDa and 8 kDa), both of which had phenoloxidase-inhibitory activity. The venom components identified in extracts from *P. hypochondriaca* (phenoloxidase, paralytic component, cytotoxic factor etc.) may be used to rapidly kill the pupal host. This would neutralize the immune defences of the host, as well as precluding any future production of host-derived hormonal substances that may interfere with development of the parasitoid. Pimpline that parasitize lepidopteran larvae, rather than the pupae (Fitton *et al.*, 1988), may also use similar parasitization strategies.

The phenoloxidase of *P. hypochondriaca* venom elutes from gel filtration as a broad peak, suggesting a wide mass range for the enzyme (Parkinson and Weaver, 1999). The genes encoding the protein in these fractions have been cloned, and have been found to represent a new form of phenoloxidase with properties that are strikingly different from those of all other arthropod phenoloxidase enzymes that have been characterized to date (Parkinson *et al.*, 2001). Three different *P. hypochondriaca* phenoloxidase genes were characterized from a cDNA library. The deduced products (*Figure 2*) are similar in sequence to those of the prophenoloxidases, a family of proteins that are normally produced as proenzymes by arthropod haemocytes, and are released (rather than secreted) to participate in immune defence to invasive challenge. A comparison of the sequences of the *P. hypochondriaca* phenoloxidase with those of various insect prophenoloxidases reveals some notable differences, including the unprecedented occurrence (for arthropod phenoloxidases) of a signal peptide sequence (indicative of a secretory product), and the absence of sequences specifying the site of proteolytic activation of prophenoloxidases. Amino-terminal analysis of a purified venom phenoloxidase has confirmed the authenticity of the signal peptide, and has indicated that proteolytic processing (i.e. activation) of the enzyme, other than to remove the signal peptide, does not occur in the venom protein. The phenoloxidases of *P. hypochondriaca* venom are thought to be constitutively active, which possibly explains the presence of phenoloxidase-inhibitory factors in the pre-secreted venom (Parkinson *et al.*, 2001). In relation to the possible harmful effects of the venom on the developing parasitoid larva, it is perhaps relevant that larvae of *P. turionellae* produce an anal secretion that is an inhibitor of phenoloxidase activity, and which also has antibiotic effects against bacteria and fungi (Führer and Willers, 1986).

The cDNA library of the *P. hypochondriaca* venom gland has yielded partial or complete sequences of a wide range of additional *Pimpla* venom constituents, including many sequences with homologies to other types of enzymes (N.M. Parkinson, personal communication). None are obvious contenders to function directly on the host hormonal system, an aspect of host manipulation that is not manifestly essential for the larval development and survival of *P. hypochondriaca* larvae in a pupal host. They are more likely to be involved in suppressing the immune defences of the host, and in preserving the host tissues as a food resource by preventing bacterial or fungal infection (Marris *et al.*, 2000).

The lateral oviduct of the ichneumon, *Campoletis sonorensis,* a solitary parasitoid of *Heliothis virescens* (tobacco budworm), and a few other noctuid lepidopteran larvae,

```
POI    MVRLRILLLSLALIASIAGDETDRDINQEILDQFTQPPR-----------------TLIV :  43
POII   MVRLRSLLLSLALIASIAGDETDRDINQEILDQFSPPSR-----------------TIIV :  43
POIII  MIRLRSLLFSLALIASIAGDETDTSVNEDILALFDPPPRPNAEPSVPFVTEPPVKAEGAV :  60

                           **
GSSVAQEENH-KNTPV---GNSQEVKNVPEQPKKLAYDIPKDLYKDLLRFDIDWAFSAFIEQHGQLA : 106
GSSGENGEGQ-K--------------IVPEQPKKLKHDIPKKLYKDLLRFEIDKPFTPFINKHSNLA :  95
GNASKAGESPPKMSPPKMSSASHTSGTKPIPPKKLTHDIPKELYRALLKFGYDEPFSAFIDTHARLA : 127

GKLIQVLLAAPTKDDISAIVKKYINTLNRALLTYAVSVTAMHRDDMRDFRIDLLSLSNPSKLVDGRQ : 173
RRLVDVLIDVPSQQDISAIVKKYVNLINRNLFTYALSVAAMHRADMRDFRTDMLTLSNPRKLVAGRH : 162
GRLIDVFMKAPGGQAISQMAERLSKIINRGLLSYALSVAILNRNDMTDIYVDKLSLKSLKKYVVGSQ : 194

ISTMQKAINVNEQVGNFSIWAKYSTATNSDDEHRLAYFREDVGINLHHWHWHLVYPFEGSLDIVKKD : 240
LSTLQKAIDANEQVGNFSIWVKYSTATNSDDEHRLAYFREDLGINLHHWHWHTVYPFGGSYKVVHKD : 229
INAAQKAVDEGDPGAPVVIAEKYSTASDTDVEHCLAYFREDIGVNLHHWHWHLVYPYEGSRDLVNKD : 261

RRGELFYYMHHQILARYNAERLSNELSLVKRFNNFREPIIEGYFPKLDSQVAGRSWPSRPAGVVLGD : 307
RRGELFYYMHHQLLARYNAERLSNELSLVKKFNNFREPIIEAYFPKLDTQVASRSWPSRPEGMVLKD : 296
RRGELFYYMHHQILARYNVERLSNGLNFVERFGNVREPIIEGYFSKLDSQLASRSWPSRPARMLLTD : 328

LHRHDINLKLDIEDLERWANRLYEAIYTGHMTDKAGALVPFDEKTGIDVLGNMIEASPVLSPNFGYY : 374
LNRRDHGLKINIEDLDRWANRLYEAIYTGHVMNEAGVLIPLNETAGIDLLGDMIEASSLLSPNFGYY : 363
VDRKDSDYYGSIHNKGLWAERLYEAIRAGHVKNEAGKPIPLDEKTGINVLGNIIGSPVLLSVNYDYY : 395

GSIHNRGHMAISYCHDPDGRYLEQFAVMGDSTTAMRDPIFYRWHAFIDEFFVTYKNTLKPYSRDELN : 441
GSLHNMGHMAISYCHDPVGRYLEDFAVMGDPATAMRDPIFYRWHAFIDEIFVTYKNTLKPYSDEELN : 430
GSIHNKGHMAIAYCHDPDGRYLEKFGVMGESSTAMRDPIFYRWHTYVDEFFVAYKNTLAPYSDSELN : 462

YEGITVTSIRVISDCGKENHLETFWTNFTVDLAKGLDFTPKSNVGAIFTHLDHCSFKYEIKATSSLA : 508
YKGIIVKSIRVISDCHKVNHLETFWANYTIDLGKGLDFKPKSNVGAIFTHLDHCPFKYEIKATSSLA : 497
CDGIEVSSIKVISKGYKDNHLKTFWGNFTVDLGKGLDFASKADVGAIFTHLDHRPFEYEIRAANTLP : 529

EAKKGVVRIFIAPRIDDNGKPYSLEDQRALMIEMDKFNYTIQPGANNIERQSINSTVTNPFETTGRE : 575
EAKKGVVRIFIAPRIDDNGDQYTLADQRALMVEMDKFDYTLQPDLNTIVRRSINSTVTNPFETTDRE : 564
KPKKAVVRIFIAPKVDKDGNVLNLERQRPLMIEMDKFSYSLKPGVNTIVRKSVKSTVTNPFARVSRG : 596

LGEKKTDPIEEFCGCGWPQYMLVPKGTEEGLLMDLFVMISDDDADISQWDSVTGFKAASTMCGLRNR : 642
LRNANIPPIEEFCGCGWPQYMLVPKGTELGLDMDLFVMISDDDADISQWDPATGFKAASTMCGLRNR : 631
LHMEHVPAVETFCGCGWPQYMLVPKGTEKGLPMDLFVMISNDGDHIAERDAITGNKVASTMCGLRNR : 663

NYPDPRAMGFPFDRPLSEGVRTLQNFLTPNMATSTIKVQFIENNPKTAAKNPQTAAP          : 699
MYPDPRAMGFPFDRPLPDDVRTLQNFLTPNMAATTIKVQFVENDPKKAEPIKRVEVIQA        : 690
TYPDARAMGFPFDRPFPRGVTTLRKFLTPNMAMSSITVQFIKKKP                      : 708
```

Figure 2. *Deduced amino acid sequences of the* Pimpla *venom phenoloxidases, POI, POII and POIII (data from Parkinson et al., 2001). The amino-terminal residues shown in bold italic represent putative signal sequences that are indicative of secretory proteins. Two conserved histidine-containing motifs thought to be involved in copper binding are underlined in bold. Asterisks indicate the position where a dibasic cleavage site (Arg-Phe) occurs in the pro-phenoloxidases of arthropods. This sequence is modified or absent from the* Pimpla *venom proteins.*

produces calyx fluid, a particulate suspension comprising both ovarian proteins and symbiotic polydnavirus (PDV), which is injected, together with the venom, as the wasp oviposits (Stoltz and Vinson, 1979a, b). Polydnaviruses, which occur only in three lineages of the Ichneumonidae and Braconidae, are divided into two major subgroups based on their morphology and genome composition. The functional biology of the PDV, and some molecular details, are discussed elsewhere in this volume (see Chapter 4). In the *H. virescens–C. sonorensis* host–parasitoid system, two primary host-regulatory effects are observed – a suppression of the cellular immune response and a marked reduction of larval growth. These are followed by a delay in host moulting and a prevention of host metamorphosis (Edson *et al.*, 1981; Vinson, 1972b). Successful parasitization of *H. virescens* by *C. sonorensis* requires only the PDV. Suppression of host larval development can be experimentally induced by injections of *C. sonorensis* calyx fluid alone, without involvement of the wasp (Dover *et al.*, 1987, 1988b). Host ecdysteroid production ceases for up to 10 days after injection and pupation of the injected last-instar larvae is prevented. Host ligation, ecdysteroid replacement therapy, and ultrastructure studies, suggest that the *C. sonorensis* PDV induces a selective, but recoverable, degeneration of the host's prothoracic glands (PG), which stops ecdysteroid production (Dover *et al.*, 1988b, 1989). If *C. sonorensis* venom alone is injected into 5th-stadium hosts (i.e. in the absence calyx fluid), it is completely lacking in effect on host larval development (Dover *et al.*, 1988b). The toxic effect of calyx fluid on PG is also absent in parasitized 3rd or 4th-instar host larvae. Host PGs only become sensitive to virally induced degeneration at around the same time as the major ecdysteroid peak in the 4th stadium. This is just before cuticle apolysis (separation of the epidermis from the old cuticle) as the host prepares to moult to the last larval stadium (Dover and Vinson, 1990; Dover *et al.*, 1995a).

A number of new mRNAs and proteins appear in the haemolymph of *H. virescens* after parasitization by *C. sonorensis*, or following injection of *C. sonorensis* PDV (CsPDV), and some have been correlated with successful parasitization of permissive hosts (Cook *et al.*, 1984). Whereas additional factors (e.g. venoms) appear necessary in some bracovirus infections (see below), the data from *H. virescens*–CsPDV strongly suggest that the alterations to the physiology of the parasitized host depend primarily on expression of the viral genome (Fleming and Krell, 1993). Expression of PDV genes shows species specificity, with some genome segments being expressed only in the adult parasitoid (Blissard *et al.*, 1986; Fleming *et al.*, 1983), some only in the parasitized larvae (Blissard *et al.*, 1989), and others in both parasitoid and host (Theilmann and Summers, 1986, 1988). Several of the CsPDV genes have now been characterized, including a family of four genes that encode cysteine-motif proteins (Blissard *et al.*, 1987; Cui and Webb, 1996; Dib-Hajj *et al.*, 1993), one that encodes a member of the repeat element (*rep*) gene family (Theilmann and Summers, 1987, 1988), and another (p12) that encodes a virion structural protein (Deng and Webb, 1999), although the expressed sequence of the CsPDV genome that brings about the degeneration of the PG is at present unknown. In relation to the venom, monoclonal antibodies raised against venom gland tissues from *C. sonorensis* are able to block viral inhibition of host growth and development, whilst Western blot studies suggest that some venom gland/oviduct proteins and some CsPDV envelope proteins may have epitopes in common, and hybridization studies indicate that some venom-related mRNAs are expressed in venom gland and oviduct, and in the parasitized host (Webb and Summers, 1990). Polydnavirus infection of *H. virescens* inhibits the translation of specific growth-associated host proteins, including storage proteins, and prevents the appearance of the pre-wandering peak of plasma juvenile hormone esterase (JHE) (Webb and Luckhart, 1996). Whether the venom and oviduct proteins of *C. sonorensis* have direct involvement as mediators of host endocrine responses to parasitism in this system has yet to be fully established.

The endoparasitic ichneumon, *Tranosema rostrale*, transmits a PDV to its host, the eastern spruce budworm, *Choristoneura fumiferana*, which inhibits host metamorphosis, but does not appear to suppress the cellular immune response (see Chapter 6). Larvae of *C. fumiferana* that are parasitized by *T. rostrale* early in the final larval stadium fail to undergo metamorphosis, and injections of wasp calyx fluid can cause dose-dependent delays in the initiation of metamorphosis (Doucet and Cusson, 1996a). Parasitization, or injection of calyx fluid containing PDV, prevents the normal rise in last-instar moulting hormone titre and lowers haemolymph JHE levels, although there is no elevation of juvenile hormone (JH) titre (Cusson *et al.*, 2000). Analysis of *T. rostrale* polydnavirus genes expressed in parasitized and virus-injected last-instar larvae reveals only one detectable mRNA. This encodes a protein of about 9.3 kDa, and is mainly transcribed in fat body and epidermis, rather than haemocytes (Béliveau *et al.*, 2000; and Chapter 6). The role of the non-virally associated venom components of *T. rostrale* in the manipulation of the host endocrine system has yet to be fully explored, and there is as yet no information on the composition of the venom for this species.

3.2 *Venoms of the Braconidae*

The Braconidae comprise one of the largest families in the Hymenoptera, in which there are more than 15 000 known species (Quicke and van Achterberg, 1990). They parasitize predominantly larvae of holometabolous insects, showing a marked preference for the Lepidoptera. Braconids are widely used in biological control programmes, and for studies on host–parasitoid interactions. With a few exceptions, they can be divided into idiobont ectoparasitoids, in which host paralysis is permanent or the host is killed (i.e. those that do not allow the host to continue functioning), and koinobiont endoparasitoids, which allow the host to continue functioning after parasitization, and which generally do not paralyse, or do so only temporarily at the time of oviposition (Shaw, 1983; Shaw and Huddleston, 1991; Wharton, 1993). The majority of ectoparasitic braconids attack concealed hosts, but the endoparasitoid species are fairly equally divided between those using concealed hosts and those that develop in hosts that remain exposed. Ectoparasitoids predominate in the Braconinae and Doryctinae, while endoparasitic groupings include the Ophiinae, the microgastrid complex (Adeliinae, Cardiochilinae, Cheloninae, Migrogastrinae etc.), Euphorinae, and Aphidinae.

Most investigations on the composition and action of the permanently paralysing venoms of ectoparasitic braconids have been limited to just three species: namely, *Bracon hebetor* (= *Habrobracon juglandis* = *Microbracon hebetor*), *Bracon brevicornis* and *Bracon gelechiae* (Beard, 1978), the venoms of which have extremely potent paralysing activities. It is estimated that in larvae of the greater waxmoth, *G. mellonella*, complete and permanent paralysis can occur at the level of one part *B. hebetor* venom to 200 million parts host haemolymph (Beard, 1952), and that the total quantity of venom produced by a single *B. brevicornis* during its lifetime has the potential to paralyse more than 1.5 million *Plodia* larvae (Beard, 1978). Early attempts to characterize the proteinaceous paralytic components were hampered by the apparently labile nature of the major paralytic factors (Spanjer *et al.*, 1977; Visser *et al.*, 1976). Two protein toxins (44 kDa and 57 kDa) were eventually purified, and partially characterized, but together represent only 2% of the original paralytic activity (Visser *et al.*, 1983). A paralytic toxin of apparent mass 18 kDa was isolated by Slavnova *et al.* (1987), whereas Quistad *et al.* (1994) characterized several larger (~73 kDa) insecticidal toxins (Brh-I–Brh-V) from *B. hebetor*, although some insecticidal activity was associated with lower mass toxins (~20–40 kDa). Three of the larger molecular weight proteins (Brh-I, Brh III and Brh-V) yielded partial amino-terminal sequences (Quistad *et al.*, 1994). Subsequent screening of a *B. hebetor* cDNA library with degenerate oligonucleotide sequences based on the Brh-I sequence gave a full length clone encoding a product of 678 amino acids that included an 18-amino acid signal sequence. The predicted MW of the 660 amino acid derived protein (77 912) was similar to that which had been estimated by SDS-PAGE for the purified venom product, and there was also matching amino-terminal sequence and total amino acid composition (Quistad and Leisy, 1996). Amino-terminal sequence comparisons indicated that Brh-III and BrH-V were related, but possibly substantially different proteins. The venom protein Brh-V is a single subunit polypeptide chain containing three cysteines. Although no similarity to any other invertebrate or non-invertebrate protein sequence was originally reported for Brh-V, subsequent analysis has shown close homology (44.3% identity, 69.7% similarity) to an aromatic storage protein, or hexamerin (CfeHex2; accession number AJ251271) that accumulates in the fat body of virgin queens and broodless workers of the ant *Camponotus festinatus* (Martínez *et al.*, 2000). Protein Brh-V has not otherwise been formally declared as a member of the hexamerin family, although comparisons with the complete hexamerin amino acid sequences (Burmester *et al.*, 1998) place it firmly within this protein group (Martínez *et al.*, 2000). The implications of this finding are possibly beyond the scope of this review, except to comment that the hexamerins are some of the most abundant proteins in the insect haemolymph and fat body, and that no other hexamerins have been reported to show paralytic activity.

The deduced amino acid sequences of two very different putative *B. hebetor* paralytic toxins, BrhTX-1 and BrhTX-2, with molecular masses in the range 87–90 kDa, were reported by Windass *et al.* (1996). Each of these proteins comprises four mainly cysteine-rich smaller proteins, with relative masses of approximately 15, 17, 21, and 34 kDa (BrhTX-1), and 17, 18.5, 21, and 34 kDa (BrhTX-2), respectively. The amino-terminal fragments and deduced amino acid sequences of the subunit proteins of BrhTX-1 and 2 show no significant similarity to any of the sequences reported by Quistad *et al.* (1994) and Quistad and Leisy (1996). One of the BrhTX-1 subunits (142 amino acids) appears unrelated to any other known protein, but the other three subunits each show significant homologies, to either phospholipases A2 (EC 3.1.1.4) or to a family of chymotrypsin-like proteases (EC 3.4.21.-) from various sources (R.J. Weaver, unpublished observations; J.D. Windass, personal communication).

A series of lower molecular weight *B. hebetor* paralytic toxins (of masses 16, 18.1, 18.2, 20, and 30 kDa, respectively) have been reported by Johnson *et al.* (1996, 1999). They are in part identical to two BrhTX-1 subunits reported by Windass *et al.* (1996), but the identities of three other proteins (17.2, 18.1, 20 kDa) have yet to be determined. The possible relationship between the proteins and subunits of Quistad and Leisy (1996), with those of Windass *et al.* (1996) and Johnson *et al.* (1996), and those identified earlier by Visser *et al.* (1983) and Slavnova *et al.* (1987), have yet to be fully resolved. Each of these proteins has been claimed to be involved in the flaccid paralysis that follows the envenomation of host insects. The mode of action of the primary active factor appears to involve the presynaptic blockage of excitatory glutaminergic transmission at neuromuscular junctions, possibly by inhibiting the release of synaptic vesicles (Piek and Mantel, 1970; Walther and Reinecke, 1983).

The likelihood of endocrine disturbances in the host as a consequence of envenomation by *Bracon* spp. appears to have been largely overlooked. An inhibitory effect on ecdysis (the physical component of shedding of the old cuticle) would seem inevitable in hosts that were completely, or even partially, paralysed, but whether any non-paralysing components that alter aspects of host hormonal control of development (host or parasitoid) in factitious hosts are also deployed has yet to be determined. Low doses of extracts of the venom glands from *M. hebetor* caused inhibition of larval growth and delayed pupation in the gypsy moth, *Lymantria dispar* (Masler and Kovaleva, 1999), a species that is not a recorded host for the wasp (Drenth, 1974). Size exclusion chromatography of *M. hebetor* venom gland extract showed the growth-inhibiting component eluting between molecular weight markers of 35 kDa and 66 kDa. Higher doses of venom (>1 ng per mg host larval weight) stopped larval growth completely, but both crude venom gland extract and the active size-separated fractions caused their effects in the apparent absence of extended paralysis (Masler and Kovaleva, 1999). The corn borer, *Ostrinia nubilalis,* the natural host of *B. brevicornis,* is totally refractory to the venom of *B. hebetor* (Beard, 1952; Drenth, 1974), perhaps indicating a degree of compositional variability between the venoms of two closely related species.

The chelonine wasps (Braconidae: subfamily Cheloninae) are solitary, endoparasitic koinobionts that attack lepidopteran eggs. Development of the parasitoid larva commences in the host egg, and continues in the host larval stages (Gauld and Bolton, 1988). The biological and biochemical interactions between chelonine wasps and their lepidopteran hosts have been reviewed (Jones, 1996). A discussion of the endocrine interactions of the species *Chelonus inanitus* with its host, *Spodoptera littoralis,* appears in Chapter 5.

At oviposition, chelonine wasps inject various secretions into the host egg, including venom, ovarian proteins and PDV, along with the single parasitoid egg. The wasp egg hatches (either inside the host egg or 1st-stage larva), and then remains as a 1st-stage larva for several host stadia, before resuming development as the host initiates metamorphosis. Host metamorphosis under these circumstances is typically a parasitoid-induced precocious metamorphosis, commencing in the penultimate larval stadium. Further host development is invariably blocked, such that the host fails to develop beyond the precocious pre-pupal stage. When the parasite emerges shortly afterwards, it kills the host and pupates nearby. The venoms of 'egg-larval' parasitoids would not be expected to possess directly-acting endocrine modulating substances, although the ovipositing wasp may inject something in addition to the egg that will later induce the premature metamorphosis (Brown and Kainoh, 1992; Jones, 1987; Reed-Larsen and Brown, 1990). An opposing view is that the 2nd-instar larva of the parasitoid is the primary cause of precocious metamorphosis in the host (Pfister-Wilhelm and Lanzrein, 1996). Contrary to both of these views is the suggestion that teratocytes (which are large cells liberated into the host haemocoel from the embryonic serosal membrane of the developing wasp larvae) are the source of the regulatory agent (Buhler *et al.*, 1985). To reconcile these different views, it has been postulated that the adult female injects one or more agents (e.g. venom/calyx fluid/PDV), which act only later to induce precocious host metamorphosis, and that the parasitoid 2nd-instar larva also releases a factor that either complements or reinforces this effect (Jones, 1987). It was suggested that the parasitoid factor might lower the threshold critical size by which the host larva 'perceives' that it has attained the stadium from which it should metamorphose (Jones, 1986, 1996). Several groups have undertaken investigations into the mechanism of the subsequent suppression of pre-pupal development using a variety of chelonine species. Ables and Vinson (1981) reported a delayed, but not prevented, pupation after injecting the host, *H. virescens,* with a combined extract of venom and calyx fluid from *Chelonus insularis.* Experiments with other chelonine associations, using pseudoparasitized hosts in which the developing parasitoid eggs were killed by treatment at low temperatures, gave clear indications that the ovipositing female wasp is the source of a regulatory agent (Ferran and Laforge, 1974; Jones, 1987; Reed-Larsen and Brown, 1990). More recently, Soller and Lanzrein (1996) have reported that injection of venom plus calyx fluid from *C. inanitus* suppresses pre-pupal development in *S. littoralis.* The suppression of pre-pupal development in larvae of the host, *Trichoplusia ni,* that have been parasitized by *Chelonus* near *curvimaculatus* may involve, in part, a reduced pre-pupal ecdysteroid peak in the last-larval stadium (Jones, 1985b; 1986, Jones *et al.*, 1986a; 1992b). Contradictory findings were reported for the same host by Lanzrein and co-workers (see below), where the pre-pupal ecdysteroid peak appeared not to be suppressed in larvae of *T. ni* that were fully parasitized by *Chelonus* sp. (Grossniklaus-Bürgin *et al.*, 1988, 1989; Grossniklaus-Bürgin and Lanzrein, 1990b). Jones (1996) has suggested that this absence of reduction of the ecdysteroid peak in fully parasitized hosts might reflect secretion of ecdysteroids by the developing parasitoid larvae, and that additional endocrine changes (e.g. reduced juvenile hormone titre) may also be involved in the suppression of host pre-pupal development. A castration effect, seen as a failure of the gonads to develop properly, is found in the parasitized hosts of some chelonine species (Ferran and Daumal, 1973; Reed-Larsen and Brown, 1990), although Brown and Kainoh (1992) suggest maternal venom is not in itself sufficient to elicit the response. Jones (1996) has suggested that there is correlative evidence to indicate that the regulatory agent for the castration effect involves a PDV.

The components of the *Chelonus* maternal secretions have been investigated for several species. Two-dimensional polyacrylamide gel electrophoresis (2-D PAGE) of the venom reservoir proteins of *C.* near *curvimaculatus* revealed the presence of about 40 (mainly acidic) proteins (Jones and Leluk, 1990). Immunological analysis indicated that some proteins were conserved in the venoms of other chelonine species and had epitopes in common with constituents of other non-chelonine venoms (Jones *et al.*, 1990a; Jones and Coudron, 1993; Leluk *et al.*, 1989). In *C.* near *curvimaculatus,* there are four major proteins, of M_r values 131 000, 53 000, 47 000 and 32 500, respectively (Jones and Leluk, 1990). All four proteins could be detected in the stung eggs of *T. ni.* Isoelectric focussing and 2-D PAGE showed that the majority *C.* near *curvimaculatus* venom proteins were acidic, and that the most abundant basic protein appeared to be the 47-kDa protein. Some acidic proteins were detected with M_r below 30 000, although no low-molecular-weight basic proteins, which are characteristic of the venoms of many non-parasitic Hymenoptera, appeared to be present (Jones and Leluk, 1990). Female chelonine wasps are estimated to inject approximately 0.02 venom reservoir equivalents into host eggs during each oviposition, an amount that was apparently without effect upon host larval development when experimentally applied (Leluk and Jones, 1989). A component of the *Chelonus* venom is apparently necessary for survival of the parasitoid in the host, perhaps acting to prevent encapsulation of the egg or developing parasitoid larva by the host's immune defence response (Leluk *et al.*, 1988). A 32.5-kDa component of the venom (subsequently called the 33-kDa protein), which comprises about 25% of the total protein content of the venom gland, was found to actively nullify the encapsulation response, with an efficacy comparable to that of the whole venom (Taylor and Jones, 1990). Screening of an expression cDNA library prepared from venom gland mRNA yielded an inferred sequence for a protein having amino-terminal sequence identical to that of the 33 kDa protein. The full cDNA encodes a 20-amino-acid signal sequence and 226-amino protein. An amino-terminal hydrophobic domain (~5000 kDa region) is followed by a central domain consisting of 12 tandem repeats of a highly charged (e.g. Glu, Asp), typically 14-amino-acid sequence motif, having a conserved central core of sequence -Ser-Gly-Ser-Xaa- (where Xaa represents a hydrophobic residue) (Jones *et al.*, 1992a). An immunologically similar protein (35 kDa) is found in the venom of *Chelonus insularis* (Leluk *et al.*, 1989). The 33-kDa protein from the venom of *C.* near *curvimaculatus* appears unrelated to any of the sequences in the genome of the *C.* near *curvimaculatus* PDV (Jones *et al.*, 1992a, 1994). Further screening of the venom gland library, using primers derived from the amino-terminus of a 52-kDa protein, yielded a clone that encoded a protein of 483 amino acids, analysis of which revealed a high degree of similarity (66%) to a 62-kDa, epidermal and gut chitinase of *Manduca sexta* (Krishnan *et al.*, 1994). The *M. sexta* chitinase is produced at each moult and is involved in digesting the chitin component of the old cuticle (Kramer *et al.*, 1993). A structural similarity to the chitinases from other organisms (e.g. bacteria, yeast, cucumber, and *Arabidopsis*) was also evident, but was weaker and more localized. Kinetic analyses of the chitinase activity of the crude venom and the purified 52-kDa protein showed that both extracts possessed chitin-lysing activities. The presence of a chitinase in the venom of a non-paralysing parasitoid wasp was unprecedented, although hyaluronidase (an enzyme that degrades hyaluronic acid, a chitin-like material) is found in the paralysing and defensive venoms of bees, ants and some predatory wasps (Jacques 1955; Kemeny *et al.*, 1984; Schmidt *et al.*, 1986). The significance of the *Chelonus* venom chitinase has yet to be resolved, and the presence of this enzyme in the venom of an egg-attacking parasitoid is somewhat enigmatic, as chitin is not known to be a component of lepidopteran eggs (Krishnan *et al.*, 1994). Despite its unspecified role in the early colonization of the host egg, it seems rather unlikely that the enzyme has a deferred effect as a contributory factor in the disruption of the host endocrine system later in parasitization.

The nature and structure of the PDV from chelonine wasps have been described for different species (Albrecht *et al.*, 1994; Jones *et al.*, 1986b; Stoltz *et al.*, 1976; Stoltz and Vinson, 1979a). The viral genome of the *C. inanitus* PDV consists of at least 10 different segments, and newly emerged adult female wasps already possess large numbers of 'encapsidated' viral DNA in their venom/ovarian calyxes (Albrecht *et al.*, 1994). Viral replication was shown to occur at a specific stage of parasitoid pupal-adult development (Gruber *et al.*, 1996). A comparison of the sodium dodecyl sulfate-polyacrylamide gel electrophoretic (SDS-PAGE) profile of *C. inanitus* calyx fluid (without virus) with that of purified PDV revealed that each was composed of numerous proteins and, although there were many similarities between the two preparations, there were also many proteins that seemed to be present only in calyx fluid (Soller and Lanzrein, 1996). Larvae of *S. littoralis*, developing from eggs injected with calyx fluid/venom of *C. inanitus*, and larvae developing from parasitized eggs in which the parasitoid egg is killed by heat treatment or X-irradiation, become developmentally arrested at the pre-pupal stage in the 6th stadium. Venom alone is ineffective in causing developmental arrest, as is calyx fluid on its own, except at high doses. When injected together, venom and calyx fluids were effective at much lower doses. Since X-ray-irradiated *C. inanitus* oviposited normally, but the wasp eggs did not hatch, it is suggested that PDV may be the active component in the calyx fluid. *S. littoralis* larvae exposed to

X-ray treated wasps (i.e. those containing PDV + venom, but no developing parasitoid) were found to have several hormonal differences from their non-parasitized counterparts (Grossniklaus-Bürgin *et al.*, 1998). The prothoracic glands of PDV/venom-containing larvae produced less ecdysteroid, and the haemolymph ecdysteroid titres were lower. There was also a delayed and reduced increase in the JH and JHE levels of pre-pupae, and a slightly modified metabolism of ecdysteroid. It seems that the PDV or venom alters the hormonal system of the host only after the pupal commitment peak of ecdysteroid (see Riddiford, 1980) in the final larval stadium, and that inhibition of PG activity is the earliest detectable response (Grossniklaus-Bürgin *et al.*, 1998). The PDV of *C.* near *curvimaculatus* has been partially purified, with 18 structural polypeptides having been separated by gel electrophoresis (Chelliah and Jones, 1990). The amino-terminal sequence of a 41-kDa component was determined, but showed no similarity to any previously reported sequences. Viral mRNA transcripts, and a novel 185-kDa parasitization protein, have been detected in *T. ni* following oviposition by *C.* near *curvimaculatus*. Synthesis of the parasitization protein (which occurs in several isoforms) is not initiated at the onset of parasitization, but occurs later, at the same time as precocious metamorphosis of the host. The 185-kDa protein is immunologically related to some of the *C.* near *curvimaculatus* PDV components and, although apparently unrelated to any known protein, it is quite possibly involved in the endocrine disruption of the host (Soldevila and Jones, 1994). The function and identities of the non-viral proteins in the calyx secretions of chelonine parasitoids have yet to be addressed (Jones, 1996).

The subfamily, microgastrinae, is the largest group of the Braconidae. They are solitary or gregarious koinobiont endoparasitoids, and include numerous species that are important in the natural regulation of lepidopteran populations (Gauld and Bolton, 1988). Many of the braconid genera (e.g. *Cotesia* [= *Apanteles*], *Microplitis, Cardiochiles*) are used in the study of endocrine and immune interactions in host–parasitoid relationships. The host–parasitoid interactions of *Cotesia congregata,* a gregarious endoparasitoid of tobacco hornworm, *M. sexta*, are discussed in detail elsewhere in this volume (see Chapter 3). Larvae of *M. sexta* parasitized by *C. congregata* in early larval stadia, develop normally until they become 5th (= last)-instar larvae, at which time they become developmentally arrested (Beckage and Riddiford, 1978, 1982a). The presence of the developing parasitoid larva, or factors introduced by the wasp at oviposition, or perhaps both, are responsible for several endocrine disturbances in the host. These hormonal disturbances include abnormally high levels of JH and lower levels of JHE in the early 5th-instar larvae, followed later by the absence of both pupal commitment and pre-moult haemolymph ecdysteroid peaks (Beckage and Riddiford, 1982a; Gelman *et al.*, 1998). The way in which *C. congregata* manipulates the haemolymph ecdysteroid levels of its host is not yet fully understood. The *C. congregata* the PDV, teratocytes (see Dahlman, 1990), and the parasite larvae, all appear to be involved to varying degrees (Beckage *et al.*, 1994; de Buron and Beckage, 1997; Dushay and Beckage, 1993; Gelman *et al.*, 1998). The venom of *C. congregata* does not seem to be required for successful parasitization, or for maturation of the wasp larvae (Beckage *et al.*, 1990, 1994; Dushay and Beckage, 1993). Injections of calyx fluid (which includes PDV), purified PDV alone, or eggs washed to remove PDV, have each been shown to delay host development and disrupt metamorphosis (Beckage *et al.*, 1994). Parasitization by *C. congregata,* or injection of *C. congregata* PDV into non-parasitized larvae, induces a group of early-expressed, parasitism-specific glycoproteins (EP1, EP2, and EP3) in *M. sexta* (Beckage *et al.*, 1987; Harwood and Beckage, 1994), at least one of which (EP1) has been shown to be a PDV gene product (Harwood *et al.*, 1994). The function of these 'early-induced' proteins has yet to be fully defined, but is perhaps more likely to involve protecting the parasitoid egg from encapsulation by host haemocytes, rather than regulating the host endocrine system.

Last-instar larvae of the armyworm, *Pseudaletia separata,* that are parasitized by *Cotesia* (= *Apanteles*) *kariyai* do not initiate metamorphosis, and pupation is blocked, apparently by disruption to host PTTH release. Prothoracicotropic hormone stimulates the PG to produce ecdysone, which in turn initiates further development (see Chapter 1). The parasitoid requires approximately 10 days to complete its larval development. This may necessitate a prolonging of the duration of the host larval stage, particularly when parasitization occurs late in host development (Tanaka *et al.*, 1987). Injections of the *C. kariyai* calyx fluid/venom fluid alone (i.e. without teratocytes) were insufficient to prolong the host larval stage for more than one or two days (Tanaka, 1987). In contrast, when host larvae were injected with teratocytes and both venom and calyx fluid, the larval-pupal moult was completely suppressed, as in normal parasitization. This indicates that teratocytes could be involved in the prolongation of host larval duration that follows parasitization by *C. kariyai* (Wani *et al.*, 1990). In early last-instar larvae of most Lepidoptera, a fall in plasma JH has been correlated with a burst of JHE activity, after which there is PTTH release. This stimulates the PG to produce an ecdysteroid pulse, which in turn initiates a cascade of events leading to pupation (Hammock, 1985; Steel and Davey, 1985; see also Chapter 1). No such burst of JHE activity is detected in early last-instar *P. separata* that are parasitized by *C. kariyai* (Hayakawa, 1990), or in *M. sexta* larvae that are parasitized by *C. congregata* (Beckage and Riddiford, 1982a). Injections of plasma from last-instar larvae of *P. separata* that were parasitized by *C. kariyai* into unparasitized larvae suppressed the normal rise in level of JHE activity, leading to a suggestion that a JHE-repressive factor was present, or induced, in the haemolymph of the parasitized host (Hayakawa, 1990). This factor, subsequently referred to as growth-blocking peptide (GBP), was shown to be a 25-amino acid, cysteine-bridged peptide having the sequence Glu-Asn-Phe-Ser-Gly-Gly-Cys-Val-Ala-Gly-Tyr-Met-Arg-Thr-Pro-Asp-Gly-Arg-Cys-Lys-Pro-Thr-Phe-Tyr-Gln-OH (Hayakawa, 1991, 1995). The GBP was subsequently purified from the plasma of PDV-infected, but otherwise unparasitized, last-instar *P. separata* larvae, and has also been shown to be present in haemolymph of unparasitized penultimate (5th)-instar larvae (Hayakawa, 1992; Hayakawa and Yasuhara, 1993). Further investigations have revealed the GBP to be a host-gene derived product, and changes in the titre of GBP have been correlated with normal larval development (Hayakawa *et al.*, 1995; Ohnishi *et al.*, 1995). The parasitism-induced elevation of GBP may occur as a result of PDV infection, which may cause a post-transcriptional elevation of the GBP-precursor, or an increase in the levels, or activity of the associated precursor-processing enzymes (Hayakawa *et al.*, 1998). An intriguing observation is that the GBP sequence shows a high degree of structural homology to a group of 23- to 24-amino acid paralytic peptides that were purified from the haemolymph of various last-instar lepidopteran larvae, but had been reported to be devoid of inhibitory effect on larval development (Skinner *et al.*, 1991, 1993). Both the GBP, and the paralytic peptides are, in turn, closely homologous to the plasmatocyte-spreading peptide (PSP) of the noctuid moth *Pseudoplusia includens* (Clark *et al.*, 1997), and to a cardioactive peptide of *Spodoptera eridiana* (Furuya *et al.*, 1999). These peptides comprise a group of structurally-related,

'ENF' peptides, designated from the single-letter code for the common, first three amino acid residues (Glu-Asn-Phe). The GBP of *P. separata* and the paralytic peptide of *M. sexta* have both been found to have plasmatocyte-spreading activity, and the PSP of *P. includens* has been shown to block growth of injected larvae (Strand *et al.*, 2000; Wang *et al.*, 1999), indicating that these 'ENF' peptides are multifunctional homologues (Strand *et al.*, 2000). Whether the parasitism-associated elevation in the haemolymph levels of GBP (and perhaps plasmatocyte spreading activity) is a directly-induced parasitism-specific response (resulting from envenomation, parasitoid, or PDV), rather than being a non-specific stress response to the presence of the developing parasitoid larva, has yet to be fully resolved.

Parasitism-associated growth retardation and blocked development are seen in both sexes of host larvae, although male *P. separata* that are parasitized by *C. kariyai* also show marked retardation in testis development and a partial inhibition of spermatogenesis (Tanaka *et al.*, 1994; Yagi and Tanaka, 1992). This occurs also in larvae of *Pieris brassicae* that are parasitized by *Cotesia glomeratus* (Junnikkala, 1985), in *M. sexta* parasitized by *C. congregata,* and in some other host-parasitoid associations (Reed and Beckage, 1997). In *P. separata*, the degree of parasitic castration in larvae varies with the age of host when parasitized, and can be mimicked by injections of unpurified *C. kariyai* PDV plus venom (Tanaka *et al.*, 1994). Some early-expressed *C. kariyai* PDV genes may be involved in the parasitic castration process, but the role of venom in this response is less clear (Tagashira and Tanaka, 1998). Shimizu *et al.* (1993) report the presence of tyrosine, tryptophan, tyramine, hydroxyphenylacetic acid and kyneurinine in the venom reservoir of *C. kariyai*, but the functional relevance of these components is not known.

The effects of the venom from *Apanteles* (=*Cotesia*) *glomeratus* have been investigated *in vivo* and *in vitro* using its usual host *Pieris rapae crucifera* (Kitano, 1982, 1986; Wago and Kitano, 1985). Mature wasp eggs inserted with calyx fluid into host larvae became encapsulated, but eggs inserted with calyx fluid plus venom did not. Venom and egg materials may act synergistically to inhibit the encapsulation response at the egg surface, and may also suppress the ability of haemocytes to react against the inserted egg. The venom had no discernible effect upon the morphology or contractility of cytoplasmic processes of two types of host haemocytes, nor did it affect the phenoloxidase-dependent melanization response to foreign substances (Kitano, 1986). A reduction in haemolymph phenoloxidase is a common response to parasitism in several lepidopteran larvae (Beckage *et al.*, 1990; Sroka and Vinson, 1978; Stoltz and Cook, 1983) and may be mediated by teratocytes (Kitano *et al.*, 1990; Tanaka and Wago, 1990).

The venom of *Cotesia melanoscela*, a parasitoid of the tussock moth *Orygia leucostigma,* may be required for successful parasitism, since the presence of PDV alone is insufficient to produce immune suppression in the host (Guzo and Stoltz, 1985). The venom promotes uncoating of the viral DNA, and enhances its persistence *in vivo* (Stoltz *et al.*, 1988). Embryos and developing larvae of the related species, *Cotesia rubecula*, are protected against the defensive reactions of the host, *Pieris rapae*, by a combination of passive evasion, and by active suppression of the immune defences of the host (Asgari and Schmidt, 1994; Asgari *et al.*, 1996). A calyx-fluid protein that protects the egg against encapsulation has been identified on the eggs of *C. rubecula* after oviposition. The cDNA encoding this 32-kDa protein has been cloned and sequenced (Asgari *et al.*, 1998). The deduced protein is not obviously related to any known protein, but does contain an identifiable hyaluron binding motif (Yang *et al.*, 1994). This is a feature that is commonly found in proteins involved in modifying cellular adhesion, motility, and growth. A transiently-expressed gene of *C. rubecula* PDV has been shown to encode a 45-kDa glycoprotein that inactivates the cellular immune-defence responses of *P. rapae* (Asgari *et al.*, 1996, 1997).

Microplitis croceipes (Microgastinae), a solitary endoparasitoid of *H. virescens*, undergoes its larval development without destroying the host tissues. Parasitization results in an inhibition of last-instar larval growth of the host, and a failure to pupate (Dahlman *et al.*, 1990a). The haemolymph ecdysteroid titre in parasitized 5th (last)-instar host larvae shows an early (small) pupal commitment peak, which then stays at the same level for the remainder of the stadium. Levels of JHE are also abnormally low throughout the 5th stadium in parasitized hosts, and a major peak of enzyme activity, normally associated with the latter part of the feeding period, does not occur (Dahlman *et al.*, 1990a). Zhang and Dahlman (1989) showed that the pupation of *H. virescens* could be prevented if unparasitized larvae were injected with *M. croceipes* teratocytes. The teratocyte-injected larvae had abnormally low levels of haemolymph JHE activity, and the titres of ecdysteroid were similar to those found in parasitized larvae (Zhang *et al.*, 1992). The influence of the teratocytes of *M. croceipes* on unparasitized *H. virescens* larvae was shown to be independent of any involvement from venom or calyx fluid (Zhang *et al.*, 1992). This contrasted with the results of Wani *et al.* (1990), who showed that teratocytes from *C. kariyai* prevented larval-pupal ecdysis in the host *P. separata* only in the presence of both venom and calyx fluid. Fathpour and Dahlman (1995) reported that *H. virescens* larvae that were injected with either *M. croceipes* calyx fluid or with sucrose-gradient purified PDV showed lower larval weight gains than did saline-injected larvae, and that the injected larvae had reduced haemolymph titres of both JHE and arylphorin (a storage protein). Jones and Lewis (1971) had reported that *M. croceipes* calyx fluid on its own was capable of causing reduced weight gain in the host *Helicoverpa zea*, and Tanaka and Vinson (1991b) had shown that a mixture of calyx fluid and venom from *M. croceipes* caused delayed larval development in *H. virescens*, and produced many larval-pupal intermediates. To rationalize these different findings, Fathpour and Dahlman (1995) suggest that since the teratocytes originate from the wasp embryonic serosa, it can be assumed that the PDV genome is incorporated in the DNA present in these cells, and that the teratocyte-induced reduction in titres of JHE, ecdysteroid, and storage protein might result from products encoded by the teratocyte PDV genes and expressed in the teratocytes. Subsequent studies (Schepers *et al.*, 1998) using teratocytes of *M. croceipes* cultured *in vitro*, have revealed the existence of four teratocyte-produced proteins, with molecular weights between 8 and 29 kDa, that could be involved in regulating the production of host JHE and storage proteins in larval fat body. The relationship of these teratocyte products to any PDV-encoded protein, venom, or calyx fluid component has yet to be determined.

The solitary migrogastrine endoparasitoid *Cardiochiles nigriceps* has been much studied in relation to host-parasitoid interactions (see Chapter 7). The habitual host of *C. nigriceps* is *H. virescens* (Lewis and Brazzel, 1966). Early studies showed that weight-gain and development of host larvae were substantially reduced by parasitism, and that injections of calyx fluid and venom mimicked these effects (Guillot and Vinson, 1972; Vinson and Barras, 1970). Tanaka and Vinson (1991a) showed that PDV/venom combinations reduced the ecdysteroid titres in 5th (= last)-instar

H. virescens, which then failed to pupate. Pennacchio *et al.* (1992a) reported that teratocytes would also inhibit pupation when injected into hosts, but analysis of haemolymph ecdysteroid titres revealed higher, rather than lower levels of ecdysteroid, although predominantly as polar conjugates (Pennacchio *et al.*, 1994a, b). The reduction in haemolymph ecdysteroid titres in parasitized and in venom/calyx-injected hosts was shown to be due largely to a reduced biosynthesis of ecdysone caused by a failure of the PG to respond to PTTH (Pennacchio *et al.*, 1997; Tanaka and Vinson, 1991a). A further reduction in the 20-hydroxyecdysone (20E) titre results from metabolic conversion to inactive polar metabolites, for which teratocytes seem to be partly involved (Pennacchio *et al.*, 1994b). Calyx fluid and venom do not seem to alter PTTH production by the host brain (Tanaka and Vinson, 1991a). The calyx fluid of *C. nigriceps* contains viral particles (PDV), which (along with those of *Chelonus texanus* and *M. croceipes*) were the first discovered members of the Polydnaviridae (Stoltz *et al.*, 1976; Vinson and Scott, 1975). Tanaka and Vinson (1991b) showed that the *C. nigriceps* PDV required the presence of venom to inhibit host pupation. Injection of *H. virescens* larvae with *C. nigriceps* calyx fluid mixed with *C. nigriceps* venom, either inhibited pupation, or produced larval-pupal intermediates. In contrast, *C. nigriceps* calyx fluid mixed with venom of either *M. croceipes* or *C. sonorensis*, failed to inhibit host pupation, even though all three parasitoid species (*C. nigriceps, M. croceipes, C. sonorensis*) can successfully parasitize the same host. Separation of whole venom by SDS-PAGE revealed five major bands and several minor bands. When venom was fractionated using a molecular weight cut-off filter, the synergistic activity of the venom was found in the fraction greater than 30 kDa. Purification of the >30-kDa fraction by high-performance liquid chromatography and SDS-PAGE revealed that the synergistic component was a major protein of approximately 66 kDa (Tanaka and Vinson, 1991b). It may be involved in enhancing the uncoating of the viral genome in infected host cells, as seems to occur in other braconid PDVs (Krell, 1991a; Stoltz *et al.*, 1988).

Aphids parasitized by endophagous braconids of the genus *Aphidius* (sub-family Aphidinae) show various degrees of ovarian degeneration and reproductive suppression depending on the stage attacked (Campbell and Mackauer, 1975; Pennacchio *et al.*, 1995; Polaszek, 1986; Soldan and Stary, 1981). Extracts of ovaries and venom glands from *Aphidius ervi* microinjected into non-parasitized 4th-stadium pea aphid, *Acyrthosiphon pisum*, caused developmental arrest and host death (Digilio *et al.*, 1998). Venom alone was as effective as combined ovarian and venom gland extract. The bioactive component of the venom is both heat and protease sensitive, indicating its proteinaceous nature. Analysis of venom gland proteins from *A. ervi* by SDS-PAGE revealed two major bands of approximately 21 kDa and 35 kDa, respectively. The bioactivity was localized in fractions enriched in these two major components of the venom (Digilio *et al.*, 2000). These so-called 'castration proteins' are possibly subunits of a single oligomeric protein, and are not glycosylated – as commonly found in venom proteins of other parasitic Hymenoptera (Leluk *et al.*, 1989).

4. Venoms of the parasitic Aculeata

The superfamily Aculeata comprises three of the most widely known families within the Hymenoptera, the ants (Formicidae), vespid wasps (Vespidae) and the bees (Apidae), along with several families, of perhaps lesser known solitary and mainly predatory wasps (e.g. the Bethylidae, Drynidae, Chrysidae, Pompilidae, Mutilidae, Scoliidae, Tiphidae, and the Sphecidae). Many of the latter family members can be classified as 'provisioning predators', but some aculeate families also contain true parasitoids, in the sense that only a single host is provided for the developing offspring and the adult wasps do not move the host after they have stung it, prior to egg-laying (Godfray, 1994). Recent phylogenetic studies of the Hymenoptera have placed the superfamilies of the Aculeata (e.g. Chrysidoidea, Vespoidea and Apoidea) between the Ichneumonidea (Terebrantia = Parasitica) and those of the other major parasitoid assemblage, the Proctotrupomorpha, which includes amongst others, the Chalcidoidea (see Whitfield, 1998). This phylogenetic division (see *Table 1*), which places the chalcid line ancestrally quite remote from that of the ichneumonid/braconid lines, may be of considerable significance when comparing the biochemical compositions and modes of action of the Ichneumonid/Braconid and Chalcid-type venoms (both paralytic and non-paralysing), particularly in relation to the developmental endocrinology of the host. Ancestral divergence of the parasitoid-containing hymenopteran superfamilies may indicate that host hormonal control almost certainly evolved independently on more than one occasion, and in more than one group.

The venoms of the non-parasitic Aculeate Hymenoptera, particularly those of the honey bee, and some social and solitary predatory wasps, have been the subject of numerous studies (see reviews by Banks and Shipolini, 1986; Nakajima, 1986; Piek, 1985, 1993; Piek and Spanjer, 1986; Schmidt, 1982, 1986; and references therein). The majority of these studies fall outside the scope of the present chapter, and are not further reviewed here. The predominantly defensive venoms of the aculeate species tend to contain low-molecular-weight basic proteins and peptides (kinins), having cytolytic or pain-inducing activities (Leluk *et al.*, 1989), although some species have low-molecular-weight proteins with paralytic activities (Piek, 1985; Piek and Spanjer, 1986). The targets for these venoms are not always invertebrates (sometimes rarely so), and there have been few studies that have described any effects on the prey endocrine system or development, other than as secondary responses in relation to a central neurotoxic action.

Aculeate wasps of the family Bethylidae are ectoparasitic species associated with lepidopteran or coleopteran hosts in cryptic habitats. They sting the prey several times, and lay several eggs on the outside of the host (Godfray, 1994; Rathmayer, 1978). Analysis of whole body extracts of the solitary Bethylid, *Goniozus leneri*, a species that has been used for biocontrol of navel orangeworm (*Amyelosis transitella*) and pink bollworm (*Pectinophora gossypiella*) shows that >70% of the paralytic component of the venom resides in small molecular weight compounds, the majority of which have yet to be identified, but may include putrescine, proline and dopamine (Skinner *et al.*, 1990). The venom itself has not been analysed, nor is there information available concerning the possible effects on host endocrinology and development.

5. Venoms of the Chalcidoidea

The Chalcidoidea are a very large, extremely diverse group of ecologically and economically and important wasps, comprising many hymenopteran families (Gibson *et al.*, 1999). The majority of these are parasites of other arthropods, although some, such as the fig wasps, are entirely phytophagous. Chalcids attack an extremely broad spectrum of hosts, of all stages, and can be ectoparasitic, endoparasitic, and solitary or gregarious (Gauld and Bolton, 1988). Hosts may be paralysed, but often transiently or incompletely (Piek, 1985). The temporary paralysis of the host may prevent it from avoiding or repelling attack. It may preclude damage to the wasp, or stop the host from removing parasitoid eggs (see also Chapters 8 and 9).

5.1 *Venoms of the Eulophidae*

The Eulophidae is the largest chalcid family, having an extremely diverse biology among its members. Many eulophids do not paralyse their hosts, or do so temporarily, but others, such as *Hyssopus pallidus,* a parasite of the codling moth *Cydia pomonella,* can induce sustained paralysis (Zaviezo and Mills, 1999). Bouèek (1988) separates the Eulophidae into four sub-families (Eulophinae, Euderinae, Entendoninae and Tetrastichinae), with a further subdivision of the Eulophinae into six tribes, two of which are the Euplectrini and Eulophini (see below). A more recent phylogeny of the Eulophidae (formulated by sequence analysis of the D2 region of the 28S rDNA gene of 87 species) shows only three tribes for the Eulophinae, with Eulophini, Euplectrini and Elachertini being combined (Gauthier *et al.*, 2000). Many have been studied in relation to the developmental effects, constituents, and modes of action of their venoms. The Euplectrini (*Euplectrus, Platyplectrus* spp.) are gregarious ectoparasites of (mainly) noctuid and geometrid lepidopteran larvae. Female euplectrines possess prominent spines on the hind tibia and have a highly flexible abdomen, both of which may facilitate the laying of eggs onto an actively moving host (Efremova, 1997). Host paralysis in this group is transient or non-existent, perhaps being less essential because of the attack-oriented morphological adaptations. Larval lepidoptera that are parasitized by euplectrines will continue to feed, at least in the short term, but do not subsequently moult or undergo further development (Chatterjee, 1945; Coudron and Puttler, 1988; Gerling and Limon, 1976; Neser, 1973; Smith, 1927; Thomsen, 1927; Uematsu, 1986; Wall and Berberet, 1974).

Euplectrus kuwanae is a gregarious external parasitoid of the noctuid, *Argyrogramma albostriata* (Uematsu, 1981a, b). Host larvae that are attacked by *E. kuwanae* fail to moult or pupate (Uematsu, 1986). Injection of *E. kuwanae* venom into non-parasitized larvae gives similar results (Uematsu and Sakanoshita, 1987). The reservoirs of the venom gland comprise paired pouches that are united at their bases. Numerous thin rhomboidal crystals, that are readily soluble in saline, can be seen through the transparent reservoir membrane. Tests showed that 0.1 venom gland equivalents (VE) was sufficient to prevent moulting in 85% of 4th instar-larvae that were tested, although apolysis occurred in a few cases. By contrast, injection of 0.05 VE caused only partial inhibition of moulting in about 40% of larvae tested, with apolysis taking place but not ecdysis, whereas injection of 1/40th VE was completely without effect (Uematsu and Sakanoshita, 1987).

Of the other euplectrines that have been examined, the endocrine and developmental consequences of ectoparasitism by *Euplectrus plathypenae* and *Euplectrus comstockii* on their host *T. ni* have been described in the greatest detail (Coudron and Brandt, 1996; Coudron and Puttler, 1988; Coudron *et al.*, 1990; Kelly and Coudron, 1990; see also Chapter 9, this volume). Several of the known and factitious hosts of *E. plathypenae* were tested for their developmental response to injections of extracts of the lower reproductive tract of the wasp that contained the venom apparatus (Coudron and Puttler, 1988). Developmental arrest was seen in all of the natural hosts and in most, but not all, of the mainly lepidopteran larvae that were known to be outside the natural host range of the parasitoid.

Coudron *et al.* (1990) showed that parasitism by *E. plathypenae* inhibited larval-larval ecdysis of *T. ni* through of the action of a non-paralytic factor from the venom of the adult female wasp. The presence of the developing parasitoid larvae was not required for the expression of moulting arrest after parasitization. Injections of ecdysteroid into parasitized larvae failed to produce apolysis and moulting, suggesting that an 'arrestment' factor was interfering with the response of epidermal target tissues to the moulting hormone. Isolated abdomens of abdominal-ligated host larvae failed to moult after parasitism, whilst anterior portions moulted normally, providing support for the theory that the arrestment factor acts as an anti-ecdysteroid (Coudron *et al.*, 1990). Analysis of haemolymph ecdysteroids indicated that the natural onset of elevated ecdysteroid titre in penultimate instars of the host was significantly delayed following parasitism by *E. plathypenae*. The ecdysteroid composition of parasitized larvae was qualitatively different from that of control larvae, being predominantly 20,26-dihydroxyecdysone (85% as estimated by radioimmunoassay) in the parasitized insects, rather than a mixture of 20E and 20,26-dihydroxyecdysone as seen in unparasitized controls (Kelly and Coudron, 1990). A subsequent examination of the immunoreactive ecdysteroids in larvae of *T. ni* parasitized by *E. plathypenae* or *E. comstockii* (Kelly and Coudron, 1996), indicated that the major ecdysteroid in 4th-stadium parasitized larvae is a glycosyl conjugate of 20E, rather than 20,26-dihydroxyecdysone as suggested earlier (see Kelly and Coudron, 1990). This may indicate the presence or action of an ecdysteroid-glycosylation enzyme, such as ecdysteroid UDP-glycosyl transferase (EGT), a factor not previously associated with this stage of larval lepidopteran development, except as a consequence of infection by baculovirus (O'Reilly and Miller, 1989; O'Reilly *et al.*, 1992; see also Chapter 14, this volume). See Chapter 9 for a fuller discussion of the possible involvement of an EGT-like enzyme in parasitization by euplectrine wasps.

Crude extracts of *E. comstockii* venom were shown to cause development arrest in unparasitized 4th and 5th larval instars and pupae of *T. ni* (Coudron and Brandt, 1996). Dose–response studies showed that 0.1 VE stopped development in all treated larvae, 0.05 VE caused developmental arrest in 80% of treated larvae, and 0.02 VE blocked development in only 20% of the larvae. The arrestment factor was shown to be water soluble, salt precipitable and heat sensitive, and to have a molecular size in excess of 30 kDa, all indicating that it was a protein. Electrophoretic analysis of the *E. comstockii* venom gland proteins yielded distinct bands, in a pattern that was unique from that typical of venom proteins from several other Hymenoptera (Jones and Coudron, 1993). Native PAGE, which is not always a good indicator of molecular size, produced a region at approximately 66 kDa that caused moulting arrest in treated insects (Coudron and Brandt, 1996). Activity was lost when the 66-kDa protein was denatured by heat in the presence of 2-mercaptoethanol and SDS. Analysis by SDS-PAGE yielded two weakly stained proteins at approximately 31 kDa and 33–36 kDa, neither of which was biologically active. The results suggest that the arrestment factor in the venom from *E. comstockii* is a protein of approximately 66 kDa in native form. The disappearance of the protein and bioactivity under denaturing conditions, and the appearance of new proteins at 31 kDa and 33 kDa, indicate that the factor may be composed of subunits.

Differential responses to *E. comstockii* venom were obtained upon injection into larva stages of the European corn borer, *O. nubilalis*, which is not a natural host for the wasp (Gelman *et al.*, 1997). Venom-injected 3rd and 4th stadium-larvae underwent apolysis, produced a new cuticle, but did not moult, whereas 5th-stadium larvae failed to initiate apolysis. A pre-moult ecdysteroid peak was not apparent in venom-treated 5th-stadium larvae, but injections of 20E (1 μg/larva) caused apolysis and new cuticle formation. Venom-treated 4th-stadium larvae showed a pre-moult ecdysteroid peak, but with a significant delay. The haemolymph ecdysteroid in venom-treated, pharate 5th-stadium larvae (i.e. non-moulting 4th) was predominantly 20E, and there was no evidence of glycosylation. The results indicate that the arrestment factor fails to prevent sufficient ecdysteroid synthesis for apolysis to occur in the 4th-stadium *O. nubilalis*, and yet still blocks moulting, whereas in 5th-instar larvae the effect of the venom is to prevent ecdysteroid synthesis. A more extensive discussion of the nature and biochemical mode of action of the developmental arrestment factor of the venom of *E. comstockii* is presented in Chapter 9.

Eulophids of the tribe Eulophini (e.g. *Necremnus sp., Dicladocerus* sp.), with the exception of the genus *Eulophus*, are parasitoids of 'mining' Lepidoptera, and of tortricids that develop in rolled or aggregated leaves. Many of these paralyse their cryptic hosts at least transiently prior to oviposition (Efremova, 1997). Wasps from the genus *Eulophus* occupy an intermediate niche, in that they attack open-living hosts (which are usually noctuids), but without the benefit of the euplectrine morphological adaptations (Efremova, 1997), and using a non-paralysing venom (Shaw, 1981a, Weaver *et al.*, 1997).

The non-paralytic venom of *Eulophus larvarum*, a gregarious ectoparasitoid of *Orthosia stabilis* (Lepidoptera: Noctuidae), induces delayed inhibition of its host's larval development. A similar effect was observed with *Eulophus pennicornis*, a gregarious ectoparasitoid of tomato moth, *Lacanobia oleracea* (Marris and Edwards, 1995; Marris *et al.*, 1996). Parasitized, penultimate-stadium larvae of *L. oleracea* fail to grow as quickly as unparasitized larvae, and do not moult to the 6th (final) larval stadium. Unparasitized larvae show a moderate ecdysteroid peak at approximately 48 h into the 5th stadium, and a much larger peak commencing on day 5 of the last larval stadium (Marris *et al.*, 2001; Weaver *et al.*, 1997). Larvae that are parasitized by *E. pennicornis* at the beginning of the 4th larval stadium, or injected with extracts of venom, show a much reduced, abolished, or possibly postponed 48 h peak, and a new peak that appears at 120 h post-treatment in the prolonged 5th stadium of the development-arrested larvae (Marris *et al.*, 2001). Analysis shows that both the pre-moult peak in unparasitized 5th-instar larvae, and the new peak in the non-moulting parasitized larvae, are predominantly 20E, together with smaller amounts of ecdysone and unidentified polar ecdysteroid (Marris *et al.*, 2001). In contrast to the report by Kelly and Coudron (1996), in which the major immunoreactive ecdysteroid in penultimate-instar larvae of *T. ni* parasitized by *E. plathypenae* or *E. comstockii* was apparently a glycosyl conjugate of 20E (see also above), we find no evidence of glycosylated ecdysteroids in larvae of *L. oleracea* that are parasitized by *E. pennicornis*, or injected with *E. pennicornis* venom, nor could an EGT-like enzyme activity be detected in haemolymph of parasitized larvae (R.J. Weaver, unpublished observations). Coudron *et al.* (1990) suggest that inhibition of host ecdysis in larvae that are parasitized by euplectrine wasps results from reduced sensitivity to moulting hormone, possibly in combination with a change in host ecdysteroid metabolism. By contrast, in larvae of *L. oleracea* that are parasitized by *E. pennicornis*, we find that the primary reason for the failure to moult is the absence of the 48-h ecdysis-initiating 5th-stadium ecdysteroid peak, which is possibly delayed beyond a critical period to initiate the moult. The venom of *E. pennicornis* may alter the sensitivity of the host PG to the ecdysiotropic action of PTTH, and thereby down-regulate ecdysteroid production (Marris *et al.*, 1999, 2001; see also Chapter 8, this volume). The venom is a complex mixture of several proteins. At least 45 different components can be readily detected using SDS-PAGE, of which 9 to 10 proteins are comparatively abundant (*Figure 3*). Gel filtration chromatography had previously distinguished at least 9–12 major venom components (Weaver *et al.*, 1997). Injection of size-fractionated components into day 0, 5th-instar *L. oleracea* larvae (0.7 VE/larva) had revealed the presence of one or more proteins within the range 43–158 kDa that had anti-moulting activity. There is presently no additional information on the further identity of the anti-moulting factor from *E. pennicornis* venom, nor of its possible relationship to the 66-kDa protein from the venom of *E. comstockii* (see Coudron and Brandt, 1996; Chapter 9, this volume).

Colpoclypeus florus is an oligophagous ectoparasitoid eulophid that injects a non-paralysing venom into the head of its tortricid host that prevents ecdysis and causes a developmental arrest (van Veen and van Wijk, 1985, 1987). Larval tortricid hosts that have been stung by *C. florus* are prevented from moulting even if parasitoid eggs are removed. No information is available on the composition or mode of action of the *C. florus* venom, although it has been suggested, but not yet substantiated, that collateral damage to the cerebral ganglia, caused by the sting, may contribute to the anti-ecdysis effect (van Veen and van Wijk, 1985).

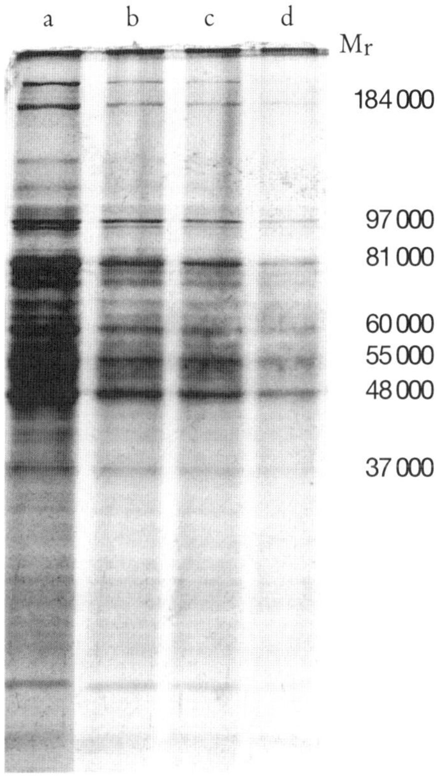

Figure 3. *SDS-PAGE of* Eulophus pennicornis *venom gland proteins. Lanes a – d represent 1, 0.5, 0.5 and 0.25 venom reservoir protein equivalents, respectively. Proteins were visualized by silver staining. Mr, estimated molecular size of main protein components.*

5.2 *Venoms of other Chalicidae*

The Eupelmidae are ecto- or endoparasitoids having a wide host range. The solitary external parasitoid *Eupelmus orientalis* attacks larvae of the bruchid beetle, *Callosobruchus maculatus* – a common West African pest that develops within seeds of the cowpea. The wasp injects a venom that causes permanent paralysis and host developmental arrest (Doury and Rojas-Rousse, 1994). Injections of diluted venom as low as 0.0025 VE were sufficient to cause developmental arrest, although higher doses (e.g. 0.1 VE) were required to produce paralysis (Doury *et al.*, 1995). First-instar larvae of the wasp were also able to induce permanent paralysis and developmental arrest in the host, possibly through the action of components from their salivary secretions (Doury *et al.*, 1995). Ecdysteroid levels do not appear to have been measured in developmentally arrested host larvae. The venom of *E. orientalis* contains several proteins by SDS-PAGE, some having antigenic similarity to those from the 1st-instar larvae salivary secretions (Doury *et al.*, 1997). There was evidence also for antigenic conservation between some of the *E. orientalis* venom proteins, and those of the honey bee, *Apis mellifera.* Assays of *E. orientalis* venom and larval salivary secretion for enzymes commonly found as venom constituents in many bees, ants and vespid wasps (e.g. hyaluronidase, phospholipase, lipase), showed that hyaluronidase was specific to venom, that phospholipase activity occurs in both, and that lipase activity was not detected (Doury *et al.*, 1997). Doury *et al.* (1997) suggest that hyaluronidase from *E. orientalis* venom may have a similar 'spreading factor' action to that of the same enzyme from non-parasitoid Hymenoptera (see Banks and Schipolini, 1986), and that the venom phospholipase may be used for both defence and host cell lysis (Schmidt, 1982). Little is known of the nature or mode of action of the paralytic or developmental-blocking components from either venom or larval secretion (Doury *et al.*, 1997).

The gregarious Pteromalid ectoparasitoid, *Nasonia vitripennis,* attacks pupae and pharate adults from several families of higher Diptera (Rivers and Denlinger, 1994a; Whiting, 1967). It was thought that the wasp would have no need to paralyse a non-motile host (Evans, 1933), although Beard (1964) showed that venom was injected at oviposition, and Ratcliffe and King (1967) showed that factors associated with stinging, apart from mechanical damage, were lethal to the host. Rivers and Denlinger (1994a) report that pupae and pharate adults of the flesh fly, *Sarcophaga bullata,* either died immediately, or entered developmental arrest, after parasitism by *N. vitripennis.* Hosts envenomed as pupae stayed developmentally arrested as pupae; whereas hosts envenomed as pharate adults matured slowly but still failed to complete their development within the puparium. The suppression of host development was caused by the venom, and was independent of the presence of the parasitoid larvae (Rivers and Denlinger, 1994a). Developmental arrest could not be reversed by injection of 20E. The mechanisms of action of the venom are as yet unknown, but have recently been examined using cultured cells of five insect and two vertebrate species (Rivers *et al.*, 1999). The venom caused cells from the insect cells to 'round up', swell, and eventually die. In contrast, cells from the frog and goldfish showed little sensitivity to the venom. The effects on the insect cells were consistent with those associated with insect-selective toxins produced by the bacterium *Bacillus thuringiensis.* It is suggested that these toxins may share a similar mode of action involving changes in cell-membrane permeability (Rivers *et al.*, 1999).

The non-paralysing venom of the ectoparasitic pteromalid, *Catolaccus grandis*, causes reduced growth and arrested development of the boll weevil host, *Anthonomous grandis*. There is evidence that venom may be active topically, as survival of non-parasitized host larvae was reduced when venom glands and wasp ovipositors were applied directly to the host integument (Morales-Ramos *et al.*, 1995). Venom of the hyperparasitoid pterolmalid, *Asaphes lucens*, induces an immediate paralysis and stops the development of its host, the aphid parasitoid, *Aphidius smithi* (Braconidae: Aphidiinae). In contrast, the venom of another hyperparasitoid, *Dendrocerus carpenteri* (Evaniomorpha: Megaspilidae), acts much more slowly on the same host (Boccino and Sullivan, 1981). Nothing is known of the venom constituents of either of these hyperparasitoid wasps.

5.3 *Venoms of Scelonidae*

The solitary egg parasitoid, *Telenomus heliothidis* (Proctotrupoidea: Scelonidae), arrests development and prevents emergence of its host, *H. virescens* (Strand *et al.*, 1980). A rapidly acting factor of maternal origin causes the effect. The factor is produced by the exocrine cells of the wasp oviduct, but its mode of action is unknown (Strand *et al.*, 1986).

6. Summary and conclusions

The study of parasitoid venoms is very much in its infancy, and only a handful of venom constituents have yet been identified. Except for the paralytic factor(s) from *B. hebetor*, the function of the venom components that have been characterized is not always clear. The phenoloxidases of the pupal parasitoid, *P. hypochondriaca*, and other venom proteins from the same source (e.g. paralytic, cytotoxic, phenoloxidase-inhibitory factors etc.), are likely to be involved in suppressing the immune defences of the host. They may also kill the host, and preserve host tissues as a food resource by preventing bacterial or fungal infection. They are unlikely to be involved in controlling the hormonal system of the pupa, as this may constitute a relatively endocrine-benevolent environment. The function of the 'chitinase' from the venom of the egg-larval parasitoid, *Chelonus insularis*, has yet to be confirmed, as is the role of the 52-kDa *C. insularis* venom protein with the novel internal repeating structure, although this may act as an anti-immune defence substance. Even with *Bracon* venom, it has yet to be confirmed whether any, or all, of the characterized components are directly responsible for the paralysis of the host, and if so, by what means, and at what site of action. As to the nature of venom components that may act to interfere directly with the host endocrine system, there is little known for any of the host–parasitoid associations. The possibility that hormonally-disruptive effects may be caused by paralytic components, or by non-paralysing components from paralysing venoms, whether acting independently or in concert, has not widely been considered. Nor has the possibility that the endocrine-disrupting actions of venoms may have arisen as new functions for former paralytic proteins. In many endoparasitoid associations the endocrine-disrupting effects of venom have often been difficult to dissociate from those caused by other maternally derived factors (e.g. PDV, teratocytes, and ovarian proteins), or from the presence of the developing parasitoid offspring. The venoms of endoparasitoids may function primarily to enhance the actions of these other components. Ectoparasitoids, such as *Euplectrus* spp. and *E. pennicornis*, are perhaps more likely to require a venom that acts directly on the host endocrine system. Progress in the characterization of endocrine-disrupting components may come from a further analysis of the moult-inhibiting actions of these eulophid wasps. The venoms of parasitic wasps offer a rich resource of endocrine-disrupting, immune-suppressing, and host-manipulating substances. The challenge for the future will be to characterize more of these molecules, and to identify their modes of action.

Acknowledgement

The authors acknowledge the support and funding from the Pesticides Safety Directorate of the Department for Environment, Food and Rural Affairs, UK.

3

Parasitism of *Manduca sexta* by *Cotesia congregata*: a multitude of disruptive endocrine effects

Nancy E. Beckage and Dale B. Gelman

1. Introduction

Many endoparasitoids have a dramatic influence on the growth and development of their host, often invoking developmental derangement and alterations in metamorphic programming (Beckage, 1985, 1997a; Lawrence and Lanzrein, 1993). Some parasitoids appear to regulate the development of their host by secreting hormones, whereas others appear to be conformers that respond to hormonal cues emanating from the host (Lawrence, 1986a). Many endoparasitoids use both strategies, alternating their strategies of hormonal offence and defence during different stages of their association with the host.

The interactions of the tobacco hornworm, *Manduca sexta*, with the braconid wasp, *Cotesia congregata*, have served as a useful model system for studying the physiological dynamics of host–parasitoid interactions (Beckage, 1993a,b, 1997a; Beckage and Templeton, 1986). The wealth of physiological and biochemical data we now have in hand describing the normal growth, development, and endocrine physiology of unparasitized tobacco hornworms (Riddiford, 1994; Riddiford *et al.*, 1999) facilitates ready comparisons with parasitized animals. Moreover, the relatively large size of the host permits the investigator to carry out experimental manipulations that would not be feasible using smaller species. The ease with which the host can be reared and staged accurately within the moult cycle also facilitates large-scale physiological studies using synchronously staged animals. One minor complication stems from the gregarious nature of *C. congregata*, since large numbers of hosts must be reared to generate sufficient numbers of synchronously staged animals containing similar numbers (and sizes) of parasitoids.

Endocrine Interactions of Insect Parasites and Pathogens, edited by J.P. Edwards and R.J. Weaver.

Like other endoparasitic braconid and ichneumonid wasps (Shelby and Webb, 1999; Stoltz, 1993;) *C. congregata* injects a polydnavirus (PDV) into host larvae during parasitization (de Buron and Beckage, 1992). The virus plays a critical role in inducing changes in the host which are required for successful parasitism (Beckage, 1997b). The PDV has an immediate devastating effect on the immune system (Beckage, 1998a; Lavine and Beckage, 1995, 1996; Strand and Pech, 1995) and then sits cryptically within the host, acting as a 'stealth pathogen' which later affects host endocrine programming prior to metamorphosis (Dushay and Beckage, 1993; Beckage *et al.*, 1994).

Our goal here is to summarize the major findings elucidating how parasitism by *C. congregata* alters the hormonal physiology of host tobacco hornworm larvae. Indeed, the parasitized insect is so different from the non-parasitized insect, that relatively few aspects of its endocrine physiology appear to remain unchanged by parasitism. Haemolymph titres of juvenile hormone (JH) (Beckage and Riddiford, 1982a) and ecdysteroids (Gelman *et al.*, 1998), as well as levels of neuropeptides in the central nervous system (Zitnan *et al.*, 1995a) and gut (Gelman and Beckage, 1995; Kingan *et al.*, 1997; Zitnan *et al.*, 1995b) are all affected. The response of endocrine targets such as the prothoracic glands (PG) to tropic hormones (i.e. PTTH) is also altered during the final stages of parasitism (Kelly *et al.*, 1998). Indeed, many phenomena seen in parasitized tobacco hornworm larvae, such as parasitic castration of male hosts (Reed and Beckage, 1997; Reed *et al.*, 1997) or metabolic effects of parasitism (Alleyne and Beckage, 1997; Alleyne *et al.*, 1997) may well be regulated by heretofore unidentified endocrine factors.

2. Effects of parasitism on host development

The effects of *C. congregata* on host growth and developmental programming are complex due to the gregarious nature of the parasitoids, but nevertheless offer fascinating insights into the dynamics of host–parasitoid relationships. The host's parasitoid load, as well as the host instar parasitized, determines the instar from which the parasitoids will emerge. Following parasitization of 1st-instar larvae, depending upon parasitoid load, wasp emergence may occur from 3rd, 4th, 5th, or supernumerary 6th-instar larvae (*Figure 1*; Beckage and Riddiford, 1978). The 5th instars weigh approximately 3 g when their parasitoids emerge, and are hence significantly smaller than unparasitized larvae of the same chronological age, which weigh 8–10 g prior to wandering (*Figure 2*). In contrast, parasitization of newly ecdysed 4th-instar larvae is followed invariably by emergence from 5th-instar hosts, and these animals weigh almost 6 g when their wasps emerge (Adamo *et al.*, 1997; Alleyne and Beckage, 1997; Washburn *et al.*, 2000). Host age at wasp oviposition also influences the rate of parasitoid development, with the wasps requiring a shorter development time in older hosts (Beckage and Riddiford, 1983a).

In hosts parasitized as 1st-instar larvae, the parasitoids moult to their 2nd instar after the host ecdyses to its final larval instar. Once host ecdysis is completed, the parasitoids ecdyse 48–72 h later (Beckage and Riddiford, 1983a; Gelman *et al.*, 1999). This relatively tight moult synchronization suggests that endocrine cues might co-ordinate these events (see below). Presumably, the 1st-instar parasitoids, like other insects (Nijhout, 1994), have a minimum critical size required for moulting, which is not attained until after the host completes its last larval moult. Thus, *C. congregata* parasitoids do not moult until they have attained a length of 2.8 mm (Gelman *et al.*, 1999).

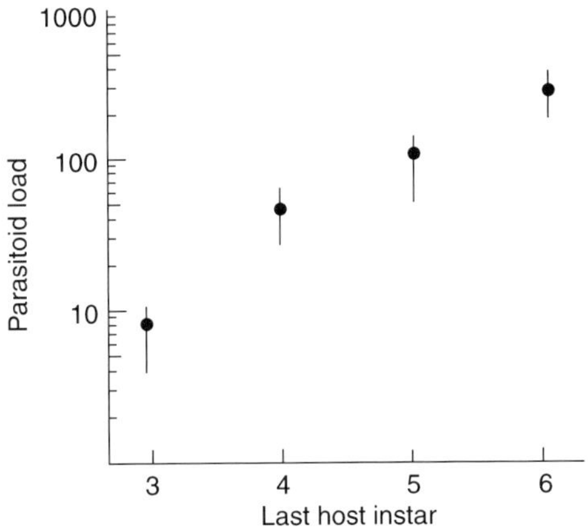

Figure 1. Mean parasitoid loads (± SD) of host larvae that were parasitized in the 1st instar and later had emerged wasps at the end of the 3rd, 4th, 5th, or supernumerary 6th instar. The parasitoid load represents the total number of emerged wasps plus unemerged parasitoids dissected from the haemocoel. (Figure redrawn from Beckage and Riddiford, 1978.)

Figure 2. Photograph of day 3 non-parasitized 5th instar larva (above) and parasitized 5th-instar larva of M. sexta *with emerged wasps. When parasitization occurs in the first instar, 5th-instar host larvae weigh 2-3 g when the wasps emerge (Beckage and Riddiford, 1978). In contrast, when parasitization occurs after ecdysis to the 4th instar, larvae weigh approximately 6 g at the time of wasp emergence (Alleyne and Beckage, 1997).*

Usually many parasitoids die within the host following emergence of the successful parasitoids (Alleyne and Beckage, 1997; Beckage and Riddiford, 1978). This is presumably due to the limited nutrients available within a 'crowded host.' The weight of the host–parasite complex at emergence is directly proportional to the total number of parasitoids present within the host; more heavily parasitized hosts weigh more at emergence than hosts with fewer parasitoids (Alleyne and Beckage, 1997; Beckage and Riddiford, 1983a). From very heavily parasitized hosts (>200 parasitoids) sometimes very few (<20) wasps emerge. One might ask why the host is not stimulated to continue to grow to achieve a size sufficient to allow the emergence of more parasitoids, since enhanced host growth might 'rescue' many doomed parasitoids. We do not yet know the answer to this question.

Compared to the numbers emerging from animals parasitized during earlier instars, dramatically fewer parasitoids emerge following parasitization of 5th-instar hosts (Alleyne and Beckage, 1997; Beckage and Riddiford, 1983a). This trend seems counterintuitive, since larger hosts offer more abundant resources to developing parasitoids. Several factors contribute to the reduced suitability of 5th instars as hosts. The adult female wasp appears to have difficulty penetrating the host's cuticle with its ovipositor. Subsequently, even in cases of successful oviposition, the host's cuticle progressively increases in thickness as the larva continues to grow, and thus interferes with emergence of the wasp's progeny. The parasitoids develop to mature 2nd instars, but then fail to emerge, and characteristic melanized spots form in integumental areas where the wasps have attempted to exit but failed (Alleyne and Beckage, 1997). The parasitoids die trapped within the host, while the host is developmentally arrested; neither the host nor its parasitoids survive.

Even in cases where the parasitoids are encapsulated, the host fails to complete normal development, and frequently succumbs as a supernumerary 6th-instar larva, pre-pupa or larval-pupal intermediate (Beckage and Riddiford, 1978). Possibly, the dosage of PDV injected by *C. congregata* during oviposition plays a role in inducing hormonal disruption in these instances, since the ultimate fate of hosts with encapsulated eggs (Dushay and Beckage, 1993) appears similar to that of non-parasitized larvae injected with PDV (Beckage *et al.*, 1994; Washburn *et al.*, 2000). The larvae grow, and attain unusually high pre-metamorphic weights (10 g or more) before they die. The endocrine implications of these events are obvious (see below) since the hormonal trigger for metamorphosis (the prewandering peak of ecdysteroid) is normally induced when a non-parasitized larva reaches a weight just exceeding 5 g (Nijhout, 1975, 1994; Nijhout and Williams, 1974a).

3. Host behavioural alterations

Many parasites and parasitoids induce changes in host behaviours including feeding activity, thermal preferences, reproduction, and geo- or phototropisms (Hurd, 1999; Moore, 1984, 1993, 1995). Certainly, our understanding of the physiological control of insect behaviour could be improved by studying how parasitoids are able to influence their host's behaviour. However, few studies have targeted the elucidation of the mechanisms by which parasitoids induce behavioural changes in their host insects (Adamo, 1997).

Following parasitization by *C. congregata*, the host initially continues to feed normally, and grows at the normal rate (Alleyne and Beckage, 1997). The duration of feeding intervals, bite rate, and spontaneous locomotory activity levels are similar in parasitized and non-parasitized insects (Adamo *et al.*, 1997). However, approximately 24 h before the parasitoids emerge from the host, the rate of food intake declines (Beckage and Riddiford, 1978; Miles and Booker, 2000), as does locomotion (Adamo *et al.*, 1997). The host fails to feed and shows no spontaneous locomotion once its parasitoids emerge (Beckage and Templeton, 1986). This parasitoid-induced quiescence is irreversible and continues until the developmentally arrested host larva dies about 2 weeks later (Adamo *et al.*, 1997; Beckage and Riddiford, 1978).

During emergence, the parasitoids exit through the host epidermis and cuticle, causing significant damage to the host integument. Despite the massive injury caused by emergence of up to a hundred or more parasitoids, the host retains normal reflexive behaviours (e.g. the righting reflex) despite the absence of feeding and spontaneous locomotion (Adamo *et al.*, 1997).

Importantly, the decline in feeding is not influenced by the number of parasitoids present in the host. In hosts parasitized at the beginning of the 1st instar by 10 or fewer parasitoids, emergence occurs from 3rd-instar host larvae (Beckage and Riddiford, 1978; *Figure 1*), and the feeding inhibition is complete even if only one wasp emerges. When late 1st- and early 2nd-instar parasitoids are implanted into the haemocoel of non-parasitized 4th-instar larvae, the parasitoids complete their development and emerge several days later despite the absence of teratocytes and PDV. Even if only one wasp successfully emerges, the cessation of host food consumption is complete and irreversible and the host is developmentally arrested (Adamo *et al.*, 1997; Lavine and Beckage, 1996).

When the parasitoids emerge from the host, the host shows no spontaneous locomotion although its reflexes remain intact (Adamo *et al.*, 1997). Beckage and Templeton (1986) showed that if the host brain was removed prior to parasitoid emergence, the resultant activity level was equivalent to that of wandering stage non-parasitized larvae. However, in these larvae the dorsal vessel does not become exposed (as it does in non-parasitized wandering larvae) – presumably because of the high haemolymph titre of JH. Since debrained larvae cannot feed, we do not know if their tendency to feed is altered by this operation. Nevertheless, the hyperkinesis that occurs in debrained parasitized larvae suggests that the reduction in feeding and locomotion that is characteristic of the final stages of parasitism is not due to tissue damage caused by the parasitoids, but rather to brain inhibitory signals that suppress feeding in the intact animal.

Octopamine, too, might conceivably have an influence on host behaviour, since high-performance liquid chromatography (HPLC) analysis of the haemolymph of parasitized larvae reveals that octopamine levels begin to increase a few days prior to emergence, and then peak at 125 pg μl^{-1} level when the wasps emerge from the host (Adamo *et al.*, 1997). Levels remain high for two days before exhibiting a sharp drop (Adamo *et al.*, 1997). In contrast, octopamine levels in non-parasitized, pre-wandering larvae are approximately 50 pg μl^{-1}. The octopamine elevation is temporally coincident with the decline in locomotory and feeding behaviours. Therefore, the increased haemolymph titres of octopamine could be responsible for suppressing feeding in parasitized last-instar *M. sexta* larvae.

Recent neurophysiological evidence suggests that the rising octopamine level might indeed be responsible for causing inhibition of feeding. The frontal ganglion regulates contractions of the foregut and ingestion of food, and octopamine inhibits firing of this ganglion and slows the rate of peristaltic contraction by the muscles lining the foregut (Miles and Booker, 2000). Injection of haemolymph from post-emergent parasitized larvae inhibits firing of the ganglion and contraction of the oesophageal musculature of non-parasitized larvae in a manner similar to injection of D L-octopamine. Moreover, the effects of either blood from post-emergence parasitized insects, or of octopamine injection, are blocked by the octopamine antagonists phentolamine or mianiserin (Miles and Booker, 2000). Thus, the high level of haemolymph octopamine seen in hosts with emerging wasps may inhibit their feeding, although feeding does not resume when the octopamine level falls significantly a few days later (Adamo *et al.*, 1997). Instead, feeding behaviour is irreversibly suppressed.

Like *C. congregata*, many species of parasitoids induce an inhibition in their host's feeding behaviour in preparation for parasitoid emergence. This 'anorexia' prevents the host from consuming the parasitoids as they emerge from the host, which non-parasitized feeding stage *M. sexta* larvae readily do when *C. congregata* cocoons are artificially glued to their dorsum (Adamo *et al.*, 1997; Adamo, 1998). Thus, the suppression of host feeding at this critical time may greatly enhance parasitoid survival by preventing predation of emerging parasitoids by the host.

Octopamine activates glycogen phosphorylase in non-parasitized *M. sexta* (Meyer-Fernandes, 2000) suggesting it may also have a metabolic role in parasitized tobacco hornworms. Many metabolic changes occur during the final stages of parasitism during the parasitoid's rapid growth phase preparatory to emergence (Thompson, 1993). Glycogen deposition appears to be enhanced in the fat body tissues of 5th-instar *M. sexta* larvae parasitized by *C. congregata* (Thompson and Dahlman, 1998), so octopamine may affect glycogen metabolism in parasitized *M. sexta* beginning midway through the 5th instar when the level begins to increase.

Dopamine and *N*-β-alanyldopamine (NBAD), which play critical roles in melanization and sclerotization of insect cuticle (Hopkins and Kramer, 1992), also increase significantly in the haemolymph of the 5th-instar hosts prior to parasitoid emergence (S. Adamo, personal communication; Hopkins *et al.*, 1998). In contrast, tyrosine and tyrosine glucoside (the main storage form of tyrosine) are present in only trace amounts in the haemolymph of parasitized 5th-instar larvae (Hopkins *et al.*, 1998). Monophenoloxidase activity, which converts tyrosine to DOPA, is very low (Beckage *et al.*, 1990). The pre-emergent parasitoid larvae accumulate high levels of tyrosine, dopamine, and NBAD, either from direct uptake or synthesis from precursors obtained from the host (Hopkins *et al.*, 1998), and presumably utilize these molecules during melanization and sclerotization of the adult cuticle following pupation.

Notably, Noguchi *et al.* (1995) reported that a haemolymph dopamine increase occurs in *Pseudaletia separata* larvae parasitized by *Cotesia kariyai*, and that the elevation precedes parasitoid emergence. Dopamine also increases in concentration in the ventral nerve cord at this time. The increase in haemolymph dopamine levels and induction of developmental disruption was mimicked by injection of polydnavirus or growth-blocking peptide into non-parasitized larvae, suggesting that PDV may alter the developmental programme of the host, in part by manipulating dopamine levels (Noguchi *et al.*, 1995). Interestingly, dopamine levels are known to affect developmental programming of non-parasitized insects, such as *Drosophila* (Neckameyer, 1996). Since dopamine also serves as a neuromodulator of nervous system function and behaviour (Chapman, 1998), another hormone-behaviour link could exist in addition to the one proposed above which may involve octopamine.

4. Effects of parasitism on host haemolymph JH titres

Beckage and Riddiford (1982a) reported that, in contrast to the JH titre of non-parasitized *M. sexta* larvae, which declines following the animal's ecdysis to the 5th instar (Fain and Riddiford, 1975; Riddiford *et al.*, 1999), the JH levels of parasitized 5th-instar larvae increase steadily following moulting to the last instar. In hosts initially parasitized as 1st instars, haemolymph JH attains a peak level (8 ng JH I equivalents/ml) on day 3 of the 5th instar and then declines slightly prior to parasitoid emergence on day 7 (*Figure 3*).

After parasitoid emergence, 4th- and 5th-instar hosts initially parasitized as 1st instars with different parasitoid loads (*Figure 1*) have similar JH titres (Beckage and Riddiford, 1982a), indicating that in parasitized larvae the stage of parasitism determines the host's JH titre, whereas in non-parasitized larvae the JH titre is tightly linked to larval weight. It would be useful to monitor the JH titres of host larvae with different numbers of parasitoids, to determine if the titres change in relation to the number of parasitoids present in the host, and vary according to the time when emergence is destined to occur.

Several factors are likely to contribute to the increased JH titre of parasitized larvae, including decreased hormone catabolism (see below) and possibly increased synthesis and secretion of JH III by the wasps (Lawrence *et al.*, 1990). JH III is indeed present in the haemolymph of parasitized 5th-instar *M. sexta* larvae (Beckage and Riddiford, 1982a), and may be of wasp origin. Probably one of our most important observations is that of the disappearance of haemolymph JH-specific esterase (JHE) activity during the 5th instar prior to parasitoid emergence (*Figure 4*; Beckage and Riddiford, 1982a). In contrast to non-parasitized larvae, in which the level of haemolymph JHE increases dramatically as the larva grows, in parasitized larvae, the level of haemolymph JHE activity declines following ecdysis to the 5th instar, and is absent from day 3 onward until the host larva dies (*Figure 4*). Since JHE is not detectable in the blood of hosts from which wasps have emerged, JH levels remain high, and therefore hosts are developmentally arrested in the larval stage. Even surgical allatectomy of parasitized 4th-instar larvae fails to provoke their metamorphosis (Beckage and Riddiford, 1982a).

The level of haemolymph JHE detected in parasitized 5th-instar larvae is far lower than levels detected in starved non-parasitized larvae, which also have an elevated JH titre (Cymborowski *et al.*, 1982). If non-parasitized 5th-instar larvae are starved after achieving weights in the range of 3.5 to 4.5 g, they usually pupate, but form abnormally small pupae. Animals starved after achieving weights in the range of 2 to 3.5 g generally moult to larval-pupal intermediates, and larvae starved after attaining weights in the range of 1.0 to 2.5 g or lower die without initiating a moult.

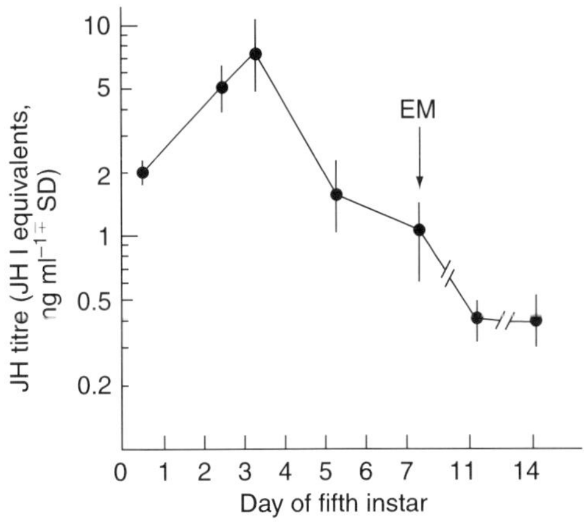

***Figure 3**. The haemolymph juvenile hormone titres of parasitized 5th-instar* M. sexta *larvae, which were previously parasitized as 1st instars by* C. congregata. *Titres were determined using the black* Manduca *bioassay for JH (Fain and Riddiford, 1975) and hormone concentration is expressed as ng JH I equivalents/ml haemolymph (± SD). Parasitoid emergence occurred on day 7 (EM). (Figure redrawn from Beckage and Riddiford, 1982a).*

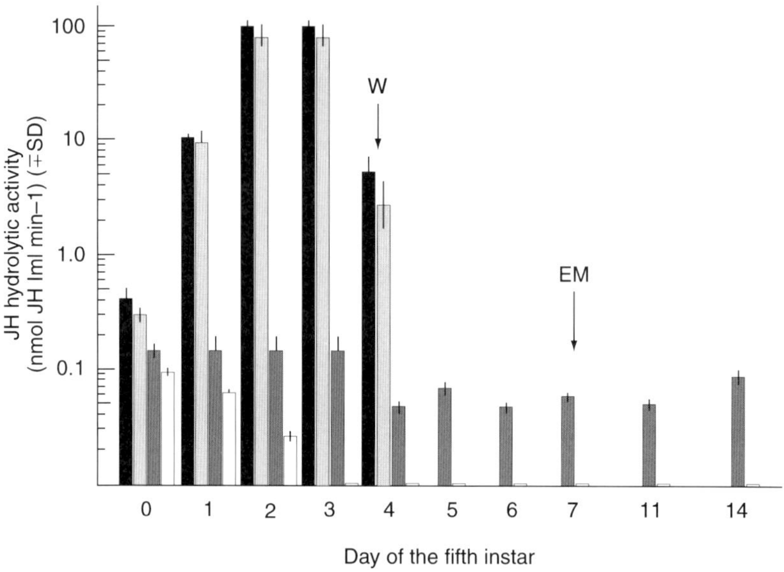

Figure 4. *Haemolymph juvenile hormone esterase activity in non-parasitized and parasitized 5th-instar* M. sexta *larvae. Host larvae were parasitized as 1st instars. Graph plots JH hydrolytic activity (expressed as nmol JH I converted to JH acid per ml haemolymph per minute) on various days of the 5th instar. See Beckage and Riddiford (1982a) for methods used to analyse total and JH-specific esterase activity. Bars represent total haemolymph esterase activity in non-parasitized larvae (solid bars), JH-specific esterase activity in non-parasitized larvae (light shading), total esterase activity in parasitized larvae (dark shading) and JH-specific esterase activity in parasitized larvae (open bars). Note that JH-specific esterase activity was undetectable in parasitized larvae beginning on day 3, when the JH titre peaks midway through the instar (Figure 3). Wandering occurred on day 4 (W) in non-parasitized larvae and parasitoids emerged on day 7 (EM) from parasitized larvae. (Figure redrawn from Beckage and Riddiford, 1982a).*

In contrast, larvae parasitized as 4th instars moult to the 5th instar and weigh 3.5–8.0 g following wasp emergence; they are invariably developmentally arrested as 5th instars (Alleyne and Beckage, 1997). Hence, the normal size-detection cues, which trigger the animal to initiate metamorphosis, are not operational in the parasitized animal, which suffers irreversible developmental arrest in the larval stage. Interestingly, larvae from which wasps have emerged are capable of responding to exogenous ecdysteroids and initiate larval cuticle synthesis if either infused or injected with 20-hydroxyecdysone (20E) (Beckage and Riddiford, 1982a).

An elevation of haemolymph JH titres coupled with reduction of haemolymph JH esterase activity during the last instar has also been reported to occur in *Pseudeletia includens* larvae parasitized by *Microplitis demolitor* (Balgopal *et al.*, 1996), in larvae of *Anastrepha suspensa* parasitized by *Diachasma-mimorpha longicaudata* (Lawrence *et al.* 1990), in *Heliothis virescens* larvae parasitized by *Microplitis croceipes* (Dong *et al.*, 1996) and in *Lymantria dispar* larvae parasitized by *Glyptapanteles liparidis* (Schopf *et al.*, 1996), indicating that JH enhancement commonly accompanies developmental arrest in parasitized larvae. Parasitism by *Tranosema rostrale* causes a decline in JHE activity in host *Choristoneura fumiferana* larvae, and JHE is also depressed in non-parasitized larvae injected with PDV, but the JH titre of parasitized larvae is not elevated (Cusson *et al.*, 2000). Importantly, Shelby and Webb (1994, 1997) observed an inhibition of translation of the mRNA encoding JHE in parasitized last-instar larvae that was caused by the wasp's PDV. Parasitism induced inhibition of translation of several insect plasma proteins typically seen in the last instar, including JHE and arylphorin (Dong *et al.*, 1996; Shelby and Webb, 1994, 1997).

In contrast, in *Spodoptera exigua* larvae parasitized by the egg-larval parasitoid *Chelonus inanitus*, and in *Trichoplusia ni* larvae parasitized by *Chelonus* sp., a precocious metamorphosis of the host begins before the parasite emerges in what is normally the penultimate instar (Grossniklaus-Bürgin *et al.*, 1994; Jones *et al.*, 1981). Juvenile hormone disappears from the blood one instar earlier than normal in truly parasitized larvae (Lanzrein and Hammock, 1995) and in pseudoparasitized larvae, which contain polydnavirus and venom, but no developing parasitoids (Jones, 1985a; Jones *et al.*, 1986a). The haemolymph JHE level rises prematurely to trigger wandering prior to the induction of host developmental arrest in the pre-pupal stage (Jones *et al.*, 1981; Lanzrein and Hammock, 1995).

5. The effects of parasitism on host ecdysteroid production, levels and metabolism

Parasitoids have evolved amazingly complex strategies for manipulating host ecdysteroid titres to favour their own development. Host moulting is often suppressed at various times during the life cycle; thus, the ecdysteroid that is physiologically active in triggering the moult (typically 20E) never reaches threshold levels. *Manduca sexta* parasitized by *C. congregata* neither wander nor pupate, processes which have been shown to be triggered by increases in haemolymph ecdysteroid levels (see reviews by Gilbert *et al.*, 1980a, b; Safranek and Williams, 1980). Indeed, although there is a small peak in host haemolymph ecdysteroid titre (3- to 4-times higher than the wandering peak, Gelman *et al.*, 1998) that occurs on day 3–4 post-moult to the last instar, neither a wandering peak nor a pre-moult ecdysteroid peak is observed (Beckage and Riddiford, 1982a; Gelman *et al.*, 1998). Ecdysteroid production in lepidopteran larvae has been shown to be regulated by a brain hormone (PTTH) (Bollenbacher and Granger, 1985; Gilbert *et al.*, 1980a, b; Warren *et al.*, 1988) as well as by JH (Fain and Riddiford, 1975; Kimura, 1974; Nijhout and Williams, 1974; Rountree and Bollenbacher, 1986), neural signals (Okajima and Watanabe, 1989; Richter and Gersch, 1983), a haemolymph factor produced in the fat body (Gruetzmacher *et al.*, 1984; Watson *et al.*, 1985), the concentration of ecdysone in the haemolymph (Bollenbacher *et al.*, 1975; Hanaoka and Ohnishi, 1974; Lafont *et al.*, 1975; Sakurai and Williams, 1989) and the responsiveness of the PG to stimulation by PTTH and other regulatory factors (Bollenbacher, 1988; Gelman *et al.*, 1991; Thyagaraja *et al.*, 1993). Thus, the parasitoid could compromise any one or more of these regulatory systems in order to maintain haemolymph ecdysteroid titres at optimum levels for its own development. In this section we review the various mechanisms used by *C. congregata* to manipulate its host's moulting hormone titres.

5.1. *Effects of PDV and venom on host haemolymph ecdysteroid levels*

At the time of parasitization, the wasp injects calyx fluid containing *C. congregata* PDV (Buron and Beckage, 1992; *Figure 5(a), (b)*), which delays wandering and interferes

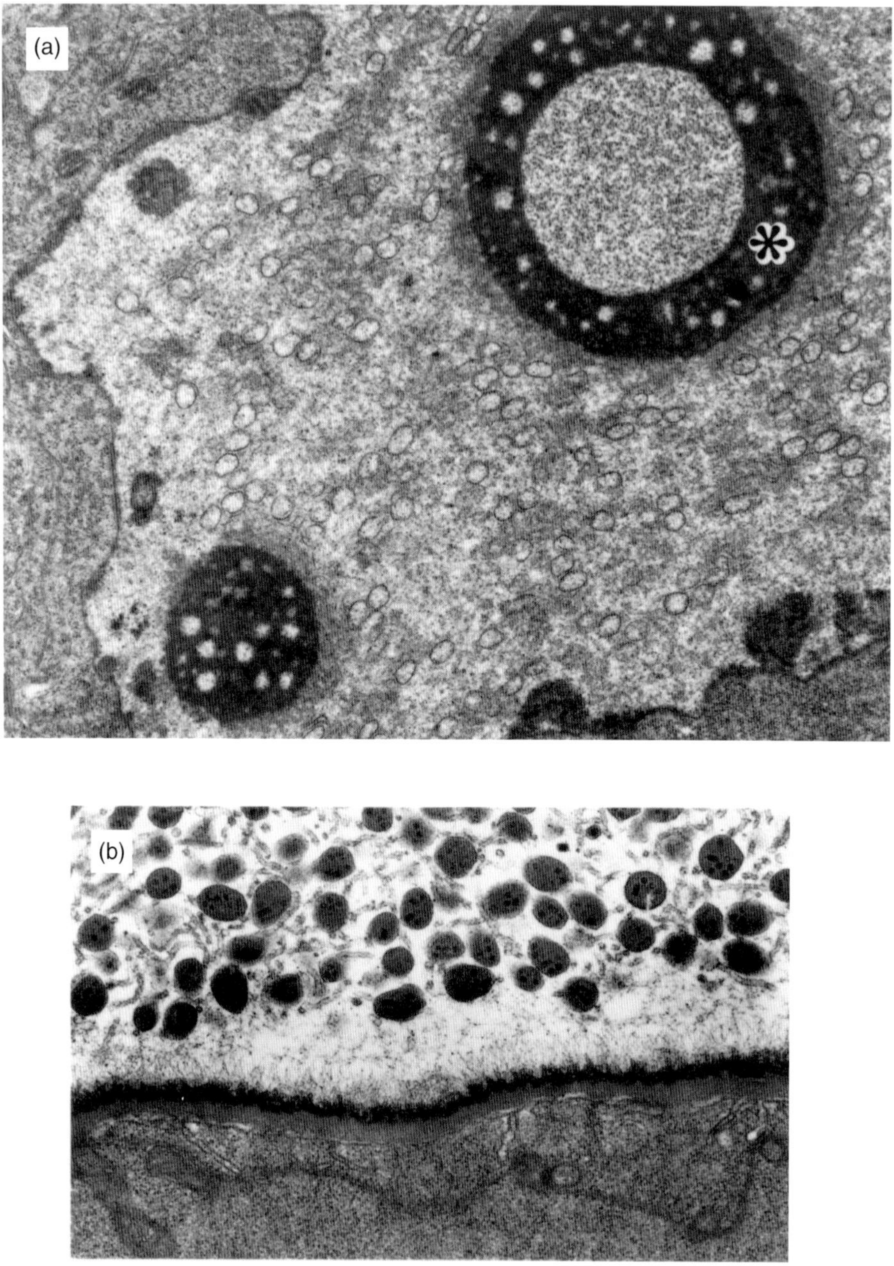

Figure 5. *(a) Transmission electron micrograph showing that in the pharate adult female wasp, the PDV virions undergo replication and assembly in the nucleus of an ovarian calyx cell. Various stages of virion envelope synthesis can be seen at the periphery of the virogenic stroma (asterisk) and nucleocapsids are being inserted into the nascent virions in the nucleoplasm. (b) Transmission electron micrograph showing fully formed PDV virions in the lumen of the calyx (above) and the surface of a* C. congregata *egg (below). Note that PDV virions are not impregnated in the surface of the egg. Each virion contains 1–15 nucleocapsids and has a tail. Photos provided by Isaure de Buron (College of Charleston, Charleston, SC). (Reproduced from Buron and Beckage, 1982,* J. Invertebr. Pathol., *with permission from Harcourt Brace.)*

with host pupation (Beckage *et al.*, 1994; Dushay and Beckage, 1993). The accompanying venom does not appear to enhance the effects of PDV or affect host growth and development (Beckage *et al.*, 1994). The terminal stage achieved by larvae artificially injected with calyx fluid containing PDV may be pre-wandering, pre-pupal or larval-pupal intermediate (Beckage *et al.*, 1994; Dushay and Beckage, 1993). The effect of wasp PDV on host haemolymph ecdysteroid titres has not been determined, but since the abnormal pupae formed by PDV-treated hosts display larval characteristics, it is likely that the virus disrupts the normal hormonal milieu, particularly with regard to JH levels. A moult would require that ecdysteroid has been produced and released into the haemolymph in supra-threshold levels. However, the normal maximum titre may never be achieved and, in those injected larvae that fail to pupate, titres may be too low to induce a moult. The effects of PDV and venom on the host in other host-parasitoid systems vary. The calyx fluid and venom of the wasps, *Apanteles kariyai* and *Chelonus inanitus,* interact to inhibit host PG activity in *Pseudaletia separata* and *Spodoptera littoralis*, respectively, thus causing host development arrest (Grossniklaus-Burgin *et al.*, 1998; Soller and Lanzrein, 1996; Tanaka, 1987). In the *C. inanitus-S.littoralis* system, reduced levels of haemolymph ecdysteroid were observed (Grossniklaus-Burgin *et al.*, 1998). Similarly, calyx fluid of *Campoletis sonorensis* and calyx fluid and venom of *Cardiochiles nigriceps* inhibit host (*Heliothis virescens*) pupation by preventing an increase in last-instar haemolymph ecdysteroid levels (Dover *et al.*, 1987; Tanaka and Vinson, 1991a). In contrast, the percentage of *Helicoverpa zea* larvae parasitized by *Microplitis croceipes* that pupate is not affected by parasitoid calyx fluid or venom (Gupta and Ferkovich, 1998).

5.2. *Effects of teratocytes on host haemolymph ecdysteroid and JH levels*

The role of teratocytes, which are released from the serosal membrane during hatching of 1st-instar wasps (de Buron *et al.*, 1997; Dahlman and Vinson, 1993), in altering *M. sexta* haemolymph ecdysteroid levels also has not been investigated. The teratocytes of *C. congregata* grow to an extremely large size (150–200 μm in diameter) in the last instar of the host, are covered initially with abundant microvilli (*Figure 6(a)*) and then develop large secretory blebs prior to the emergence of the wasps (*Figure 6(b)*).

Injection of several hundred teratocytes into non-parasitized larvae induces formation of giant (12 g) larvae, which are paler and pinker than non-parasitized larvae (Adamo *et al.*, 1997). A pale, pinkish colour in the cuticle of *M. sexta* larvae is often indicative of excess JH (Staal, 1977a). Non-parasitized 4th-instar larvae injected with 350-450 teratocytes harvested from day-zero 5th-instar host larvae, showed developmental anomalies in that it took the majority at least twice as long as normal to show any pre-metamorphic symptoms (de Buron and Beckage, 1997). These teratocyte-injected larvae often displayed faint dorsal vessel exposure, but did not show wandering behaviour. Our observations indicated that the timing of dorsal vessel exposure and the onset of wandering behaviour were uncoupled in these teratocyte-injected 'hosts'. Moreover, none of the larvae pupated normally. Instead, they either died as pre-wandering larvae or as larval-pupal intermediates, forming primarily larval cuticle despite their large size. These animals grew progressively more pale following the moult to the 5th instar and developed jet black spiracles prior to their death. The injected teratocyte cells were not encapsulated and persisted in the haemolymph as freely floating cells until the larvae died.

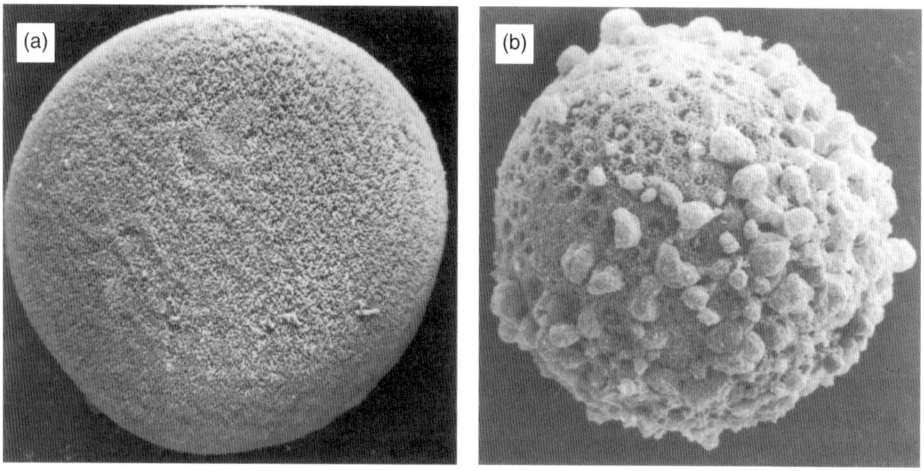

Figure 6. *Scanning electron micrographs of teratocytes isolated from 5th-instar* M. sexta *hosts examined a few days prior to parasitoid emergence (a) and on the day the wasps emerged (b). Large secretory blebs appeared interpersed among the abundant microvilli on the cell surface when the parasitoids began emerging from the host. Since each parasitoid egg serosal membrane gives rise to approximately 160 teratocytes (Buron et al., 1997), and as many as 150 parasitoid eggs are deposited in a single oviposition, several hundred thousand cells may be present in the haemolymph of the host. Photos provided by Isaure de Buron (College of Charleston, Charleston, SC). (From Beckage 1997a.)*

These observations are not inconsistent with teratocytes having a pronounced inhibitory effect on the production or activity of JHE as reported in other species (Dahlman and Vinson, 1993), although the direct physiological route whereby the cells and/or their secretions might invoke this effect remains unclear. Thus, teratocytes alter the host hormonal milieu, and may contribute to the increased level of JH observed in parasitized *M. sexta* 5th-instars. Since teratocytes alone cannot inhibit a moult, haemolymph ecdysteroid titres in teratocyte-injected larvae must increase sufficiently to stimulate apolysis. In other host–parasite systems (e.g. *H. virescens* parasitized by *M. croceipes*), injections of teratocytes also result in incomplete larval-pupal ecdysis or delayed pupation (Zhang and Dahlman, 1989) and inhibit ecdysteroid production, possibly secondarily via the inhibition of JHE release (Zhang *et al.*, 1992).

Teratocytes from *C. congregata*, like those of *M. croceipes* (Schepers *et al.*, 1998), when cultured *in vitro* secrete many polypeptides, the functions of which are unknown (de Buron *et al.*, 1997). In hosts with pre-emergent wasps, the teratocytes begin to show dramatic symptoms of apoptosis midway through the 5th instar. The teratocytes secrete large numbers of exocytotic vesicles or blebs, the nucleus forms dendritic-like projections, and the chromatin adheres to the inner nuclear membrane, and there is extensive vacuolization of the cytoplasm (de Buron *et al.*, 1997). Since about 150 eggs are injected per oviposition, and each egg gives rise to 150–170 cells (de Buron *et al.*, 1997), the exocytotic cells of hosts with maturing wasps could be a major source of haemolymph macromolecules, some of which could conceivably function in endocrine regulation in addition to serving a nutritive role for the parasitoids (Dahlman and Vinson, 1993).

Heliothis virescens injected with teratocytes from *C. nigriceps*, exhibit decreased haemolymph ecdysteroid levels late in day 5 of the 5th instar compared with control larvae (Pennacchio *et al.*, 1994a). However, on day 7, an ecdysteroid peak is observed in teratocyte-injected larvae. Ecdysone and polar ecdysteroids (including polar conjugates of 20E – which are not able to stimulate a moult) were the most prevalent ecdysteroids making up this peak (Pennacchio *et al.*, 1994a). Therefore, teratocytes from *C. nigriceps* appear to play a role in inhibiting host pupation through the inactivation (conjugation) of 20E. In contrast, teratocytes do not appear to play a role in inducing development arrest in the *C. inanitus–S. littoralis* system (Grossniklaus-Bürgin *et al.*, 1994).

5.3 *Fluctuations in ecdysteroid levels in parasitized 5th-instar* M. sexta*: the origin of the peak*

As mentioned earlier, in 5th-instar *M. sexta* hosts, neither a wandering nor a pre-moult haemolymph ecdysteroid peak is observed. However, haemolymph ecdysteroid levels do begin to increase on day 2 and peak (approximately 350 pg μl^{-1}) by the morning of day 3 (Gelman *et al.*, 1998; *Figure 7*). Titres remain high until the morning of day 5, at which time they have returned to relatively low levels (50 pg μl^{-1}). The increase in ecdysteroid titre precedes the parasitoid's moult from the 2nd to the last instar and concomitant emergence from the host, and appears to be associated with one or both of these events. These events are normally observed on day 4 of the host's 5th instar (Gelman *et al.*, 1998). The levels of 20E in host haemolymph are sufficiently high to induce wandering. However, high levels of JH (Beckage and Riddiford, 1982a) probably prevent the onset of wandering.

In parasitized insects, host PGs remain active, i.e. produce and release ecdysteroid (Beckage and de Buron, 1993; Kelly *et al.*, 1998). However, in contrast to glands from unparasitized 5th instars, PG from parasitized 5th instars are refractory to stimulation by PTTH (Kelly *et al.*, 1998; *Figure 8*). Interestingly, glands from these parasitized last instars, when cultured *in vitro*, continue to produce ecdysteroid after 3 hours of incubation rather than producing low basal levels, as do PGs from similarly aged unparasitized *M. sexta* larvae (Kelly *et al.*, 1998) (*Figure 8*). Enhanced secretory activity was observed for day-3 and day-4 host glands, a time that is coincident with the host haemolymph ecdysteroid peak associated with the wasp's final moult and emergence from its host. Thus, it appears that the parasitoid has uncoupled the normal mechanisms

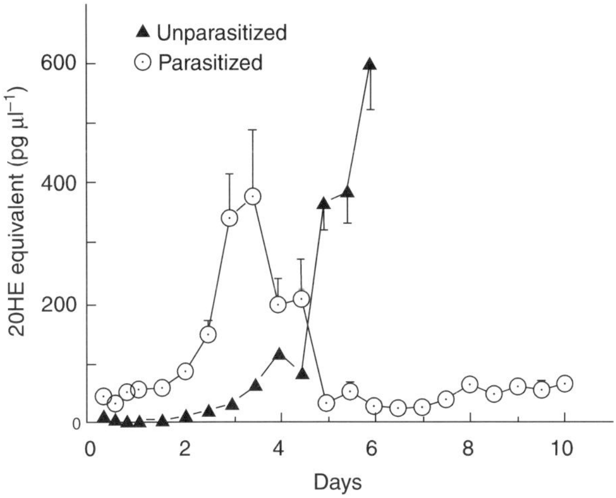

Figure 7. Fluctuations in haemolymph ecdysteroid levels of unparasitized and parasitized 5th-instar M. sexta, *which previously were parasitized as 1st instars. Haemolymph was sampled every 6 h on day 0, and in the morning and afternoon on subsequent days. Results are expressed as PG 20E equivalents/µl haemolymph. (From Gelman* et al., *1998, Elsevier Science Ltd.)*

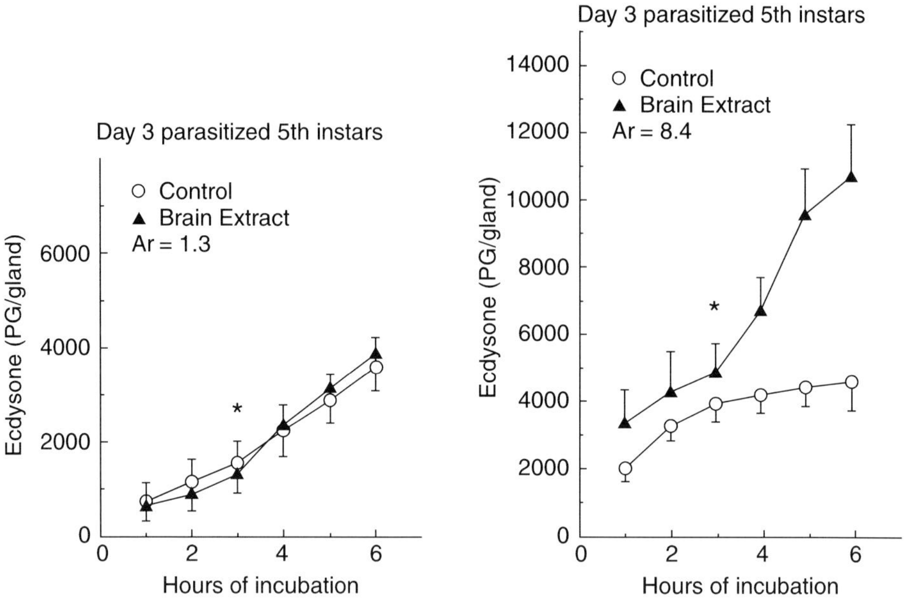

Figure 8. *Typical time course of ecdysteroid production by prothoracic glands from day-3 parasitized and unparasitized 5th-instar* M. sexta*. The asterisk denotes the time of brain extract addition. Ar (activation ratio) was determined by dividing the amount of immunoreactive ecdysteroid secreted into the medium for the 3–6 h period for a brain extract-stimulated prothoracic gland by the amount secreted during the same period for the unstimulated contralateral (control) gland from the same animal. (From Kelly* et al.*, 1998, Elsevier Science Ltd.)*

by which the brain regulates PG activity. Prothoracicotropic hormone cannot stimulate ecdysteroid production, and the parasitoid, by some unknown mechanism, regulates PG activity so that ecdysteroid is synthesized and released at the appropriate time in relation to parasitoid development.

Although host PGs appear to be responsible for the observed haemolymph ecdysteroid peak, evidence also exists for a contributory role by the parasitoid. Thus, host haemolymph ecdysteroid levels increase similarly in anterior and posterior portions of host larvae that have been ligated behind the first pair of abdominal prolegs (Gelman *et al.*, 1998) (*Figure 9(a)*). The ratio of parasitoids in anterior portions (containing PG) and posterior portions (PG absent) of ligated hosts is 1:5, and yet the haemolymph ecdysteroid levels in the sections in front of and behind the ligation are not significantly different (Gelman *et al.*, 1998), attesting to the exceptional ability of the parasitoid to fine-tune the levels of ecdysteroid in host haemolymph. Feedback mechanisms must exist so that parasitoids can maintain the host haemolymph ecdysteroid concentration at approximately 350 pg μl^{-1} – a concentration similar to that reported for haemolymph collected from 2nd-instar parasitoids one day prior to emergence (Gelman *et al.*, 1998).

In parasitized *M. sexta*, ligated behind the first proleg, haemolymph ecdysteroid profiles, as determined by reverse phase HPLC (RPHPLC), for portions anterior and posterior to the ligation are similar. Ecdysone (E), 20E and a peak containing polar ecdysteroids are the prominent ecdysteroids, and small quantities of 20,26-dihydrox-

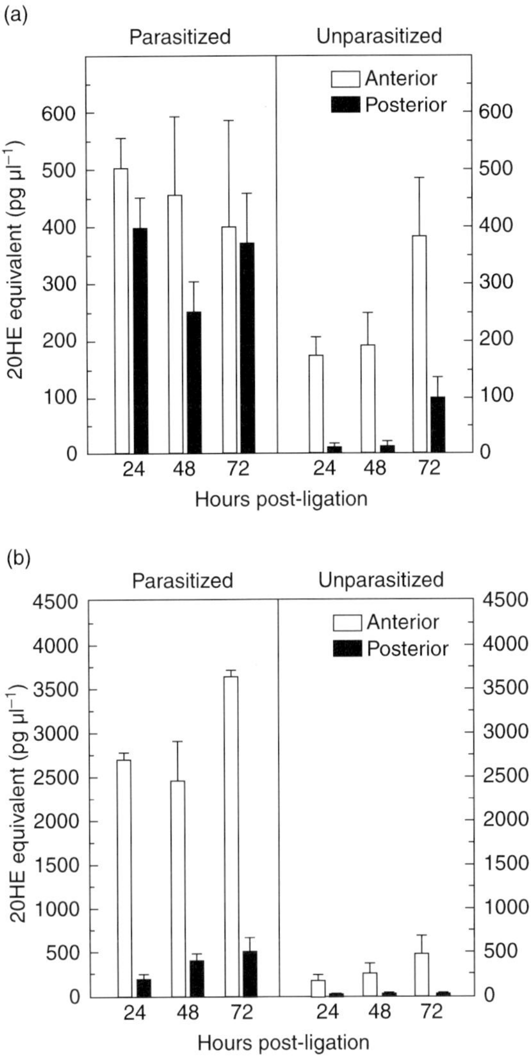

Figure 9. *Haemolymph ecdysteroid levels of anterior and posterior portions of parasitized and unparasitized* M. sexta *5th instars that had been ligated on day 2. (a) Larvae ligated behind the first abdominal proleg. (b) Larvae ligated behind the last thoracic segment. Results are expressed as PG 20E equivalents/µl haemolymph. (From Gelman* et al.*, 1998, Elsevier Science Ltd.)*

yecdysone (20,26E) and 26-hydroxyecdysone (26E) are also present (Gelman *et al.*, 1998). When parasitoids are excluded from the anterior portion of the host (when ligation is placed between the last thoracic and first abdominal segments), ecdysteroid titres increase dramatically to approximately 3000 pg μl^{-1} (Gelman *et al.*, 1998), indicating that the parasitoid must be present to prevent host PG from synthesizing and releasing too much ecdysteroid (*Figure 9(b)*). As determined by RPHPLC, E and 20E are the predominant ecdysteroids in the section anterior to the ligation (Gelman *et al.*, 1998). In follow-up experiments, ligations were placed behind the first abdominal segment providing larvae with varying numbers of parasitoids in the anterior portion. Haemolymph ecdysteroid titres in anterior portions containing from 0.5 to 2 parasitoids were approximately one half as high as titres in anterior portions with no parasitoids (Gelman *et al.*, unpublished), further supporting a role for *C. congregata* in regulating ecdysteroid production by host PG.

Ecdysteroids have been reported to enhance the growth of parasitoids cultured *in vitro* (Lawrence, 1986b; Nakahara *et al.*, 2000). Thus, when young 1st-instar *Diachasma-mimorpha longicaudata* are removed from their hosts (*Anastrepha suspensa*) and cultured *in vitro*, 20E is required for the parasitoids to moult to the 2nd instar (Lawrence, 1986b). For the hymenopteran parasitoid *Venturia canescens* the growth promotional effects of 20E increase in a concentration-dependent manner and peak at a concentration of 1ng μl^{-1} (Nakahara *et al.*, 2000). When higher concentrations are used, parasitoid larvae are malformed. When JH is added to the medium, the developmental time is shortened, but for *V. canescens*, the addition of JH does not appear to be essential for parasitoid development.

In other host–parasite systems, mechanisms of manipulating host ecdysteroid titres vary, but typically, the host does not exhibit a pre-pupal ecdysteroid peak containing threshold levels of 20E capable of stimulating a moult. When *C. sonorensis* injects PDV into 5th-instar *H. virescens*, host PGs degenerate (Dover *et al.*, 1989). In *H. virescens* parasitized by *C. nigriceps*, host PGs are structurally intact, but inactive – the block in ecdysteroid synthesis occurring at the protein phosphorylation step (Pennacchio *et al.*, 1994a). Prothoracicotropic hormone is able to stimulate the production of the second messenger, cAMP, but the cAMP-dependent protein kinase is never activated. As mentioned earlier, wasp-injected calyx fluid (containing PDV) and venom are believed to be responsible for inactivating this host's PG (Tanaka and Vinson, 1991a). However, host PGs only become refractory to PTTH stimulation after the critical period associated with pupal commitment, while PGs of parasitized *M. sexta* are refractory during the entire last instar (Kelly *et al.*, 1998). *Trichoplusia ni* parasitized by *Chelonus* species also do not exhibit a pre-moult ecdysteroid peak, but like *M. sexta* parasitized by *C. congregata*, a small parasitoid-induced host-haemolymph ecdysteroid peak is observed just prior to parasitoid emergence (Grossniklaus-Bürgin *et al.*, 1989; Jones *et al.*, 1992b). Since pseudoparasitized *T. ni*, as well as parasitized *T. ni* from which the parasitoid has been removed, do not exhibit this small haemolymph ecdysteroid peak, it appears that the peak observed in truly parasitized hosts originates from the endoparasitoid (Jones *et al.*, 1992b). *Cydia pomonella* parasitized by *Ascogaster quadridentata* and *Adoxophyes* sp. parasitized by *Ascogaster reticulatus* also exhibit haemolymph ecdysteroid peaks prior to the emergence of the parasitoid (Brown and Reed-Larsen, 1991). In these hosts, however, the ecdysteroid peak contains sufficient 20E to trigger apolysis, a condition that the authors suggest may facilitate parasitoid egression. Similarly, in *M. sexta* parasitized by *Hyposoter exiguae* or *H. fugitivus*, apolysis of the host cuticle occurs just prior to parasitoid emergence (Beckage and Templeton, 1985). Interestingly, in pseudoparasitized *Adoxophyes* sp., haemolymph ecdysteroid levels remain low and apolysis does not occur (Brown and Reed-Larsen, 1991).

Wasp parasitoids cultured *in vitro* have been reported to synthesize and release ecdysteroids (Brown and Reed-Larsen, 1991; Brown *et al.*, 1993; Gelman *et al.*, 1999), which may contribute to, or may be entirely responsible for, the small peak in haemolymph ecdysteroid observed in many host insects just prior to the emergence of the parasitoid(s). For *C. congregata*, in order to be able to synchronize parasitoid larvae with respect to their stage of development, Gelman *et al.* (1999) developed a staging system based on parasitoid larval body length, cuticle shedding and ballooning of the anal vesicle. These authors reported that *in vitro* ecdysteroid synthesis is characterized by two peaks of activity, one that occurs at the time of the moult to the 2nd instar and the 2nd that precedes the moult to the 3rd instar (*Figure 10*). Release of ecdysteroid

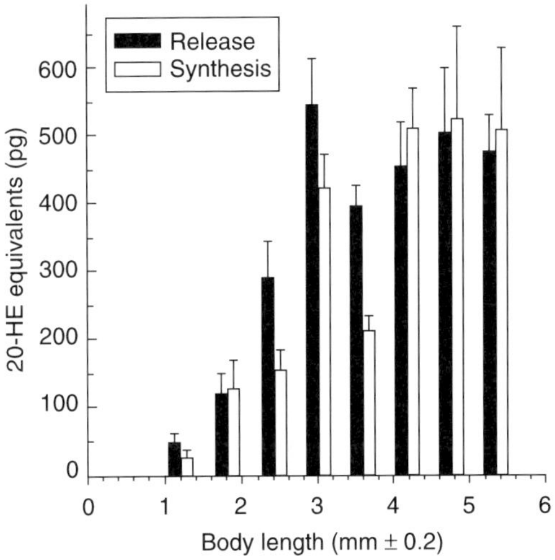

Figure 10. Ecdysteroid synthesis and release per parasitoid graphed as a function of parasitoid body length. Five to 10 parasitoids were placed in 25 μl drops of Grace's medium and incubated for 6 h. (From Gelman et al., *1999, Wiley-Liss, Inc., a subsidiary of John Wiley & Sons Inc.)*

into the medium shows a similar pattern, but the decline after the parasitoid's moult to the 2nd instar is not as pronounced (*Figure 10*).

Significant amounts of both E and 20E as well as of polar ecdysteroids are released into the medium, as are small quantities of 20,26E and 26E. *Ascogaster* sp. also release E, 20E and polar ecdysteroids when incubated *in vitro* (Brown and Reed-Larsen, 1991; Brown *et al.*, 1993).

Surprisingly, although synthesis and release of ecdysteroid by *C. congregata* is relatively high just prior to and during the parasitoid's moult to the 2nd instar, host haemolymph titres do not show a significant increase at this time (Gelman *et al.*, 1998). Thus, for its first larval moult, *C. congregata* appears to be primarily dependent on its own internal ecdysteroid resources, while for its second larval moult, an increase in host haemolymph ecdysteroid levels (to which the parasitoid contributes) is probably required. As mentioned earlier, in *Cotesia congregata* the moult to the 2nd instar is never observed until its host moults to the last instar. A similar trigger is observed for other host-parasitoid systems, including *C. inanitus–S. littoralis* (Grossniklaus-Bürgin *et al.*, 1994) and *C. nigriceps–H. virescens* (Pennacchio *et al.*, 1994a). For the latter system, *C. nigriceps* cultured *in vitro* moulted from the 1st to the 2nd instar in an ecdysteroid-free medium supporting the view that a mechanism not involving ecdysteroid triggers this parasitoid's moult from the 1st to the 2nd instar. Thus, a trigger other than ecdysteroid appears to be responsible for *C. congregata* and *C. inanitus* and, perhaps, for *C. nigriceps* to moult to the 2nd instar. In contrast, for the *A. suspensa–D. longicaudata* system, the ecdysteroid peak accompanying host pupariation appears to be required for *B. longicaudatus* to moult to the 2nd instar (Lawrence, 1986b). Similarly, in four other host–parasite systems, the parasitoid's moult to the 2nd instar is linked to the increase in ecdysteroid titre that occurs in 5th-instar hosts or at the end of diapause (Plantevin and Grenier, 1990).

5.4 *Effects of parasitism on the metabolism of host ecdysteroids*

Haemolymph ecdysteroid profiles of parasitized and unparasitized *M. sexta* are similar (Gelman *et al.*, 1998). Twenty hydroxyecdysone, 20,26E and a peak containing polar ecdysteroids were the most prevalent ecdysteroids in host haemolymph sampled on day 3 post-ecdysis to the 5th instar and in haemolymph from unparasitized *M. sexta* sampled on days 4 and 5 post-ecdysis to the last instar. Thus, at the time when host haemolymph ecdysteroid titres peak, physiologically active 20E is present. Interestingly, the profile of ecdysteroids in the haemolymph of host and parasitoid are almost identical (Gelman *et al.*, 1998) (*Figure 11*). Haemolymph ecdysteroid profiles for ligated hosts (whether behind the first pair of prolegs or behind the thorax) and unligated hosts are similar with the exception that increased amounts of E are present in anterior portions of thorax-ligated hosts (Gelman *et al.*, 1998), perhaps reflecting a reduced rate of conversion of E to 20E (Beckage and Templeton, 1986). Fat body and midgut from parasitized as compared to unparasitized *M. sexta* 5th instars, when cultured *in vitro*, show reduced levels of 20-monooxygenase activity (Beckage and Templeton, 1986). However, host haemolymph ecdysteroid profiles do not show a build-up of ecdysone. Reduced levels of ecdysteroid production by host PG may be responsible for not triggering host ecdysteroid 20-monooxygenase activity. In the *T. ni–Chelonus* system, ecdysteroid profiles are also similar for parasitized and unparasitized *T. ni*. However, polar metabolites (including ecdysteroid acids) and apolar metabolites are present at slightly higher levels in parasitized larvae (Grossniklaus-Bürgin *et al.*, 1989). Results suggest that the parasitoid is involved in the conversion of host ecdysteroids to apolar metabolites that are released into host haemolymph.

While in parasitized *M. sexta* 5th instars and *T. ni* precocious 5th instars, ecdysteroid production/release by the host is prevented from attaining levels observed in unparasitized pre-pupae, in other host–parasite systems ecdysteroid is produced, but converted to inactive metabolites. As mentioned previously, teratocytes of *C. nigriceps*

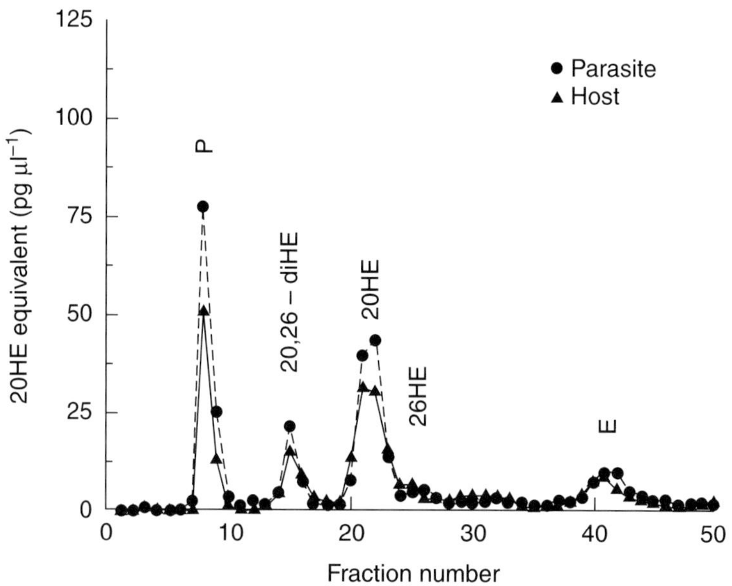

Figure 11. *Profiles of ecdysteroids in haemolymph of day-3 hosts and their respective parasitoids. Profiles were determined by radioimmunoassay of eluate fractions from HPLC. P, peak containing polar ecdysteroids; 20,26-diHE, 20,26-dihydroxyecdysone; 20HE, 20-hydroxyecdysone; 26HE, 26-hydroxyecdysone; E, ecdysone. (From Gelman* et al.*, 1998, Elsevier Science Ltd.)*

appear to be responsible for converting 20E to polar conjugates, the latter being responsible for the increased haemolymph ecdysteroid titre (as compared to unparasitized larvae) observed in parasitized larvae on day 7 of the last instar (Pennacchio *et al.*, 1994d). Similarly, the ectoparasitoid, *Euplectrus plathypenae,* induces a large haemolymph ecdysteroid peak (7200 pg 20E equivalents/μl) in its host *T. ni.* The major component of this peak is reported to be an ecdysteroid that has the same elution time (on RPHPLC) as 20,26E (Kelly and Coudron, 1990). Additional investigations revealed that this peak contains a glucoside conjugate of 20E (D.B. Gelman, *et al.* unpublished results). When a close relative of *E. plathypenae, E. comstockii,* parasitizes 4th-instar *Ostrinia nubilalis,* host haemolymph ecdysteroid titres increase and the host exhibits apolysis but not ecdysis (Gelman *et al.*, 1997). Thus, when *O. nubilalis* is the host, *E. comstockii* does not induce the conversion of 20E to inactive metabolites, but, rather, inhibits an endocrine event regulating the development of the pharate 5th instar and/or ecdysis. When 5th rather than 4th-instar *O. nubilalis* are parasitized, host haemolymph ecdysteroid levels remain low and the host does not undergo apolysis (Gelman *et al.*, 1997). Perhaps, as in the *H. virescens–C. sonorensis* system (Tanaka and Vinson, 1991), *O. nubilalis* PGs do not become sensitive to parasitoid inhibition until the last instar.

6. The effects of parasitism on host neurosecretion

The synthesis and release of a variety of endocrine and neuroendocrine secretions in conjunction with fluctuating tissue receptivity regulate a multitude of physiological events, and are therefore probable targets for parasitoid activity. Several studies have been conducted to compare the production and accumulation of important neuropeptides/peptides including PTTH, bombyxin, allatotropin, allatostatin, diuretic hormone, eclosion hormone, proctolin and FMRFamide in parasitized and unparasitized insects (Zitnan *et al.*, 1995a, b).

Since parasitism is so often associated with an alteration in ecdysteroid titres, the effect of parasitism on the regulation of ecdysteroid production by PTTH is of special interest. Prothoracicotropic hormone is released five times during the 5th larval instar of *Bombyx mori* (Shirai *et al.*, 1993) and, thus, may trigger a number of important events in the last instar. In *Bombyx mori,* Sakurai *et al.* (1998) reported that prior to gut purge, gradually increasing haemolymph ecdysteroid titres accompanied by daily fluctuations appear to be important for the sequential occurrence of developmental events associated with larval-pupal metamorphosis. Two peaks of PTTH activity have been observed in *M. sexta* 5th instars (Bollenbacher *et al.*, 1987), although it is possible that additional small peaks have yet to be detected.

When parasitized and unparasitized *M. sexta* are compared, immunocytochemical techniques reveal that levels of PTTH and bombyxin are increased in the brain neurosecretory cells and in associated axonal pathways of parasitized insects (Zitnan *et al.*, 1995a). Accumulation is most pronounced in brains and axons of older hosts, those from which the parasitoids have already emerged, and the level of accumulation appears to increase with time following emergence. With antibodies to *B. mori* PTTH and to bombyxin, respectively, cells, axons and axon terminals of the CA, and the latter, plus axon terminals of the corpora cardiaca, also stain more intensely in older hosts from which the parasitoids have emerged. In experiments in which the *in vitro* incubation of PG was used to determine relative amounts of PTTH activity in brains of parasitized and unparasitized *M. sexta* larvae, when differences in brain size are considered, ecdysiotropic activity is more concentrated in brains from day-7 hosts (wasps had emerged) than in brains from similarly aged unparasitized larvae (Kelly *et al.*, 1998). Therefore, after parasitoid emergence, PTTH appears to be accumulating in host brains. In younger brains of parasitized and unparasitized larvae, even though concentrations of PTTH are not significantly different (Kelly *et al.*, 1998), it is possible that processes associated with the synthesis and release of PTTH are operating at different levels. Thus, in host brains, production and release of PTTH could be occurring at reduced rates, especially since host PGs are not responsive to this brain hormone. Experiments to measure turnover rates in these brains would shed light on this issue. Importantly, one form of the PTTH isolated from brains of parasitized larvae is larger than 29 kDa, and this suggests that a unique form of PTTH (precursor) is present and that parasitization may inhibit the processing of the PTTH precursor (Kelly *et al.*, 1998). Previously, for unparasitized larvae, Kelly *et al.* (1996) reported a molecular weight of 11.5 kDa for the large form of *M. sexta* PTTH.

From an evaluation of staining intensity, and in some cases, number of cells stained, other neuropeptides, including allatotropin, allatostatin, diuretic hormone, eclosion hormone, proctolin and FMRFamide also increase in host brain corpora allata–corpora cardiaca complexes examined 7 days after wasp emergence (Zitnan *et al.*, 1995a). Similarly, parasitism induces the accumulation of FMRFamide-like peptides in endocrine cells located in the *M. sexta* gut, as well as in the nerves innervating the gut (Kingan *et al.*, 1997; Zitnan *et al.*, 1995b). As compared to the condition in unparasitized larvae, both the number of immunopositive cells and the intensity of staining increases in these cells as well as in cells of the frontal ganglion. Processing of FMRFamide peptides also appears to be inhibited in the midgut endocrine cells of parasitized larvae (Kingan *et al.*, 1997). As determined by an *in vitro* assay of PG activity, parasitism leads to an increase in the concentration of the low-molecular-weight ecdysiotropin(s) that is present throughout the hindgut (Gelman and Beckage, 1995). Thus, it appears that in developmentally arrested parasitized larvae, neuropeptides are synthesized in neuroendocrine cells of the brain, CC–CA complexes and frontal ganglia, and that peptides are synthesized in endocrine cells of the gut, perhaps at an elevated rate, but that release is suppressed. This inhibition of neuropeptide/peptide release may play a major role in the induction and maintenance of developmental arrest in host insects.

7. Effects of hormonally active insect growth regulators (IGRs) on parasitized larvae

Despite the fact that *C. congregata* larvae normally develop in a high JH milieu (i.e. the host's haemolymph), they are extremely sensitive to JH analogues, such as methoprene, administered to the host. Parasitoids fail to emerge from hosts parasitized in the 1st instar and then treated with 100 μg methoprene prior to the time of expected parasitoid emergence in the 5th instar (Beckage and Riddiford, 1982b). Those parasitoids that do emerge following treatment of the host with low doses (1–5 μg), often fail to eclose from their cocoons due to disruption of parasitoid metamorphosis after emergence from the host.

Recently we have repeated the experiments of Beckage and Riddiford (1982a) using host larvae parasitized in the 4th instar. Application of 200μg methoprene on day 0 of the 5th instar delays parasitoid emergence by 2 days, and fewer than the normal number of parasitoids emerge from methoprene-treated host larvae. The later in the 5th instar the host is treated, the more profound is the suppressive effect on emergence, and no parasitoids emerge from hosts treated early on day 4 just prior to the beginning of wasp emergence on day 5.

Parasitized *M. sexta* larvae appear less sensitive to anti-JHs compared to non-parasitized larvae of the same age (Beckage and Riddiford, 1983b), which may be due to their higher than normal JH titre. Fluoromevalonate (FMev) acts as an anti-JH in a variety of insects and inhibits the synthesis of JH I, II, and III by the corpora allata of a wide range of insects (Edwards *et al.*, 1983, 1985; Quistad *et al.*, 1981; Schooley and Edwards, 1996). In parasitized 4th-instar *M. sexta,* larger doses of FMev are required to induce formation of larval-pupal intermediates which are induced in non-parasitized larvae with much lower doses (Beckage and Riddiford, 1983b). Nevertheless, the fact that parasitized larvae can be induced to form larval-pupal intermediates with this compound is worthy of note. As noted above, even surgical allatectomy of parasitized 4th instars fails to induce their metamorphosis, perhaps because the parasitoids continue to produce JH (presumably JH III) in the allatectomized host. Treatment of the host with FMev, however, may curtail JH production by the parasitoids as well as inhibiting JH secretion by the CA of the host, resulting in the formation of larval-pupal intermediates. No parasitoids emerge from these intermediates.

The anti-JH ETB (ethyl 4-[2-{*tert*-butyl carbonyloxy} butyoxy]benzoate) has a complex mode of action, acting as an anti-JH in low doses and as a JH-like compound at high doses in non-parasitized *M. sexta* larvae (Schooley and Edwards, 1996; Staal, 1977b). Regardless of the dose of the anti-JH ETB applied to 3rd-instar larvae (which previously were parasitized in the 1st instar), the compound does not induce formation of larval-pupal intermediates, or even black 5th-instar larvae (Beckage and Riddiford, 1983b). A black 5th-instar cuticle normally forms in response to a lowering of the JH titre during the moult (Fain and Riddiford, 1975). The parasitoids emerge at the normal time and in normal numbers from ETB-treated hosts, indicating that parasitized larvae are refractory to ETB, although they are paler and pinker than normal compared to solvent-treated control larvae (Beckage and Riddiford, 1983b). This pigmentation effect may reflect the JH-like effect of high ETB dosages, since the epidermis of non-parasitized larvae treated with high doses of ETB also turns pink (Staal, 1977b).

Another class of insect growth regulators, the benzyl-1,3-benzodioxole derivatives, has major disruptive effects on development of *C. congregata* parasitoids in treated host larvae (Beckage *et al.*, 1987). Large pockets of moulting fluid accumulate in the ecdysial space between the epidermis and cuticle during moulting and the parasitoids subsequently fail to emerge. The head and thoracic region of the newly moulted 5th-instar hosts also show morphological abnormalities, which interfere with feeding (Beckage *et al.*, 1987). Since wasps fail to emerge from host larvae that are starved beginning on day 0 or day 1 of the 5th instar (Beckage and Riddiford, 1983a), the induction of morphological anomalies, which prevent host feeding during the critical period early in the 5th instar, may interfere subsequently with parasitoid emergence from the host.

In contrast to methoprene, which disturbs the wasps' second larval moult, azadirachtin interferes with the first larval ecdysis of the parasitoids developing within the host haemocoel (Beckage *et al.*, 1988). The 2nd-instar cuticle is synthesized, but the 1st-instar cuticle is not shed, so the larvae die encased in the 1st-instar exuvial cuticle, within the host haemocoel. A dose of 10 μg azadirachtin injected into a day-0 5th-instar hosts completely inhibits emergence of its developing parasitoids (*Figure 12*). As expected, despite the parasitoids' demise, the host does not wander or pupate – probably due to the action of the PDV-containing calyx fluid injected by the female wasp at the time of oviposition. Why the parasitoids fail to shed the exuvial 1st-instar cuticle is not known. The compound may interfere with the release of eclosion hormone (Horodyski, 1996) or ecdysis-triggering hormone (Zitnan *et al.*, 1996) during parasitoid moulting. Notably, azadirachtin has been reported to interfere with the moulting process and ecdysis in a wide variety of insects (Beckage, 1998b; Kubo and Klocke, 1982; Schmutterer, 1990). Yet azadirachtin treatment typically leads to reduced ecdysteroid titres (Kubo and Klocke, 1982; Schmutterer, 1990), and a reduction in ecdysteroid levels is required for the release of eclosion hormone (Truman *et al.*, 1983) and ecdysis-triggering hormone (Zitnan *et al.*, 1999) in lepidopterans.

Recently, we have examined effects of the non-steroidal ecdysone agonists (Beckage, 1998b; Dhadialla *et al.*, 1998), which are potent compounds that induce apolysis and initiation of supernumerary larval moulting of *M. sexta* (Wing *et al.*, 1988). These compounds appear to specifically target Lepidoptera (Sundaram *et al.*, 1998). It appears likely that these compounds act by directly binding to ecdysteroid receptors in epidermis and other target tissues, since ecdysteroid titres are suppressed in animals treated with these agonists and yet moulting is initiated (Wing, 1988; Wing *et al.*, 1988). When the compounds RH-5849 and RH-2485 are administered to last-instar hosts, the injected larvae are stimulated to undergo apolysis and form a slipped head capsule, and the parasitoids die as pre-emergent wasps in the host haemocoel (Beckage *et al.*, unpublished results).

In summary, the effects of IGRs can be classified as to whether they have direct or indirect effects in causing disrupted development of endoparasitoids (see summary in Beckage *et al.*, 1988). Following their administration to the host, the effects of IGRs may be interpreted as being indirect if they appear to be due to induction of anomalies in the host which prevent maturation of the parasitoids. The induction of precocious metamorphosis in parasitized larvae treated with FMev probably has indirect effects

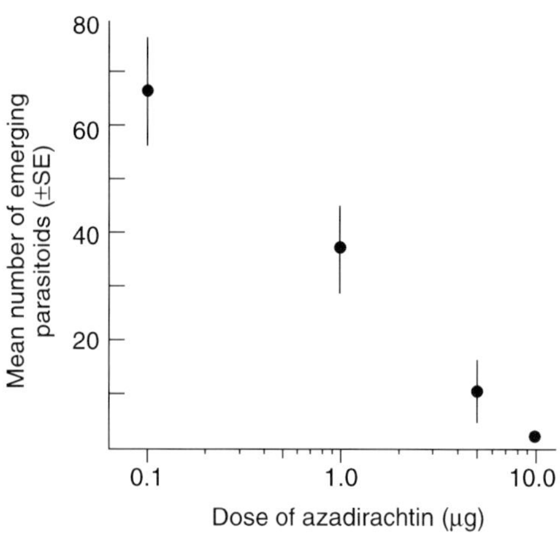

Figure 12. *Dose–response relationship showing how different doses of azadirachtin inhibit emergence of* C. congregata *parasitoids from treated host larvae. The mean number of emerging wasps per host (± SD) is plotted as a function of the dose of azadirachtin injected into day-0 5th instar larvae. (Figure redrawn from Beckage* et al.*, 1988, with permission from Wiley-Liss Inc, a subsidiary of John Wiley & Sons, Inc.)*

on the parasitoid in that it appears likely that parasitoids cannot emerge since the host fails to feed for the period necessary to assure adequate growth of the parasitoid, and a hardened pupal cuticle is formed which prevents parasitoid emergence. Similarly, the administration of benzyl-1,3-benzodioxole derivatives to parasitized larvae induces formation of morphological abnormalities in the head and thorax, which suppress host feeding. In contrast, administration of methoprene or azadirachtin to 5th-instar host larvae appears to have direct effects on the parasitoids, interfering with their larval ecdyses, emergence and metamorphosis. Hormonally active IGRs have been shown to affect development of many species of parasitoids (Beckage, 1985). Thus, their potential implementation in insect control must be assessed carefully to eliminate the possibility of their use having unanticipated detrimental effects on beneficial species.

4

Endoparasitoid-mediated disruption of host endocrine systems: common themes through uncommon means

Bruce A. Webb, Rico L. Rana and Douglas L. Dahlman

1. Introduction

Although some hymenopteran endoparasitoids evade immune detection, most parasitoid eggs and larvae are recognized as foreign by host immune systems, but escape immune reactions by actively suppressing host immunity. Not only is immunity actively suppressed, parasitized insects are usually maintained in a physiological and developmental state that supports endoparasitoid development. To establish an environment permissive for endoparasitoid survival and development, factors from the reproductive tract of the female parasitoid and from the developing parasitoid larva are employed to disrupt immunity and alter other aspects of host physiology. These factors include proteins synthesized by the female wasp and introduced into hosts during oviposition (venoms, ovarian proteins), symbiotic viruses (polydnaviruses (PDV), entomopoxviruses, ascoviruses) and cells released from the parasitoid egg at the completion of embryonic development (teratocytes) (Dahlman and Vinson, 1993; Strand and Pech, 1995; Webb, 1998). Some parasitoid eggs and larvae also secrete proteins into the host, although the physiological significance of these secretions has not been established (Soldevila and Jones, 1993). Because endoparasitoids encounter similar physiological challenges, it is perhaps not surprising that similar physiological effects are associated with the several factors introduced during parasitization. Venoms, ovarian proteins, PDV and teratocytes have been linked to immune suppression as well as to alterations of host growth and development (Dahlman and Vinson, 1993; Jones and Coudron, 1993; Stoltz, 1993). Parasitized hosts often exhibit physiological distress characterized by a slowing or cessation of growth and development (Vinson *et al.*, 1979; Webb and Dahlman, 1985), which prevents host metamorphosis and allows successful

Endocrine Interactions of Insect Parasites and Pathogens, edited by J.P. Edwards and R.J. Weaver.

parasitization by hymenopteran parasitoids that culminates in host death. Our focus here is on those factors that alter host growth and development. Specifically, our objective is to compare the similarities and distinctive roles of teratocytes and PDV. We will concentrate on two systems that have been investigated over several years. Teratocytes from the braconid parasitoid, *Microplitis croceipes*, suppress host development when injected into the permissive host, *Heliothis virescens*, larvae (Zhang and Dahlman, 1989) in a manner similar to that seen in parasitized larvae. Similarly, parasitization or injection of purified PDV from the ichneumonid, *Campoletis sonorensis*, inhibits host growth and development (Vinson *et al.*, 1979). Some of the mechanisms through which PDV and teratocytes alter host developmental physiology are known, and some intriguing similarities exist. The data suggest that teratocytes and PDV may have similar mechanisms for disrupting host developmental physiology. They also encode genes that exhibit structural similarities (D.L. Dahlman, B.A. Webb *et al.*, unpublished observations) suggesting that some elements of their evolutionary background may be shared.

The development of larval insects is regulated primarily by changing titres of haemolymph juvenile hormone (JH) and ecdysteroids (Nijhout, 1994; see also Chapter 1 this volume). Therefore, parasitoid-mediated developmental alterations are likely to be associated with abnormalities in the host endocrine physiology. Parasitization by both *M. croceipes* and *C. sonorensis* blocks host larval development prior to pupation. Specifically, larvae parasitized by *C. sonorensis* do not initiate premetamorphic behaviour (e.g. wandering), while parasitization by *M. croceipes* blocks host development at a post-wandering stage (see *Figure 2*; Webb and Dahlman, 1985). Based upon the developmental stage of host larvae at the completion of endoparasitoid development, *C. sonorensis* parasitization produces the high JH and low ecdysteroid titres characteristic of intermoult, penultimate-instar larvae. Hosts parasitized by *M. croceipes* exhibit developmental arrest after the initiation of metamorphic development, but before the surge of ecdysteroids leading to pupation (Dahlman *et al.*, 1990a). These host larvae exhibit low JH and ecdysteroid titres, but pupal development can be induced by injection of 20-hydroxyecdysone (20E) (Webb and Dahlman, 1986). Introduction of teratocytes or PDV alone will cause many of the developmental and physiological abnormalities observed in parasitized larvae. Specifically, injection of teratocytes or PDV delays or blocks host development in a dose-dependent manner (Fathpour and Dahlman, 1995; Zhang and Dahlman, 1989). Normal endocrine function in the injected larva is affected as indicated by a reduction in ecdysteroid titre and by alteration of juvenile hormone esterase (JHE) activity (and presumably, JH levels) in a manner similar to that observed in parasitized insects (Pennacchio *et al.*, 1994a, b; Tanaka and Vinson, 1991; Zhang *et al.*, 1992). Thus, teratocytes and PDV appear to have evolved, in part, as the means to deliver genes and proteins that inhibit host growth and development. These strategies maintain parasitized larvae in a developmental and physiological state that permits parasitoid growth and development.

Teratocytes are cells that result from the dissociation of the serosal membrane of the parasitoid egg upon hatching (*Figure 1(a)*). These cells circulate in the haemolymph of the parasitized insect. Typically, teratocytes do not divide after they have been released, but do increase significantly in size. Prior to the 1980s, teratocytes were viewed primarily as supplementary nutritional sources for the developing parasitoid larva (Vinson and Iwantsch, 1980). However, more recent data have associated teratocytes with disruption

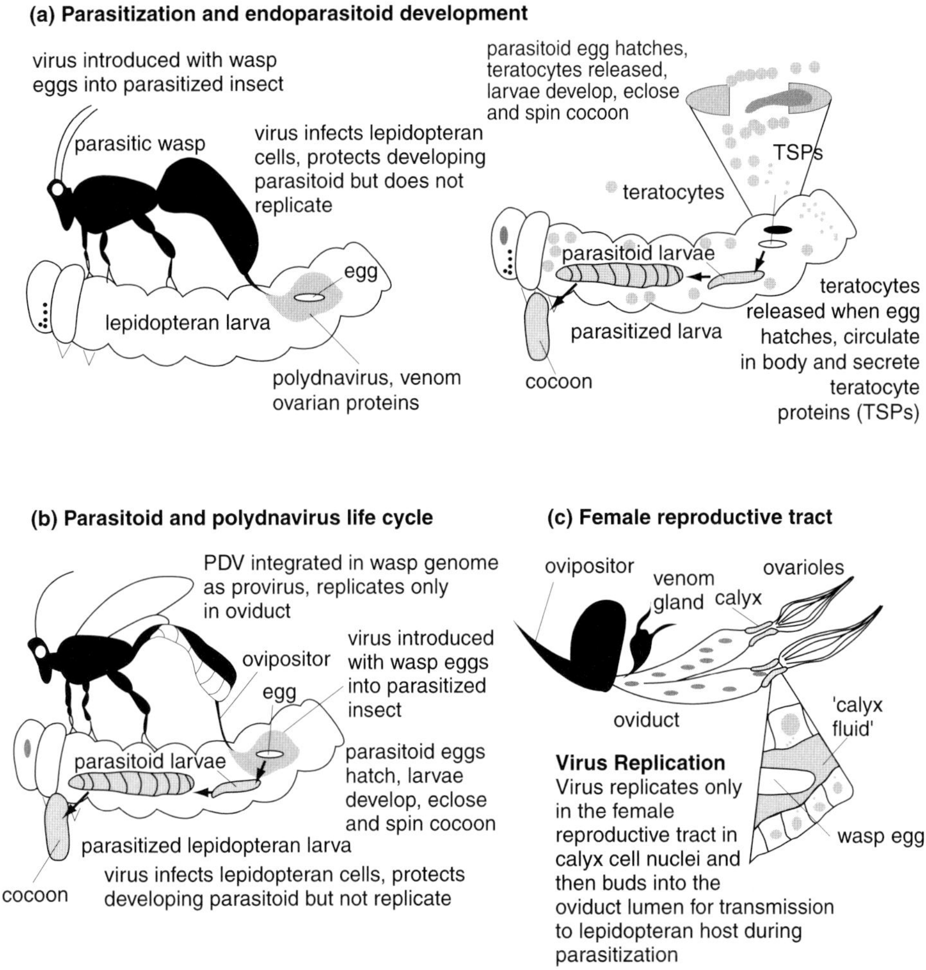

Figure 1*. Schematic diagram showing parasitization and introduction of maternal factors, responsible for suppressing host immune system while supporting growth and development of endoparasitic wasp, into lepidopteran host larvae. (a) The dissociation of teratocytes from the serosal membrane enveloping the parasitoid embryo and subsequent release into host haemolymph after egg hatch. (b) and (c) detail the life cycle and the site where polydnaviruses replicate prior to their transmission into the host during parasitization.*

of immune responses and inhibition of host growth and development (Dahlman, 1990, 1991; Dahlman and Vinson, 1993; Strand *et al.*, 1988). While it is likely that teratocytes have largely nutritional roles in some wasps, in others, teratocytes are now viewed as parasitoid factors that actively alter their hosts to support the nutritional and developmental needs of endoparasitoid larvae.

Polydnaviruses are symbiotic viruses produced in the calyx cells of the female parasitoid oviduct and bud into the oviduct lumen (*Figure1(b)* and *(c)*). Polydnaviruses are introduced into parasitized larvae during oviposition, whereupon they infect host cells (primarily haemocytes), and express a host-specific complement of viral genes. Polydnaviruses, alone or in conjunction with other factors, suppress host immune responses and alter host development (Webb, 1998). There are two phylogenetically unrelated groups of PDV: the ichnoviruses and the bracoviruses that are present in ichneumonid and braconid wasps, respectively. Our focus here is to compare the ichnovirus from *C. sonorensis* with the teratocytes from the braconid wasp, *M. croceipes*.

2. Teratocytes

When braconid wasps lay eggs into their host larvae, several endoparasitoid-derived substances are introduced into the host haemolymph. These include calyx fluid (which may contain PDV), venom, and ovarian proteins (*Figure 1(a)*) (Dahlman and Vinson, 1993). When the eggs of some braconid parasitoids hatch, the serosal membrane enveloping the developing embryo dissociates into individual cells called teratocytes. After release from the wasp egg, teratocytes circulate in the host haemolymph, grow rapidly, become greatly enlarged and secrete large amounts of biologically-active proteins (*Figures 1(a)* and *2*). These cells do not undergo further cell division, but may become polyploid (Strand and Wong, 1991). Dahlman and Vinson (1993) reviewed the physiological and biochemical functions of teratocytes. Newly released teratocytes have surface microvilli (*Figure 3*) that are thought to absorb nutrients from the host haemolymph. The rapid growth of teratocytes suggests that these cells efficiently take up and process host metabolites. Subsequently, they may serve as a nutrient source for the developing parasitoid, or synthesize proteins or other substances for release into the host haemolymph where they may perform a more active role in parasitization. Some parasitoid larvae are known to ingest teratocytes as demonstrated by the finding of complete cells or cell fragments in the digestive tract of parasitoid larvae (Kadono-Okuda *et al.*, 1998). Teratocytes have also been implicated in suppression of host immunity, as both encapsulation of parasitoid larvae and host haemolymph phenoloxidase activity are reported to be inhibited (Kitano *et al.*, 1990; Vinson, 1972). Teratocytes are known to secrete proteins and have a cellular architecture indicative of this capability. For example, large, electron-dense vesicles are seen in teratocytes from the egg parasitoid, *Telenomus heliothidis*. The contents of these vesicles are presumably released into the host egg, and may contain enzymes capable of digesting host tissues, thereby providing nutrients for the developing parasitoid (Strand *et al.*, 1988). So far, the scelionid egg parasitoid, *T. heliothidis*, is the only parasitoid in which teratocyte extracts show detectable levels of digestive enzymes (e.g. acid phosphatase, esterase, and leucine aminopeptidase) capable of digesting host tissues to provide a nutrient source. Additionally, Dahlman and Vinson (1993) postulated that teratocyte-secreted proteins (TSP) caused the reduction in haemolymph JHE levels, which may indirectly impact the overall JH, ecdysteroid and other hormonal titres in the host haemolymph. Schepers *et al.* (1998) reported that TSP inhibited *in vitro* synthesis of JHE in *H. virescens* fat body cultures.

In the *H. virescens–M. croceipes* host–parasitoid model, we are interested in the role of teratocytes in synchronizing host and endoparasitoid development, and in understanding the mechanism(s) through which teratocytes alter host developmental physiology. These questions have been approached by investigating the influence of teratocytes harvested from parasitoid eggs cultured *in vitro* and assaying the effects of cultured teratocytes and of proteins secreted from these cells. Reports in the literature have largely concentrated on the effects of teratocytes on host physiology (Dahlman and Vinson, 1993). Unequivocal evidence for the precise role(s) of teratocyte-secreted

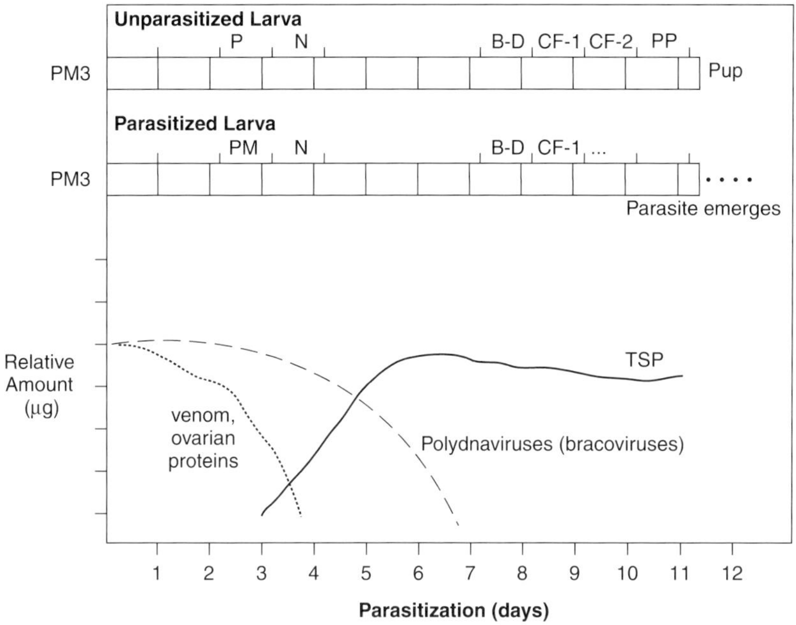

***Figure 2**. Hypothetical depiction of the relative quantities of maternal factors introduced from a braconid endoparasitoid into its lepidopteran host at parasitization. Numbers in each box located at the top two bars represent the developmental stage of the larval host, Heliothis virescens. Pre-moult third- (PM3) instar larvae are used as the starting developmental stage for parasitization. PM, pre-moult stage; N, newly moulted; PP, pre-pupal stage; D, digging stage; B-D, burrowing-digging stage; CF-1 and 2, cell formation days 1 and 2; TSP, teratocyte-secreted proteins.*

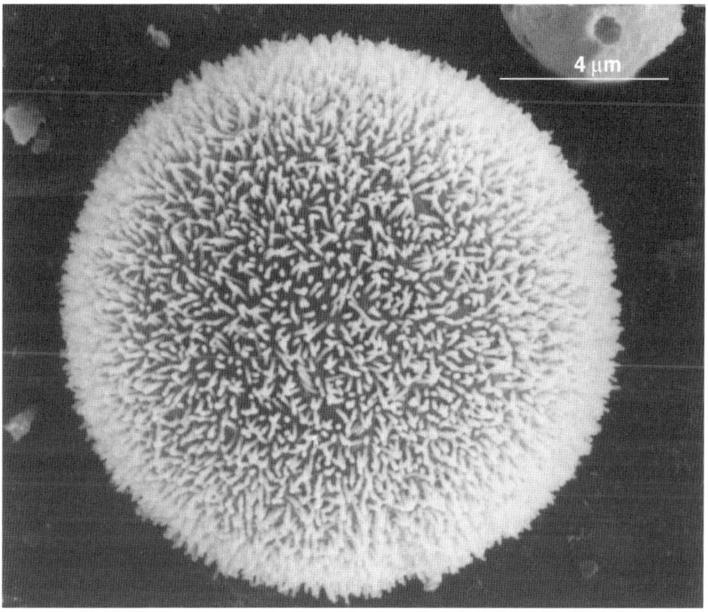

***Figure 3**. Scanning electron micrograph of a 1-day-old teratocyte showing short and relatively sparse microvilli at the outer surface.*

proteins, and for their interaction with components of the calyx fluid (PDV and venom) and other ovarian proteins (which are also introduced to the host during parasitization) is not yet available. Zhang and Dahlman (1989) investigated the influence of teratocytes on host physiology, in the absence of parasitoid larvae, by injecting various numbers and ages of *M. croceipes* teratocytes into its larval host, *H. virescens*. They observed that introduction of approximately 750 three-day-old teratocytes into 4th stadium, day-two larvae, produced incomplete larval-pupal ecdysis or delayed pupation. In addition, teratocyte injections significantly compromised the growth of host fat body, and depressed JHE levels in the host haemolymph to about 5% of the levels found in control larvae (Zhang *et al.*, 1992). Thus, teratocytes released into the host haemolymph at parasitoid egg hatch may be responsible for the reduction in JHE levels observed in parasitized larvae (Dahlman *et al.*, 1990). Dong *et al.* (1996) suggested that reduced levels of JHE and storage proteins (arylphorins) in fat body and haemolymph of teratocyte-injected larvae could result from post-transcriptional regulation or modification of the JHE and arylphorin mRNAs. Although the levels of JHE and arylphorin proteins were reduced in teratocyte-injected host larvae, the amounts of JHE and arylphorin mRNAs were similar to those in control larvae.

Similarly, injections of teratocytes obtained from another braconid endoparasitoid, *Cardiochiles nigriceps*, into last-instar larvae of its host, *H. virescens*, resulted in host developmental arrest and altered ecdysteroid titres (Penacchio *et al.*, 1992). The hormonal alterations seen in parasitized hosts were attributed, at least in part, to the conversion of 20E into inactive polar metabolites by teratocytes, as well as to the depression of secretory activity of the prothoracic glands by calyx fluid/venom (Pennacchio *et al.*, 1994a). These hormonal changes are thought to induce an increase in the suitability of the nutritional milieu in which the parasitoid larvae develop (Pennacchio *et al.*, 1993; see also Chapter 5 in this volume).

Normally, teratocytes function in parasitized larvae, and these also contain ovarian proteins, venom, PDV, and a developing parasitoid larva. However, teratocytes can significantly impact host physiology in the absence of other maternal factors and can independently support successful development and egression of the parasitoid from its host larva. Whether one or more specific proteins or other teratocyte secretions interact to cause the observed physiological abnormalities seen in parasitized hosts remains to be elucidated. The results from the *M. croceipes* system suggest that a specific protein (TSP14) is partially responsible for at least some of the aberrant physiological and developmental symptoms evident in parasitized and teratocyte-injected insects.

3. Polydnaviruses

Polydnaviruses are symbiotic viruses that replicate only in the oviduct calyx cells of some ichneumonid and braconid wasps (*Figure 1*). As suggested by their name (i.e. poly-DNA-viruses), they are the only virus group having genomes comprised of multiple, circular DNA molecules or segments. Polydnaviruses are delivered to parasitized insects with the parasite egg during parasitization. In parasitized larvae, PDV infect several host tissues, but especially haemocytes, and express a host-specific complement of viral genes. Viral gene expression induces changes in host immunity and alterations in host development, but the virus itself does not replicate (Webb and Summers, 1990). The virus is transmitted vertically because viral DNA segments are integrated at several loci within the genomes of their parasitoid host. The PDV replicate only in specialized 'calyx' cells of the female reproductive tract, and do so in response to the hormonally induced maturation of the oviduct during adult parasitoid development (Webb and Summers, 1990). The PDV–parasitoid system is an unusual example of a mutualistic virus association, in which the expression of viral genes is essential for wasp survival in the parasitized host.

One of the earliest effects described in calyx fluid- (and/or PDV-) injected larvae, was a dose-dependent reduction in the growth of the host larvae (Vinson *et al.*, 1979). Injection of low (subphysiological) doses of virus resulted in transitory effects, while higher doses prevented the pupation of injected larvae (Cui *et al.*, 2000), and increased the level of viral gene expression. To establish a bioassay for the developmental and physiological effects of PDV injection, Shelby and Webb (1994) investigated the levels of haemolymph storage proteins in parasitized and in PDV-injected larvae. As well as showing reduced feeding, virus-injected larvae exhibited reduced haemolymph titres of storage proteins (*Table 1*). Surprisingly, the observed reductions in storage protein synthesis were not accompanied by a concomitant reduction in mRNA levels – suggesting that arrest of host development may involve regulation of host gene expression at a post-transcriptional level (Shelby and Webb, 1997). A factor that inhibited protein synthesis in organ culture assays was characterized and named the haemocyte translation inhibiting factor (HTIF). The HTIF was sensitive to proteolysis and heat treatments, suggesting that it was proteinaceous. Retention on concanavalin A columns suggested that the protein was glycosylated and, based on size-exclusion chromatography, had a molecular weight of approximately 45 kDa (K.S. Shelby and B.A. Webb, unpublished results). The HTIF is present in the haemolymph only following virus infection and expression, but it has not been demonstrated that this gene is virally encoded. Efforts to clone the gene encoding HTIF are in progress.

On a parallel research track, the *C. sonorensis* PDV (CsIV) genome is being sequenced. The genes that are expressed in parasitized larvae are largely characterized. At least 37 CsIV genes are known or predicted to be expressed in parasitized insects (B.A. Webb, unpublished results). Thirty-six of these genes belong to one of three gene families, each having from three to 25 gene family members. One of the gene families with potential relevance to host–parasitoid interactions is the cysteine-motif gene family (Blissard *et al.*, 1987; Dib-Hajj *et al.*, 1993). Cysteine-motif genes are

Table 1. *Comparison of the main features and effects of teratocytes and polydnaviruses (PDV) in ichneumonid and braconid wasps*

	Ichneumonidae	Braconidae
Polydnavirus	Yes (ichnovirus)	Yes (bracovirus)
Teratocyte	No	Yes
Host juvenile hormone	High	Low
Ecdysteroid	Low	Low
Host larvae wander	No	Yes
Host storage protein reduced	Yes	Yes
Host storage protein mRNA reduced	No	No
Inhibitory protein	HTIF	TSP14

abundantly expressed, and encode proteins that are subsequently glycosylated and secreted. The majority of cysteine-motif genes are encoded on viral segments that produce progeny segments from a parental segment. Interestingly, this production of progeny segments increases the copy number of some cysteine-motif genes. The resultant increase in gene copy number increases the titre of cysteine-motif proteins in parasitized insects, and this increase in titre is required for survival of the parasitoid. Experimental manipulations that reduced the virus infectious dose, lowered the haemolymph titre of cysteine-motif proteins and did not protect parasitoid eggs from encapsulation (Cui and Webb, 1997; Cui *et al.*, 2000). Some cysteine-motif proteins are known to bind to insect haemocytes, and have been associated with a reduction in the encapsulation response of haemocytes (Li and Webb, 1994). Whether the various cysteine-motif gene family members have other activities is not known.

4. Teratocyte-secreted protein (TSP)

There have been relatively few studies on the physiological effects of teratocytes, and even fewer have addressed the functional significance of individual teratocyte proteins. As evidence accumulated that teratocytes inhibited host growth and development, *in vitro* assays were developed to assess teratocyte function in the absence of other host or parasitoid factors. Pre-moult 3rd-stadium larvae parasitized by *M. croceipes* exhibit wandering behaviour, but further development is blocked at the CF-1 stage. These larvae have a reduced weight, and are without the high levels of storage protein associated with normal pre-metamorphic behaviour in the final stadium. Teratocyte injection was found to lower haemolymph storage protein titres in a manner similar to the reduction observed in parasitized larvae and analogous to that described above for CsIV. To determine if proteins or other factors synthesized by and released from teratocytes inhibited storage protein synthesis, fat body was explanted and the incorporation of [^{35}S]-labeled methionine into newly synthesized proteins was measured *in vitro*. These experiments suggested that proteins secreted by teratocytes inhibited protein synthesis by the host fat body, but the data were variable, possibly because of the irregular structure of fat body tissue. By comparison with fat bodies, the paired and symmetrical nature of the testes (which also synthesize storage proteins) appeared to provide an ideal alternative candidate tissue with which to perform carefully controlled organ culture assays for studying the effects of TSP on protein synthesis. In an *in vitro* culture of teratocytes, (D.L. Dahlman, unpublished results) demonstrated that TSP released into the culture medium inhibited protein synthesis in testes bioassays. Testes bioassays were then used to evaluate different fractions obtained from TSP purification procedures. A teratocyte protein (TSP14) subsequently purified from teratocyte culture media was one of the primary factors causing inhibition of protein synthesis, and it is also possible that this protein may be associated with the blockage of host development. In the following sections, we outline the experimental findings implicating the *M. croceipes* teratocyte protein, TSP14, as a candidate agent in disrupting the developmental physiology of its host, *H. virescens*.

4.1 In vitro *culture of teratocytes*

To evaluate the effects of TSP on host physiology, Schepers *et al.* (1998) described an *in vitro* culture system for isolating *M. croceipes* teratocytes and teratocyte proteins. Endoparasitoid eggs were collected from superparasitized, pre-moult 3rd-instar

H. virescens larvae, and reared aseptically at 28°C in ExCell-400 medium supplemented with gentamicin. Three days after hatching, the medium was separated from the developing parasitoid larvae, teratocytes and other cell debris by centrifugation, and stored at -80°C until use. The resulting supernatant served as a source for preparation of crude TSP. As much as 0.7 μg of TSP are secreted from about 900 teratocytes (the average number per egg) per day. Crude preparations of TSP inhibited the incorporation of [^{35}S]-methionine into proteins synthesized *in vitro* by host larval fat body (Schepers *et al.*, 1998) or testes (D. Dahlman, unpublished results). In addition, a dose of 3.5 larval equivalents (LE) (1 LE = TSP secreted *in vitro* by teratocytes from one parasitoid egg over 3 days of *in vitro* culture) caused a reduction of about 50% in the incorporation of radiolabeled methionine in a rabbit reticulocyte lysate translation assay, using mRNAs from *H. virescens* testes (D.L. Dahlman, unpublished results). Likewise, the same preparations caused a reduction in the amount of JHE produced from fat body cells (Schepers *et al.*, 1998). The striking influence of crude TSP preparation on these physiological and biochemical phenomena warranted further investigation. Fractionation of the crude TSP preparation with molecular weight filters produced a 3–30 kDa fraction containing biological activity associated with four low molecular weight proteins (Schepers *et al.*, 1998). The predominant protein, estimated (by SDS-PAGE) to have a molecular weight of 14 kDa, was isolated and sequenced (D.L. Dahlman, unpublished results).

4.2 *Isolation and identification of a TSP and the TSP gene from teratocytes*

Functional analyses of the TSP required separation and purification of the predominant TSP component from crude preparations. About 1500 LE of TSP, obtained from crude preparations, was used as starting material for purification. The material was fractionated using molecular weight filters and the 3–30 kDa fraction was concentrated, resuspended and separated by C18 reversed-phase high-performance liquid chromatography (RP-HPLC). The fraction corresponding to the 14 kDa protein was collected and subjected to Lys-C proteolysis then re-chromatographed on the C18 column. Two peptides with short retention times were sequenced using Edman degradation. One of the fragments provided a sequence of only five amino acids, while the other fragment produced a sequence extending for 28 residues. Both peptides were suitable for the design of degenerate oligonucleotide primers (D.L. Dahlman, unpublished observations).

Degenerate primers were synthesized and used to amplify and clone both the 3′ and 5′ ends of TSP14 mRNAs by rapid amplification of cDNA ends (RACE). The resulting 540bp cDNA encoded a known amino acid sequence that was not primer-specified, and an open reading frame of approximately the correct size to encode the TSP14. The complete cDNA contains an open reading frame of 426 nucleotides that encodes a 22-amino-acid signal peptide, and a sequence encoding 129 amino acids which would produce a protein with an estimated molecular weight of 14 012 (D.L.Dahlman *et al.*, unpublished results). The predicted peptide sequence of the 14-kDa cDNA encoded a cysteine-motif protein similar to those previously described from PDV (Blissard *et al.*, 1987; Cui and Webb, 1996; Dib-Hajj *et al.*, 1993), possibly indicating an evolutionary (and perhaps, functional) similarity between PDV and teratocyte genes (*Table 2*).

Table 2. Comparison of C. sonorensis *polydnavirus (PDV) N-terminal (N) and C-terminal (C) cysteine-motifs of the VHv1.1 and VHv1.4 gene products* with a similar deduced sequence motif in the M. croceipes *TSP14 protein. Cysteines and conserved residues are indicated in bold. Introns are represented by an asterisk.* Campoletis sonorensis *PDV genes are named by the segment on which the gene resides, the host in which it is expressed and the size of the mRNA. For example, VHv1.4 is a gene encoded on segment V, that is 1.4 kb in size, expressed only in the* H. virescens *host.*

TSP14			
TSP14	CAPASGI*CHRVGLEITKPCCNK—FDRCFASVSEPVSRC		
PDV motifs			
VHv1.4(N)	CIGNYQP*C	IESTKPCCRLEDRTSVQFGRKEYIC	DRFFGGLC
VHv1.4(C)	CIPNYEL*C	VNSKRPCCWENKLFAGSSKPRNFVC	GLHGRSYC
VHv1.1(N)	CIGNFQP*C	IKSTKPCCRLEDRTSVQFGREEYIC	DRFLGGLC
VHv1.1(C)	CIGHYQK*C	VNADKPCCSKTVRYGDSKNVRKFIC	DRDGEGVC
WHv1.6	CIDNWKY*C	RGINKPCCGQQLMEDGTLGPKHFVC	FELGQGIC
WHv1.0	CMANWDY*C	LGFGKPCCDQ—HSIC	FKFGEGIC

4.3 *Recombinant expression of TSP*

The 14-kDa cDNA encoding TSP14 was expressed in baculoviruses as a fusion to a poly-histidine tag to simplify the purification of the recombinant TSP14 (D.L. Dahlman *et al.*, unpublished results). The purified recombinant TSP14 caused a reduction in protein synthesis in organ cultures of testes in a dose-dependent manner (E.J. Schepers and D.L. Dahlman, personal communication), and in a rabbit reticulocyte lysate assay. Unfortunately, we recovered a very low yield of recombinant proteins from the baculovirus expression system (between 20–50 $\mu g\ l^{-1}$ in lepidopteran *Spodoptera frugiperda* (Sf9) suspension cell cultures). We are now evaluating TSP14 expression using recombinant fusion vectors in yeast (*Saccharomyces cerevisiae*) cells.

Taken together, we have established that *M. croceipes* teratocytes injected into *H. virescens* larvae inhibit the synthesis of proteins in host larvae and block host development. These abnormalities can probably be attributed to a reduction in the synthesis of host proteins necessary for normal host development. We also hypothesize that the inhibition of protein synthesis is directed at the post-transcriptional/translational levels as indicated by the results from reticulocyte lysate assays. These results indicate that either the teratocytes themselves, or their secretions, or both, are probably responsible for the developmental abnormalities observed in host insects.

5. Common themes in the effects of polydnaviruses and teratocytes on host development

Teratocytes are cells released from endoparasitoid eggs that persist through the remainder of parasitoid development, while PDV are highly modified virus particles that enter host cells and, although their genes are expressed, they do not replicate. When injected into larvae, teratocytes and PDV suppress larval growth, inhibit host development and reduce the amount of storage protein synthesized by host fat body, but do so without reducing the storage protein mRNA levels (*Table 1*). A teratocyte cDNA encoding a glycosylated, secreted protein that inhibits storage protein synthesis in organ culture bioassays has been isolated. Surprisingly, this protein has a cysteine-rich region that is similar to those described from a PDV cysteine-rich gene family. Although the cysteine-rich PDV gene family encodes secreted glycosylated proteins, these proteins have not yet been associated with the inhibition of host growth, development or storage protein synthesis. Interestingly, all of the PDV cysteine motifs are interrupted by an intron just before the second cysteine residue in the motif. Preliminary characterization of the TSP14 gene indicates that an intron is also present at this position, suggesting that not only are the cysteine residues conserved, but also that at least some aspects of gene organization are shared. The possibility exists that TSP14 and the PDV cysteine-motif proteins have a common evolutionary origin. Polydnaviruses have an unusually close relationship with their host endoparasitic wasps. There is evidence that at least one other viral gene family, the vinnexin gene family, is evolutionarily related to insect innexins (gap junction proteins) (M.W. Turnbull, personal communication), and is thought to be derived from the parasitoid wasp genome. If a member of the PDV cysteine-motif gene family is isolated, and is subsequently shown to inhibit storage protein synthesis in fat body and testis, then there will be evidence of conservation of both function and structure. However, the similar mechanisms through which PDV and teratocytes inhibit host growth and development may also represent the exploitation of a limited number of regulatory points in a conserved developmental programme. Endoparasitic insects must overcome similar host defences and regulatory networks. It remains possible that the introduction of proteins encoded by PDV- and teratocyte-derived genes throughout endoparasite development, and regulation of host physiology so as to block host larval development is an exquisite example of convergent evolution. But only a more complete understanding of the structure and function of polydnavirus and teratocyte genes and proteins will tell.

5

Effects of an egg-larval parasitoid and its polydnavirus on development and the endocrine system of the host

Beatrice Lanzrein, Rita Pfister-Wilhelm and Felix von Niederhäusern

1. Introduction

Parasitoids display a great variety of life-styles, which are classified on the basis of the host stage they attack and from which they exit (egg parasitoids, larval parasitoids, egg-larval parasitoids and so on), as well as on the site of development (ectoparasitoids versus endoparasitoids) and the number of larvae developing per host (solitary versus gregarious). Even within a life-style each parasitoid–host interaction appears to be unique. Originally it was thought that the major effects on the host are caused by the parasitoid larva, but it soon became clear that in Hymenopteran parasitoids, other host regulatory factors, such as polydnaviruses (PDV), venom and teratocytes, also have an essential role. Interestingly, the relative importance of these factors, and of the parasitoid larva, is different in the various parasitoid/host systems (Coudron, 1991; Lawrence and Lanzrein, 1993; Stoltz, 1986; Vinson and Iwantsch, 1980a). Parasitoids of larval insects lay their eggs into fully immunocompetent hosts, and the primary aim of the adult parasitoid and its PDV is the prevention of encapsulation of the parasitoid egg (Lavine and Beckage, 1995; Stettler *et al.*, 1998; Strand and Pech, 1995). Growth and development of hosts are affected differently by various parasitoids (Vinson, 1990b) but often a prolongation of feeding and a developmental arrest in the last instar is observed, particularly when the parasitoid eggs are laid into late instars of the host (Lawrence and Lanzrein, 1993).

In egg-larval parasitoids, the egg is placed into the egg stage of the host, which is believed not to be immunocompetent (Salt, 1968). Thus, the immediate early

Endocrine Interactions of Insect Parasites and Pathogens, edited by J.P. Edwards and R.J. Weaver.

protection of the parasitoid egg in the host's egg stage may not be important. Later on, however, once the host has become a larva, inhibition of encapsulation of the parasitoid larva appears to be essential (Stettler *et al.*, 1998). We are working with the wasp *Chelonus inanitus* (Braconidae) and one of its natural hosts *Spodoptera littoralis* (Noctuidae), the latter being a major pest of cotton, apple, alfalfa, grapes and tomatoes (Gerling, 1971). All members of the subfamily Cheloninae described up to now are solitary egg-larval endoparasitoids of Lepidoptera. Hosts parasitized by Cheloninae enter metamorphosis precociously and prepare a pupation retreat. Then, the last (= 3rd)-instar parasitoid larva leaves the host and consumes the host remains (Shaw and Huddleston, 1991; Wharton, 1993).

We have already collected data on various aspects of the *C. inanitus–S. littoralis* system. The morphology and development of this parasitoid and its effect on growth and development of the host has been analysed (Grossniklaus-Bürgin *et al.*, 1994). The PDV of *C. inanitus* has been characterized (Albrecht *et al.*, 1994) and we have shown integration of viral DNA into the *C. inanitus* genomic DNA as well as time-specific excision and circularization of viral DNA (Gruber *et al.*, 1996). Furthermore, we have demonstrated that the *C. inanitus* PDV, synergized by venom, plays an essential role in causing developmental arrest of the host in the pre-pupal stage (Soller and Lanzrein, 1996) and that the quantities of viral transcripts increase mainly in the last instar (Johner *et al.*, 1999). Polydnavirus/venom are also essential in preventing the encapsulation of the parasitoid larva (Pfister-Wilhelm and Lanzrein, 1996; Stettler *et al.*, 1998). The parasitoid larva, in the presence of PDV/venom, was shown to play an essential role in inducing the precocious onset of metamorphosis (Pfister-Wilhelm and Lanzrein, 1996) and to release proteins into the host which are possibly involved in this process (Hochuli *et al.*, 1999). Polydnavirus/venom of *C. inanitus* was seen to affect ecdysteroids, juvenile hormones (JH) and juvenile hormone esterase (JHE), mainly in the last instar (Grossniklaus-Bürgin *et al.*, 1998), while the effects of successful parasitization were seen much earlier, and included both the induction of a precocious drop in production and titre of JH, as well as a precocious increase in JHE and ecdysteroids (Pfister-Wilhelm and Lanzrein, 1996; Steiner *et al.*, 1999).

Here we directly compare how PDV/venom alone and in combination with the parasitoid larva influence growth, development and the endocrine system of the host. We do this by combining already published data, together with new data, in composite figures. We also show conditions under which super-parasitism occurs, to which extent it is tolerated by the host, and how it affects host growth. Furthermore, we present data on the quantity of viral DNA injected into the host in relation to the degree of parasitism. The effects of PDV/venom alone, and in combination with the parasitoid larva, are discussed in relation to viral transcription and release of proteins by the parasitoid larva into the host.

2. Experimental details

2.1 *Insects, parasitization and X-ray-irradiation*

Spodoptera littoralis (Noctuidae) was mass reared at 27 ± 1°C under a 14L:10D photoperiod and fed on artificial diet. Adult *S. littoralis* and diet were kindly donated by Novartis AG, Basle. The solitary egg-larval endoparasitoid *C. inanitus* (Braconidae) was reared on *S. littoralis*; the parasitoid larva emerges from the developmentally arrested precocious pre-pupa, eats its host, spins a cocoon and then pupates. Details on the biology, morphology and rearing of the parasitoid and its host are given in Grossniklaus-Bürgin *et al.* (1994).

For experimental purposes, 27- to 32-h-old *S. littoralis* egg patches (kept at 20°C) were divided and some were reared under normal conditions (non-parasitized control larvae), whilst others were parasitized with either normal *C. inanitus* (parasitized larvae), or with X-ray-irradiated *C. inanitus* (X-ray-parasitized larvae). X-ray-irradiation was carried out as described by Soller and Lanzrein (1996). X-ray-irradiated wasps oviposit normally and inject PDV and venom, but the eggs they lay are infertile, and no parasitoid larva hatches. Host larvae developing from eggs parasitized with X-ray-irradiated wasps become developmentally arrested in the pre-pupal stage, as do larvae developing from eggs injected with calyx fluid/venom or those larvae developing from parasitized eggs in which the parasitoid embryo was killed by heat-treatment (Soller and Lanzrein, 1996). Parasitization with X-ray-irradiated wasps is thus an elegant way to study the effect of PDV/venom in the absence of a developing parasitoid larva. Earlier investigations have indicated that the active component of calyx fluid in *C. inanitus* is the PDV (Soller and Lanzrein, 1996 and unpublished observations). Thus it is likely that most if not all observations made with X-ray-parasitized larvae reflect PDV/venom-mediated effects. For the sake of simplicity, the term PDV/venom will therefore be used instead of calyx fluid/PDV/venom throughout the rest of this chapter. The degree of parasitism was monitored by dissection of some host eggs at various time points after the eggs were offered to the female wasps. Parasitization was continued until 95–100 % of the eggs were parasitized with one or two eggs.

2.2 *Induction of super-parasitization and its evaluation*

As an additional approach to investigate the role of PDV/venom and the parasitoid on host development, super-parasitization and X-ray super-parasitization was induced. To obtain super-parasitism, two or three females were offered a single small egg patch for parasitization. In order to monitor the degree of parasitism and to study its effect, a patch of *S. littoralis* eggs was subdivided into a group of 10 eggs, and an adjacent group of 20 eggs, and then offered for parasitization to three females for a period of 2 hours. The group of 10 eggs was then dissected to determine the degree of parasitism and the remaining 20 eggs were used for experimental purposes. Extensive preliminary observations had revealed that for each group of three females the degree of parasitism in the 10 eggs was representative for the other 20 eggs.

2.3 *Measurement of content of viral DNA in host eggs*

Spodoptera littoralis eggs (3–8-h-old) were parasitized as described in Section 2.2 and, after determination of the degree of parasitism in the group of 10 eggs, the group of 20 eggs was flooded with liquid nitrogen. As a control, 20 non-parasitized eggs from the same egg patch were used. The frozen eggs were transferred into an Eppendorf tube and ground with a fine pestle. Egg DNA was isolated with the QUIamp Tissue Kit (Qiagen), according to the manufacturer's protocol. In addition to the egg DNA, 10μg of calf thymus DNA was loaded as carrier onto the QUIamp column. Calyx fluid was collected and the PDV was purified by sucrose gradient centrifugation as described in Soller and Lanzrein (1996). Isolation of PDV DNA was as described for the eggs, but no carrier DNA was used. Viral DNA concentrations in parasitized eggs were measured by Southern blot. Egg DNA, along with a dilution series of PDV DNA, was denatured, dotted and UV-crosslinked to a positively charged nylon membrane as described (Gruber *et al.*, 1996). The blots were pre-hybridized for 2–3 h at 42°C in 5 × SSC (0.75 M NaCl, 75 mM sodium citrate), 20 g l^{-1} blocking reagent (Boehringer), 1 g l^{-1} laurosylsarkosin, 0.2 g l^{-1} sodium dodecyl sulfate (SDS), 200 μg ml^{-1} denatured calf thymus DNA and 50% formamide. The blots were hybridized at 42°C overnight in the same buffer, which in addition contained 15 ng ml^{-1} [α–^{32}P] dCTP-labelled *Hin*dIII digested PDV DNA. Labelling was done with the Random Primed DNA Labeling Kit (Boehringer). The blots were washed in 2 × SSC, 0.1% SDS at room temperature and in 0.2 × SSC, 0.1% SDS at 68°C. The nylon membranes were exposed to a Storage Phosphor Screen (Molecular Dynamics) and quantitatively analysed on a PhosphorImager (Molecular Dynamics).

2.4 *Measurement of JH, JHE and ecdysteroids*

Juvenile hormones were quantified using the physico-chemical method developed by Rembold and Lackner (1985), with minor modifications as described in Schopf *et al.* (1996). Juvenile hormone esterase activities were measured using a slight modification of the partition assay developed by Hammock and Sparks (1977). The data on JH II and on JHE are from Grossniklaus-Bürgin *et al.* (1998) and Steiner *et al.* (1999). Ecdysteroids were extracted and purified by C18 Sep-Pak (Waters) as described (Grossniklaus-Bürgin *et al.*, 1998). The ecdysteroid data are a combination of results from Grossniklaus-Bürgin *et al.* (1998) and Steiner *et al.* (1999), together with some new measurements.

3. Results

3.1 *Growth curves of non-parasitized, PDV/venom-containing and parasitized larvae*

A comparison of the growth curves (*Figure 1*) shows that PDV/venom only very weakly affected weight gain up to the stage of pupal cell formation, while parasitism caused a reduction in weight gain at earlier stages. The maximum weight reached by parasitized larvae was no more than one-tenth that of non-parasitized or X-ray-parasitized larvae. The course of development indicated at the bottom of *Figure 1*, shows that non-parasitized and X-ray-parasitized larvae passed through the various stages almost synchronously until the late pupal cell-formation stage. Subsequently, non-parasitized larvae went through the pre-pupal phase and pupated while X-ray-parasitized larvae slowed down development and eventually became developmentally arrested (for details see Soller and Lanzrein, 1996 and Grossniklaus-Bürgin *et al.*, 1998). Such insects are termed stationary pre-pupae (SPP). Parasitized larvae developed synchronously with the non-parasitized and X-ray-parasitized larvae up to the 4th instar and then precociously entered metamorphosis in the 5th instar. The parasite larva remains as a 1st instar until the host has moulted to the 5th instar, and then passes through a short 2nd instar and moults to the 3rd instar before leaving the host, which is now a precocious pre-pupa (for details see Grossniklaus-Bürgin *et al.*,

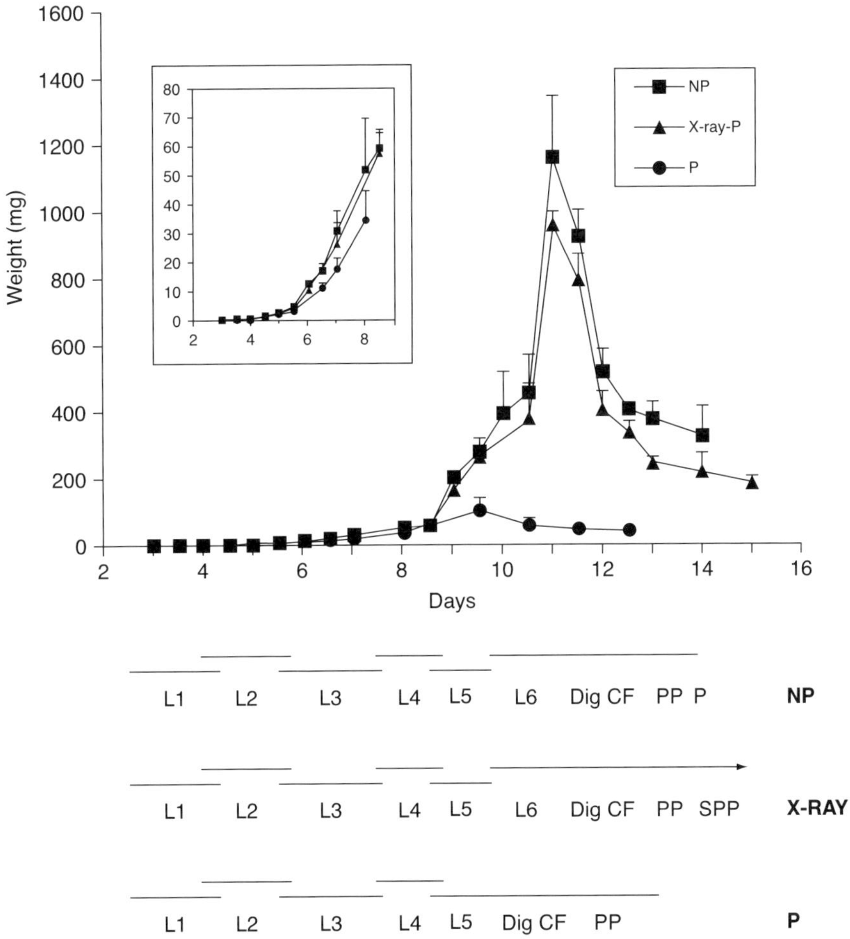

Figure 1. *Growth of non-parasitized, X-ray-parasitized and parasitized* S. littoralis. *On the bottom, the development of non-parasitized (NP), X-ray-parasitized (X-RAY) and parasitized (P) larvae is shown. L1–L6, first to sixth larval instar; Dig, digging; CF, pupal cell formation; PP, pre-pupa; SPP, stationary pre-pupa; P, pupa. Data are means with standard deviations. For L1–L3, 5–8 larvae were weighed together and 3–7 measurements were made for each stage. For L4–L6, single larvae were weighed and the number of determinations was 20–40. The inset shows L1–L3 at a higher magnification.*

1994). When non-parasitized or X-ray-parasitized larvae were kept in small groups they occasionally passed through five larval instars instead of the usual six. In this case the head capsule width of the 4th-instar larvae was broader than normal, and the duration of the instar longer than that of normal 4th-instar larvae. In the last larval instar, the head capsule width and the weight curve were similar regardless of whether or not the larvae were in the 5th or the 6th instar. For experimental purposes, we preferred to use larvae that underwent the normal six-instar development, and this could be achieved by selecting those 4th-instar larvae with head capsule widths below 1.15 mm.

3.2 *Super-parasitism and its effect on host development*

We also investigated whether and under which conditions super-parasitism occurs and how it affects host development. When a single *C. inanitus* female was offered 80–150 *S. littoralis* eggs for 2 h, it always left a few eggs not parasitized, laid one egg per host egg in most cases, two eggs in 0.5% of the eggs and three eggs in 0.02% of the eggs (average of 12 females). Reduction of the size of the egg patch to only 10 eggs did not lead to super-parasitism. Only when two or three females were offered a small egg patch did super-parasitism occur. The degree of parasitism obtained when three females were offered 30 eggs during 2 h is shown in *Figure 2*. The data reveal that approximately 25% of the host eggs contained one egg, 5% no eggs and the remainder up to 17 eggs. We then determined the maximum level of parasitism that is tolerated by the host. For this purpose, we made a series of experiments where 10 eggs of a prepared clutch of 30 eggs were dissected to determine the degree of parasitism, while the remainder were placed on a diet to observe the development of the host. From the same egg patch, another 30 eggs were submitted to the same preparatory treatment, but were not parasitized (control clutch). The data in *Figure 3* show that, from the control clutches (*Figure 3,* top), a majority of larvae hatched. In the few eggs that larvae did not hatch from, this failure was probably caused by the preparatory treatment of the eggs, since hatching rates between 95 and 100% were observed from non-treated eggs. The data from the parasitized eggs (*Figure* 3, bottom) show that 5–6 eggs can be tolerated, but with higher egg loads hatching of the host larvae seldom occurred. Further observation of the hatched hosts revealed that most of them survived and that only one parasitoid eventually emerged from each host.

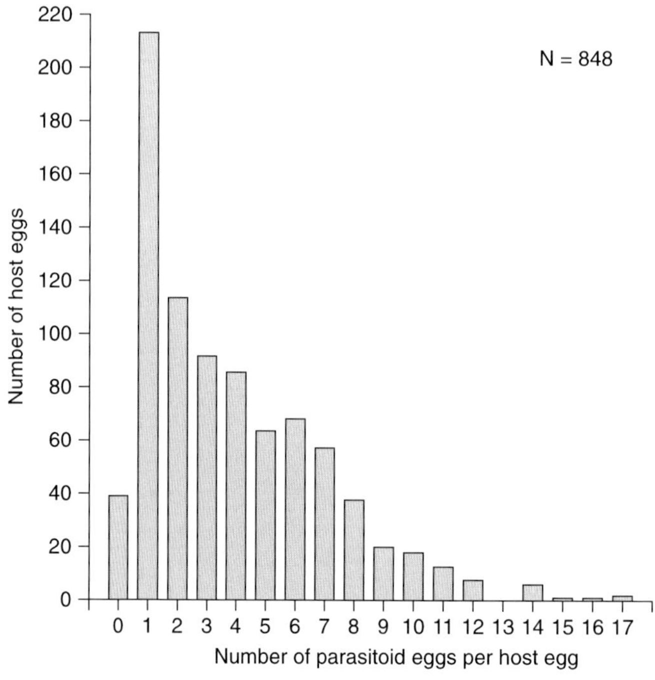

Figure 2. *Super-parasitism of* S. littoralis *eggs by* C. inanitus. *Three females were offered 30 eggs during 2 hours; then each egg was dissected and the number of parasitoid eggs was counted.*

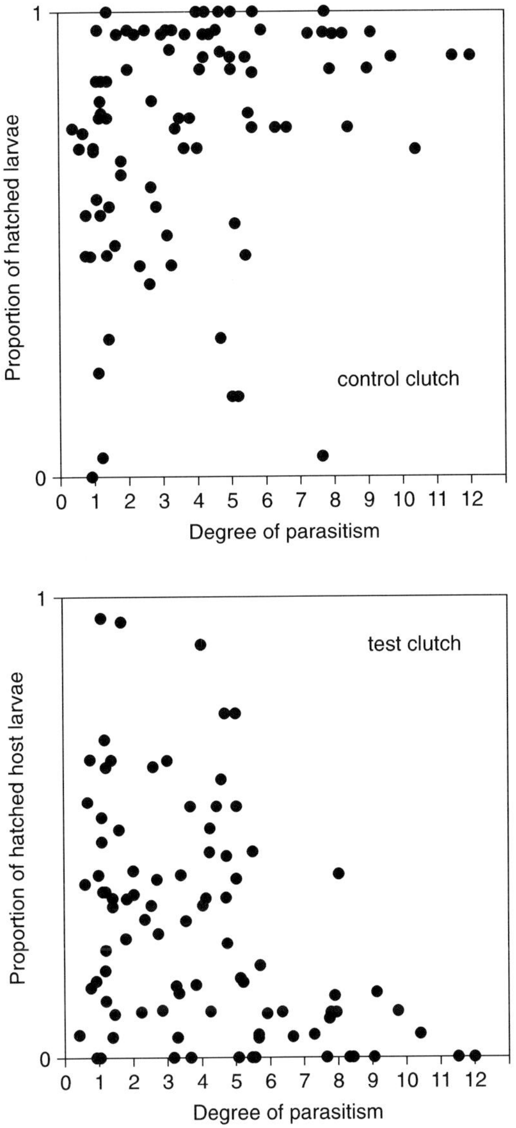

Figure 3. *Influence of the degree of parasitism on the development of the host. Thirty eggs were prepared, as described in Section 2.2, and offered to three* C. inanitus *females during 2 h. Then 10 eggs were dissected to determine the degree of parasitism and the remaining 20 eggs were kept for observation (test clutch, bottom). From the same egg patch another 30 eggs were treated the same way, but not offered for parasitization (control clutch, top).*

Occasionally, earlier instar host larvae were dissected after super-parasitism and it was found that even these younger host larvae contained only one parasitoid larva – indicating that the reduction to only one developing larva is accomplished very early.

3.3 *Amount of PDV DNA injected per host egg in relation to the degree of parasitism and effect of super-parasitism and X-ray-super-parasitism on host growth*

We then measured the amount of viral DNA injected by the wasps under conditions of normal parasitism and super-parasitism. The experimental set-up to induce and monitor super-parasitism was the same as described in Section 3.2, but after parasitization the eggs were analysed for their content of PDV DNA by Southern dot blot. For quantification, a dilution series of PDV DNA was also blotted onto the

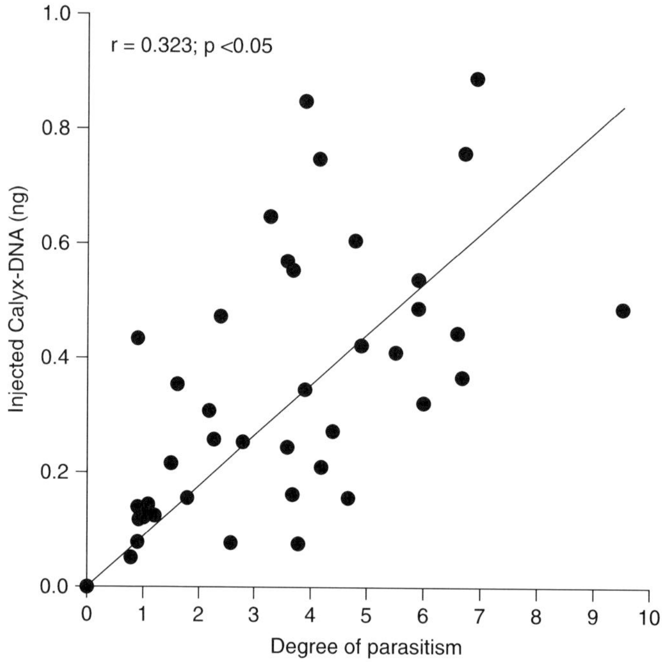

Figure 4. *Amount of polydnavirus DNA injected per host egg in dependence of the degree of parasitism.*

membrane. The results are given in *Figure 4* and show that quantities of viral DNA increased with the degree of parasitism and that, on average, around 0.1 ng PDV DNA was injected along with each parasitoid egg. We do not know how to explain the great variations observed in super-parasitized eggs, but they are certainly not due to variations in the yield of collected DNA as this was very constant.

We then investigated how super-parasitism and X-ray super-parasitism affect host growth as compared to the effect of normal parasitism and X-ray-parasitism. *Figure 5* gives an overview for selected stages of development and the weights are given as percentages of the corresponding non-parasitized larvae. The latter had the greatest weight at all stages investigated, followed by X-ray-parasitized, X-ray-super-parasitized, parasitized, and lastly, super-parasitized larvae. The X-ray-parasitized larvae had a weight of around 80–85% of that of non-parasitized larvae until the 6th instar, but this dropped to around 60% in the stationary pre-pupae. For X-ray-super-parasitized larvae the values were slightly lower. Obviously the presence of PDV/venom has a continuous effect on growth and the even lower weights of X-ray-super-parasitized larvae suggest a certain dose dependence. The reduction in weight gain of parasitized and super-parasitized larvae was much greater than that in X-ray- and X-ray-super-parasitized larvae and increased up to the 3rd larval instar. This weight differential then remained at a similar level in the 4th and early 5th instar, but became much greater upon the precocious onset of metamorphosis. Thus, the parasitoid larva plays a major role in reducing host weight gain.

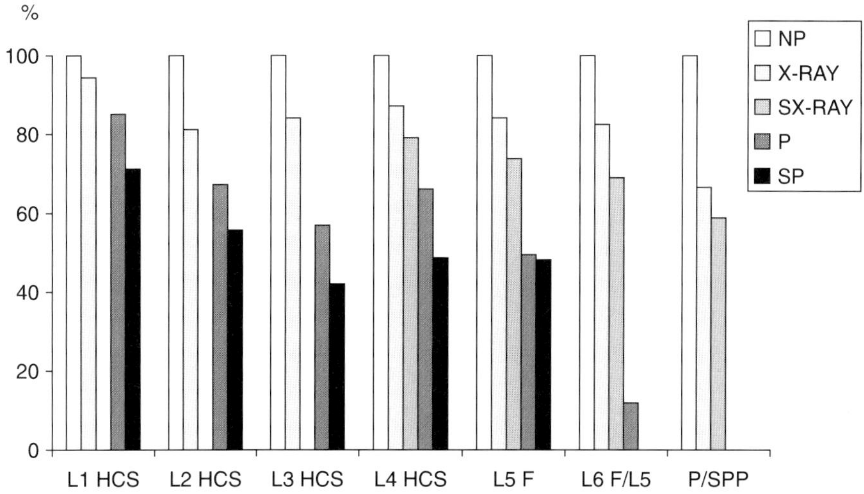

Figure 5. *Effect of X-ray-parasitization (X-RAY, 1–2 eggs per host egg), X-ray-superparasitization (SX-RAY, 4–5 eggs per host egg), parasitization (P, 1–2 eggs per host egg) and superparasitization (SP, 4–5 eggs per host egg) on weight gain of* S. littoralis. *A portion of an egg patch was always kept non-parasitized (100%) and the weight at various developmental stages was calculated as a percentage of the non-parasitized status. HCS, head capsule slippage; F, feeding; P, pupa; SPP, stationary pre-pupa. Data are means of 1–4 determinations.*

3.4 *JH II, JHE and ecdysteroids in non-parasitized, X-ray-parasitized and parasitized larvae*

In earlier investigations it was shown that JH II is the predominant JH in *S. littoralis* larvae, whether non-parasitized, X-ray-parasitized or parasitized (Grossniklaus-Bürgin *et al.*, 1998; Steiner *et al.*, 1999). A comparative analysis of fluctuations in the haemolymph titres of JH II, JHE and ecdysteroid is shown in *Figure 6* for the time from the 4th instar until pupation (non-parasitized) or until developmental arrest as stationary pre-pupae (X-ray parasitized) or until parasitoid emergence (parasitized). Titres of JH II (*Figure 6(a)*) were very similar in non-parasitized and X-ray-parasitized larvae until the stage of pupal cell formation. Then, in non-parasitized larvae, JH II slightly increased for a short period of time, while in X-ray-parasitized larvae the increase was delayed and was of longer duration. In parasitized larvae, the pattern of JH II titre was again totally different, dropping to very low or undetectable levels from the end of the 4th stadium onwards.

The JHE data (*Figure 6(b)*) show the following. In non-parasitized larvae, two peaks of JHE activity were seen, the first in the feeding/digging phase and the second in the late pupal cell formation/pre-pupal stage. In X-ray-parasitized larvae, the first increase of JHE occurred at the same time as that in non-parasitized larvae, but values were slightly lower. The second peak, however, of JHE, was delayed by a similar extent to the delayed increase of JH II in the pre-pupae. In parasitized larvae, JHE increased much earlier than in non-parasitized and X-ray-parasitized larvae and the values were much lower. The first increase was seen in the feeding/digging phase and the second in the late pupal cell formation/pre-pupal stage, i.e. at developmentally similar stages to those in last instar non-parasitized and X-ray-parasitized larvae.

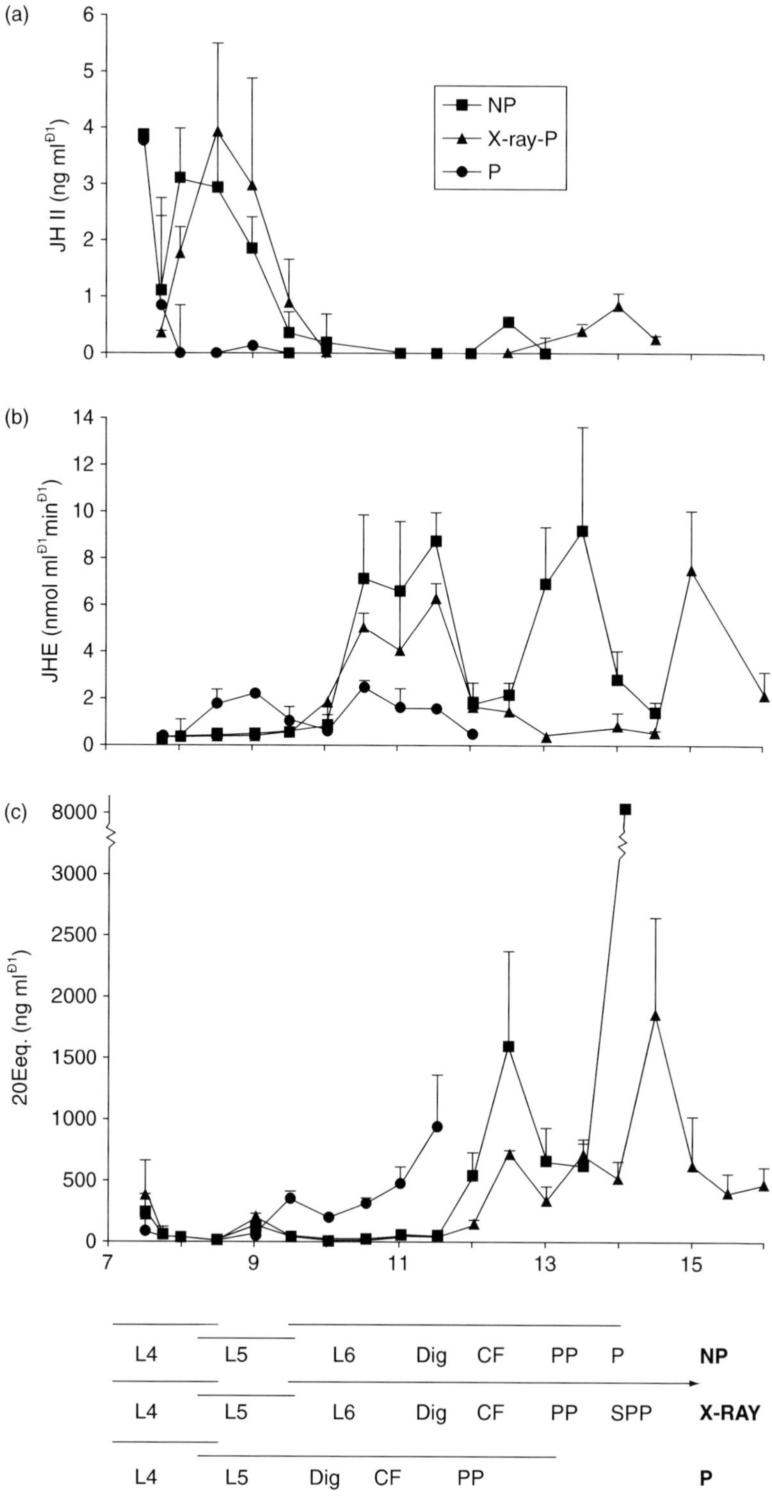
(a)
JH II (ng ml^Ð1)
NP
X-ray-P
P
(b)
JHE (nmol ml^Ð1 min^Ð1)
(c)
20Eeq. (ng ml^Ð1)
7
9
11
13
15
L4 L5 L6 Dig CF PP P NP
L4 L5 L6 Dig CF PP SPP X-RAY
L4 L5 Dig CF PP P

Titres of ecdysteroids in haemolymph of non-parasitized, X-ray-parasitized and parasitized larvae are shown in *Figure 6(c)*. We always analysed the 60% methanol fraction of the C18 Sep Pak purification, and we have shown earlier that this fraction contains predominantly 20E together with ecdysone (E) and some 20,26-dihydroxy-ecdysone (Grossniklaus-Bürgin *et al.*, 1998). In non-parasitized and X-ray-parasitized larvae, the titres of ecdysteroids were similar up to the stage of digging. Thereafter, titres in non-parasitized larvae increased rapidly, reached a peak at the pupal cell formation stage, decreased intermittently and increased to very high values in early pupae. In X-ray-parasitized larvae, the increase of ecdysteroids in the pupal cell formation stage was comparatively delayed, values were lower, and ecdysteroid levels increased at a later time, when stationary pre-pupae were seen to be producing patches of pupal cuticle. It appears that the lower peak end extended increase in ecdysteroids, which is seen in X-ray-parasitized larvae after the formation of the pupal cell, is related to their developmental deficiencies (for details see Grossniklaus-Bürgin *et al.*, 1998). In parasitized larvae, ecdysteroids increased earlier than in non-parasitized and X-ray-parasitized larvae, notably in the digging stage of the precocious last (5th) instar, but values were lower than those in X-ray-parasitized larvae of the corresponding stages. Towards the end of parasitization, the titre of haemolymph ecdysteroids increased. In summary, the first effect caused by PDV/venom is seen in the 6th instar when JHE concentrations in the feeding/digging stage are reduced. Thereafter, the increase of ecdysteroid in the pupal cell formation stage was reduced and delayed, and was followed by a delayed increase in JH II, and a subsequent delayed increase in prepupal JHE. The additional presence of a parasitoid larva affected the endocrine system of the host much earlier as it led to a precocious decline of JH II at the end of the 4th instar, followed by a precocious increase in JHE and ecdysteroids. Thus, the sequence of endocrine events preparatory to metamorphosis that is seen in X-ray-parasitized larvae begins one instar earlier in the presence of a parasitoid larva.

4. Discussion

4.1 *The contributions of PDV/venom and of the parasitoid larva in manipulating host growth and development*

Injection of calyx fluid/venom has been shown to have a very similar effect to that of X-ray parasitization on host development, whereby a quantity of 0.05–0.1 *Chelonus* equivalents per 100 eggs was found to be effective (Soller and Lanzrein, 1996). From one female wasp approximately 1 μg of calyx DNA can be collected (S. Wyder and S. Zumbach, personal communication) and thus the amount injected corresponds to 0.5–1 ng per egg, i.e. slightly above the average quantity of 0.1 ng which a wasp injects (*Figure 4*). Analysis of the effects of PDV/venom alone, or in combination with a

◀ ***Figure 6.*** *Comparison of endocrine parameters between non-parasitized, X-ray-parasitized and parasitized larvae from the 4th instar until pupation for non-parasitized larvae (NP) or until formation of stationary prepupae (SPP) for X-ray parasitized larvae or until parasitoid emergence for parasitized larvae (P). The abscissa gives the age in days and underneath the development for each status is indicated. For abbreviations see Figure 1. The data are from Grossniklaus-Bürgin et al., 1998, Steiner* et al., *1999 or newly generated. (a) Titre of JH II in haemolymph; (b) juvenile hormone esterase (JHE) activity; (c) ecdysteroids in the 60% methanol fraction of C18 Sep-Pak, expressed in 20-hydroxyecdysone (20E) equivalents.*

developing parasitoid larva, on growth and development of the host revealed the following. In comparison to non-parasitized larvae, X-ray-parasitized larvae had a slightly lower weight throughout development, but the difference in weight became significant only in the 6th instar (*Figures 1* and *5* and Grossniklaus-Bürgin *et al.*, 1998). Head capsule widths were significantly smaller from the 4th instar onwards (Grossniklaus-*Bürgin et al.*, 1998). In X-ray-super-parasitized larvae the weight difference was even more pronounced (*Figure 5*), and as the amount of viral DNA injected into a host egg was seen to increase with the degree of parasitism (*Figure 4*) this suggests that PDV/venom affects larval growth in a dose-dependent way. It is not known whether this effect is due to a reduced uptake of food or to altered metabolism. After pupal commitment, PDV/venom markedly affected host development and led to the formation of stationary pre-pupae (for details see Grossniklaus-Bürgin *et al.*, 1998). An additional effect of PDV/venom was inhibition of the development of testes (Grossniklaus-Bürgin *et al.*, 1998). A serosal membrane could be seen adhering to the embryo and young 1st-instar larva, but no teratocytes could be seen in the haemolymph of later larval instars; thus, teratocytes do not appear to be involved in manipulating host development in this parasitoid (Grossniklaus-Bürgin *et al.*, 1994). For a comparison with the effects of other Cheloninae see Jones (1996).

The weight of parasitized larvae, and even more of super-parasitized larvae, was always lower than that of X-ray-parasitized larvae (*Figures 1* and *5*), and head capsule widths of parasitized larvae were significantly smaller than those of non-parasitized larvae from the 3rd instar onwards (Grossniklaus-Bürgin *et al.*, 1994). Thus, the presence of a parasitoid larva in addition to PDV/venom further reduces weight gain, as previously reported in the system *C. insularis–Heliothis virescens* (Ables and Vinson, 1981). Deposition of more than five parasitoid eggs per host egg prevented development of the host larva in most cases (*Figure 3*); similar observations have been made in other solitary egg-larval parasitoids (Biliotti and Daumal, 1969; Jackson *et al.*, 1979; Moreno and Jimenez, 1993). It is not known whether failure to develop is due to multiple perforation of the egg, an overdose of PDV/venom and/or the excessive volume occupied by the many parasitoid eggs as they are relatively large (approximately 200 μm in length, Grossniklaus-Bürgin *et al.*, 1994) in relation to the diameter of the host egg (500 μm). It was interesting to note, that the reduction in weight gain of parasitized, and even more of super-parasitized larvae, increased progressively until the 3rd instar compared with X-ray-parasitized and non-parasitized larvae (*Figure 5*). Then, in the 4th and early 5th instar, when the host larva grows rapidly and the parasitoid larva is still a 1st-instar larva (Grossniklaus-Bürgin *et al.*, 1994), the difference in weight gain was less apparent, but became very large when the parasitoid larva moulted into the 2nd instar and started to consume host haemolymph (*Figure 1* and Grossniklaus-Bürgin *et al.*, 1994). The strong host-regulating effect emanates from the parasitoid larva, as transplantation of parasitoids into X-ray-parasitized larvae induced the precocious onset of metamorphosis (Pfister-Wilhem and Lanzrein, 1996). Parasitoid-removal experiments have shown that the parasitoid needs to be present only up to the moult of the host to the 5th instar to induce the precocious onset of metamorphosis (Pfister-Wilhelm and Lanzrein, 1996), and indirect evidence suggests that protein(s) released by the late 1st-instar parasitoid larva into a PDV/venom containing host play an important role in this process (Hochuli *et al.*, 1999). Thus, PDV/venom appears to have only subtle effects on the host during larval development, but a more profound effect after the host has commenced metamorphosis. When a parasitoid larva is present in addition to PDV/venom, metamorphosis is induced precociously, and the effects of PDV/venom appear earlier. It is interesting to note that the intensity of PDV transcription increased predominantly in the 6th-instar in X-ray-parasitized larvae, and in the 5th instar in parasitized larvae, although some differences between different viral segments were observed (Johner *et al.*, 1999). It is not known whether the presence of a parasitoid larva, in addition to PDV/venom, directly influences the increase in viral transcription, or whether this is mainly dependent on the state of development of the host.

4.2 *The contributions of PDV/venom and of the parasitoid larva in manipulating the host's endocrine system*

The effects of PDV/venom were apparent mainly in the last larval instar (*Figure 6*) when viral transcripts also increased (Johner *et al.*, 1999). At the feeding/digging stage, JHE values were slightly lower in X-ray-parasitized than in non-parasitized larvae and, at the stage of pupal cell formation, ecdysteroids increased more slowly and the peak was delayed. At this stage, PDV/venom was shown to suppress ecdysone release by prothoracic glands (Grossniklaus-Bürgin *et al.*, 1998). It appears that this suppression is the result both of an impaired production of prothoracicotropic hormone and a reduction in responsiveness of prothoracic glands to this hormone (R. Pfister-Wilhelm, unpublished). At the late cell formation/early pre-pupal stage, there was a delayed increase in JH II levels in X-ray-parasitized larvae. The rise in pre-pupal JHE was also delayed, and the increase in level of ecdysteroids was reduced. It was shown earlier that the composition of ecdysteroids was very similar in non-parasitized and X-ray-parasitized larvae and that the same metabolites were formed – indicating that PDV/venom mainly affects the production and rate of metabolism of E and 20E (Grossniklaus-Bürgin *et al.*, 1998). Likewise, JH II was shown to be the predominant JH, together with some JH III, in both non-parasitized and X-ray-parasitized late-instar larvae, indicating that PDV/venom does not influence JH homologue composition (Grossniklaus-Bürgin *et al.*, 1998).

When a developing parasitoid larva was present in addition to PDV/venom, host endocrinology was altered earlier and in a different manner compared to PDV/venom alone. Parasitization had no effect on JH homologue composition, and titres were not altered in the embryonic stages (Steiner *et al.*, 1999). This is in accordance with the observation that JH degradation in eggs of *S. exigua* was not altered by parasitization with *C. inanitus* (Lanzrein and Hammock, 1995). In newly hatched 1st-instar parasitized larvae, titres of all three JHs were lower than in non-parasitized larvae (Steiner *et al.*, 1999). This could be due to the fact that newly hatched parasitized larvae begin to feed less voraciously than non-parasitized larvae and this is reflected by the lower weight reached in the first instar (Grossniklaus-Bürgin *et al.*, 1994). In the 2nd and 3rd stadium, no difference in JH titre was seen between non-parasitized and parasitized larvae. However, at the end of the 4th stadium, JH II and JH III dropped in parasitized larvae, but remained higher and increased again in the 5th stadium in non-parasitized and X-ray-parasitized larvae (*Figure 6* and Steiner *et al.*, 1999). This precocious drop in JH II titre in parasitized larvae appears to be due to an inhibition of JH production by the corpora allata (Pfister-Wilhelm and Lanzrein, 1996) and to a precocious increase in JHE (*Figure 6* and Steiner *et al.*, 1999). The decrease in JH titre seems to be directly related to the presence of the parasitoid larva as implantation of a late 1st-instar parasitoid larva into young 5th-instar X-ray-parasitized larvae led to a drop in JH titre within 24 h (Steiner *et al.*, 1999). The level of the precocious JHE peak found in feeding/digging parasitized larvae was much lower than that in X-ray-parasitized larvae of the corresponding stage. In addition, the ecdysteroid levels at the cell formation stage in parasitized larvae were lower than those in X-ray-parasitized larvae (*Figure 6*). Obviously, a parasitoid larva together with PDV/venom has a stronger effect on host endocrinology than PDV/venom alone. The parasitoid larva strongly targets the JH system and enhances some of the effects of PDV/venom. Interestingly, the parasitoid's own JH (JH III) reaches highest values in the 2nd instar, when the host's JH is at very low levels or undetectable (Steiner *et al.*, 1999).

Peaks of feeding/digging stage JHE, ecdysteroids, pre-pupal JH II and JHE were seen to alternate in the timing of their appearance in non-parasitized, X-ray-parasitized and parasitized larvae (*Figure 6*). This suggests that these parameters are dependent on each other, but it is not known how they are interconnected. Thus, it is not clear whether the delay in the appearance of pre-pupal peaks of JH II and JHE seen in X-ray-parasitized larvae are a consequence of the reduced and delayed increase in ecdysteroids, or are mediated independently by PDV/venom. Likewise, it is not clear whether the precocious increase in JHE in parasitized larvae is linked to the precocious drop in JH, or is directly mediated by the parasitoid.

Taken together, the data accumulated for the *C. inanitus–S. littoralis* system are indicative of the following sequence of events. Polydnavirus/venom injected into the host egg at oviposition exerts only subtle effects on the host in the early larval instars, one of these effects being the important protection of the parasitoid larva from encapsulation. Only small quantities of viral transcripts are found in this phase. At the end of the 4th larval instar, parasitoid-derived substances, in the presence of PDV/venom, initiate the precocious onset of metamorphosis by inhibiting the corpora allata and by inducing JHE. Interestingly, the quantity of viral transcripts increases at this stage. After the moult of the host into the precocious last instar, the parasitoid moults into its 2nd instar, has a high titre of JH III, and grows rapidly. The weight of the host then decreases and it burrows into the ground, constructs a pupal cell and enters the pre-pupal stage; in this phase ecdysteroids increase. The parasitoid then moults into the 3rd instar, emerges from the host, which is now arrested in a precocious pre-pupal state, and eats it.

4.3 *Comparison with larval parasitoids*

None of the solitary or gregarious 'larval' endoparasitoids investigated up to now affect host development and endocrinology in a way that is comparable to that described above for *C. inanitus–S. littoralis*. In several cases, hosts attacked as larvae either failed to gain weight or gained weight at a reduced rate, and in some cases this effect has been attributed to calyx fluid/venom (Vinson, 1990b). The dose of 0.1 ng of viral DNA injected per egg by *C. inanitus* (*Figure 4*) is lower than the 2.4 ng or 6.5 ng injected per host larva by the solitary larval parasitoids *Microplitis demolitor* and *Campoletis sonorensis* (Strand *et al.*, 1992; Theilmann and Summers, 1986). In addition, the level of viral transcription appears to be much lower in *C. inanitus*, and follows a different pattern (Johner *et al.*, 1999) than that seen in these two larval parasitoids (Strand *et al.*, 1992; Theilmann and Summers, 1988). Some larval parasitoids cause a prolongation of the host's last instar by preventing metamorphosis. This occurs particularly when late instars are parasitized, and is often associated with an elevation in the JH titre and/or repressed JHE activity (Balgopal *et al.*, 1996; Beckage and Templeton, 1986; Cusson *et al.*, 2000; Dong *et al.*, 1996; Dover *et al.*, 1995b; Schopf *et al.*, 1996). It is not clear in these particular associations to what extent the parasitoid larva contributes to the endocrine manipulation of the host. However, measurements of JHE mRNA in the *M. croceipes–H. virescens* system indicate that the parasitoid larva plays a role, as do calyx fluid and teratocytes (Dong *et al.*, 1996). The larva of *M. demolitor* has also been shown to contribute to endocrine manipulation of the hosts *H. virescens* or *Pseudoplusia includens* (Dover *et al.*, 1995b; Balgopal *et al.*, 1996). Obviously, the effects of the larval parasitoids investigated up to now are, with respect to the JH system, opposite to those seen in the egg-larval Cheloninae. The former often cause an increase in JH titre and a suppression of JHE leading to a prolongation of the feeding phase in the last larval instar and prevention of metamorphosis, while the latter causes a decrease in JH titre and an increase in JHE in the penultimate stadium leading to a precocious onset of metamorphosis. By contrast, inhibition of pupation and suppression of ecdysteroids is seen in the Cheloninae as well as in several larval parasitoids. This effect is associated with calyx fluid or PDV alone, or in combination with venom and/or teratocytes, in different parasitoid–host systems (Cusson *et al.*, 2000; Dover *et al.*, 1995a; see review by Lawrence and Lanzrein, 1993; Pennacchio *et al.*, 1994a, 1998a). Thus, our current knowledge suggests that each parasitoid–host interaction is characterised by a specific life style and endocrinology, and that manipulation of the host can be achieved by various means. The diversity of host hormonal manipulation displayed by parasitic wasps is certainly fascinating, and will hopefully continue to be challenging to many scientists.

Acknowledgements

We should like to express our thanks to Novartis AG, Basle, for providing us with adult *S. littoralis* and diet for rearing the larvae, and to Dr. D. Gerling, Tel Aviv, for having collected the progenitors of our *C. inanitus* colony. We also thank S. Wyder for invaluable assistance in preparing the figures. Financial support from the Swiss National Science Foundation (grants 3100-041828.94 and 3100-52399.97) is also gratefully acknowledged.

6

Hormonal alterations and molecular mechanisms underlying the induction of host developmental arrest by endoparasitic wasps

Michel Cusson, Catherine Béliveau, Marlène Laforge, Guy Bellemare, Anic Levasseur and Don Stoltz

1. Introduction

Suppression of host metamorphosis is a recurring theme in the literature on interactions between endoparasitic wasps and their lepidopteran hosts. While it is difficult to identify the precise constraints that have led to this adaptation, several hypotheses have been put forward. For example, the hardened cuticle of the host pupa may represent an obstacle to the successful egression of the parasitoid from the host (Tanaka *et al.*, 1987). Similarly, if pupation takes place underground, it may be difficult for the adult wasp to make its way out of the soil (Beckage and Riddiford, 1982a). With respect to the appropriateness of the host environment for development of the wasp larva, it has been suggested that the hormonal changes associated with host metamorphosis may be detrimental to a parasitoid that is responsive to host hormones (Beckage and Templeton, 1986). In addition, maintenance of a 'larval' nutritional milieu may assure the parasitoid access to haemolymph nutrients that would otherwise undergo sequestration (Beckage and Riddiford, 1982a).

Depending on the parasitoid and host species involved, developmental arrest can be induced at different stages of the last or (normally) penultimate host instar, suggesting important differences in the physiological milieu of affected hosts, as well as different strategies and endocrine targets used to suppress host metamorphosis. In this chapter, we provide a brief survey of the literature on host-parasitoid hormonal interactions, with special emphasis on the endocrine events that are targeted by parasitic wasps to

Endocrine Interactions of Insect Parasites and Pathogens, edited by J.P. Edwards and R.J. Weaver.

block metamorphosis in their hosts, and how these vary as a function of the stage at which developmental arrest is induced. We then review the limited data available on the molecular mechanisms postulated to explain some of the physiological alterations that accompany developmental arrest, with particular attention being given to the role played by the wasp-associated polydnaviruses (PDV). Lastly, we provide a summary of our work on the role played by the PDV of the ichneumonid *Tranosema rostrale* (TrPDV) in the suppression of metamorphosis in its host *Choristoneura fumiferana*, and the search for TrPDV genes that may be involved in bringing about the observed developmental arrest.

2. Endocrine targets: influence of the stage at which developmental arrest is induced

Many parasitized caterpillars moult to their last stadium and become developmentally arrested while in the feeding stage. This would indicate that the larval-pupal transformation is blocked at a very early stage of the process, and that one of the first hormonal events that sets metamorphosis in motion, such as the drop in the juvenile hormone (JH) titre and the associated release of prothoracicotropic hormone (PTTH; Rountree and Bollenbacher, 1986), is probably disrupted.

Examples of wasp species whose habitual hosts, when parasitized, exhibit none or few of the characteristics normally associated with metamorphosis include *Campoletis sonorensis* (Shelby and Webb, 1997), *Tranosema rostrale* (Cusson *et al.*, 2000), *Cotesia congregata* (Beckage and Riddiford, 1982a), *Cotesia kariyai* (Hayakawa, 1990), and *Microplitis demolitor* (Dover *et al.*, 1995b). Parasitism by three of these species, namely *C. sonorensis* (Shelby and Webb, 1997), *C. congregata* (Beckage and Riddiford, 1982a), and *M. demolitor* (Balgopal *et al.*, 1996), has been reported to promote the maintenance of elevated JH titres in the last instar of their respective hosts, *Heliothis virescens, Manduca sexta* and *Pseudoplusia includens*. In larvae of *Pseudaletia separata* parasitized by *C. kariyai*, JH titres have not been determined, but indirect evidence suggests that high JH levels may not be required for the observed suppression of host metamorphosis; instead, it has been suggested that PTTH synthesis or release is disrupted by parasitism (Tanaka *et al.*, 1987). More recent work, however, has shown that the activity of JH esterase (JHE), a JH-specific degradative enzyme, is inhibited in last-instar hosts parasitized by *C. kariyai* or injected with its PDV (Hayakawa, 1990, 1995), suggesting that JH titres may indeed be higher than normal in these insects. On the other hand, no detectable difference in the JH titre, 3 days post-parasitization, was observed between last-instar *C. fumiferana* larvae parasitized by *T. rostrale* and control insects, although the activity of JHE was inhibited by parasitism (Cusson *et al.*, 2000). This observation calls into question the functional significance of the inhibition of JHE activity reported for several lepidopteran species in which parasitism-associated developmental arrest has been documented (Balgopal *et al.*, 1996; Beckage and Riddiford, 1982a; Dahlman *et al.*, 1990a; Dover *et al.*, 1995b; Hayakawa, 1990; Shelby and Webb, 1997; Strand *et al.*, 1990; Zhang *et al.*, 1992).

The ultimate consequence of an inhibition of the early endocrine events of metamorphosis, as described above, is the suppression of ecdysone secretion by the prothoracic glands (PG), which prevents both the metamorphic commitment peak of ecdysteroid, and the subsequent (larger) peak that induces the pupal moult. However, whether this type of hormonal interference is sufficient to maintain the host in a state of arrested development is unclear as work on the effects of *C. sonorensis* calyx fluid (the main component of which is a PDV (CsIV)) on *H. virescens* indicates that this wasp-borne factor, when injected at a dose of two female-equivalents, causes a rapid degeneration of the PG and that the effect may be direct, as suggested by ligation-injection experiments (Dover *et al.*, 1987; Dover *et al.*, 1988a). Likewise, injection of 0.5 female-equivalent of *T. rostrale* calyx fluid was observed to cause partial degeneration of *C. fumiferana* PG, but the effect was not noticeable until several days after injection (Cusson *et al.*, 2000). Differences in the impact of these two viruses on host PG may be partly due to differences in injected doses, as subsequent work on CsIV indicated that its effect on the PG was dose-dependent (Dover *et al.*, 1988b). Interestingly, the PGs of *M. sexta* do not degenerate in response to parasitism by the gregarious parasitoid *C. congregata,* or in response to injection of its PDV (Beckage and de Buron, 1993). In this system, however, parasitism inhibits the 20-monooxygenase-driven conversion of ecdysone into the active moulting hormone, 20E, and parasitoid egression requires a transient increase in host ecdysteroid titres (Beckage and Riddiford, 1982a; Beckage and Templeton, 1986).

Unlike the aforementioned species, the braconid *Cardiochiles nigriceps* induces developmental arrest in final-instar larvae of its host, *H. virescens*, immediately after the pupal commitment peak of ecdysteroids, with 20E titres reaching commitment levels but being suppressed thereafter (Tanaka and Vinson, 1991a). Not surprisingly, JHE activity is not inhibited in these parasitized insects, and JH titres are suspected to undergo their normal pre-metamorphic decline, since treatment of parasitized larvae with a JH analogue inhibits the moult of the parasitoid (Pennacchio *et al.*, 1992a). In this system, the wasp PDV has a direct effect on host PG, as demonstrated by *in vitro* treatment of *H. virescens* PG with *C. nigriceps* PDV/venom, which inhibits secretion of ecdysone by the host PG. Interestingly, the PGs become sensitive to the PDV/venom combination only after pupal commitment (Pennacchio *et al.*, 1998a, b). However, the PGs do not undergo any apparent morphological degeneration, although they become functionally inactive and unresponsive to PTTH (Pennacchio *et al.*, 1997). In addition, the teratocytes of *C. nigriceps* have been shown to convert host (*H. virescens*) 20E into inactive polar metabolites, thus making a contribution to the suppression of 20E titres following pupal commitment (Pennacchio *et al.*, 1992a).

The effects of parasitism by the braconid, *Microplitis croceipes,* on the development of its host, *H. virescens*, appear similar to those described above for *C. nigriceps* in that parasitized larvae proceed normally to reach the pupation-cell-formation stage (which immediately precedes the pre-pupal stage; Webb and Dahlman, 1985), and their 20E titres appear to reach commitment levels. However, the PGs of the parasitized insects are not unresponsive to PTTH, and JHE activity remains low throughout the last stadium of parasitized larvae (Dahlman *et al.*, 1990a). These effects are mediated both by the PDV (Fathpour and Dahlman, 1995) and teratocytes (Zhang *et al.*, 1992). Thus, the *M. croceipes–H. virescens* system provides an example of host endocrine disruption that follows a pattern that is intermediate between that described for *C. nigriceps* and that which is observed in hosts that do not appear to initiate pre-metamorphic behaviours.

Some endoparasitic wasps arrest the development of their hosts at an even later stage of the metamorphic programme. The egg-larval parasitoid, *Chelonus inanitus,* first induces precocious metamorphosis in what would normally be the penultimate instar of its host, *Spodoptera littoralis*, and then causes developmental arrest in the precocious pre-pupa (Grossniklaus-Bürgin *et al.*, 1994; Soller and Lanzrein, 1996). Similar alterations are observed in the hosts of other chelonine wasps, including the noctuid *Trichoplusia ni* parasitized by *Chelonus curvimaculatus* (see Jones, 1996 for a review). The hormonal events that underlie the precocious initiation of metamorphosis have been shown to parallel those normally observed in the last stadium of unparasitized larvae, i.e. inactivation of the corpora allata (CA) followed by a rise in the activity of JHE, and the occurrence of the ecdysteroid pupal commitment pulse (Grossniklaus-Bürgin *et al.*, 1998). Obvious differences in the endocrine profile of metamorphosing parasitized and control larvae only become manifest at the pupation-cell-formation stage, when ecdysteroid titres of parasitized larvae are lower than those of controls. Parasitized larvae also display reduced levels of pre-pupal JH and JHE, the peaks of which occur later than in controls. These hormonal changes have been shown to be caused by the PDV/venom injected into the host egg during oviposition (Grossniklaus-Bürgin *et al.*, 1998), whereas the precocious induction of metamorphosis requires the presence of proteins secreted by the larval parasitoid, in addition to the PDV and venom (Hochuli *et al.*, 1999).

The foregoing discussion and examples show that the endocrine events altered by parasitism, for the purpose of achieving host developmental arrest, vary as a function of the stage at which metamorphosis is blocked. These differences should be taken into account when comparing data on hormonal disruptions among different host-parasitoid systems, as they are likely to represent different underlying molecular mechanisms.

3. Molecular mechanisms underlying developmental arrest

Various factors transmitted to the host by the parasitic wasp during oviposition have been shown, either alone or in combination, to bring about the hormonal and biochemical changes that effect developmental arrest. Among braconids, these factors include PDV, venoms, teratocytes and, possibly, larval parasitoid secretions, whereas among ichneumonids, PDV alone appear to be responsible for most of the physiological alterations observed in the parasitized hosts.

Among the wasp-associated factors known to be involved in suppressing host metamorphosis, the PDVs have received the most attention. These viruses are obligate symbionts of many ichneumonid and braconid endoparasitoids, and have a segmented genome that consists of several superhelical dsDNA circles, generated from linear copies integrated in the wasp genome. Replication and virion assembly take place in the calyx cells of the wasp ovary, and virus particles are released in the lumen of the oviduct where they form the particulate fraction of the calyx fluid. Typically, a small quantity of this fluid is injected into the caterpillar host during oviposition. Although PDVs do not replicate in the host caterpillar, expression of some of their genes induces changes in host physiology that would appear to be beneficial to the developing wasp embryo and/or larva. Among these changes, immune suppression and developmental disruption are the two best-documented effects of PDV (Stoltz, 1993; Webb, 1998).

Convincing evidence for a specific role in developmental regulation has not yet been provided for any of the genes isolated from PDV and known to be expressed in lepidopteran hosts; instead, most investigations have pointed to an involvement of the identified genes, transcripts and protein products in immune suppression (Webb, 1998; Béliveau *et al.*, 2000). In fact, it is not yet clear whether the same or different sets of genes are responsible for the induction of the two types of host pathologies. Certainly, it is possible to conceptually envisage common mechanisms for immune dysfunction and developmental regulation. For example, the promoter of the prophenoloxidase (PPO) gene in the mosquito *Anopheles gambiae* contains an ecdysone response element (Ahmed *et al.*, 1999a). Because PPO is a component of the humoral immune system that is often inhibited by PDV infection (Shelby and Webb, 1999), one can see that inhibition of ecdysone synthesis in a host that has a similar PPO gene organization would probably result in the concurrent inhibition of PPO transcription. Alternatively, several investigations have pointed to a role for PDV in the inhibition of translation of various proteins, including growth-associated proteins such as arylphorins (Dong *et al.*, 1996; Shelby and Webb, 1994, 1997), JHE (Béliveau *et al.*, 2000; Dong *et al.*, 1996; Shelby and Webb, 1997), and regulatory proteins involved in ecdysone synthesis (Pennacchio *et al.*, 1997). Shelby and Webb (1999) have suggested that a common mechanism through which host development and immune responses are regulated by inhibiting the expression of critical host proteins at a post-transcriptional level, would be a rapid and effective means to redirect host physiology. If some PDV-encoded proteins turn out to be translational inhibitors secreted into the haemolymph by infected cells, the mechanism whereby the translation of other major plasma proteins appears unaffected (Shelby and Webb, 1997, 1999) will need to be examined. Since most plasma proteins are synthesized in the fat body, the functional differentiation of this tissue (Blissard, 1982; Haunerland and Shirk, 1995) and the possibility that only specialized regions of it may be responsive to a given PDV gene product should be considered. Alternatively, permissiveness for PDV infection and/or gene expression could be restricted to certain regions of the fat body.

Despite the lack of direct evidence for a role of specific PDV-encoded proteins in developmental regulation, and the possibility that the two principal alterations of host physiology ascribed to these viruses may be effected by the same gene products (see above), recent studies on PDV gene expression in lepidopteran hosts have provided data on temporal patterns and tissue-specific transcription of PDV genes that are consistent with a function for the encoded proteins in developmental regulation. For example, the level of transcription of genes encoded by segments CiV12 and CiV14 of *C. inanitus* PDV was observed to increase substantially in the last instar of parasitized hosts, coincident with the induction of developmental arrest (Johner *et al.*, 1999). Similarly, Jones and Wache (1998) identified a 2.7-kb PDV transcript in arrested *T. ni* pre-pupae parasitized by *C. curvimaculatus*. In *H. virescens* larvae parasitized by *C. nigriceps*, Varricchio *et al.* (1999) identified a PDV gene that is transcribed in the host PG, coincident with the loss of functional integrity of these glands.

Similar efforts in our laboratory have led to the identification of a PDV gene whose pattern of transcription suggests that the protein it encodes may be involved in the induction of developmental arrest in *C. fumiferana* larvae parasitized by *T. rostrale* (Béliveau *et al.*, 2000). A summary of this work is provided below, preceded by an introduction to this host–parasitoid system and a description of some of its distinctive features.

4. Disruption of *C. fumiferana* metamorphosis by TrPDV

4.1 *Developmental perturbation versus immune dysfunction*

Parasitization of early 5th (penultimate)- or 6th (last)-instar *C. fumiferana* larvae by *T. rostrale* prevents the initiation of host metamorphosis, with development of the 5th-instar larvae being apparently unaffected (Doucet and Cusson, 1996a). Similarly, injection of *T. rostrale* calyx fluid into 2-day-old unparasitized 6th-instar larvae causes a dose-dependent delay in the initiation of metamorphosis. However, this effect can be abolished by prior treatment of the calyx fluid with UV and the DNA cross-linker, psoralen. This observation strongly suggests that the developmental arrest observed in parasitized larvae is the result of PDV infection (Doucet and Cusson, 1996a).

Although the effect of *T. rostrale* calyx fluid on host development is typical of that which has been documented for other host–parasitoid systems and their associated PDV, its effects on host immunity appear uncommonly weak, particularly with respect to the host cellular immune response. In one experiment, small glass rods, about the size of a parasitoid egg, were implanted in *C. fumiferana* larvae three days following parasitization by *T. rostrale*. When the rods were examined 24 h later for signs of encapsulation, they were all encapsulated, irrespective of whether they had been implanted in parasitized or control larvae. However, wasp eggs and larvae had all escaped encapsulation and the haemocyte-coated rods retrieved from parasitized insects were not melanized, unlike those dissected from control caterpillars (Doucet and Cusson, 1996b). These results suggested that the ability of *C. fumiferana* haemocytes to spread and adhere to foreign surfaces, other than wasp eggs and larvae, was not altered by parasitism (Doucet and Cusson, 1996b). Such observations were in contrast with those reported for other host–parasitoid systems in which the behaviour of host haemocytes was altered to an extent that prevented their adherence to foreign surfaces (e.g., Stoltz and Guzo, 1986), a pathology that has often been attributed to PDV infection (Stoltz 1993; Webb 1998). We later confirmed the ability of haemocytes from parasitized *C. fumiferana* larvae to spread onto glass microscope slides by comparing their behaviour in haemolymph samples drawn from larvae at different time-points following parasitization with their behaviour in samples obtained from unparasitized caterpillars of comparable ages (D. Doucet, personal communication). Thus, although parasitism inhibited melanization (a pathology that resulted from an inhibition of phenoloxidase activity and which could be duplicated by injection of *T. rostrale* calyx fluid into unparasitized larvae), the cellular immune response appeared to remain unaffected by parasitism, with the possible exception of an inhibition of the normal increase in total haemocyte counts, which we speculated was an indirect consequence of the developmental arrest (Doucet and Cusson, 1996b).

Microscopic examination of *T. rostrale* eggs revealed the presence of fine chorionic projections within which TrPDV virions appeared to be entrapped, resulting in the formation of a particulate coat on the egg surface (Cusson *et al.*, 1998). A similar feature had previously been described for the eggs of the ichneumonid *Venturia canescens* (Rotheram, 1973a,b), although the particulate coat in that case was found to be made of virus-like particles (VLP) that are devoid of DNA (Bedwin, 1979). In *V. canescens*, these chorion-bound VLP apparently confer passive protection to the egg against encapsulation, possibly because structural proteins from the particles share antigenic determinants with a protein from the host, *Ephestia kuehniella*, thus keeping the egg from being recognised as foreign by the host (Theopold *et al.*, 1994).

By analogy with this system, we have proposed that the eggs of *T. rostrale* evade the host immune response by a similar passive mechanism (Cusson *et al.*, 1998), although this hypothesis still awaits formal testing.

Thus, since the available evidence suggests that TrPDV infection causes no apparent cellular immune dysfunction – which should minimise the number of viral genes, if any, involved in immune suppression – the *T. rostrale–C. fumiferana* host–parasitoid system appears to be particularly well suited to the identification of PDV genes that may have functions in host developmental regulation. In this respect, the PDV of *C. inanitus* may have similar 'virtues' since it does not alter haemocyte spreading and fails to prevent the encapsulation of latex beads; however, its presence is required to protect the parasitoid larva from encapsulation (Stettler *et al.*, 1998).

4.2 *Hormonal alterations induced by* T. rostrale *in its host,* C. fumiferana

As mentioned in Section 2, *C. fumiferana* hosts parasitized by *T. rostrale* shortly after the last larval moult do not exhibit external signs of metamorphosis. Juvenile hormone esterase activity, which reaches maximal levels on day 4 in control insects, is suppressed by both parasitism and injection of calyx fluid. Nonetheless, JH titres drop from measurable levels (126 pg JH II equivalents/ml) immediately after the moult to almost undetectable levels (~25 pg JH II equivalents/ml) on day 4 in both control and parasitized larvae. Interestingly, *in vitro* rates of JH biosynthesis by isolated CA follow the same pattern, with the CA becoming completely inactivated by day 4 in parasitized, calyx fluid-injected, and control insects (Cusson *et al.*, 2000). Thus, in the absence of sustained JH biosynthesis, inhibition of JHE activity does not appear sufficient to insure the maintenance of elevated JH titres in parasitized insects. However, we cannot discount the possibility that our analytical method for measuring JH titres was not sufficiently sensitive to detect titre differences in the low pg range. However, the JH titres that we reported for control *C. fumiferana* larvae, using a radioimmunoassay (Cusson *et al.*, 2000), are in good agreement with recently published JH titres obtained for the same species using physico-chemical analysis (Palli *et al.*, 2000).

Parasitism of newly moulted last-instar *C. fumiferana* larvae, or injection of 0.5 female-equivalents of calyx fluid, completely suppressed the rise in haemolymph 20E titres normally observed in unparasitized larvae between day 4 (pupal com-mitment peak) and the larger ecdysteroid peak initiating the pupal moult that is observed on day 7 (Cusson *et al.*, 2000). Whether this inhibition is functionally related to the suppression of JHE activity is unclear as a reduction in JHE level does not appear to affect JH titre. However, the possibility that this enzyme plays another, yet undescribed, role (beside the reduction in JH titre) in the initiation of lepidopteran metamorphosis (see below) should be examined (Cusson *et al.*, 2000). In relation to the modulation of ecdysteroids, this might be particularly pertinent, as we have no convincing evidence that TrPDV directly affects host PG. Recent work by Ismail *et al.* (2000) suggests that JH acid may be required to induce metamorphic competence in certain tissues of *M. sexta*. In this context, it appears possible that the primary function of an inhibition of JHE activity may be to block the formation of JH acid and, as a consequence, to prevent acquisition of metamorphic competence by certain tissues. Thus, a more thorough screening of the potential mechanisms underlying the TrPDV-induced suppression of 20E titre is required to shed new light on these questions. Indirect strategies using recombinant viral proteins and other biotechnological approaches may facilitate such an analysis.

4.3 *TrPDV gene expression in* C. fumiferana

In an effort to identify TrPDV genes that may be involved in blocking *C. fumiferana* metamorphosis, we conducted Northern analysis of RNA extracted from parasitized and calyx fluid-injected last-instar larvae using, as a probe, PCR-amplified DNA generated from *Hin*dIII and *Sph*I TrPDV genomic libraries. Using this approach, we detected a single transcript of ~650 bp, which was most abundant 24 h after injection in calyx fluid-injected larvae, or 3–4 days after oviposition in parasitized insects (Béliveau *et al.*, 2000). We then screened two TrPDV genomic libraries with ^{32}P-labelled cDNAs generated from mRNAs extracted from calyx fluid-injected *C. fumiferana* larvae, which led to the selection of several positive clones. Restriction enzyme mapping of these clones revealed the presence of four different families, one of which represented 86% of the positive clones. We focused our attention on this particular family and localized coding sequences in *Pvu*II-digested clones using the same cDNA probe used for library screening. One clone was chosen for sequence analysis and found to encompass the entire gene sequence (Béliveau *et al.*, 2000).

This gene encodes a 103-amino-acid protein, designated TrV1, which has a predicted molecular mass of 9.3 kDa after cleavage of a putative signal peptide (first 21 amino acids). An initial BLAST (Altschul *et al.*, 1997) search revealed limited similarity with the *C. sonorensis* VHv1.4 protein expressed in CsIV-infected *H. virescens* larvae (Cui and Webb, 1996), but mostly within the signal peptide region of the two proteins (Béliveau *et al.*, 2000). Unlike VHv1.4, TrV1 displays no cysteine-motif, and has no potential glycosylation site; the latter are considered essential for binding of CsIV cysteine-motive proteins to haemocytes (Soldevila *et al.*, unpublished results, cited in Shelby and Webb, 1999).

Using an RT-PCR approach, we cloned the cDNA corresponding to the TrV1 gene. Sequence analysis of this cDNA revealed a predicted transcript size identical to that observed on the initial northern blot. Subsequent northern analysis of the same RNA used in the first blot, using a probe specific to the TrV1 gene, generated the same transcription pattern and thus confirmed our conclusion regarding the origin of the transcript detected using the first probe (Béliveau *et al.*, 2000).

The availability of the TrV1 cDNA made it possible to produce a recombinant protein in a bacterial expression system and to raise TrV1-specific antibodies. This allowed the analysis of tissue-specific expression, the results of which were then compared with tissue-specific transcription using the TrV1-specific probe. Northern analysis of RNA extracted from haemocytes, midgut, fat body and cuticular epithelium, dissected from parasitized larvae, revealed strong signals in the latter two tissues, but very weak signals in haemocytes and midgut. Western analysis of the proteins extracted from the same tissues confirmed the presence of the protein in fat body and cuticular epithelium, but indicated that the bulk of the protein was found in the plasma (haemocyte-free) fraction of the haemolymph, with no signal detected in haemocytes. These results suggested that, unlike that which has been observed for other PDV genes (see Webb, 1998 for a review), there is little TrV1 transcription in haemocytes. In addition, the TrV1 protein is secreted into the haemolymph (Béliveau *et al.*, 2000). The strong signal observed in the cuticular epithelium may, however, have originated from fragments of fat body and muscle tissue that we could not completely remove from the epithelium during dissections; *T. rostrale* females typically lay their eggs in the host fat body, just under the integument (Cusson *et al.*, 1998). The fact that

TrPDV virions are bound to the egg surface and apparently tend to enter tissues directly in contact with the egg in the caterpillar host (Cusson *et al.*, 1998) may help to explain the observed pattern of tissue-specific transcription.

Since the publication of the above results, we have completed the analysis of the other three families of positive clones selected in our initial library screening, and sequenced the associated genes (*Table 1*). One of them, designated TrV2, encodes a 102 amino acid protein that displays a high degree of similarity to TrV1 (the similarity is even greater at the nucleotide sequence level). This homology suggests that the transcription signal that we identified as specific to TrV1 could, in fact, be contributed by both TrV1 and TrV2. We are currently designing oligonucleotide probes specific to each of the two genes, in order to examine this possibility. Another gene, termed TrV4, encodes a somewhat larger protein displaying limited similarity to TrV1 and TrV2. Lastly, we have identified a TrPDV gene that is homologous to the repeat element protein (REP) gene family isolated from CsIV (Theilmann and Summers, 1987, 1988). We are in the process of analysing these genes and their protein products to ultimately determine whether they play roles in the inhibition of *C. fumiferana* metamorphosis or in other pathologies.

Table 1. *List of cloned cDNAs and genes from TrPDV, with some of their characteristics*

Clone no.	Name	cDNA	No. of exons	No. of aa[a]	Signal peptide	N-Glyco. site[b]	Similarity with other proteins
114	TrV1	Yes	2	103	Yes (21 aa)	No	TrV2 (79/103)[c] TrV4 (29/52) VHv1.4 (22/46)
208	TrV2	Yes	2	102	Yes (21 aa)	No	TrV1 (79/103) TrV4 (21/30) VHv1.4 (23/46)
9	TrV3	Yes	1	230	No	No	REP (101/211)
89	TrV4	No	2	159	Yes (15 aa)	Yes	TrV1 (29/52) TrV2 (21/30)

[a]aa, amino acids. [b]*N*-glycosylation site. [c]Ratios: number of similar amino acids/number of amino acids in portion of sequence displaying significant similarity with other protein; according to BLAST (Altschul *et al.*, 1997).

In the process of conducting preliminary Northern analysis for the newly isolated TrPDV genes, using specific probes, some of the transcription signals observed were much stronger than expected, given that they were not detected at all by our first northern analysis using PCR-amplified DNA as a probe (see above). The decision to use this type of probe (as opposed to labelled, undigested viral DNA) had been made necessary due to the very limited availability of TrPDV DNA (*T. rostrale* cannot be reared in the laboratory for more than two or three generations). Recently, we repeated this analysis using ^{32}P-labelled TrPDV (undigested, not amplified) as a probe, and observed other transcripts in addition to that attributed to TrV1 (*Figure 1*). This suggests that the amplification of the DNA present in the genomic library introduced

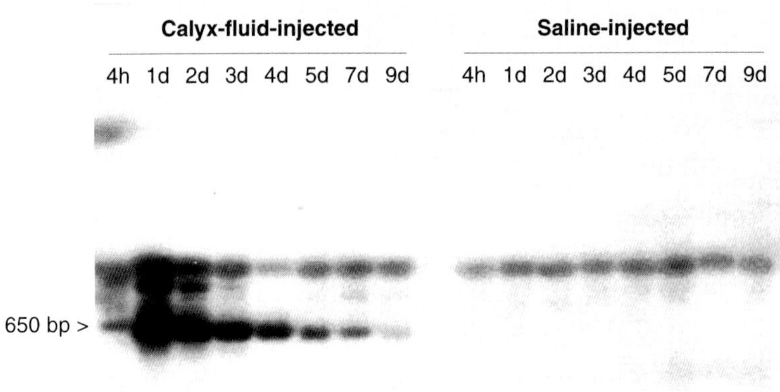

Figure 1. *Northern analysis of total RNA extracted from last-instar* C. fumiferana *larvae at various times (4 hours to 9 days) following injection with either* T. rostrale *calyx fluid or saline; ^{32}P-labelled undigested TrPDV DNA was used as a probe. The 650-bp signal is believed to be contributed by TrV1 and, possibly, TrV2. Hybridization conditions were as described in Béliveau* et al. *(2000).*

a bias leading to an over-representation of TrV1-related sequences. Nevertheless, the 650-bp signal contributed by TrV1 (and possibly TrV2) remains the strongest signal observed (*Figure 1*). Interestingly, a signal was detected in the RNA extracted from both calyx fluid- and saline-injected larvae (*Figure 1*). This band has the same mobility as the rRNA band seen on the stained blot. Whether this observation is of any biological significance remains to be seen.

In summary, TrV1 encodes what appears to be the most abundant transcript in TrPDV-infected last-instar *C. fumiferana* larvae. Its overall pattern of transcription is temporally associated with host developmental arrest, which appears to be the principal pathology induced by TrPDV in *C. fumiferana*. Its tissue-specific expression differs significantly from that observed for other known PDV genes (Webb, 1998), with only trace levels of transcription detected in haemocytes. The TrV1 protein shows no significant similarity with any other known protein, except for two other putative TrPDV proteins, the genes of which have recently been cloned, and the VHv1.4 protein encoded by CsIV. In the latter case, most of the similarity between the two proteins is seen in the signal peptide regions, and TrV1 lacks the cysteine-motif and the glycosylation site that appear necessary for interference with haemocyte behaviour. For these reasons, we believe that TrV1 (and possibly the homologous TrV2) is a good candidate for a PDV-encoded protein that may play a role in the induction of host developmental arrest. We are currently planning experiments to test this hypothesis.

5. Conclusion

Endoparasitic wasps have evolved various mechanisms to interfere with the metamorphosis of their lepidopteran hosts. Endocrine disruptions associated with the induction of developmental arrest appear to vary as a function of the exact stage at which metamorphosis is blocked. However, among hosts that display apparently similar symptoms, differences in hormonal perturbations have been documented, particularly with reference to the maintenance of high JH titres in last-instar parasitized animals. These observations raise questions regarding the functional significance of the often-reported inhibition of JHE activity in parasitized insects.

Factors transmitted to the host by the wasp during oviposition appear to be responsible for most of the observed endocrine alterations and the associated developmental arrest. Among these factors, the wasp PDVs have received the most attention and hold the most promise for the identification of genes that are instrumental in inducing the observed developmental disturbances. Some PDV genes appear to be good candidates, although it remains to be seen whether putative functions in immune suppression and developmental regulation are associated with the same, or separate, groups of genes.

Acknowledgements

This work was funded by a grant from the National Biotechnology Strategy Fund (Government of Canada) to Michel Cusson, by post-graduate scholarships from the Natural Sciences and Engineering Research Council of Canada (NSERC) and the Canadian Forest Service to Marlène Laforge, and by an NSERC operating grant to Guy Bellemare.

7

Regulation of host endocrine system by the endophagous braconid, *Cardiochiles nigriceps*, and its polydnavirus

Francesco Pennacchio, Carla Malva and S. Bradleigh Vinson

1. Introduction

Parasitoid insects exhibit a considerable diversity of adaptive strategies to parasitic life. These range from a loose association with the host, in the case of idiobionts, to a more complex physiological integration seen in koinobionts (Gauld, 1988). In the case of koinobionts, the juvenile stages of the parasitoid do not kill the host. For most of their development, they show a remarkable degree of adaptation to parasitic life, which is comparable to that of specialized metazoan parasites, and is characterized by an astonishing diversity of subtle physiological interactions (Godfray, 1994; Quicke, 1997). Among these interactions, the endocrine relationship between the host and the parasitoid is one of the most intriguing issues. A tentative classification of the different types of endocrine interaction has led to the definition of 'conformers' or 'regulators' to distinguish different categories of parasitoids (Lawrence, 1986a). The development and hormonal balance of conformers is linked to that of the host, whereas the regulators actively disrupt host endocrinology in order to produce a more suitable environment for their pre-imaginal stages. These categories, even though conceptually valid, do not fully cover the heterogeneity of hormonal strategies exhibited by parasitoid insects. This is mainly because any rigid, dichotomous classification of biological phenomena cannot possibly encompass all of the astonishing bio-diversity that is the result of the intense adaptive radiation process that has occurred during the evolution of parasitic Hymenoptera.

Endocrine Interactions of Insect Parasites and Pathogens, edited by J.P. Edwards and R.J. Weaver.

Host regulation is a developmental strategy that is more common in the parasitic Hymenoptera than in most other parasitic insects (Quicke, 1997). This is partly due to a series of host colonization and regulatory factors that are produced by the female reproductive and accessory tissues, and are delivered through the ovipositor (Quicke, 1997). Host colonization by endophagous parasitoids is associated with severe pathological alterations affecting both the immune and neuro-endocrine system of the host (Lavine and Beckage, 1995; Lawrence and Lanzrein, 1993; Strand and Pech, 1995). While the adaptive significance of immune defence suppression is clearly evident, the significance of alterations to host endocrinology and development are perhaps less obvious. In addition, the functional constraints on evolutionary driving forces, of changes to the endocrine system, are ill defined and more difficult to analyse. This complexity is due to the great diversity of pathological mechanisms that are observed in different host–parasitoid associations (Beckage, 1985; Lawrence and Lanzrein, 1993). The biochemical basis of immune suppression in parasitized hosts has received a great deal of attention, and some basic molecular aspects have been elucidated in a few selected host–parasitoid model systems (Shelby and Webb, 1999; Strand and Pech, 1995). In contrast, many of the studies on host neuroendocrine and developmental alterations are largely descriptive, and only a small amount of molecular information is currently available (Beckage, 1993a,b; Lawrence and Lanzrein, 1993).

Cardiochiles nigriceps Viereck, an endophagous braconid parasitoid of the tobacco budworm, *Heliothis virescens* (F.), is one of the few parasitoid species that has been studied in depth, particularly with reference to the physiological and molecular bases of host endocrine alterations. The majority of studies on this model system have been conducted in our laboratories over the last ten years. Many descriptive aspects of host manipulation have been reported, the role and the mechanism of action of host-regulatory factors have been elucidated in part, and a few genes – putatively involved in the host neuroendocrine disruption – have been cloned and sequenced. This chapter reviews the currently available information on the endocrine alterations to *H. virescens* mature larvae that are induced by *C. nigriceps*. It also suggests a tentative interpretation of the adaptive significance of the redirection, by the parasitoid, of host development and metabolism, both of which are triggered, and maintained, by dramatic alterations to the host hormonal balance.

2. Host endocrine alterations

Cardiochiles nigriceps is a solitary endoparasitoid of tobacco budworm larvae. When host *H. virescens* larvae are parasitized, they develop as far as the last larval instar, but then fail to pupate (Lewis and Vinson, 1968; Pennacchio *et al.*, 1992a). These developmentally arrested last-instar larvae are completely consumed by the mature parasitoid larva, which eventually emerges from the host to spin a cocoon. This developmental disruption is associated with a substantial modification of the total ecdysteroid titre in host last-instar parasitized larvae (Pennacchio *et al.*, 1992a; Tanaka and Vinson, 1991a). These larvae, unlike developmentally concurrent unparasitized controls, show no detectable pupal commitment peak of 20-hydroxyecdysone (20E), or subsequent burst of ecdysteroid, which would normally be triggering the larval to pupal moult. Instead, they show a slow and gradual increase of total haemolymph ecdysteroids during the final part of parasitoid development. The reversed phase high-performance liquid chromatography (RP-HPLC) separation of ecdysteroids, and their quantification by radioimmunoassay, has provided evidence that compared to developmentally concurrent unparasitized controls there is a substantial reduction of the total ecdysteroid titre in parasitized host last-instar larvae and a completely different ecdysteroid profile. The parasitized hosts have very low levels of 20E, and an accumulation of inactive polar metabolites that appear during the final part of parasitoid larval development (Pennacchio *et al.*, 1994a). Enzymatic digestion of these polar metabolites with β-glucuronidase and sulfatase enzymes releases free ecdysteroids, especially 20E (Pennacchio *et al.*, 1994a). Thus, in these late-stage parasitized host larvae, there is both a reduction of ecdysteroid biosynthesis by the PG, and a greater conversion to inactive metabolites, which occurs through a parasitoid-controlled degradative pathway, of the reduced amounts of 20E that are produced.

Measurements of juvenile hormone (JH) titre in the host haemolymph have never been made. However, levels of juvenile hormone esterase (JHE) activity in the haemolymph of parasitized *H. virescens* larvae do not show any significant difference from those seen in developmentally concurrent unparasitized controls, and a normal peak of JHE activity is seen on day 2 of parasitized last-instar host larvae (Pennacchio *et al.*, 1994b). This suggests that any subsequent decrease of JH titre in mature parasitized host larvae would be unlikely to be prevented, although this hypothesis is not yet supported by experimental evidence.

The parasitoid-mediated interference with host ecdysteroid levels as a strategy for altering development and reproduction is a common theme in several host–parasitoid associations. However, the alteration of ecdysteroid titre in *H. virescens* larvae parasitized by *C. nigriceps* appears to be rather peculiar, and different from that of other systems. In many of these other cases, a low 20E titre in parasitized larvae is found to be associated with, and in some cases may be induced by, a reduced rate of degradation and/or an increased rate of synthesis of JH (Beckage and Templeton, 1986; Cusson *et al.*, 2000; Dahlman *et al.*, 1990a; Hayakawa, 1995; Jones *et al.*, 1986a; Lanzrein and Hammock, 1995; Noguchi *et al.*, 1995; Pfister-Wihelm and Lanzrein, 1996; Tanaka *et al.*, 1987; Zhang *et al.*, 1992).

3. Host regulation factors

When *C. nigriceps* oviposits, it injects into the host body, along with the egg, a venom and a symbiotic polydnavirus (*C. nigriceps* polydnavirus; CnPDV) (*Figure 1*). This was one of the first members of the Polydnaviridae to be discovered (Stoltz and Vinson, 1979a; Vinson and Scott, 1975). After the egg hatches, the parasitoid larva emerges from the serosal membrane, which then dissociates into single rounded cells known as teratocytes (*Figure 2*). These cells float freely in the haemocoel of the host and, as the parasitoid larva develops, increase in both size and level of ploidy, but do not undergo cell division (Dahlman and Vinson, 1993; Pennacchio *et al.*, 1994c; Vinson, 1970; Vinson and Scott, 1974). The CnPDV and teratocytes are the two major parasitoid-derived host-regulating factors, and are maternal and embryonic in origin, respectively. It is very likely that both of them are involved in the regulation of many aspects of host physiology, which is important for successful parasitism (Vinson, 1972a). Hereafter, only the role of these parasitoid-derived factors in the regulation of host endocrine system will be considered.

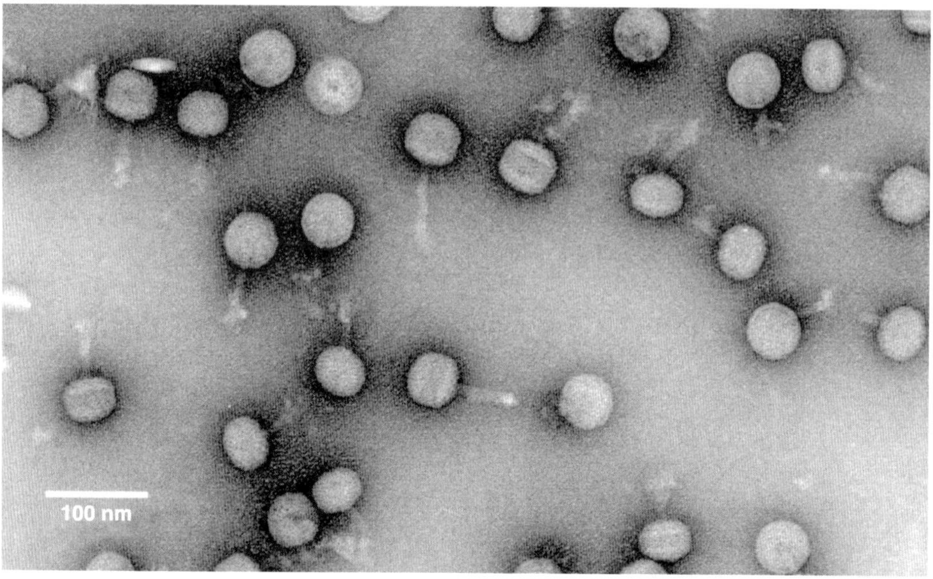

Figure 1. Negative staining of liberated nucleocapsid of Cardiochiles nigriceps *polydnavirus.*

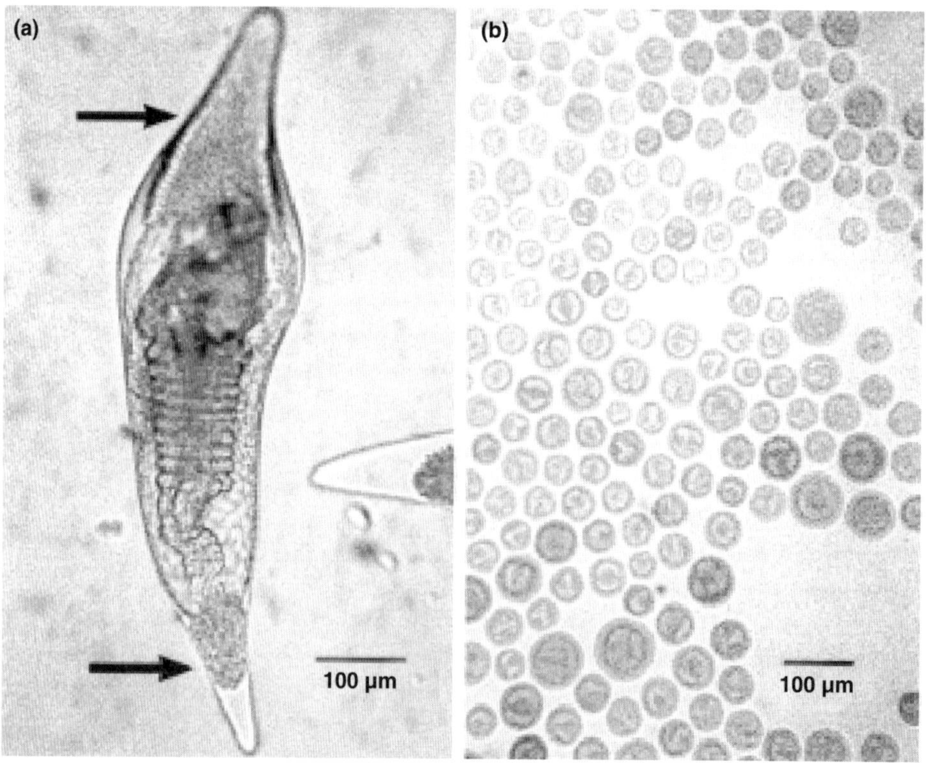

Figure 2. (a) Cardiochiles nigriceps *embryo; at hatching, the indicated terminal poles (arrows) of the embryonic serosa dissociate into single cells called teratocytes (b).*

In last-instar *H. virescens* larvae parasitized by *C. nigriceps*, the synthesis of ecdysone by prothoracic glands (PG) is strongly depressed. This reduced synthesis of ecdysone is also evident when PGs are treated *in vitro* with female wasp reproductive secretions, comprising calyx fluid and venom (Tanaka and Vinson, 1991a). Recent studies (Pennacchio *et al.*, 1998a) have demonstrated that this *in vitro* treatment is effective only when the CnPDV is transcriptionally active, suggesting that viral infection of the PG might be responsible for the depression of ecdysone biosynthesis.

The inhibitory effect of CnPDV on host PG activity is not associated with either a gross morphological degeneration of gland cells, or signs of necrosis or apoptosis (Tanaka and Vinson, 1991a; Pennacchio *et al.*, 1997). The strategy, or mechanism of inactivation of the PG in the same host (*H. virescens*), adopted by a different parasitoid, the ichneumonid *Campoletis sonorensis* (Cameron), appears completely different. In this case, the PG of last-instar host larvae are completely disrupted by the action of the PDV from *C. sonorensis* (Dover *et al.*, 1988a, b). The functional inactivation of PG induced by *C. nigriceps*, however, does not seem to be unique. An apparently similar situation is reported for the host–parasitoid system *Manduca sexta* L.–*Cotesia congregata* (Say), where PGs do not show signs of structural degeneration, even though their activity is severely impaired (Beckage and de Buron, 1993; Gelman *et al.*, 1998; Kelly *et al.*, 1998). A reduced functional activity of host PGs has been recently reported for another braconid species, *Chelonus inanitus* L. (Grossniklaus-Bürgin *et al.*, 1998).

The teratocytes of *C. nigriceps*, when injected into unparasitized *H. virescens* larvae, have a negative affects on both host weight and pupation. The effects are dose-dependent and vary according to the developmental stage of the host at the time of injection (Pennacchio *et al.*, 1992a; Vinson, 1970). The most pronounced effect is seen with injections into mature host larvae before the pupal commitment peak of ecdysone. Such treatment always results in the complete failure to pupate (Pennacchio *et al.*, 1992a). However, the total ecdysteroid titre in teratocyte injected host larvae is much higher than that in synchronous unparasitized controls (Pennacchio *et al.*, 1992a), even though the levels of 20E are comparatively lower. This suggests that the presence of teratocytes may be altering metabolism of ecdysteroids (Pennacchio *et al.*, 1994a). This ability of teratocytes of *C. nigriceps* to metabolize host ecdysteroids can be demonstrated by incubation with radiolabelled ecdysteroids *in vitro*, which shows that significant amounts of polar conjugates are produced only when 20E is used as a substrate (Pennacchio *et al.*, 1994a). Then, the unexpected higher ecdysteroid titre in host larvae that receive a teratocyte injection can be interpreted as a result of the elimination of the feedback regulation normally exerted by 20E on prothoracicotropic hormone (PTTH) production and release, as well as on PG biosynthetic activity (Steel and Davey, 1985; Watson *et al.*, 1989).

In conclusion, *H. virescens* last-instar larvae parasitized by *C. nigriceps* show developmental arrest in response to severe alteration of their hormonal balance. The overall physiological picture can be summarized as follows (*Figure 3*). The brain of mature parasitized host larvae appears to produce normal levels of PTTH (Tanaka and Vinson, 1991a). However, the inhibitory effect of CnPDV infection on PGs of host mature larvae dramatically reduces their biosynthetic activity. Then, the subsequent reduced amounts of 20E produced are readily converted by *C. nigriceps* teratocytes to inactive polar metabolites. The integration of parasitoid-derived host regulation factors, both of maternal and embryonic origin, nearly completely suppresses the

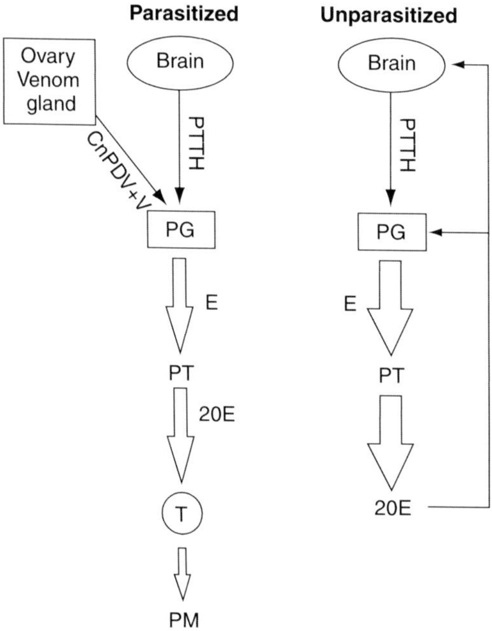

Figure 3. Schematic representation of the functional integration between Cardiochiles nigriceps *polydnavirus and teratocytes, and their effect on ecdysone synthesis and metabolism in parasitized hosts, compared to unparasitized controls. CnPDV,* Cardiochiles nigriceps *polydnavirus; E, ecdysone; 20E, 20-hydroxyecdysone; PM, polar metabolites; PT, peripheral tissues; PTG, prothoracic glands; PTTH, prothoracicotropic hormone.*

20-hydroxyecdysone titre in parasitized host mature larvae, acting on both synthesis and metabolism of ecdysteroids.

4. Functional and molecular bases of host prothoracic gland inactivation

The PGs of last-instar *H. virescens* larvae parasitized by *C. nigriceps*, even though not structurally disrupted, are functionally inactivated (Pennacchio *et al.*, 1997; 1998b; Tanaka and Vinson, 1991a). The glands are responsive to PTTH stimulation, but the PTTH signal transduction pathway is interrupted. Several biochemical changes at cellular level are observed. Basically, a lower level of ecdysone biosynthesis is associated with reduced levels of general protein synthesis. Protein synthesis is blocked at translational level, although transcriptional activity appears to be unaffected (Pennacchio *et al.*, 1997). The decrease of general protein biosynthesis may well account for the reduced ecdysteroidogenesis. In fact, as is the case with steroidogenesis in higher animals (Orme-Johnson, 1990; Vinson, 1987), ecdysone biosynthesis in insects is triggered by a rapid increase in the rate of general protein synthesis. This is necessary for the production of specific proteins, which regulate rate-limiting steps in the ecdysone biosynthetic pathway (Keightley *et al.*, 1990; Rybczynski and Gilbert, 1994). The observed translational block in the PGs of parasitized last-instar *H. virescens* larvae is a possible consequence of the marked under-phosphorylation of the cAMP dependent ribosomal S6 protein. Phosphorylation/activation of ribosomal S6 is one of the critical steps in the PTTH signal transduction pathway in the PG of lepidopteran larvae (Rybczynski and Gilbert, 1994; Song and Gilbert, 1994).

Furthermore, phosphorylation of β-tubulin also plays a pivotal role in the signal transduction pathway of PTTH, by enhancing the cytoskeleton-mediated transport of ecdysone precursors and intermediate metabolites (Rybczynski and Gilbert, 1994, 1995). Phosphorylation of ß-tubulin is an important cellular biochemical event, and is involved in many other responses to hormonal stimulation (De Loof and Vanden Broeck, 1995). Interestingly, in PG of last-instar *H. virescens* larvae parasitized by *C. nigriceps*, a reduced level of phosphorylation has been observed for a protein having the same molecular mass of β-tubulin in response to PTTH stimulation (Pennacchio *et al.*, 1997) and this may further contribute to the reduced ecdysteroidogenesis. The concurrent decrease in the phosphorylation level of the key regulatory proteins controlling the PTTH signal transduction pathway, the rate of general protein synthesis, and that of ecdysteroid biosynthesis, provide further support to the hypothesis that these cellular activities are functionally linked (Pennacchio *et al.*, 1998a).

Even though the phosphorylation level of key regulatory proteins, such as ribosomal S6 and β-tubulin, is markedly reduced in intact PG cells, the expression level of these phosphorylatable proteins in *H. virescens* PG is not affected by *C. nigriceps* parasitism (Pennacchio *et al.*, 1997). Furthermore, in RP-HPLC fractions of PG cytosolic extracts no substantial change of cAMP-dependent protein kinase activity (measured as ribosomal S6 kinase activity) is detected in response to parasitism (Pennacchio *et al.*, 2000). These pieces of evidence suggest that cAMP-dependent protein kinase and/or phosphatase activities are very likely altered *in vivo* in intact PG cells of parasitized last-instar host larvae.

The functional inactivation of PG in parasitized larvae is induced by CnPDV infection (Pennacchio *et al.*, 1998a, b). This expression of viral genes in host PG is likely to be of crucial importance in mediating the suppression of ecdysteroidogenesis. The CnPDV is a typical polydnavirus (Webb, 1998), integrated as a pro-virus in the genomic DNA of the parasitoid, and having a segmented genome made of circular double-stranded DNA molecules, which range in size from 2.5 to more than 23 kb (Varricchio *et al.*, 1999). Between 12 and 48 h after oviposition, at least 6 major poly A^+ transcripts of CnPDV are detected in parasitized host larvae undergoing PG inactivation (Varricchio *et al.*, 1999). Genomic libraries of CnPDV have been prepared, as has a cDNA library of mature parasitized host larvae showing developmental arrest. The cDNA library was screened with ^{32}P-labelled total DNA of CnPDV, resulting in the isolation of several clones containing viral sequences expressed in parasitized hosts. The isolated clones were then used to screen the CnPDV genomic libraries. Using this approach, two cDNA clones, and the corresponding genomic clones, have been fully sequenced and the structure of the viral genes determined (Varricchio *et al.*, 1999; and personal communication). These genes both contain an intron, and the mature transcriptive sequences in each case code for small proteins having molecular masses around 15 kDa. These proteins show no significant sequence homologies with other known proteins. *In situ* hybridization experiments with labelled viral cDNAs indicate that the two CnPDV genes are expressed in the PGs of last-instar *H. virescens* larvae parasitized by *C. nigriceps*. Whether these genes are responsible for the disruption of the PTTH signal transduction pathway in host PGs remains to be demonstrated.

Based on data gathered so far on the parasitoid-induced inactivation of host PG, both host protein kinases and phosphatases are potential candidates for study. Disrupting the balance between phosphorylation and dephosphorylation of regulatory proteins could be an effective way of interfering with several cellular pathways in different host tissue, which results in a complex pathological syndrome observed in parasitized hosts. This proposed action of CnPDV-encoded molecules on conserved pathways in different host tissues is an intriguing possibility, and appears to be supported by our recent experimental data gathered on CnPDV gene expression in specific host tissues. Northern blot and *in situ* hybridization experiments have demonstrated that the two CnPDV genes cloned so far are expressed both in host PG and haemocytes, presumably playing an important role in the alteration of both the endocrine and immune systems of the host (Falabella, P., Varricchio, P., Graziani, F., Malva, C., Pennacchio, F., personal communication). This corroborates the hypothesis that PDV induction of seemingly disparate phenomena in parasitized hosts may be due to a cascade of related events precipitated by a direct action upon a few primary targets (Stoltz, 1993). The identification of these primary targets in different host tissues and organs will probably shed new light on the interpretation of the complex array of pathological symptoms induced by parasitoid insects on their hosts. It would be also interesting to characterize and compare the parasitoid-induced pathologies on specific host tissues in different host–parasitoid associations. The parasitoid polydnavirus-induced reduction of PG biosynthetic activity has been reported in other host–parasitoid systems (Beckage and de Buron, 1993; Gelman *et al.*, 1998; Grossniklaus-Bürgin *et al.*, 1998; Kelly *et al.*, 1998); however, detailed functional and molecular data are not yet available.

5. Endocrine alterations and host nutritional suitability

The alteration of 20E titre and metabolism in last-instar *H. virescens* larvae parasitized by *C. nigriceps* is responsible for host developmental arrest. However, even though this is the most prominent pathological symptom, other metabolic changes are also induced by this hormonal disruption. The titre of proteins in the haemolymph of last instar lepidopteran larvae is hormonally controlled (see Kanost *et al.*, 1990). The peak of 20E that initiates the larval-pupal transition also regulates the activity of the fat body by reducing the biosynthesis of proteins, and in most cases triggering their sequestration as storage granules in fat body cells (Kanost *et al.*, 1990). The absence of 20E from the parasitized host last-instar larva can be considered as the key endocrine change contributing to the maintenance of a high level of haemolymph proteins, which results from a failure by fat body cells to reabsorb storage proteins (Pennacchio *et al.*, 1993, 1994b). This change in metabolism occurs in correspondence to the exponential growth phase of the parasitoid larva, which is triggered and supported by the larva being able to feed continuously on the nutritionally enriched host haemolymph (Pennacchio *et al.*, 1993). As a consequence, the parasitoid larva undergoes its first moult only after the protein titre in the haemolymph of the host is elevated, and is sustained at a sufficient level that allows the parasitoid to attain a critical body mass. In this nutritionally enriched milieu, the major part of parasitoid larval size increase is attained in about one-third of its total developmental time (Pennacchio *et al.*, 1993; 1994b). Ligation of parasitized last-instar larvae at different times during the last instar has further demonstrated an important link between the increase in protein titre in the host haemolymph and the development of the parasitoid larva (Pennacchio *et al.*, 1992a, 1993). These studies also provide circumstantial evidence in favour of the hypothesis that host hormones are not directly involved in the regulation of parasitoid moulting and development.

It is evident that *C. nigriceps* redirects host development and metabolism, in order to increase the nutritional suitability of the host as a resource ensuring the optimal growth and development of the parasitoid larva. This is accomplished through a dramatic alteration of the host endocrine balance that allows the parasitoid to exploit the tremendous metabolic energy that normally would be expended by the host in preparation of metamorphosis. This is a common host exploitation strategy in hymenopteran parasitoids of lepidopteran larvae (Coudron, 1991; Lawrence, 1990).

Optimizing the nutritional exploitation of the host is undoubtedly one of the most stringent constraints determining evolutionary processes in parasitic Hymenoptera. In the case of koinobiont parasitoids, this is accomplished through a series of complex host-regulatory mechanisms, which aim to manipulate the host as a living microenvironment for optimal exploitation by the parasitoid. The central role of this evolutionary pattern in endophagous Hymenoptera can be fully appreciated by comparing how taxonomically related parasitoids might exploit different types of insect hosts. Interesting insights are offered by a comparative analysis of *C. nigriceps* and *Aphidius ervi* Haliday, two endophagous braconids that parasitize lepidopteran larvae and aphids, respectively. Using completely different mechanisms, these two parasitoids achieve similar control of host nitrogen metabolism, synchronizing the most food-demanding part of their larval development with a parasitism-induced increase of nitrogen-derived nutrients. This is accomplished by preventing pupation, in the case of *C. nigriceps,* whilst *A. ervi* severely limits host reproduction (Pennacchio *et al.*, 1995). Host castration, mediated by both parasitoid venom (Digilio *et al.*, 1998, 2000) and teratocytes (Falabella *et al.*, 2000), is part of a nutritional strategy based on complex biochemical interactions between the host aphid, its bacterial endosymbiont, *Buchnera,* the parasitoid larva, and the teratocytes (Falabella *et al.*, 2000; Pennacchio *et al.*, 1999). Thus, the host exploitation strategies of these two braconids are ultimately the result of similar evolutionary processes, both of which are driven by an important functional requirement, such as the need to provide the exact amount and type of food at the appropriate time. In both cases, the common theme is the competition between the nutritional needs of the parasitoid and the primary metabolic requirements of the host. The disruption of these key host functions, and the subsequent metabolic redirection of the host, results in a dramatic increase in its nutritional suitability for the developing parasitoid larva. Basically, the parasitoid larva waits for the most favourable biochemical changes taking place in the host to support reproduction and/or development, and then out-competes the host target tissues for the available resource. The metabolic redirection in favour of the parasitoid is frequently regulated by subtle endocrine mechanisms (*Figure 4*). For a further discussion of the nutritional aspects of parasitoid–host interactions see Chapter 11.

6. Conclusions

The host–parasitoid association of *H. virescens–C. nigriceps* is probably one of the best-studied model systems for the analysis of functional and molecular bases of host developmental alterations. Many aspects of the endocrine regulation of the host have been elucidated, but several intriguing molecular questions remain unanswered. In particular, the study of the molecular mechanisms of action of CnPDV-encoded molecules in the induction of inactivation of host PG is a challenging issue that may provide

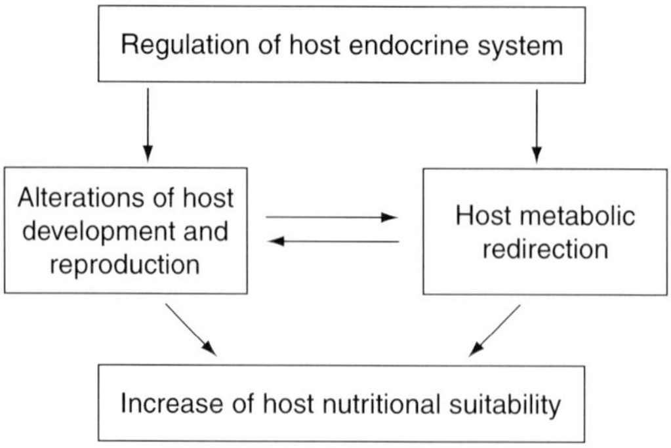

Figure 4. Schematic representation of the relationship between host endocrine alterations and host nutritional suitability.

interesting new insights into both PDV biology and the molecular physiology of ecdysteroidogenesis in insects.

The analysis of the nature of host–parasitoid interactions and of the complex biochemical and molecular processes leading to the disruption of neuroendocrine control of host physiological homeostasis will considerably contribute to the advancement of both basic and applied entomology. We predict that unravelling the molecular bases of the astonishing diversity of host–parasitoid associations in insects might offer valuable new opportunities for exploiting natural biodiversity in the development of innovative, biotechnology-based insect control strategies.

Acknowledgements

The Ministry of University (MURST, Programma di ricerca scientifica di rilevante interesse nazionale, 2000–01) has financially supported part of the experimental work here described.

8

Endocrine interactions of ectoparasitoid wasps with their hosts: an overview

Gay C. Marris, Robert J. Weaver and John P. Edwards

1. Introduction

Ectoparasitoid wasps occur in several hymenopteran families, and comprise a large and varied collection of insects. Just as they are many in number, so they have adopted a diverse array of lifestyles within the ectophagous habit. Britain's largest ichneumonid, *Rhyssa persuasoria*, uses the sharp cutting ridges of its ovipositor to drill several centimetres through tree-trunks to reach grubs of the wood-wasp, *Sirex* sp. (Askew, 1971; Gauld and Bolton, 1988). Tiny pteromalids are ectohyperparasitoids of other Hymenoptera (Gauld and Bolton, 1988), while males of the aphelinid *Coccophagoides utilis* are obligate autoparasitoids, which can only develop as external parasites of their conspecifics (Broodryk and Doutt, 1966). *Cedria paradoxa* is unique amongst all parasitic wasps (ecto- or endo-) in that the female guards the parasitized host caterpillar until her brood reaches adulthood (Beeson and Chatterjee, 1935). The extraordinary aquatic ichneumonid, *Agriotypus armatus*, literally walks into the water, remaining submerged for up to half an hour while it places its eggs onto pre-pupae of case-making caddis flies (Askew, 1971; Clausen, 1940). Irrespective, however, of the particular natural history of any given species, certain basic similarities exist between the biology of these ectoparasitoids. The adult female parasitoid wasp must locate and successfully oviposit onto a suitable host. This host is invariably another arthropod, and is often an insect in its pre-adult stages of development. The progeny of the wasp are then confined on the outside of the selected host, from where they consume the body fluids and tissues of the host, for most, if not all, of their immature lifespan. These fundamental traits – which define the ectoparasitic habit, particularly the requirement to develop on the exposed outer surfaces of an insect that may have not yet completed its development, present ectoparasitoids with a concomitant set of challenges. The different ways in which wasps respond to these challenges, in turn, define the complex physiological relationships that occur between ectoparasitoids and their hosts.

Endocrine Interactions of Insect Parasites and Pathogens, edited by J.P. Edwards and R.J. Weaver.

2. The need for ectoparasitoids to regulate the host

The most immediate problem encountered by many ectoparasitoids is the need to contend with an active, and sometimes aggressive, host. Not only is the ovipositing female herself open to attack, but, unlike the progeny of endoparasitoid species, the parasitoid eggs and larvae will remain exposed to the threat of physical damage long after the moment of oviposition. The writhing, mobile host larvae eject saliva or inflict lethal bites on adult wasps and their offspring (Neser, 1973; Takasu and Overholt, 1996; Vinson and Iwantsch, 1980a), and there are numerous examples in the literature of very precise egg placement, in very specific oviposition sites, which provide some refuge against such defences (Baltensweiler and Moreau, 1957; Clausen, 1940; Gerling and Limon, 1976; Godfray, 1994; Kasparayan, 1981; Mason, 1967; Neser, 1973; Quicke, 1997; Wall and Berberet, 1974). These various physical and behavioural adaptations do not, however, necessarily afford reliable protection to the developing ectoparasitoid offspring, and even the most carefully placed eggs are sometimes destroyed by the physical actions of the host. Many ectoparasitic Hymenoptera therefore require some means of subduing their hosts to allow oviposition and permanent attachment of the growing wasp brood.

Assuming successful oviposition has been achieved, ectoparasitoids that use immature insects as hosts face further challenges posed by the developmental condition of the host. A host that is parasitized during its larval stages will undergo gross morphological changes if it continues to grow and develop. Although a very few ectoparasitoid species, such as some tryphonine ichneumonids, can remain attached to their host during the process of ecdysis (Kasparyan, 1981), the vast majority are in danger of being shed should the host moult or metamorphose. Ectoparasitoids of larval hosts must either complete their own development very rapidly, i.e. within the duration of a single host stadium, or they must employ some regulatory mechanism to slow down or arrest host development. Apart from the potentially lethal physical consequences of unregulated host growth, ectoparasitoids may also have to contend with the dangers of direct physiological incompatibility with the endocrine system of their hosts. A host that continues to grow and develop has a changing hormonal status. By the same token, developing parasitoid larvae use comparable (but not identical) endocrine processes to regulate their own development. This means that parasitoid offspring, which come into contact with host hormones, may be susceptible to the effects of these hormones.

It is accepted that endoparasitic larvae, developing directly in the haemocoel of larval insect hosts will be exposed to biologically active endocrine substances. Moreover, it has been widely recognized that this exposure poses a threat to endoparasitoid survival when the host's hormonal status is incompatible with the needs of the developing wasp (Beckage, 1985, 1993b; Lawrence, 1986; Strand, 1986; Thompson, 1983; Vinson and Iwantsch, 1980a). Clearly, ectoparasitoids do not share such an intimate association with their host's physiology, but it is wrong to assume (as has often happened in the past) that ectoparasitoids will never have to contend with the effects of potent hormones that have the potential to disrupt their own development. In endoparasitic associations, the primary courses of endocrine communication between host and wasp larvae are via the cuticle of the parasitoid (Beckage, 1985, 1993b), by ingestion (Lawrence, 1986; Lawrence and Hagedorn, 1986; Riddiford, 1975), and possibly via the wasp's anal vesicle (Edson and Vinson, 1977). In ectoparasitic associations, it would seem that oral uptake is probably the most likely route of transfer. Only a very limited number of insect species have been examined for their susceptibility to ingested endocrine products. However, it has been shown that while some insects do not exhibit any developmental disruptions after ingesting dietary ecdysteroids (Blackford *et al.*, 1996; Blackford and Dinan, 1997c; Robinson *et al.*, 1987) other species certainly do experience adverse effects after consuming ecdysone or 20-hydroxyecdysone (Blackford and Dinan, 1997b). In some ectoparasitoid Hymenoptera, a specialized corrugated plug of cells on the egg-surface forms a pedicel that penetrates the host's cuticle (Neser, 1973; Gerling and Limon, 1976). This structure may not simply provide secure attachment to the host, but could also allow the passage of 'physiologically active' substances, including hormones, between host and wasp (Gerling and Limon, 1976).

Just as hosts present a minefield of physiological challenges to each parasitoid and her progeny, so have parasitic wasps developed different strategies to allow them to exploit their hosts effectively. Extensive studies have shown that many endoparasitic Hymenoptera manipulate the endocrinology of their hosts to make them compatible with their own developmental requirements, and the mechanisms and effects of host regulation have been described for several endoparasitoid species (Beckage, 1985; Coudron, 1991; Lawrence, 1986a; Stoltz, 1986; Strand, 1986; references cited in Quicke, 1997). However, the parallel regulatory mechanisms that exist in ectoparasitic relationships are less well documented, and have not been previously reviewed.

3. Non-paralysing ectoparasitic associations: the scope of this review

Parasitic wasps (both ecto- and endo-) have been variously categorized according to the effects that they have on their hosts' developmental biology. Species that allow hosts to continue to grow in size after parasitism have been described as koinobionts, as opposed to idiobionts, which survive on no more than the host resources present at the time of oviposition (Askew and Shaw, 1986; Haeselbarth, 1979). A more useful definition in terms of the physiological, especially endocrine, relationships that parasitoids share with their hosts is provided by Lawrence (1986a), who describes those species which do not manipulate the levels of hormone in the host, but make use of host hormones to regulate their own development and ensure that it is synchronized with that of the host, as 'physiological conformers'. By contrast, 'physiological regulators' actively manipulate the endocrine system of their hosts. According to the latter definition, this review is primarily concerned with those ectoparasitoids that are physiological regulators. However, for the purposes of this review, we make an additional important distinction between those ectoparasitoids that completely and permanently paralyse their hosts during, or just prior to, the act of oviposition, and those that do not cause paralysis, or paralyse only temporarily.

In terms of the various challenges associated with ectoparasitism described in the previous section, complete, permanent paralysis offers an immediate solution to the risk of the ectoparasitoid being physically attacked by the host, either as an adult or as ectoparasitoid progeny. It also has the indirect effect of modifying the endocrine system of the host, thus alleviating many of the dangers to the parasitoid associated with host moulting and/or metamorphosis. Many paralysing ectoparasitoid species (particularly *Bracon* spp.) have profound effects on the developmental biology of their hosts, but these effects could be viewed as secondary 'side-effects' of the total immobilization of the host, caused by the general effects of wasp venom on neuromuscular transmission (Beard, 1978; Coudron, 1991; Fukushima *et al.*, 1990; Piek, 1990; Piek *et al.*, 1974; Piek and Spanjer, 1978, 1986; Quistad *et al.*, 1994; Slavnova *et al.*, 1989; Spanjer *et al.*, 1977; van Marle, 1977; Visser *et al.*, 1976, 1983; Walther and Reinecke, 1983). Although parasitoids that do not inflict permanent paralysis may partially address the perils encountered during the act of oviposition, they must have some other means of regulating the endocrinology of their hosts if they are to counteract the subsequent dangers posed by host hormones and host development. This review therefore makes a distinction between those ectoparasitoids that permanently paralyse their hosts, and hence disrupt the host endocrine system, and those that do not paralyse their hosts, but subsequently disrupt host endocrine systems in a more precise and specific way. The following sections discuss only those ectoparasitoids of the second type, which, for whatever reasons, require their hosts to remain unparalysed after parasitism. In order to survive, these ectoparasitoids engage in highly specific and often very complex interactions with the endocrine system of their hosts, in order to render the host more compatible with the developmental needs of the parasitoid.

4. Case studies of the effects of ectoparasitism on host endocrinology

Examples of wasp–host associations in which the host is no more than temporarily paralysed, and yet development of the host is subsequently disrupted, can be found within several different families of ectoparasitic Hymenoptera which, in turn, use a variety of different arthropod hosts. The Dryinidae characteristically paralyse their leafhopper hosts for only a few minutes after oviposition, but although parasitized hoppers remain active during much of the subsequent development of the wasp larvae, they do not moult (Askew, 1971). Pre-pupae of the halictid bee, *Lasioglossum zephyrum*, which have been stung by the mutilid *Myrmosula parvula* or *Pseudomethoca frigida*, continue to wriggle in response to mechanical stimulation, but do not moult to the pupal stage. Pupal development of the host continues within the last, unshed last larval cuticle (Brothers, 1972, 1978). Spiders that have been attacked by the members of the ichneumonid genus, *Polysphincta*, remain active, but fail to moult (Askew, 1971; Clausen, 1940; Cushman, 1926). Larvae of *Aphidius smithi* that have been stung by the ectohyperparasitoid, *Dendrocerus carpenteri*, are not paralysed at the time of parasitism and void a meconium before entering a greatly prolonged pre-pupal stage (Bocchino and Sullivan, 1981). The non-paralysing venom of the pteronmalid *Catolaccus grandis* reduces growth and arrests the development of host boll weevil larvae (Morales-Ramos and Cate, 1992; Morales-Ramos *et al.*, 1995). Stings inflicted by the solitary ectoparasitic eulophid *Dimmockia javana* inhibit larval development of the coconut leafminer, but do not prevent pre-pupal hosts from pupating (Taylor, 1937). Perhaps the developmental arrestant of *D. javana* only affects certain components of the host larval endocrine system, and is therefore no longer effective once the host is committed to pupation. Larvae of *Oenophthira pilleriana* and *Nacoleia octasema* quickly recover from the stings inflicted by the bethylid *Goniozus* spp., but subsequent moulting is inhibited and they only continue their normal activities until the hatching of the parasite eggs, whereupon they spin a scanty cocoon (Taylor, 1933; Voukassovitch, 1925). The gregarious pteromalid ectoparasitoid *Nasonia vitripennis* attacks pupae and pharate adults of several families of Diptera, injecting a non-paralytic venom at the time of oviposition (Beard, 1964; Darling and Werren, 1990; Dawei and Dingxi, 1987). *Sarcophaga bullata* hosts that are still pupae inside the puparium when first stung by *N. vitripennis* retain their pupa-like morphology, and never initiate pharate adult development (Rivers *et al.*, 1993; Rivers and Denlinger, 1994a). While this could be a consequence of various wasp-induced alterations to host metabolic reserves (Rivers and Denlinger, 1994b, 1995a; Yoder *et al.*, 1996), diapause-like arrest could reflect a parasitoid-induced deficiency in host ecdysteroids (Denlinger, 1981; Rivers and Denlinger, 1994a).

Although such examples provide tantalizing indications of endocrine regulation on the part of ectoparasitoids, the physiological bases for these effects have only been studied in any detail in an extremely narrow range of insects: the vast majority of accounts of host-regulation by non-paralysing ectoparasitoids are provided by the Eulophidae. These accounts are collected from comparatively few species, from still fewer genera. In almost all cases, these ectoparasitoids utilize penultimate or final instar lepidopteran larvae as their hosts. The critical endocrine events that normally take place during these developmental stadia in healthy, non-parasitized caterpillars have been reviewed by many authors (Bollenbacher and Gilbert, 1981; Gilbert *et al.*, 1980; Riddiford, 1980; see also Chapter 1, this volume), and can be summarized as follows: the immature feeding stages of lepidopteran development, including the penultimate instar, are punctuated by a series of moults, which take place in response to the release of ecdysteroid by the prothoracic glands (PG). When levels of juvenile hormone (JH) are high, the presence of a pre-moult peak of ecdysteroid in its biologically active form (20E) results in larval-larval ecdysis. During their final larval instar, caterpillars experience a rapid and marked decline in JH titre. This, in part, reflects a reduction in JH synthesis by the corpora allata, and also a rise in the levels of JHE in the haemolymph. In the absence of JH, PTTH is released by the brain, providing the endocrine milieu necessary to initiate metamorphosis. At this point, larvae typically cease feeding, void their gut contents, and commence 'wandering' in search of a pupation site (Truman and Dominick, 1983). Prothoracicotropic hormone stimulates the PG to release the ecdysteroid precursor for 20E, thus committing the insect to a pupal moult (Bollenbacher and Granger, 1985; Smith, 1993). For more detailed accounts of the interplay between the components of the lepidopteran endocrine system during specific developmental events, the reader is referred to Chapter 1. Those ectoparasitoids that act to disrupt their hosts during the latter stages of their larval development can achieve this by interfering with the synthesis, release, accumulation, regulation, or sites of action of ecdysone, 20E, JH, JHE, and/or PTTH.

Colpoclypeus florus is a gregarious eulophid ectoparasitoid of various lepidopteran larvae (Bouèek and Askew, 1968). Immediately prior to oviposition, the female wasp repeatedly stabs her ovipositor directly into the head capsule of the host. The stung host subsequently displays a brief period of trembling, but remains unparalysed, and rapidly reassumes its normal repertoire of behaviour (van Veen and van Wijk, 1985, 1987). When penultimate (4th)-instar *Adoxophyes orana* caterpillars are attacked within 48 h of their previous moult, their development is invariably arrested and they never enter ecdysis. By contrast, when 4th-instar hosts are attacked later within the same stadium (48–66 h post-moult) they do not always show complete developmental arrest: some initiate a moult, but subsequently fail to complete ecdysis; others, although they have been 'stung', moult normally. Similarly, only those 5th-instar larvae that are attacked during the first few days of their final stadium are prevented from undergoing pupation (van Veen and van Wijk, 1985, 1987). The wasp-induced disruptions to host endocrinology, which surely underlie these symptoms, have not been investigated. However, the fact that all stabbing activity is targeted solely at the head region of the host could reflect the possibility that the venom of *C. florus* acts directly on the host's brain to prevent the release of PTTH, thus removing the signal for ecdysone release by the PGs. This explanation is consistent with the observation that *C. florus* is only effective in halting ecdysis when initial attack takes place early in a stadium, perhaps before PTTH is normally released to induce the pre-moult surge of ecdysteroid. Stings inflicted after the host has already released PTTH may be too late to prevent ecdysteroid production, and thus ecdysis is able to take place.

Another Eulophid genus, *Euplectrus,* includes some well-studied examples of species that inhibit the larval development of their lepidopteran hosts. Noble (1936, 1938) noted that caterpillars that were attacked by *Euplectrus agaristae* rapidly became exhausted as they struggled to evade parasitism. Although these hosts soon recovered, they never moulted, indicating the operation of a non-paralysing developmental arrestant. Observations of a very similar nature have been made for *Euplectrus bicolor* (Bischoff, 1929; Thomsen, 1927; Zimin, 1930), *Euplectrus kurandaensis* (Temperley, 1930), *Euplectrus parvulus* (= *plecopterae*) (Chatterjee, 1945), *Euplectrus* species near *laphygmae* (Neser, 1973), *Euplectrus laphygmae* (Gerling and Limon, 1976), *Euplectrus kuwanae* (Uematsu, 1981a,b, 1986; Uematsu and Sakanoshita, 1987), *Euplectrus ronnai* (Yamamoto *et al.*, 1998) and *Euplectrus melanocephalus* (Jones and Sands, 1999). However, host-regulation has only been closely studied in just two *Euplectrus* species: *Euplectrus plathypenae* and *Euplectrus comstocki* (Coudron *et al.*, 1990, 1994, 1997; Coudron and Brandt, 1996; Coudron and Kelly, 1985, 1990).

Euplectrus plathypenae is a gregarious ectoparasitoid of a variety of lepidopteran larvae (Bottrell, 1968; Coudron and Puttler, 1988; Muesebeck *et al.*, 1951; Wall and Berberet, 1974). The female wasp stings the host prior to oviposition, but the venom only inactivates the host temporarily (Melin, 1978). Early observers noted that although parasitized hosts subsequently regain mobility, attack by *E. plathypenae* does result in their arrested development (Smith, 1927; Wilson, 1933), and more recent investigations have considered the endocrine basis for this effect (Coudron *et al.*, 1990; Coudron and Kelly, 1985; Kelly and Coudron, 1990; Vinson, 1990). While unparasitized *Trichoplusia ni* show an increase in total ecdysteroid titre in the haemolymph 24 h after entering their penultimate larval instar, parasitized caterpillars do not show any equivalent rise in their ecdysteroid titre until approximately 5 days later (Kelly and Coudron, 1990). The peak of ecdysteroid recorded in parasitized hosts coincides with (but is not thought to be caused by) the onset of parasitoid feeding, and is much greater than the pre-moult ecdysteroid surge that occurs in non-parasitized caterpillars. Moreover, the ecdysteroid measured by radioimmunoassay (RIA) in parasitized individuals is qualitatively different from that found in unparasitized caterpillars. Parasitized individuals have one major RIA-positive ecdysteroid in their haemolymph, identified as 20,26-dihydroxyecdysone (20,26E), while unparasitized caterpillars apparently contain both 20E and 20,26E (Kelly and Coudron, 1990). These authors initially speculated as to whether 20,26E, which seems to comprise more than 80% of the total ecdysteroid measured in the haemolymph of parasitized insects, might be necessary for the success of parasitism, or whether it is present as a catabolite of 20E. However, subsequent data from the same research team suggests that the major immunoreactive steroid of late-stage parasitized larvae may, in fact, be a glucose conjugate of 20E (Gelman, unpublished data; Kelly and Coudron, 1996). This would imply that the moult-arrestant produced by *E. plathypenae* operates by conjugating sugar(s) to 20E, thus rendering the hormone biologically inactive. Coudron *et al.* (1990) found that, although the injection of healthy *T. ni* larvae with 20E caused them to moult, injection of parasitized caterpillars did not initiate either apolysis (separation of the epidermis from the new cuticle) or ecdysis. Together these observations could indicate that it is not only the absence (or conjugation) of 20E that prevents moulting in *T. ni*, but also a failure of the parasitized host larvae to respond to the effects of moulting hormone. This would suggest a double-edged approach to the prevention of host moulting by *E. plathypenae*. Not only is moulting hormone absent or deactivated, but the capacity of target tissues to respond to the hormone is also inhibited.

The gregarious ectoparasitoid *E. comstockii* (Ferrière, 1941) is not easily distinguished from its close relative, *E. plathypenae*, other than by the pattern of sensory pores on the antennal scape of the adult male wasp, by comparison of particular allozyme loci, or by analysis the cuticular hydrocarbons of each species (Coudron *et al.*, 1993). There are some similarities between *E. plathypenae* and *E. comstockii* in respect to the endocrine relations that exist between these parasitoids and their hosts. *Euplectrus comstockii*, like *E. plathypenae*, injects a non-paralysing venom into each host during a stinging process that precedes oviposition (Coudron *et al.*, 1993; Coudron and Brandt, 1996). Utilizing larvae of the European corn borer, *Ostrinia nubilalis* (an atypical host), as their model, Gelman *et al.* (1997) showed that the venom of *E. comstockii* prevents larval-larval ecdysis, or inhibits pupation, depending upon the developmental stage of the host larva at the time of treatment. Envenomed larvae in their 3rd or 4th instar show apolysis, but subsequently fail to moult, whereas envenomed 5th instars fail to undergo apolysis or subsequent ecdysis (Gelman *et al.*, 1997). Envenomed penultimate-instar larvae do produce a slightly delayed, small peak of ecdysteroid, but this is not comparable, either in its timing or its size, to the late, very marked surge in ecdysteroid titre measured in the *E. plathypenae–T. ni* association (Kelly and Coudron, 1990). Haemolymph ecdysteroid titres measured in envenomed 5th-instar *O. nubilalis* are reduced to a level lower than that normally required to allow moulting to proceed in this host (Gelman and Woods, 1983). Further analysis revealed that the ecdysteroid peak in 4th-instar hosts that have been envenomated by *E. comstockii* contains 20E, but only at very low levels (Gelman *et al.*, 1997).

These observations suggest that it may not be the conjugation or deactivation of 20E that prevents moulting in the *E. comstockii* system, but the lack of sufficient quantities of biologically active ecdysteroid at the critical stage in host development. Gelman *et al.* (1997) speculate that *E. comstockii* venom suppresses ecdysteroid titres either by inhibiting PTTH release or by preventing the production/release of ecdysteroid by the PG (or both), possibly by acting on the pathways that regulate PG activity (Okajima and Watanabe, 1989; Richter and Gersch, 1983). The fact that injection of 20E into envenomed 5th-instar larvae stimulates apolysis shows that that *E. comstockii* does not cause *O. nubilalis* to become completely refractory to the effects of ecdysteroid. However, injection of 20E into envenomed 5th-instar larvae does not result in the formation of normal pupae. Interestingly, Gelman *et al.* (1997) note that removal of the old cuticle from 20E-injected, envenomed larvae revealed a newly formed cuticle that was intermediate in form between larval and pupal. Together, these observations suggest that adverse effects on host development caused by *E. comstockii* are not simply a result of suppressed levels of moulting hormone, but may also arise because the venom exerts some inhibitory effect on some other developmental regulators, such as JH, JHE, eclosion hormone or ecdysis triggering hormone.

In addition to regulating the endocrine control of ecdysis, both *E. plathypenae* and *E. comstockii* have profound effects on the titres of storage proteins found in the haemolymph of their hosts. These proteins are also developmentally regulated (Coudron *et al.*, 1994, 1997). Storage proteins supply insects with essential peptides and amino acids, which allow them to complete the latter stages of their larval development, and to undergo metamorphosis (Wyatt and Pan, 1978). The fat body is the primary site of synthesis and secretion of these proteins in the immature, feeding stages of developing insects (Levenbook, 1985; Riddiford and Law, 1983). Storage proteins can accumulate to comprise as much as 80% of the total haemolymph protein content in final-instar larvae and their sequestration by the fat body normally occurs just prior to the larval-pupal moult. However, when hosts are parasitized in their early instars, large amounts of late larval storage proteins are produced prematurely (Coudron *et al.*, 1994, 1997). Electrophoretic analysis reveals that haemolymph samples collected from parasitized hosts contain multiple prominent bands corresponding to proteins in the region of 65–85 kDa, which are characteristic of the banding pattern found in non-parasitized larvae in their final stadium (Coudron *et al.*, 1994, 1998). Immunoblotting with specific antibodies confirmed the presence of one acidic and two basic storage proteins in early-stage parasitized *T. ni* (Coudron *et al.*, 1994).

Production of storage proteins in unparasitized lepidopteran larvae generally occurs when endogenous levels of JH are low (Tojo *et al.*, 1981), and is suppressed when caterpillars are exposed to JH or a JH analogue (Jones *et al.*, 1988a, 1993; Tojo *et al.*, 1985). Injection of non-parasitized caterpillars with 20E, or *in vitro* applications of 20E to fat body tissues, both result in a decline in storage protein mRNA transcription (Kumaran *et al.*, 1987; Ray *et al.*, 1987). Similarly, mRNA transcripts of storage proteins are low when endogenous levels of 20E are high (Ray *et al.*, 1987). Together these observations suggest that the production of storage proteins in lepidopteran larvae may be negatively regulated by the presence of ecdysteroids and/or JH. Although the effects of parasitism by *Euplectrus* spp. on host JH titres have yet to be examined, the wealth of evidence of a disruption to the levels of active ecdysteroids in parasitized hosts may indicate that a parasitoid-induced absence of 20E at a critical stage of host development could account for the precocious production of storage proteins.

As for the adaptive significance of a wasp-induced increase in storage protein titre in younger larvae, the most plausible explanation would seem to be that it enhances the nutritional value of comparatively small caterpillars which are unable to develop beyond the stage of development at the time of initial parasitism (Coudron *et al.*, 1994; see also Chapter 11, this volume). In the case of *E. comstockii*, the production of storage proteins by envenomed hosts coincides with the time at which parasitoid eggs would hatch and, thus, with the onset of feeding by parasitoid larvae. This synchrony certainly supports the concept that the presence of storage proteins may be critical to the nutritional requirements of feeding larvae. However, the fact that some storage proteins are naturally present in non-permissive hosts, and yet these larvae cannot sustain parasitoid growth, contradicts the idea that storage proteins are released in envenomed hosts for the benefit of the parasitoid (Coudron *et al.*, 1998).

In addition to being under endocrine regulation, the fat body is itself an important source of endocrine regulating substances. In particular, the fat body is the main site of JHE production in larval Lepidoptera (Hammock, 1985). Clearly, anything that alters the normal status of the fat body, such as premature production of storage proteins, may in turn affect the production and release of JHE. Juvenile hormone esterase regulates the levels of JH present in the host haemolymph, and the levels of JH in turn influence the production of storage proteins. This causal link makes it very difficult to ascertain whether the host hormones are disrupted as a result of conditioning of the host as a food resource, or whether they are disrupted in order to condition the host as a resource. When considered in the context of *Euplectrus* spp., it is possible to come up with the following sequence of events. The pre-moult peak of 20E in the host is either deactivated or delayed; this affects the fat body, enabling premature production of storage proteins, but perhaps also inhibiting JHE release. As a consequence, JH levels remain high, a late peak of ecdysteroid then occurs, which (in concert with a persistently high level of JH), together have the effect of preventing moulting, even if non-conjugated 20E is present or is artificially introduced. If host *T. ni*, parasitized by *E. plathypenae* already contain abnormally high levels of JH, then this could explain why they do not respond to the topical application of JH (Coudron *et al.*, 1990). However, under normal circumstances the combination of high JH and high 20E might be expected to have a juvenilizing effect, if anything, resulting in a supernumerary larval-larval moult. At present, it is impossible to know whether this highly speculative scenario has any bearing whatsoever on the true course of events in ectoparasitic associations, but it does underline how the nutritional status of host with respect to the ectoparasitoid cannot easily be separated from its endocrine status. For a more detailed discussion of the interactions of parasitoids of the genus *Euplectrus*, with those of their hosts, see Chapter 9.

Apart from *E. plathypenae* and *E. comstockii*, the only other non-paralysing eulophids that have been studied in the context of host endocrine regulation are members of the genus *Eulophus*. Shaw (1981a) provides a detailed record of the oviposition behaviour of the gregarious ectoparasitoid *Eulophus larvarum* on host larvae of the noctuid, *Orthosia stabilis*. He noted that this wasp always rejects hosts unless they are about to moult. Hosts are stung and then parasitized immediately after moulting takes place, the timing of host-attack being ‘exceptionally acute’ (Shaw, 1981b). Parasitized hosts subsequently resume feeding (at a reduced level), but do not moult again, although they may survive for many weeks after attack. The observation by Shaw (1981a) that attack by *E. larvarum* prevents apolysis led him to be one of the very first researchers to conclude that a non-paralysing venom produced by an ectoparasitic wasp disrupts physiological events normally under endocrine control. The gregarious ectoparasitoid *Eulophus pennicornis* displays similar oviposition behaviour, and has comparable effects on the developmental biology of its habitual host, the tomato moth, *Lacanobia oleracea.* Penultimate-instar host larvae feed voraciously, gain weight rapidly, and moult to their final (6th) stadium within 3 days of their previous moult. However, larvae that have been parasitized by *E. pennicornis* cease feeding, do not moult, and eventually die (Marris and Edwards, 1995; Marris *et al.*, 1996). Comparisons of ecdysteroid levels between healthy and parasitized *L. oleracea* larvae reveal marked differences in their respective patterns of ecdysteroid titre. Two days after entering their 5th (penultimate) stadium, unparasitized larvae exhibit a distinct peak in haemolymph ecdysteroid level, which is subsequently followed by a prolonged period where ecdysteroid levels are very much lower (Weaver *et al.*, 1997). By contrast, *L. oleracea* of the same age, which have been parasitized or envenomated by *E. pennicornis*, do not show any comparable rise in their ecdysteroid titres until much later, normally about 5 days after initial attack.

This scenario bears some obvious parallels with the *Euplectrus* systems, in so far as a required pre-moult ecdysteroid surge is delayed or absent. However, whereas the 'late' peak of ecdysteroid, detected in the haemolymph of *T. ni* parasitized by *E. plathypenae*, consists almost entirely of either 20,26-dihydroxyecdysone (Kelly and Coudron, 1990) or a glycosyl conjugate of 20E (Kelly and Coudron, 1996), reversed-phase liquid chromatography separation and RIA analysis of the equivalent peak in *L. oleracea* (5 days post-parasitism), revealed a pattern of RIA-detectable material (mainly 20E) that was strikingly similar to that recorded in unparasitized 5th-instar caterpillars at peak titre, which occurs some 3 days earlier at 48 h post-moult (Marris *et al.*, 2001). The reasons for this discrepancy between the two outwardly similar host–ectoparasitoid relationships are at present unknown. On the strength of available evidence, it seems that *L. oleracea*, which have been parasitized by *E. pennicornis*, continue to produce biologically active moulting hormone (20E), but that this appears in the haemolymph at the wrong time to initiate the moulting process (Marris *et al.*, 1999, 2001). The hypothesis (Marris *et al.*, 1999, 2001), however, that developmental arrest in *L. oleracea* is solely due to an absence of a critically-timed peak of ecdysteroid is not consistent with the finding that envenomed larvae, which are subsequently injected with 20E at 48 h post-moult still do not moult (R.J. Weaver, G.C. Marris, unpublished results). In this respect, *E. plathypenae* and *E. pennicornis* are alike, for both species disrupt the timing of host ecdysteroid release and also alter the ability of the host to respond to the moulting hormone. Thus, these two eulophids from different genera both appear to use a dual approach to the inhibition of host moulting.

The mechanism by which *E. pennicornis* causes an apparent delay in host ecdysteroid production has yet to be fully resolved (Marris *et al.*, 1999, 2001; Weaver *et al.*, 1997). There is currently no evidence that normal PTTH synthesis is affected in parasitized *L. oleracea*, but parasitism may be acting to inhibit PG function, possibly by damaging PG cells, or by interfering with their sensitivity to PTTH (Marris *et al.*, 1999, 2001). Prothoracic glands excised from hosts parasitized by *E. pennicornis*, at a point in time when unparasitized caterpillars are known to produce an ecdysteroid surge (48 h post-previous-moult), release no measurable level of ecdysteroid *in vitro*, and neither do they respond to the artificial stimulator forskolin (Marris *et al.*, 2001). In larval Lepidoptera, the PTTH signal transduction pathway in the PG induces an increase in the endocellular level of the second messenger cyclic AMP (cAMP), which in turn, plays a critical role in activating the biosynthetic pathway that culminates in the production and secretion of ecdysteroids from the PG (Gilbert *et al.*, 1988; Rybczynski and Gilbert, 1994). Forskolin mimics this action, since it is a stimulator of the enzyme adenylate cyclase, which synthesises cAMP from ATP. The fact that the PG obtained from *E. pennicornis*-parasitized caterpillars at 48 h post-moult do not respond to forskolin could indicate that cAMP production is prevented, perhaps through the inhibition of adenylate cyclase. Alternatively, forskolin-treated PG may produce cAMP, but ecdysteroid synthesis is then interrupted further downstream in the process. Prothoracic glands excised from *Heliothis virescens* hosts parasitized by the endoparasitoid *Cardiochiles nigriceps*, while still intact and viable, show a severely depressed biosynthetic activity, which cannot be stimulated by either the application of forskolin or PTTH (Pennacchio *et al.*, 1992a, 1994a, 1997; Tanaka and Vinson, 1991a). In this particular endoparasitoid–host system, the PG explanted from parasitized host larvae were still able to produce the cAMP as a normal response to PTTH. However, the downstream activation of the cAMP-dependent protein kinase did not appear to occur. Taken together, these findings point to a parasitoid-induced block of protein synthesis translation in host PG cells, possibly determined by the under-phosphorylation of key target regulatory proteins (Pennacchio *et al.*, 1997, 1998a). It is conceivable that *E. pennicornis* causes a similar disruption to the PTTH transduction signal pathway in the PG of *L. oleracea*, thus inhibiting ecdysteroid production. This hypothesis could be explored by using recognised techniques to examine any build-up of intracellular cAMP in response to PTTH or forskolin in the PG of parasitized larvae. However, the apparent insensitivity to forskolin implies of PG from *L. oleracea* that have been attacked by *E. pennicornis* to produce a required pre-moult surge of ecdysteroid, possibly reflects the insensitivity of the PG to PTTH, rather than any wasp-induced reduction in the availability of PTTH.

The fact that the number of viable PG cells is reduced in caterpillars which have been parasitized by *E. pennicornis* could suggest that cell-destruction alone prevents hormone release (Marris *et al.*, 2001). However, it seems improbable that the extent of tissue damage actually measured (<30% reduction in cell viability) is sufficient to account for the complete absence of ecdysteroid output by PG dissected from parasitized hosts at 48 h post-moult. Moreover, PGs removed from parasitized hosts at 120 h also show a reduction in cell viability, and yet produce greater quantities of ecdysteroid than intact PGs removed from unparasitized larvae of the same age (Marris *et al.*, 2001). In some other host–parasitoid systems, the host PGs tend to exhibit reduced biosynthetic activity, without obvious signs of gross morphological disruption (Pennacchio *et al.*, 1997, 1998a, b) and, although little is known about the regenerative capacity of PG tissues, they do change in size and ultrastructure during the normal course of insect development (Herman and Gilbert, 1966). One cannot, therefore, assume that tissue damage is permanent, or that it necessarily accounts for depressed ecdysteroid output in parasitized *L. oleracea*. At present, one can only speculate as to the way in which the venom of *E. pennicornis* acts to regulate ecdysteroid release. Tanaka and Vinson (1991a) showed that, when the PGs of *H. virescens* were incubated with both venom and calyx fluid from *Cardiochiles nigriceps*, the release of ecdysteroid was significantly reduced. Pennacchio *et al.* (1998b) elaborated on this observation, by reporting that the PDV of *C. nigriceps* was the major parasitoid-derived factor involved in host PG inactivation. They showed that the developmental stage of the host at the time of dissection was also critical in determining the sensitivity of PG to parasitoid-derived factors. Although this demonstrates that wasp maternal factors can act directly on PG tissues to depress their function, this does not appear to be the case with *E. pennicornis*, since neither cell viability nor ecdysteroid release are altered when *E. pennicornis* venom is applied directly to *L. oleracea* PG *in vitro* (Marris *et al.*, 2001). The available evidence suggests that attack by *E. pennicornis* inhibits moulting through the indirect effects of wasp venom on the PG of its larval host. This effect is not achieved, however, through the destruction of PG cells, but more probably reflects the interruption of an innate cycle in the PG of *L. oleracea*, such that at the critical point in insect growth, the PG are temporarily switched-off, and lose their ability to respond to their normal cue to produce an essential peak of ecdysteroid.

While these findings may account for why moulting does not take place at the expected point in time, they do not explain why parasitized or envenomed caterpillars do not moult later, at a time (120 h) when they experience the abnormally high levels of biologically active ecdysteroid. In other words, why does parasitism by *E. pennicornis* arrest the development of *L. oleracea* totally, rather than just delaying it? Just as ecdysteroid levels have been shown to be disrupted by endoparasitism, so have JH and the JHE (Balgopal *et al.*, 1996; Cusson *et al.*, 2000; Joiner *et al.*, 1973). It is therefore possible that other components of the endocrine system of *L. oleracea* may also be affected by *E. pennicornis,* such that the hormone milieu becomes incompatible with moulting, irrespective of the amount of ecdysteroid that is present. Preliminary analyses of haemolymph samples collected from *L. oleracea* indicate that JHE levels may be suppressed by *E. pennicornis* (G.C. Marris, unpublished results). However, it is unclear whether the observed absence of JHE in 5th-instar hosts is actually parasitoid-induced, or simply reflects the fact that parasitized caterpillars do not moult, and therefore might not be expected to show the fluctuations of JHE normally only recorded in the final stadium (M. Cusson, personal communication). This remains the subject of ongoing research.

Unlike in the *E. comstockii* and *E. plathypenae* systems, the effect of *E. pennicornis* on host storage protein profiles has not been formally investigated, but recently it has been shown that parasitism by *E. pennicornis* is associated with an initial increase in total plasma protein concentration of *L. oleracea,* and in the appearance of a parasitism-specific protein (PSP) (Richards and Edwards, 1999a). The PSP is absent in unparasitized 5th-instar *L. oleracea*, and in hosts which have been stressed by starvation, heat and cold shock, or infected with a nematode parasite (*Steinernema feltiae*). Although the PSP identified in the *E. pennicornis–L. oleracea* system, has an estimated molecular weight of 27 kDa, and is certainly smaller than typical storage proteins, electrophoretic analysis of haemolymph collected from *T. ni* parasitized by *E. comstockii* (Coudron *et al.*, 1994) appeared to show proteins of a similar size. However, current understanding is that the PSP found in parasitized *L. oleracea* is induced, or introduced, by feeding *E. pennicornis* larvae in order to disrupt host haemocytes, implying that it plays some role in the modification of the parasitized hosts cellular immune response system (Richards and Edwards, 1999a, 2001).

5. Sources of endocrine-regulatory substances in ectoparasitic relationships

The factors that are responsible for endocrine regulation in endoparasitic associations originate from a variety of sources (Vinson and Iwantsch, 1980a). These endoparasitoid host-regulatory substances include ‘venoms’ derived from various components of the reproductive system of the adult female wasp, as well as from the poison gland itself (references cited in Quicke, 1997), cells released from the endoparasitic eggs (teratocytes) (Grossniklaus-Bürgin and Lanzrein, 1990; Joiner *et al.*, 1973; Pennacchio *et al.*, 1992a, 1994b,c; Zhang and Dahlman, 1989, Zhang *et al.*, 1992), PDV produced in the calyx region of the ovaries (Dover *et al.*, 1988a; Pennacchio *et al.*, 1998a b; Tanaka and Vinson, 1991a), and secretions produced by feeding wasp larvae (Brown *et al.*, 1993; Iwantsch and Smilowitz, 1976; Schneider, 1950, 1951). By contrast, only some of these sources are known to be important in ectophagous associations.

The venom gland is by far the most widely documented source of endocrine-regulating factors in ectoparasite–host relationships. Many species of *Euplectrus* are recorded as either 'stinging' their larval hosts, 'piercing' their host's larval integument, and/or as injecting some 'fluid' immediately prior to oviposition, and it seems very likely that these wasps are all using some kind of venom to cause developmental arrest in larval lepidopteran hosts (Gerling and Limon, 1976; Neser, 1973; Noble, 1936, 1938; Yamamoto *et al.*, 1998). Coudron and Puttler (1988) were the first to identify the source of the developmental arrestant from *E. plathypenae* as originating from the lower reproductive tracts of the female wasp, and Vinson (1990) subsequently pinpointed its origin to the poison gland. *Euplectrus kuwanae* produces venom containing thin, rhomboid crystals, which is responsible for developmental arrest of *Argyrogramma albostriata* larvae (Uematsu and Sakanoshita, 1986, 1987). Shaw (1981) attributed the developmental arrest caused by *E. larvarum* to a non-paralysing venom introduced into caterpillars during a pre-ovipositional sting; and all of the endocrine disruptions attributed to parasitism by *E. pennicornis* can be reproduced by envenomating non-parasitized *L. oleracea* larvae (Marris, 1996, 1999, 2001; Weaver *et al.*, 1997). Envenomation by *N. vitripennis*, or artificial injection with its crude venom extract, not only arrests larval development, but also alters lipid metabolism of dipteran hosts in a similar way to parasitism (Rivers and Denlinger, 1995a). The venom of the solitary ectoparasitoid *Eupelmus orientalis*, causes paralysis and developmental arrest in host larvae of the beetle *Callosobruchus maculatus*, and is also vital to the suppression of conspecific parasitoid hosts when *E. orientalis* is involved in ectohyperparastic relationships (Doury and Rojas-Rousse, 1994; Doury *et al.*, 1995, 1997). Some authors have traditionally regarded the Dufour's gland, in conjunction with the poison gland, as comprising an integral part of the parasitoid venom apparatus (Van Marle and Piek, 1986), but there is little or no evidence that the secretions of the Dufour's gland play any role in endocrine regulation by ectoparasitoid species. There was an initial suggestion that Dufour's gland secretions might form some component of *E. orientalis* venom, but it has since been shown conclusively that secretions from the Dufour's gland alone do not have any significant effect on host development (Doury *et al.*, 1995). Similarly, Vinson (1990) was unable to induce developmental arrest in *H. virescens* hosts injected with the contents of the Dufour's gland of *E. plathypenae*.

Since teratocytes are cells derived from the embryonic membrane of the endoparasitic egg, and are released into the host haemolymph upon hatching, it is difficult to see how they could be of relevance to the ectoparasitic habit. This does not mean, however, that eggs of ectoparasitoids play no role in host-regulation. The immature stages of *E.* nr. *laphygmae* do not appear to be responsible for the general developmental arrest observed in parasitized *P. acuta*, but the presence of egg clusters on the host integument apparently exerts some sort of local effect on the process of moulting (Neser, 1973). It was observed that, on those rare occasions when the female parasitoid laid eggs onto a 4th-instar caterpillar after host moulting had commenced, setae appeared on the new integument and developed normally everywhere except in the area where oviposition had taken place. In another example, the eggs laid by the solitary ectoparasitoid, *Dinarmus basalis*, are the source of an oviposition deterrent marker. The parasitized host acquires this marker following the passage of 'molecules' from the wasp egg to the host (Gauthier and Monge, 1999). Although there is no suggestion that this material is important in the regulation of host endocrinology, it does demonstrate that substances released by eggs can, in certain cases, influence the biochemistry of the parasitized host.

Polydnaviruses and virus-like particles have only ever been recorded in two parasitioid families, the Braconidae and the Ichneumonidae, and never in any of the ectoparasitic species within these groups (see references cited in Quicke, 1997). To our knowledge, *E. pennicornis* is the only ectoparasitic species from the family Eulophidae to have been screened for the presence of PDV. In preliminary experiments, we were unable to detect the presence of PDV in extracts of the reproductive system of female wasps (G.C. Marris, unpublished results). Cells within the calyx region of *N. vitripennis* have been shown to contain particles consisting of a protein-carbohydrate complex. These particles are discharged as secretory droplets during oviposition, but these are not thought to be important to the survival of the parasitoid egg (Rafai and King, 1972). Some ectoparasitoids may, however, employ other ovarian products in host-regulation. The ectohyperparasitoid *D. carpenteri* applies a substance to each host at the time of oviposition. This substance inhibits sclerotization and prevents the completion of host pupation, thus facilitating the feeding and development of wasp larvae (Bocchino and Sullivan, 1981; Höller *et al.*, 1994). The secretion is clearly not a venom in the true sense, since the female reproductive system of *D. carpenteri* does not include a poison gland (Höller *et al.*, 1993a). The source of the inhibitory substance appears to be the ovaries of the female wasp. Höller *et al.* (1994) provided evidence that the host-regulatory substance produced by *D. carpenteri* is JH. Comparatively high levels of JH III were measured in tissue samples (especially ovaries) dissected from *D. carpenteri* females. It was also observed that the arrested development in wasp-exposed late-instar or pre-pupal hosts could be mimicked by application of synthetic JH III. Moreover, the oviposition-deterrent property of tissue extracts from *D. carpenteri* could be inhibited by exposure to JH esterase. These findings led Höller *et al.* (1994) to speculate that other species of parasitic wasps, i.e. those that regulate the endocrinology of their hosts, may also secrete JH (as a host-regulatory substance).

It has been known for a long time that ectoparasitic larvae secrete potent chemicals. When more than one egg of the solitary ectoparasitic chalcid *Elachertus* sp. are deposited onto a single host caterpillar, the hatching of one causes 'the immediate cessation of development of the remainder', suggesting that the oldest wasp larva produces a toxin which poisons its conspecifics (Clausen, 1940). More recent work has shown that early instar larvae of the eulophid, *Melittobia digitata*, the larvae of which do not fully penetrate the cuticle of their host, may secrete a toxin or an enzyme that opens pore canals in the integument of the host, thus facilitating the passage of material from the haemolymph (S.B.Vinson, personal communication). Current evidence also suggests that the PSP found in *L. oleracea* parasitized by *E. pennicornis* (Richards and Edwards, 1999a) is produced by the feeding *E. pennicornis* larvae (Richards and Edwards, 2001). In another example, the extraordinary behavioural modifications observed in the orb weaving spider *Plesiometa argyra* (see Section 7) have been attributed to a chemical secretion produced late by larvae of the solitary ectoparasitoid *Hymenoepimecis* sp. (Eberhard, 2000). In spite of such clear instances of ectoparasitoid larvae acting as the source of biologically active substances, there are still comparatively few documented cases of ectoparasitoid larvae playing a key role in host regulation. Indeed, in some host–parasitoid relationships it has been demonstrated that the larva is not capable of exerting any influence on host endocrinology. Transplantation of ectohyperparasitic larvae of *Asaphes lucens* and *D. carpenteri*, onto previously non-parasitized larvae of *Aphidius smithi* had no adverse effect on host development (Bocchino and Sullivan, 1981). However, the view that ectoparasitoid larvae are not generally sources of host-regulatory substances almost certainly reflects a lack of research in this area, rather than the true state of affairs.

Adult females of the pteromalid, *C. grandis,* do not have the capacity to inject a full dose of venom into all boll weevil larvae that they encounter, especially as the wasps get older, but even though the majority of hosts are not paralysed by *C. grandis*, envenomation is not necessary for the survival of larval parasitoids (Morales-Ramos *et al.*, 1995). It has been demonstrated that 1st-instar *C. grandis* larvae can survive on non-venomated, non-paralysed hosts. The saliva of the parasitoid larvae appears to cause changes in the haemolymph amino acid composition of the host, which are similar to those caused by the venom of adult *C. grandis*, but it is not known whether the same chemicals are involved (Morales-Ramos *et al.*, 1995). The authors suggest that the saliva produced by *C. grandis* larvae regulates and redirects host metabolism, thus increasing the nutritional value of the host to the benefit of the developing parasitoid offspring. The developmental arrest observed in bruchid beetle *Callosobruchus maculatus* parasitized by *E. orientalis* is not only attributed to the venom injected into hosts by adult female wasps, but also to the feeding activity of 1st-instar wasp larvae (Doury and Rojas-Rousse, 1994; Doury *et al.*, 1995, 1997). Newly-hatched *E. orientalis* larvae merely have to bite their hosts for five minutes to induce delayed paralysis accompanied by developmental arrest, and injection of the saliva of *E. orientalis* larvae into *C. maculatus* has the same effects on the host (Doury *et al.*, 1995). In fact, after the parasitoid eggs have hatched, the sting of the adult female wasp is not necessary to allow the development of the parasitoid offspring, or to induce the observed perturbations to mobility and development (Doury *et al.*, 1995).

Although Neser (1973) ruled out the role of *E.* nr. *laphygmae* larvae in causing developmental arrest in host caterpillars of *Plusia acuta*, it was found that they were responsible for the observed cessation in the feeding activity of the host. This effect was not, however, attributed to any substance secreted by larvae (although this was not investigated), but to the cumulative debilitating effect of parasitism on the general health of the host. Kelly and Coudron (1990) made the interesting observation that the beginning of a marked surge in ecdysteroid titre in parasitized *T. ni* caterpillars coincided with onset of *E. plathypenae* feeding. Although a similar ecdysteroid surge was also recorded in stung hosts from which parasitoid eggs or larvae were removed, it occurred much later, indicating that the parasitoid larvae may themselves have been stimulating the synthesis of host ecdysteroids. It is unclear, however, whether this effect in fully parasitized hosts is simply the response of the host to the mechanical action of feeding, or to some substance introduced into the host from the ectoparasitoid larvae. Kelly and Coudron (1990) suggest that the causal relationship between parasite feeding and increased host haemolymph ecdysteroid titre warrants further research. Premature production of storage proteins in hosts parasitized by *E. comstockii* has been shown to coincide with the onset of parasitoid larvae feeding. This coincidence of timing may imply that the trigger for storage protein release derives from parasitoid larvae themselves. However, experiments carried out using parasitized hosts from which parasitoid eggs were subsequently destroyed, revealed that envenomation is clearly the major factor causing precocious production of storage protein in permissive hosts (Coudron *et al.*, 1994). It is interesting to note that, whereas *E. comstockii* arrests larval-larval moulting both in permissive and in non-permissive hosts, marked alterations in the production of putative storage proteins occurs only in the permissive hosts (Coudron *et al.*, 1998). Also, since eggs fail to hatch in non-permissive hosts, this suggests that although envenomation consistently disrupts moulting, the disruption to storage protein production may be at least partly attributable to the feeding parasitoid larvae.

Aside from the various secretions that adult and larval ectoparasitoids use to regulate their hosts, the physical damage inflicted on the host during attack by the wasp may also play a part in disrupting the endocrinology of the host. Van Veen and van Wijk (1985, 1987) observed that the eulophid *C. florus* always stabs the host larvae of *A. orana* directly into the head capsule. Such parasitized larvae do not moult (see Section 2). The ovipositor of *C. florus* is uniquely 'armed' with two rigid horn-like projections situated on either side of its pointed tip. These sharp barbs penetrate the head capsule region of the host, suggesting that mechanical damage caused to the cerebral ganglia may be responsible for preventing ecdysis (van Veen and van Wijk, 1985). This theory is plausible, since damage caused to the brain, which includes the site of PTTH production, might prevent activation of the PG, and thus the release of ecdysteroid.

6. The chemistry and specificity of non-paralysing venoms

A considerable volume of literature exists on the neurophysiological effects of the venoms of aculeate Hymenoptera, but there is much less information on the effects of venom and oviposition substances of the parasitoid wasps. The most thoroughly investigated venom of any of the ectoparasitoid wasps is that of *Habrobracon* (=*Bracon* or *Microbracon*) *hebetor*, which produces a paralysing venom (Beard, 1952, 1978; Quistad *et al.*, 1994; Visser *et al.*, 1976, 1983). By contrast, research into the chemistry on non-paralysing venoms is almost entirely restricted to two eulophid species, *E. comstockii* (Coudron and Brandt, 1996) and *E. pennicornis* (Weaver *et al.*, 1997). The reader is referred to Chapter 2 of this volume for a detailed review of the chemistry of parasitoid venoms.

The majority of paralytic venoms exert their effects on the neuromuscular systems of host insects. Because of the relative 'pharmacological conservancy' of these systems among different types of insects (Quicke, 1997), most paralysing venoms are non-specific. In general, ectoparasitoids that paralyse their hosts permanently are able to exploit a broad host-range. By contrast, some non-paralysing ectoparasitoids are comparatively host-specific, although others are able to exploit a variety of different host-types. This could imply a far greater range of specificity, or mode of action, of the host regulatory substances in these broader-range, non-paralysing ectoparasitoids. As an example of a species with a relatively narrow host range, *Eulophus pennicornis* has been recorded as completing its development on less than a dozen different host species (Bouèek, 1959; Shaw, 1987; Speyer, 1934; Thompson, 1955; Van de Veire, 1993). However, artificial injection with *E. pennicornis* venom has been shown to cause developmental arrest in several species of *Spodoptera*, which are not recorded as natural hosts (H.J. Matthews, personal communication). The narrow host-range shown normally by *E. pennicornis* may therefore reflect ecological constraints on the availability of certain host species rather than specificity of venom action, but to date the venom has only been shown to affect later stages of larval development in some Lepidoptera. By contrast, *E. plathypenae* is capable of completing its development on a wide range of Lepidoptera, including noctuids, sphingids and geometrids (Coudron and Puttler, 1988; Krombein *et al.*, 1979; Wall and Berberet, 1974). Coudron and

Puttler (1988) found that the venom of *E. plathypenae* caused developmental arrest in no less than 44 different lepidopteran species that were either subjected to parasitism or to injection with extracts of the reproductive system of the wasp. In addition, they found that venom from *E. plathypenae* affected moulting in certain orthopteran, coleopteran, and dipteran species. However, since comparatively few of the non-lepidopteran species tested showed any symptoms of disrupted development, it seems that *E. plathypenae* venom, like that of *E. pennicornis*, may be more or less specific for lepidopteran larvae. Up to 15 different lepidopteran species have been reported as natural hosts for *E. comstockii* (Bottrell, 1968; Muesebeck *et al.*, 1951; Peck, 1963; Wall and Berberet, 1974). Parasitism (i.e. envenomation) by *E. comstockii* arrests the development of both permissive and non-permissive hosts, and reduces their rates of growth, but alterations to putative storage proteins only occur in natural hosts (Coudron *et al.*, 1998). This indicates that the capacity of venom to regulate moulting may be independent of any effect on host protein titre. It also suggests that some components of the venom may be more host-specific than others (Coudron *et al.*, 1998). The effects of *E. comstockii* venom on host development are also age-dependent. Envenomation of different instars of the unnatural host *O. nubilalis* always inhibits moulting, but while 3rd and 4th instars undergo apolysis and produce new cuticle, 5th instars fail to do so (Gelman *et al.*, 1997). The ectohyperparasitoid, *D. carpenteri,* successfully utilizes a very broad range of host insects and stages (Chow and Mackauer, 1999; Höller *et al.*, 1993b; Mertins, 1985; Sullivan, 1988; Takada, 1973), including the later larval stages of its own species (Spencer, 1926). The most critical host-quality in determining acceptance by *D. carpenteri* is host developmental stage, rather than host species (Bennett and Sullivan, 1978; Chow and Mackauer, 1999). The fact that *D. carpenteri* favours mature larvae, pre-pupae and non-sclerotized pupae of aphidiine wasps suggests that its host-regulatory substance (which inhibits sclerotization) may be relatively non-specific in its action, and is probably developmental-stage-specific rather than species-specific (Chow and Mackauer, 1999).

7. Paralysis versus non-paralysis as a parasitoid strategy

Non-specific paralysing venoms enable ectoparasitoid wasps to exploit a very broad host range, compared to non-paralysing species. Paralytic venoms that permanently subdue the host effectively prevent any recurrent danger of deliberate attack by the host and, generally speaking, they override the need for complicated endocrine regulation to protect the parasitoid offspring from the potentially catastrophic consequences of host moulting or metamorphosis. Given that the benefits of complete and permanent paralysis seem so obvious, why do not all ectoparasitoids paralyse their hosts?

First, a non-paralysed (i.e. mobile) host can continue to seek refuge, or defend itself from predators, thus affording indirect protection to the developing ectoparasitoid larvae. For example, the last larval stage of the endoparasitoid *Dinocampus* (= *Perilitus*) *coccinellae* emerges from its ladybird host, *Hippodamia* sp., and spins a silken cocoon underneath the body of the host, with the legs of the ladybird 'entangled in the outer strands' of the parasitoid cocoon (Clausen, 1940). The ladybird, which is still alive and unparalysed, then actively guards the parasitoid by standing over it and, if disturbed, rearing up (G.C. Marris, unpublished results). If parasitized ladybirds are freed from the threads of the wasp's cocoon, they rush about wildly, apparently in search of the cocoon and, if found, they attempt to re-entangle their legs in the web of silk (Davis, 1928). An equally dramatic example of such exploitation of a non-paralysed host is provided by the solitary ectoparasitoid, *Hymenoepimecis* sp., which attacks spiders of the genus *Plesiometa* (Eberhard, 2000). The parasitized spider is stung at the time of oviposition, but soon resumes it's activities, spinning normal orb-webs to capture prey. However, on the night that the wasp larva finally kills the host (two weeks after initial attack), the spider spins an atypical web from which the parasitoid suspends its pupal cocoon (Eberhard, 2000). This 'cocoon' web lacks the sticky spirals and radial lines of silk seen in orb webs, and is much more compact, durable and resistant to heavy rains (Eberhard, 1990; Fincke *et al.*, 1990). There is, however, no similar evidence that either *E. pennicornis* or *Euplectrus* spp. is able to modify the behaviour of their parasitized hosts to a similar extent. When parasitized by *E. plathypenae,* the host *T. ni* will continue to respond to physical disturbances by looping, or curling its body (Coudron *et al.*, 1990). This could by interpreted as defensive behaviour, but in the case of *E. pennicornis*, the host larvae (*L. oleracea*), although not paralysed, rapidly become debilitated and sluggish, and do not actively seek refuge after parasitism (G.C. Marris, unpublished results).

A second potential advantage of using non-paralysing venom is that the host will be able to continue feeding. From the point of view of the ectoparasitoid, this means that the host will continue to improve as a food resource beyond the time of parasitism (see Section 3, definition of koinobionts). There is good evidence that non-paralysing ectoparasitoids manipulate hosts to ensure that the parasitoid larvae obtain essential nutrients. Thus, premature production of storage proteins following parasitism by *E. comstockii* is timed to coincide with the onset of the hatching of parasitoid eggs (Coudron *et al.*, 1994, 1997). This temporal enhancement of nutritional resource, however, might equally well arise in paralysed hosts, and there is little to suggest that insects attacked by non-paralysing ectoparasitoids actually do feed and grow significantly after they have been parasitized. For example, *L. oleracea* that have been parasitized by *E. pennicornis* continue to feed for less than 24 h after initial parasitism, and cease to gain any weight from then on (Marris and Edwards, 1995). Parasitism of hosts by *E.* nr *laphygmae* (Neser, 1973) or *E. comstockii* (Coudron *et al.*, 1998) also stops caterpillars from feeding within 48 h.

A third potential advantage offered by a non-paralysed host is that the host endocrine system may be harnessed by the wasp to provide hormonal substances, which could be of direct benefit to the developing ectoparasitoid offspring. This challenges the popular view that ectoparasitoids only manipulate their hosts' endocrine systems in order to alleviate the potential threats posed by physiological incompatibility. Certainly, in some (non-hymenopteran) ectoparasitic associations, ingestion of the host's hormones may be necessary to ensure reproduction. For example, yolk deposition in the *Varroa* mite, an ectoparasite of the honeybee, appears to be triggered by the uptake of JH from the host's haemolymph (Hänel and Koeniger, 1986). Kelly and Coudron (1990) found that the temporal pattern of appearance of ecdysteroids in the ectoparasitic, non-paralysing hymenopteran, *E. plathypenae,* closely followed that of its host *T. ni*. This led them to suggest that mechanisms may exist for ectoparasitoid regulation of host haemolymph ecdysteroids for use by the developing parasitoid larvae. Perhaps the large peak of 20E that occurs in caterpillars during their final stages of parasitism by *E. pennicornis* is not a delayed peak, in the sense that it should have occurred 3 days earlier, but is instead a 'new' peak of ecdysteroid, which may be induced by the ectoparasitoid for its own purposes. In order to cast light on this interpretation of the endocrine interactions between ectoparasitoids and their hosts, further work is needed. Apart from measuring ecdysteroid (and other hormone) profiles in ectoparasitoid larvae at set time intervals during host development, it would also be useful to examine any temporal relationships that exist between significant developmental events in parasitoid larvae and disrupted endocrine events in their host.

8. Conclusions

The study of endocrine interactions between ectoparasitoids and their hosts is at least 20 years old. While it is not a very new area of research, it is very narrow. Only a few host–wasp systems have been examined in any detail and, if anything, these studies have only served to demonstrate how complex and varied each association may be. Parallels do exist between systems, but so do important differences, particularly in terms of the way in which ecdysteroids may be affected to disrupt moulting in the parasitized host. Although ecdysteroids play a key role in the control of normal insect development, this is done in concert with many other substances, including other hormones and regulatory enzymes. As research progresses, it will be interesting to see how these different components of the host endocrine system, particularly JH and JHE, may also be affected by ectoparasitism. Ectoparasitoid venoms are undoubtedly a very important source of host-regulatory substances, but their chemistry and the precise mechanisms by which they achieve their effects are still poorly understood. Furthermore, parasitoid larvae as originators of host endocrine disruption have been largely overlooked. At present, there is a tendency to interpret endocrine disruption on the part of ectoparasitic Hymenoptera as a means of protecting wasp offspring, rather than as the manifestation of a more dynamic relationship between host endocrinology and the physiological needs of the wasp. More research is needed to clarify the sources, mechanisms of action and the adaptive significance of their effects in a wider range of ectoparasitic associations.

Acknowledgements

The authors acknowledge the support of the Pesticides Safety Directorate of the Department for Environment, Food and Rural Affairs.

9

Ecological and physiological relevance of biochemical changes in a host as a result of parasitism by *Euplectrus* spp.: a case study*

Maureen Knop-Wright, Thomas A. Coudron and Sandra L. Brandt

1. Introduction

The ways in which hymenopteran parasitoids regulate the physiology and biochemistry of their hosts are diverse. Nearly every physiological system in the host can be affected by parasitism. Substances derived from the adult and immature forms of these parasitoids, including polydnaviruses (PDV), venoms, teratocytes, or parasitoid-derived or induced proteins, have been shown to accelerate, slow or arrest development, change immune responses, induce or inhibit synthesis of certain proteins and change the profile of substances in the biochemical milieu, including sugars, lipids, amino acids, peptides, hormones and biogenic amines (Beckage, 1985, 1993a; Beckage *et al.*, 1990; Coudron, 1991; Lawrence, 1990; Rivers *et al.*, 1998; Thompson, 1986, 1993; Vinson, 1990b; Vinson and Iwantsch, 1980a). Few, if any, of the mechanisms by which these parasitoid-derived substances alter the host are fully understood.

Yet, following this physiological and biochemical turbulence, the end point for every parasitoid is the same – a redirection of host resources to benefit the parasitoid's offspring through a few key physiological events which must occur in order for that host to support parasitoid development. Determination of these events is critical to

* Mention of a proprietary product in this paper does not constitute a recommendation for use by the US Department of Agriculture or indicate exclusion of other suitable products. All programmes of the US Department of Agriculture are offered on a non-discriminatory basis without regard to race, colour, national origin, religion, sex, age, marital status, or handicap.

understanding the physiological interaction between the parasitoid and the host. The hymenopteran parasitoids *Euplectrus plathypenae* and *Euplectrus comstockii* parasitize a number of species of lepidopteran larvae. Physiologically, the most obvious effects of parasitism are the failure of parasitized larvae to moult (Coudron and Puttler, 1988; Coudron *et al.*, 1990) and the premature appearance of storage proteins in haemolymph of the larval host (Coudron *et al.*, 1994). For *Euplectrus* spp., moulting is one of those critical physiological events that must be inhibited in the host to ensure survival of the parasitoid's offspring. In contrast, increased storage protein production is a physiological event that presumably must be enhanced in the host to ensure the parasitoid's survival. The need for increased storage protein production is not as evident as the need for arresting moulting. However, increased storage protein production may be a mechanism by which the parasitoid is able to access the amino acid and protein resources within the host, or it may be a mechanism by which the life of the host is extended in order to support the full development of the parasitoid (Coudron *et al.*, 1994, 1997).

The arrest of moulting and storage protein production are both normally regulated, at least in part, by the steroid hormone, 20-hydroxyecdysone (20E), and the sesquiterpenoid hormone, juvenile hormone (JH) (Gilbert *et al.*, 1996; Haunerland, 1996), leading to the possibility that *Euplectrus* venom inhibits moulting and induces storage protein production by redirecting normal endocrine regulation of the host's physiology. However, the relationship between *Euplectrus* and its hosts is far more complex, and provides an excellent framework for examining the ecological and physiological relevance of biochemical changes in the host brought about by parasitism.

2. Biology of *Euplectrus* spp.

All species of the genus *Euplectrus* known to date, are gregarious ectoparasitoids of lepidopteran larvae (Coudron and Puttler, 1988; Gerling and Limon, 1976; Jones and Sands, 1999), and *E. comstockii* and *E. plathypenae* are no exception. These two species belong to the Tribe Euplectrini in the Family Eulophidae (Krombein *et al.*, 1979) (*Table 1*). Many eulophids have been studied for their potential use as biological control agents against important pests (*Table 2*). Female *Euplectrus* usually lay between five and 12 eggs on the surface of the integument of lepidopteran larvae, generally preferring 2nd to penultimate stadia (Coudron *et al.*, 1990; Uematsu and Sakanoshita, 1987). The eggs of *E. comstockii* and *E. plathypenae* are different from the eggs of many other eulophid parasitoids in that they are attached to the host by a pedicel, which penetrates the surface of the host integument, making the parasitoid eggs difficult (if not impossible) for the host larvae to remove (Vinson, 1990b; T.A. Coudron, unpublished results).

The behavioural and physiological effects induced in permissive hosts in response to parasitization by *E. comstockii* and *E. plathypenae* are nearly identical (Coudron *et al.*, 1990, 1994, 1997; Coudron and Brandt, 1996; T.A. Coudron, unpublished results). Host larvae parasitized by *E. comstockii* and *E. plathypenae* are not paralysed, but continue to feed and respond to stimuli (Coudron *et al.*, 1990). Although arrestment of moulting (Coudron *et al.*, 1990, 1997; Coudron and Puttler, 1988) and the premature production of late larval storage proteins (Coudron *et al.*, 1994, 1997) are the predominant physiological effects observed in parasitized larvae, the growth rate of host larvae is also reduced (Coudron *et al.*, 1990, 1997). An extract of the venom

Table 1. Systematics of Eulophidae

Subfamily	Eulophinae		
Tribes	Eulophini	Euplectrini	Elachertini
Genera[a]	*Eulophus*	*Euplectrus*	*Hyssopus*
	Necremnus	*Euplectromorpha*	*Elachertus*
	Diglyphus		
Subfamily	Entedontinae		
Tribes	Entedontini	Tetrastichini	Euderini
Genera[a]	*Chrysocharis*	*Tetrastichus*	*Euderus*
	Pediobius	*Melittobia*	
	Entedon	*Winnemana*	
Subfamily	Elasminae		
Genera[a]	*Elasmus*		

[a]*Genera listed are only a sampling of those found within each subfamily and tribe as per Krombein et al. (eds), 1979 (Catalogue of Hymenoptera in America North of Mexico).*

Table 2. Examples of eulophid wasps with potential for biological control of insect pests

Parasitoid	Host	Reference
Euplectrus kuwanae	*Argyrogramma albostriata* (Noctuidae)	Uematsu and Sakanoshita, 1987
Euplectrus laphygmae	*Spodoptera littoralis* (Noctuidae)	Gerling and Limon, 1976
Euplectrus melanocephalus	*Eudocima* spp. (Noctuidae)	Jones and Sands, 1999
Euplectrus sp. nr. parvulus	*Penicillaria jocosatrix* (Noctuidae)	Nafus, 1991
Euplectrus plathypenae	*Spodoptera* spp. *Helicoverpa zea* (Noctuidae)	Wall and Berberet, 1974; Parkman *et al.*, 1982
Euplectrus puttleri	*Anticarsia gemmatalis* (Noctuidae)	Puttler *et al.*, 1980
Chrysocharis sp.	*Liriomyza* spp. larvae (Agromyzidae)	Palumbo *et al.*, 1994; Ohno *et al.*, 1999
Edovum puttleri	*Leptinotarsa decemlineata eggs* (Chrysomelidae)	Schroder and Athanas, 1995; Tipping *et al.*, 1999
Eulophus pennicornis	*Lacanobia oleracea* *Mamestra brassicae* (Noctuidae)	Marris and Edwards, 1995
Tetrastichus sp.	Insect eggs, eg. *Hypera postica* (Curculionidae) and *Periplaneta americana* (Blattoidea)	Pitcairn and Gutierrez, 1992; Pawson and Gold, 1993

reservoir of either *E. comstockii* or *E. plathypenae* also arrests moulting and induces premature storage protein production when injected into host larvae (Coudron and Puttler, 1988; T.A. Coudron, unpublished results). These observations indicate that substances derived from the female wasp are sufficient to engender these host responses (Coudron and Puttler, 1988). Polydnaviruses, commonly found in Braconidae and Ichneumonidae (Webb, 1998), have never been recorded from the Eulophidae in general, and were not detected in *E. comstockii* or *E. plathypenae* (Coudron and Brandt, 1996). Therefore, the venom is presumed to be a sufficient source of substances regulating host growth rate, moulting, and storage protein production.

Parasitization by *E. plathypenae* or artificial injection of *Euplectrus* venom caused moulting arrestment in all species of lepidopteran larvae tested (Coudron and Puttler, 1988). These included species in the families Noctuidae, Sphingidae, Arctiidae, Danaidae, Geometridae, Hesperidae, Lasiocampidae, Papilionidae, Pieridae, Pyralidae, and Yponomeutidae. However, in those factitious (unnatural) hosts that were successfully parasitized by *E. plathypenae*, parasitoid larvae completed development to the adult stage only on hosts belonging to the lepidopteran families Noctuidae, Sphingidae, and Geometridae. Successful development of the parasitoid occurred on these unnatural 'permissive' hosts, but all other hosts tested were 'non-permissive' because parasitoid eggs were laid, but did not hatch.

3. Critical events in the moulting cycle: targets for the venom of *Euplectrus* spp.

A physiologically acceptable host is one with both the correct genetic makeup, and the correct developmental programming to enable the physiological changes necessary to support the offspring of the parasitoid. Although the parasitoid cannot add or remove host genes to change the host's suitability, it can change the expression of those genes, in effect changing the programming of the host's genome. Each developmental programme in an animal is made up of a series of critical physiological events, which in turn consist of a series of biochemical reactions. These programmes may mediate simple growth of the insect, or they may mediate the extensive morphological changes encountered during metamorphosis. Although a parasitoid invariably kills or reduces the reproductive fitness of its host, its own survival does not depend upon the death of the host as much as it depends on the redirection of resources away from the host and towards the parasitoid. Consequently, a parasitoid must redirect the developmental programming of the host from one that favours survival of the host to one that favours its own survival.

One way in which the developmental programming of the host can be efficiently redirected is through changes in its endocrine system, since small changes in the timing or titre of a hormone can have profound effects on gene expression. Many insect hormones have been characterized, but the two that have the most profound effects on insect development are 20E and JH. Most developmental events in the host are, at least in part, regulated by the action of these two hormones (see Chapter 1). As with most hormones, the chemical form, the titre, the time of release, and the receptivity of the target tissues have an impact on their biological activity, as does the ratio of 20E to JH (Cherbas and Cherbas, 1996; Gilbert, *et al.*, 1996; Riddiford, 1996).

Because many critical physiological events mediated by these hormones consist of similar biochemical reactions differing only in order or magnitude, the intended redirection of one biochemical reaction by the parasitoid may result in the unintentional redirection of a number of different physiological events as the host tries to regain a state of physiological homeostasis. Thus, a cascade of physiological changes may be set in motion, even though only one of these may be physiologically relevant to the parasitoid's survival.

Moulting consists of a cascade of biochemical reactions, which can be divided into at least five critical physiological events (*Table 3*). The first critical event in the moulting process of lepidopterans is the release of prothoracicotropic hormone (PTTH), a peptide that is synthesized in neurosecretory cells in the brain and secreted into the haemolymph of larvae in response to cues such as body size or photoperiod (Gilbert *et al.*, 1996). Binding of PTTH in the haemolymph to receptors on the surface of cells in the prothoracic gland (PG) induces a biochemical cascade, which triggers the second critical event in insect moulting – the synthesis of 3-dehydroecdysone, which is converted to ecdysone by ketoreductases in the haemolymph. Subsequently, ecdysone is converted to 20E inside the cells in target tissues by ecdysone 20-monooxygenases (Gilbert *et al.*, 1996).

***Table 3**. Moulting consists of five critical physiological events*

Trigger	Events
Size, photoperiod, etc.	1. Release of PTTH
PTTH	2. Synthesis of 3-dehydroecdysone in the prothoracic gland
Rise in cellular 20E	3. Apolysis
Rise in haemolymph 20E	4. Negative feedback inhibition of the prothoracic gland by 20E
Decline in haemolymph 20E	5. Ecdysis

The increased titre of 20E inside cells triggers the third critical moulting event - apolysis. Apolysis consists of all of the biochemical processes involved in the separation of the epidermis from the old cuticle (Chapman, 1998). Within the nucleus of cells, the ecdysone receptor (EcR) and ultraspiracle (USP) proteins bind 20E and facilitate transcription of genes controlled by promoters with ecdysone response elements (EcREs) (Cherbas and Cherbas, 1996). Negative feedback inhibition of ecdysteroid synthesis in the PG by 20E (Gilbert *et al.*, 1996; Sakurai and Williams, 1989; Song and Gilbert, 1998) is the fourth critical event. The subsequent drop in haemolymph 20E titre triggers the fifth and last critical event in the moulting process – ecdysis. Ecdysis consists of all of the behavioural and biochemical events necessary to enable the insect to shed the old cuticle (Chapman, 1998; Truman, 1996). A drop in ecdysteroid titre has been shown to induce the release of one or more ecdysis-regulating neuropeptides (Hesterlee and Morton, 1996; Truman, 1985) including eclosion hormone (EH), pre-ecdysis-triggering hormone (PETH), ecdysis-triggering hormone (ETH), and crustacean cardioactive peptide (CCAP) (Chapman, 1998; Ewer *et al.*, 1997; Kingan *et al.*, 1997a; Truman, 1996; Zitnan *et al.*, 1996). These peptide hormones interact in a highly co-ordinated manner to elicit the behaviours necessary to loosen the old integument and to allow the insect to emerge from it and, thus, successfully complete the moulting cycle.

4. Determining how parasitization by *Euplectrus* spp. affects events critical to moulting in lepidopteran larvae

To arrest moulting in lepidopteran larvae, *Euplectrus* venom must either inhibit the substances or processes that trigger one or more of the five critical physiological events, or the venom must inhibit the expression or action of products of genes regulated by these events (i.e. processes within the event that occur downstream of the trigger). An examination of how venom from, and parasitization by, *Euplectrus* spp. affects all of these critical physiological events may lead to a better determination of which events are redirected by the parasitoid and which are critical to the parasitoid's survival.

The first critical physiological event in moulting, PTTH release, is triggered in larvae of the noctuid *Trichoplusia ni* when a minimal weight and head capsule width is attained within a specific photoperiodic window. Jones *et al.* (1981b), measured larval weights at the time of head capsule slippage and head capsule widths at ecdysis, and determined that the most accurate predictor of ecdysis in all larval stadia of *T. ni* was the ratio of the cube root of larval weight to head capsule width. This critical size ratio (CSR) was also proposed as an accurate index of the developmental state of the larva at the time of PTTH release.

Although *T. ni* larvae parasitized by *E. plathypenae* do not moult, an approximate CSR can be calculated from the maximum weights and the maximum head capsule widths of parasitized *T. ni* larvae. Unparasitized *T. ni* larvae attained a maximum weight of about 80 mg by 40 h into the 4th stadium (Coudron *et al.*, 1990) (*Figure 1*). Head capsule widths of unparasitized larvae increased from 1.14 ± 0.08 mm to 1.59 ± 0.06 mm, yielding a CSR value of 2.71 (Coudron *et al.*, 1990). In contrast, larvae parasitized by *E. plathypenae*, with or without viable parasitoid eggs, grew more slowly but attained a similar weight (*Figure 1*), although the width of head capsules of parasitized larvae increased only slightly to 1.17 ± 0.07 mm (Coudron *et al.*, 1990). The CSR calculated for 4th-instar parasitized larvae was 3.68. Consequently, the CSR value would indicate that parasitization does not interfere with the host's ability to reach the critical size necessary for PTTH release. Unfortunately, it is not known if a minimal head capsule size is essential to the process. Therefore, a more definitive determination of the effect of *Euplectrus* venom on the physical factors governing PTTH release will depend on the measurement of PTTH titres in the host insect.

The release of PTTH is the trigger for the second critical physiological event in moulting – induction of ecdysteroid synthesis and release from the PG. The timing of PTTH release from the brain and its titre in the haemolymph has not been measured in detail in the penultimate stadium of *T. ni*. However, in the penultimate stadium of *M. sexta*, PTTH is released over a 5–7 h period. There is a sequential pattern to the titres of PTTH and ecdysteroids in the haemolymph, with the peak of PTTH occurring after

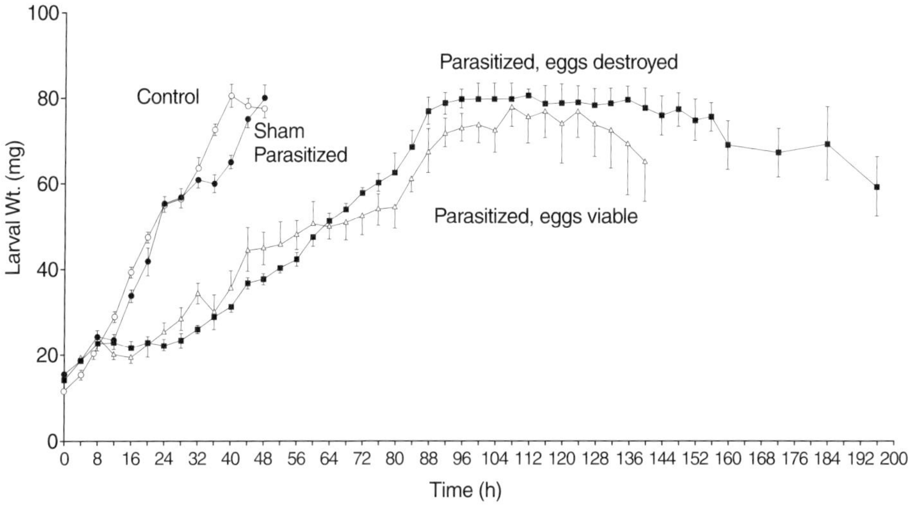

Figure 1. *Comparison of weight gain in parasitized and unparasitized 4th-instar larvae of* Trichoplusia ni. *Figure reprinted from Coudron* et al. *(1990)* Arch. Insect Biochem. Physiol. © *1990 with permission of Wiley-Liss, Inc., a subsidiary of John Wiley & Sons, Inc.*

the onset of ecdysteroid synthesis, but declining before ecdysteroid levels peak in the haemolymph (Bollenbacher and Gilbert, 1981; Nijhout, 1994). Assuming that PTTH release in penultimate instar *T. ni* larvae is similar to that in *M. sexta*, this information, in combination with data from ligation studies and haemolymph ecdysteroid titres, can be used to determine whether or not the *Euplectrus* spp. arrest moulting by interfering with the release of PTTH and subsequent ecdysteroid synthesis (or release) from the PG (Coudron *et al.*, 1990; Kelly and Coudron, 1990). To determine the approximate time of PTTH release, ligatures were applied immediately behind the heads of unparasitized 4th-instar larvae, isolating the brain from the PG. Larvae ligated earlier than 16 h after the previous ecdysis did not moult, suggesting that the ligation occurred prior to PTTH release. When larvae were neck-ligated 25 h post-ecdysis, only 43% moulted, and when ligated 35 h post-ecdysis, sufficient PTTH had been released to stimulate enough ecdysteroid production to elicit moulting in 100% of ligated larvae. Because insufficient PTTH was released in all ligated larvae by 25 h post-ecdysis, and since peak haemolymph ecdysteroid titre in *T. ni* occurs about 28–30 h post-ecdysis (Kelly and Coudron, 1990), it is likely that sufficient PTTH activity to induce moulting is released in *T. ni* between 25 and 30 h after the previous ecdysis (*Figure 2*).

To determine whether the venom of *Euplectrus* acted via the brain or the PG, parasitized 4th-instar larvae were ligated before or after PTTH and ecdysteroid release. Like unparasitized larvae, parasitized larvae neck-ligated prior to PTTH release (16 h post-ecdysis) failed to moult. However, when neck or thoracic ligation occurred after PTTH and ecdysteroid release (30–35 h post-ecdysis), and the thorax and/or abdomen were subsequently parasitized, only the parasitized regions of the larvae isolated by the ligature failed to moult (Coudron *et al.*, 1990) (*Figure 3*). Similarly, 100% of 4th-instar larvae parasitized by *E. plathypenae* after PTTH and ecdysteroid release (30–35 h post-ecdysis), but within the first 36 h of the 4th stadium, also failed to moult. This

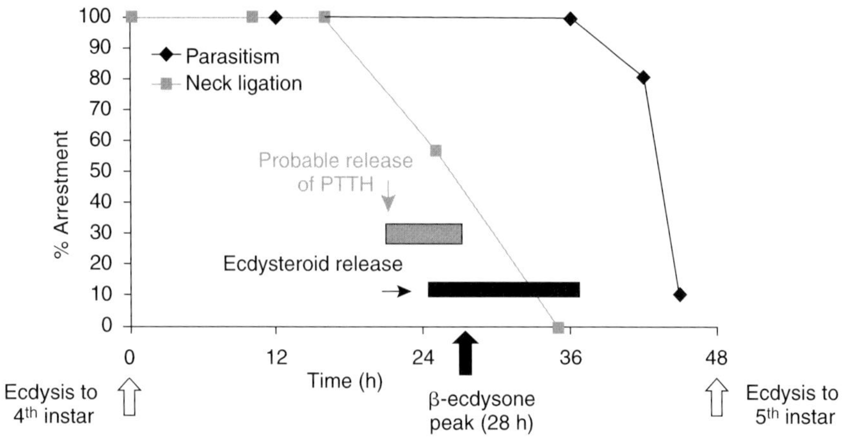

Figure 2. Effect of neck ligation and parasitism by Euplectrus plathypenae *on moulting arrestment of 4th-instar larvae of* Trichoplusia ni. *Arrows denote the probable period over which PTTH is released from the brain (based on ligation experiments) and the period over which elevated levels of ecdysteroid appear in the haemolymph. Based on data from Coudron* et al. *(1990) and Kelly and Coudron (1990).*

proportion dropped to only 11%, however, when parasitization occurred within 3 h of the moult to the 5th stadium (45 h post-ecdysis) (Coudron *et al.*, 1990) (*Figure 3*). Because a ligature that stopped the transfer of haemolymph (but may have allowed for the transfer of nervous impulses) was sufficient to localize the effect of *Euplectrus* venom (Coudron *et al.*, 1990), the haemolymph appears to be the vehicle by which the venom is delivered to target tissues. The most important finding of this research, however, was that the wasp venom was able to arrest moulting downstream of the action of PTTH and ecdysone. Although these ligation studies did not rule out the possibility that the brain and PG were targets of the venom, this research did demonstrate that action on these tissues, and inhibition of the processes they mediate, were not the only means by which the venom arrests moulting.

To determine whether the venom directly affected the major event in moulting mediated by PTTH (the synthesis of 20E or its precursors), the effect of *Euplectrus* parasitization on the timing, titre, and profile of haemolymph ecdysteroids was measured. Synchronized 4th-instar *T. ni* larvae were parasitized by *E. plathypenae* before PTTH release, and parasitoid eggs were destroyed on half of these larvae within 4 h of parasitization (Kelly and Coudron, 1990). Haemolymph samples were collected from these host larvae (and from similar-aged, unparasitized larvae) every 4 h until the larvae moulted, or the parasitoids pupated. These samples were subjected to analysis by reverse-phase, high-performance liquid chromatography (RPLC) and radioimmunoassay (RIA) for the presence of ecdysteroids. Haemolymph of 4th-instar unparasitized larvae contained very little ecdysteroid until 24 h after the beginning of the 4th stadium (*Figure 4(a)*). At this time, haemolymph ecdysteroid titre increased rapidly in unparasitized larvae, and reached a peak of 1200 pg ecdysteroid equivalents (EE) per ml at 28 h post-ecdysis (Kelly and Coudron, 1990). Larval weight at this time was 55 mg, which is about 70% of the maximum weight attained by 4th-instar larvae (Coudron *et al.*, 1990).

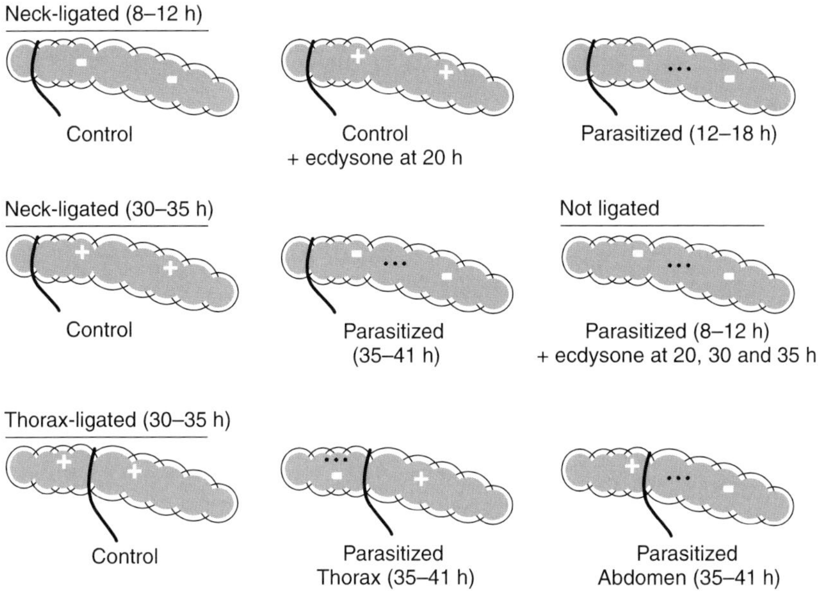

***Figure 3**. Effects of ligation and parasitism on moulting. Black lines indicate the sites of ligation. Dots indicate the sites of parasitism. (+) denotes moulting of that section, thorax or abdomen; (-) denotes failure to moult. Times refer to hours post-ecdysis from the previous stadium. Based on data from Coudron* et al. *(1990).*

In contrast, ecdysteroid levels in the haemolymph of parasitoid-envenomated larvae (parasitized host larvae on which wasp eggs were destroyed before hatching) increased only slightly, and remained below 250 pg EE/ml until about 8 days after parasitization, whereupon the ecdysteroid titre rapidly climbed to 1000 pg EE/ml. This rise in haemolymph ecdysteroid was not linked to a critical larval size because parasitoid-envenomated larvae had reached maximum weight five days earlier (Kelly and Coudron, 1990) (*Figure 4(c)*).

In parasitized larvae (those parasitized hosts on which parasitoid eggs and subsequent larvae were allowed to develop), the ecdysteroid titre increased significantly from <100 to about 400 pg EE/ml between 80 and 88 h (*Figure4(b)*). The attainment of 70% maximal weight in host larvae and the onset of parasitoid feeding coincided with this event. This ecdysteroid peak was followed by a second, much larger peak between 88 and 108 h that resulted in ecdysteroid titres rising from <100 to a peak of 1600 pg EE/ml – a titre almost 30% greater than the maximum ecdysteroid levels found in the haemolymph of unparasitized *T. ni* larvae (Kelly and Coudron, 1990). This twin-peak pattern in parasitized 4th instar larvae is similar to that found in unparasitized 5th-instar *T. ni* larvae (Grossniklaus-Bürgin and Lanzrein, 1990a), and may influence the induction and utilization of storage proteins discussed later in this chapter. A subsequent 50% drop in haemolymph ecdysteroid titre at 108 h, was followed by a rapid and continuous rise in ecdysteroid titre to a level of 7000 pg EE/ml at 140 h (approximately 6 days into the parasitized 4th stadium) – a titre five times higher than the peak ecdysteroid titre in the haemolymph of unparasitized larvae. This level then dropped sharply to nearly undetectable levels at about the time

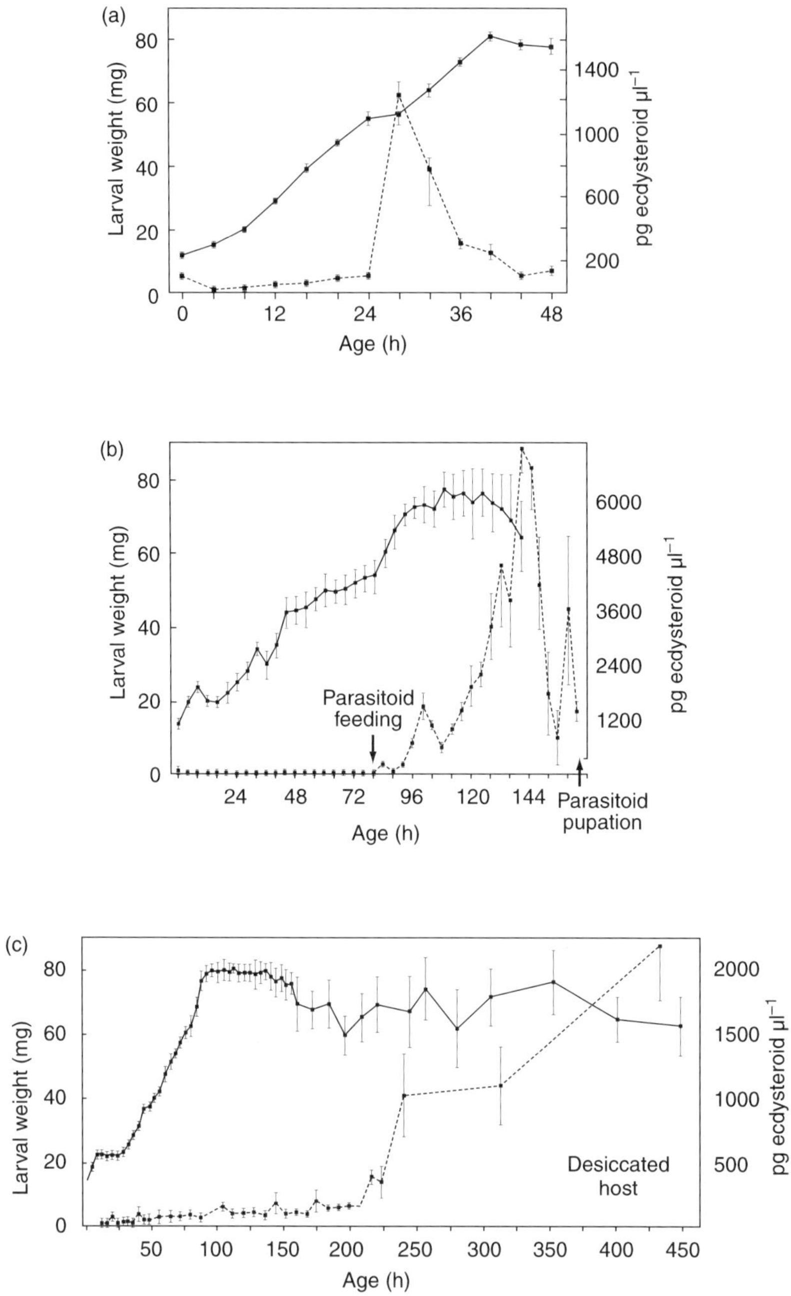

Figure 4. *Larval weights and titres of haemolymph ecdysteroids in parasitized and unparasitized 4th-instar* Trichoplusia ni *larvae. (a) Ecdysteroid profile of unparasitized larvae. (b) Ecdysteroid profile of parasitized larvae – parasitoid offspring remaining. (c) Ecdysteroid profile of parasitoid-envenomated larvae – parasitoid offspring removed. Figure reprinted from Kelly and Coudron (1990)* J. Insect Physiol. *© 1990 with permission of Elsevier Science Ltd..*

of parasitoid-induced lysis and death of the host (i.e. shortly before pupation of the parasitoid larvae (Kelly and Coudron, 1990)) (*Figure 4(b)*).

A comparison of transmission electron microscopy (TEM) images of the integument of unparasitized larvae with those of parasitoid-envenomated 4th-instar larvae at 45 h post-ecdysis, showed definite differences in the extent of apolysis (Coudron *et al.*, 1990) (*Figure 5*). Integument from the unparasitized larvae had undergone apolysis, new epicuticle had formed, and the endocuticle was partially degraded and separated from the old epicuticle by an ecdysial space. In contrast, TEM of integument removed at 45 h post-ecdysis, from larvae parasitized 30 h previously (before PTTH release), showed no signs of apolysis and appeared almost identical to integument taken from unparasitized *T. ni* larvae at 15 h post-ecdysis (Coudron *et al.*, 1990). In fact, these parasitoid-envenomated larvae showed no visual signs of apolysis two weeks after parasitism (Kelly and Coudron, 1990).

Taken together, the data from haemolymph ecdysteroid analysis and from TEM suggest that venom inhibits the first and/or second physiological event critical to moulting, namely the release of PTTH and/or the synthesis of ecdysteroid, and thereby inhibits apolysis. However, data from ligation studies indicate that the *Euplectrus* venom is capable of working independently of the host brain or PG, and is

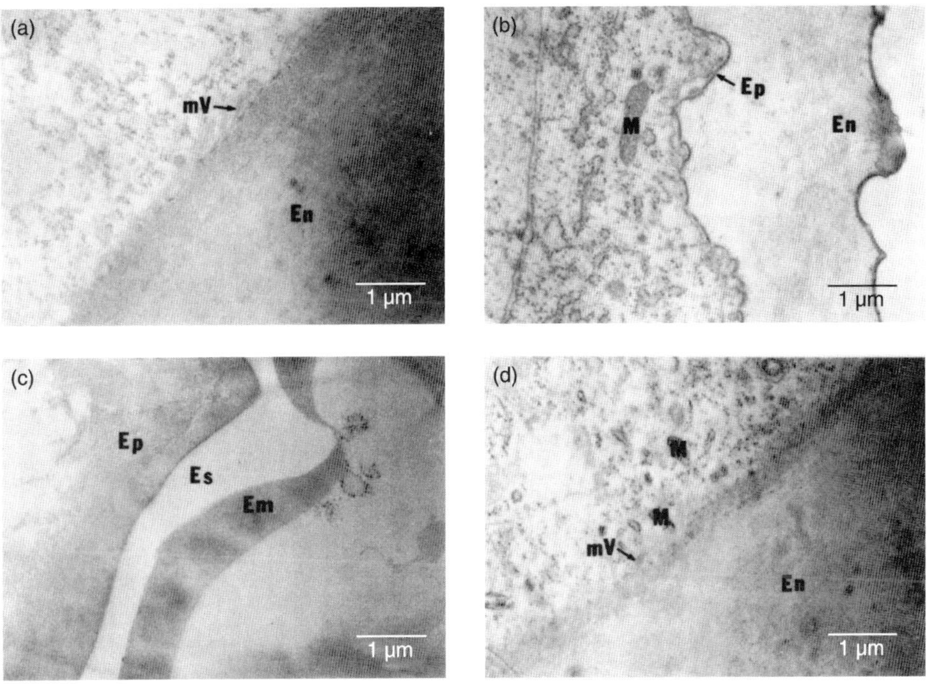

Figure 5. *Transmission electron microscopy of integument from 4th-instar* Trichoplusia ni *larvae. Panels (a), (b) and (c) show integument from unparasitized larvae 15, 35, and 45 h post-ecdysis, respectively. Panel (d) is integument sampled 45 h post-ecdysis from a larva parasitized by Euplectrus plathypenae at 15 h post-ecdysis, before PTTH release in the host. Em, ecdysial membrane; En, endocuticle; Ep, epicuticle; Es, ecdysial space; M, mitochondria; mV, microvilli. Figure reprinted from Coudron* et al. *(1990)* Arch. Insect Biochem. Physiol. *© 1990 with permission of Wiley-Liss, Inc., a subsidiary of John Wiley & Sons, Inc.*

able to act downstream of PTTH and ecdysone to arrest moulting. We can reconcile the data from these experiments by proposing that the venom has two mechanisms of action. When injected into the host before or shortly after PTTH release, the venom blocks events leading up to the third physiological event critical to successful moulting – apolysis. When injected after apolysis, the venom interferes with events leading up to the final physiological event critical to moulting – ecdysis. Subsequent studies have revealed that inhibition of apolysis, as opposed to inhibition of ecdysis, is dependent upon the developmental stage of the host at the time of parasitization (T.A. Coudron, M. Knop-Wright, unpublished results).

The mechanism by which *Euplectrus* venom arrests apolysis is unknown. The absence of a significant ecdysteroid peak until well after the end of weight gain in parasitoid-envenomated hosts (from which *Euplectrus* larvae have been removed), suggests that the inhibition of ecdysteroid synthesis alone may be sufficient to arrest apolysis. In parasitized larvae supporting the development of parasitoid offspring, however, inhibition of ecdysteroid synthesis may not be sufficient to arrest apolysis since a definite peak in haemolymph ecdysteroid titre is observed (*Figure 4(b)*). This suggests that either the venom, or the presence of the parasitoid larvae, must also alter the sensitivity of the host tissues to 20E, inactivate 20E before it reaches the target tissues, or block 20E-mediated events occurring upstream, in order to arrest apolysis in parasitized hosts.

Host tissues must become receptive to 20E before apolysis can occur (Nijhout, 1994), and current evidence suggests that the venom does decrease tissue sensitivity to 20E (Coudron *et al.*, 1990). When unparasitized 4th-instar *T. ni* larvae were neck-ligated 12 h post-ecdysis (before PTTH activation of the PG), an immediate injection of 20E (12 μg/larva) did not cause moulting. However, a similar injection made 8 h later (20 h post-ecdysis) resulted in about 90% of the injected larvae showing signs of moulting posterior to the ligature (Coudron *et al.*, 1990). In contrast, injection of 20E (12 μg/larva at 20, 30, or 35 h post-ecdysis) into 4th-instar *T. ni* larvae previously parasitized at 12 h post-ecdysis (i.e. before PTTH activation of the PG) failed to induce moulting. Assuming that parasitized larvae were at the same developmental stage as unparasitized larvae at the time of treatment, then 20E injection should have induced moulting in parasitized larvae, especially if the venom inhibits ecdysteroid synthesis, but not its action. One interpretation of these data is that venom causes target tissue to become refractory to 20E by 8 h after parasitism. Although 20E was injected into parasitized larvae at a weight (i.e. approx. 30 mg) that was below the value found to be responsive in unparasitized larvae (i.e. 40 mg), subsequent data, to be discussed later in this chapter, suggest that tissues in parasitized larvae remain refractory throughout the period of weight gain to approximately 80 mg. Other evidence for a venom-mediated reduction in the receptivity of host tissues to 20E comes from preliminary experiments evaluating the effects of parasitism by *E. comstockii* on apolysis and cuticle production in larval integument cultured *in vitro* (Hiruma *et al.*, 1991). After incubation with 20E, integument harvested from unparasitized 4th instar *M. sexta* larvae underwent apolysis and produced new cuticle *in vitro*. In contrast, integument harvested from parasitized *M. sexta* larvae did not apolyse (K. Hiruma, L. M. Riddiford and M. Knop-Wright and T. A. Coudron, unpublished results). This failure to undergo apolysis, or produce new cuticle, was not entirely due to inactivation of 20E or subsequent blocking of 20E-induced gene expression, because some ecdysteroid-responsive genes involved in moulting and cuticle synthesis were still expressed (K. Hiruma, personal communication).

Despite the presence of 20E-induced gene expression in the integument of parasitized larvae, inactivation of 20E in parasitized hosts may still play a role in the arrestment of apolysis. When haemolymph ecdysteroid profiles of parasitized and unparasitized 4th-instar *T. ni* larvae were compared using RPLC and RIA to identify the ecdysteroids, several differences were noted (Kelly and Coudron, 1990). Haemolymph taken from unparasitized larvae at the time of peak haemolymph ecdysteroid titre (about 30 h post-ecdysis) (*Figure 4(a)*) contained about 40% of the active moulting hormone (i.e. 20E) together with the (presumably) inactive ecdysteroids: ecdysone (<1%), 20,26-dihydroxyecdysone (~30%) and an unidentified polar fraction (~30%) (Kelly and Coudron, 1990). In these analyses, 3-dehydroecdysone would not have been detected since it is quickly converted to ecdysone in the haemolymph (D.B. Gelman, personal communication). In contrast, RPLC/RIA analysis of ecdysteroids in the haemolymph of parasitized larvae at 140 h post-ecdysis (the time at which haemolymph ecdysteroid titres were highest) showed that only a very small proportion of the ecdysteroid was 20E (*Figure 4(b)*). In the samples from parasitized larvae, the major ecdysteroid present co-eluted with the ecdysone metabolite 20,26-dihydroxyecdysone (Kelly and Coudron, 1990). Subsequent RPLC analysis, followed by normal-phase, high-performance liquid chromatography, revealed that the peak also co-elutes with the sugar-conjugated ecdysteroids, ecdysone-galactose (EGal) and 20-hydroxyecdysone-glucose (20EGlc), but not with ecdysone-glucose or 20-hydroxyecdysone-galactose (T.J. Kelly, T.A. Coudron, unpublished results). Enzymatic treatment of the conjugated ecdysteroid peak with β-glucuronidase/glucosidase and β-galactosidase revealed that it was largely comprised of 20E conjugated to a glucose moiety (20EGlc). Such a conjugation would probably render the 20E inactive. Glycosylated 20E was not detected in the peak that co-eluted with 20,26-dihydroxyecdysone in unparasitized *T. ni.*

The presence of glycosylated ecdysteroids in parasitized insects suggests the presence of glycosylating enzymes in these insects. Potentially, the source of the 20E-glycosylating enzyme could be the venom, the parasitoid larva, or the host. An enzyme with similar ecdysteroid-conjugating activity, ecdysone glucosyl transferase (EGT) was originally found associated with *Spodoptera frugiperda* cells infected with a baculovirus, *Autographa californica* nuclear polyhedrosis virus (AcMNPV) (O'Reilly and Miller, 1989; see also Chapter 14). Samples from parasitized larvae and from *E. plathypenae* venom were analysed for EGT activity. Four tissue samples were analysed (*Figure 6*). Extracts from two tissues, an extract from the reproductive tracts of male parasitoids (control tissue, CT) and haemolymph taken from unparasitized 3rd-star *T. ni* larvae 12–16 h post-ecdysis (control haemolymph, CH), had no arrestment activity. Extracts from two additional tissues, the reproductive tract (including venom glands) of female parasitoids (venom,V), and haemolymph taken 4 h post-parasitism from similarly aged 3rd-instar *T. ni* larvae (parasitized haemolymph, PH), were found to arrest moulting in 100% and 88% of injected 4th-instar larvae, respectively. Ecdysone glucosyl transferase activity, as determined by the migration of radiolabeled ecdysone or UDP-glucose to the position indicated (EG) in the fluorograph (*Figure 6*) was not detected in moult-arresting extracts (V and PH), or non-arresting extracts (CT and CH), unless an extract from baculovirus-infected cells containing EGT was added. Assuming that the putative 20E glycosylating enzyme is present in the haemolymph of 3rd-instar *T. ni* by 4 h post-parasitism, this TLC fluorograph suggests that this enzyme is most unlikely to be EGT, because it appears to use

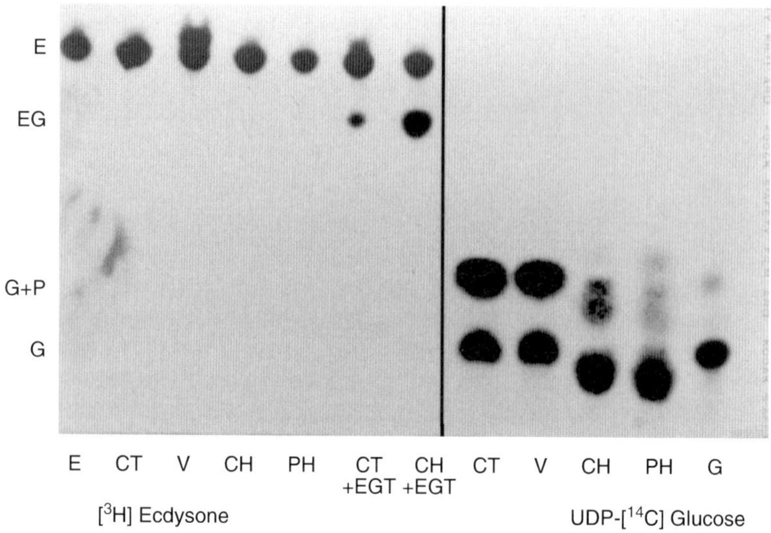

Figure 6. *Fluorograph of EGT activity in* Euplectrus *venom and haemolymph from parasitized and unparasitized 3rd-instar* Trichoplusia ni *larvae. Extracts from the venom glands of* E. plathypenae *(venom-containing parasite tissue, V) and from reproductive tissue of male parasites (control tissue, CT), as well as haemolymph from unparasitized 3rd-instar* T. ni *larvae (control haemolymph, CH) and from parasitized larvae (parasitized haemolymph, PH) were incubated 2h at 37°C with [^{3}H]ecdysone and unlabelled UDP-glucose, or UDP-[^{14}C]glucose and unlabelled ecdysone and 20E. Samples were spotted onto a silica-gel thin layer chromatography plate and run in butanol: acetone: acetic acid : 37% ammonia: water = 70:50:18:1.5:60. E indicates where ecdysone migrates. EG indicates where glycosylated ecdysone migrates. G indicates where glucose migrates, and G+P indicates where UDP-glucose migrates. As a control, CH and CT were also incubated with extracts from baculovirus-infected cells containing ecdysone glucosyl transferase (EGT). Unpublished results in collaboration with D.R. O'Reilly.*

neither ecdysone nor UDP-glucose as a substrate. Further studies are necessary in order to determine whether it is the venom, the host larva, or the parasitoid offspring that is responsible for the glycosylation of 20E seen in *Euplectrus*-parasitized larvae.

To date, ecdysteroids have been profiled only from haemolymph taken at the times of peak ecdysteroid titres in unparasitized larvae, and in parasitized host larvae with intact parasitoid progeny. In order to determine what role these glycosylating enzymes play in the physiology of the host or the parasitoid, the ecdysteroid profile must be determined at additional points in the development of the host larvae. If 20EGlc is detected shortly after parasitization in the haemolymph of both parasitoid-envenomated and parasitized larvae (with *Euplectrus* progeny intact), then both inactivation of 20E through glycosylation and inhibition of ecdysteroid synthesis/release from the PG may be critical to the arrestment of apolysis.

If glycosylation of 20E is only detected in haemolymph sampled at the time of the parasitoid-induced peak of haemolymph ecdysteroid (about 80–100 h post-ecdysis) in parasitized hosts with intact parasitoid progeny, then the developing parasitoid (and not the venom) is responsible for inducing glycosylating enzymes. In this instance, glycosylation of 20E is probably not important in the initial induction of apolysis arrestment. However, glycosylation may be important in maintaining apolysis arrestment in the face of renewed ecdysteroid synthesis and release, or an attempt on the part of the parasitoid to protect itself from exposure to the 20E produced by the host.

If the build-up of 20EGlc is only detected in the haemolymph late in the development of the parasitized host (about 140 h in a parasitized host supporting *Euplectrus* progeny and about 200 h in a parasitoid-envenomated host), then glycosylation of 20E may be a result of the natural processes linked to the deterioration of the host. In this case, a comparison of the profile of haemolymph ecdysteroids in dying parasitized host larvae and in unparasitized larvae dying from other causes may be warranted.

Alternatively, if glycosylation of 20E is only detected in the haemolymph during the time of the final increase in ecdysteroid titre (about 140 h post-ecdysis) of parasitized host larvae with intact *Euplectrus* progeny, then glycosylation of 20E may mediate programmed cell death leading to premature death of the host larvae. Towards the end of *Euplectrus* development, the host larvae turn black, and the parasitoid larvae move beneath the cadaver to spin a characteristic silken cocoon (Coudron and Puttler, 1988). This blackening occurs whether the host is supporting only one or several parasitoid larvae, but it does not occur in the absence of parasitoid larvae. One substance that is a potent mediator of programmed cell death is 20E, a build-up of which triggers degeneration of some larval tissues in preparation for pupal development (Bangs and White, 2000). Because ecdysone is converted to 20E inside cells in lepidopteran larvae, the presence of a 20E glycosylating enzyme in the haemolymph (i.e. outside the cells) might lead to increased ecdysteroid production and conversion of ecdysone to 20E (inside the cell) by preventing the negative feedback regulation of the PG (Song and Gilbert, 1998). This prolonged presence of 20E within the cell may achieve a titre sufficient to cause cell death. A further analysis of the profile of haemolymph ecdysteroids at different periods during host development must be undertaken to further delineate the role that 20E glycosylation plays in parasitized hosts. The failure of *Euplectrus* spp. venom to arrest apolysis occurs when parasitism takes place late in the development of the host larvae, or when a low dose is administered in injection treatments (Coudron *et al.*, 1990; Uematsu and Sakanoshita, 1987; T.A. Coudron, unpublished results). In most cases, however, moulting is still arrested in these host larvae due to arrestment of the final critical physiological event in successful moulting – ecdysis. The endocrine events that regulate the process of ecdysis are less well known than the endocrine events that regulate apolysis. The titre of 20E in the haemolymph, however, mediates negative feedback regulation of the PG (Song and Gilbert, 1998), which, in turn, results in a decrease in 20E titre in the haemolymph. The magnitude of the drop in 20E titre in the haemolymph (subsequent to the final haemolymph ecdysteroid peak in final instars of lepidopteran larvae) is believed to be an important determinant of ecdysis (Cherbas and Cherbas, 1996; Truman, 1985). Since *Euplectrus* venom has been shown to suppress and/or delay the appearance of ecdysteroids in the haemolymph when injected before PTTH release (Kelly and Coudron, 1990), it may also mediate the lower haemolymph 20E titre observed when injected after apolysis. In theory, this could result in an insufficient decline in haemolymph 20E to trigger the other physiological and hormonal events that regulate ecdysis.

Evidence from studies conducted in the non-permissive host, *Ostrinia nubilalis*, suggests that a drop in haemolymph ecdysteroid titre may also be associated with arrestment of ecdysis (Gelman *et al.*, 1997). Moulting was arrested in all penultimate 4th-instar *O. nubilalis* larvae parasitized by *E. comstockii*. Apolysis, however, was arrested in only 10%. The other 90% underwent delayed apolysis but failed to ecdyse. A comparison of haemolymph ecdysteroid titres in unparasitized and parasitized 4th instar *O. nubilalis* larvae showed that the peak ecdysteroid titre was delayed by 3 h and was 25% lower in parasitized larvae (Gelman *et al.*, 1997). In contrast, apolysis was arrested in 90% of final-instar *O. nubilalis* larvae parasitized by *E. comstockii*, and the haemolymph ecdysteroid titre remained at 10% of the peak titre measured in unparasitized larvae. These data are consistent with the possibility that venom-induced arrestment of apolysis and the arrestment of ecdysis only (i.e. without blocking apolysis) are both mediated by one mechanism – the level of ecdysteroid synthesized/or released from the PG. However, *Euplectrus* venom can arrest ecdysis when delivered after ecdysteroids have peaked in the haemolymph in regions of the larval host isolated by ligature from the PG (the site of ecdysteroid synthesis). Therefore, a reduction in host ecdysteroid titre cannot be the only mechanism by which the venom arrests ecdysis in the host. Clearly, the venom must also be affecting other physiological events in the host, perhaps by acting directly on the release of other ecdysis-regulating peptide hormones, including EH, ETH, PETH, and/or CCAP.

Further studies on how *Euplectrus* venom affects endocrine and other physiological events leading up to ecdysis are warranted and may clarify our understanding of the relationship between events controlling apolysis and ecdysis in insects. Current data suggest that one single substance may be responsible for both the arrestment of apolysis and the arrestment of only ecdysis. Coudron and Brandt (1996) characterized the arrestment activity in the venom of *E. comstockii*. Arrestment activity was found in aqueous extracts of crude venom and was sensitive to protease degradation, ammonium sulfate precipitation, and heat denaturation – indicating that the arrestant was a protein. Partial purification of a fraction of the venom that arrested moulting resulted in the isolation of a protein that had a molecular weight of 66 kDa. Electrophoresis of the 66 kDa protein under denaturing conditions resulted in two discernible bands with mobilities equivalent to 31 and 36 kDa, neither of which showed arrestment activity in bioassays. Injection of the 66 kDa protein into 4th-instar *T. ni* larvae before PTTH release resulted in either arrestment of apolysis or arrestment of ecdysis only (i.e without blocking apolysis), depending upon the dose delivered (M. Knop-Wright and T. A. Coudron, unpublished results).

One other physiological process that is induced by the venom of *Euplectrus* spp. is the precocious production of large quantities of haemolymph storage proteins in penultimate and earlier instars of permissive lepidopteran hosts. Normally, these proteins are closely regulated by the endocrine system and are only induced in final-instar larvae. However, *Euplectrus* venom seems to be able to mediate the same changes in storage protein induction and utilization as seen in an unparasitized final-instar larva, but under different endocrine conditions. This suggests that, as in moulting arrestment, *Euplectrus* may utilize the endocrine system of the host to induce a physiological event necessary for its own survival, but may also have a mechanism to regulate storage proteins independent of the host's endocrine system.

5. Critical biochemical events in the induction of late larval storage proteins

In lepidopteran larvae, six different classes of storage proteins have been identified on the basis of function, structure, amino-acid sequence, isoelectric points (pI), and proposed regulation of synthesis uptake and utilization (Haunerland, 1996; Telfer and Kunkel, 1991). Arylphorins, methionine-rich storage proteins, basic JH-suppressible storage proteins, acidic JH-suppressible storage proteins, and riboflavin-binding proteins are hexamerins (i.e. proteins with a total native molecular weight of around 500 kDa), which are composed of six homologous subunits of 70–85 kDa each (Telfer and Kunkel, 1991). Very high density lipoproteins (VHDL) are di- or tetrameric proteins which have a total molecular weight of 150 kDa, and have the same temporal pattern of appearance in the haemolyph as hexamerins (Haunerland, 1996).

Increased production and utilization of storage proteins in lepidopterans is normally restricted to the last larval stadium (Haunerland, 1996). All lepidopteran storage proteins share a similar pattern of appearance, typified by being low or absent in early larval stadia, but then showing marked increases and accumulation in the haemolymph of the final larval instar (Haunerland, 1996; Telfer and Kunkel, 1991). This marked increase is followed by a rapid decline in haemolymph titre during either the pre-pupal or pupal stage due to receptor-mediated resorption by the fat body, where these proteins serve as a pool of amino acids that are subsequently used for adult development and reproduction (Haunerland, 1996). Although the temporal occurrence of these proteins is similar, the endocrine events that mediate the induction of storage protein synthesis and release may be quite different.

Arylphorins and VHDL are commonly found in lepidopteran larval haemolymph prior to moulting to the final stadium (Ray *et al.*, 1987; Wang and Haunerland, 1992; Webb and Riddiford, 1988). Resorption of these proteins by the fat body begins at the pre-pupal stage (Tojo *et al.*, 1981; Wang and Haunerland 1992), and normally ceases by the end of the pupal stage (Kanost *et al.*, 1990). The sites of synthesis and resorption of VHDL and arylphorin are located in different areas of the fat body (Haunerland and Shirk, 1995; Wang and Haunerland, 1991, 1992). In *Helicoverpa zea*, arylphorin and VHDL are synthesized in the peripheral fat body (which lies between the outer muscle layer surrounding the haemocoel and the integument), but are resorbed via receptor-mediated endocytosis by the perivisceral fat body (which surrounds the gut) (Haunerland, 1996; Haunerland and Shirk, 1995; Wang and Haunerland, 1994). The identities of the molecules that trigger increased synthesis and/or resorption of arylphorin and VHDL are not known; however, JH titre alone does not appear to affect the level of arylphorin mRNA in *M. sexta* (Riddiford and Hice, 1985) or in *T. ni* (Jones *et al.*, 1993). Resorption of these proteins by the fat body may be induced by a decline in 20E in the haemolymph just prior to larval ecdysis in penultimate and earlier larval stadia. In the final stadium, however, the pupal commitment peak of 20E after the decline in JH may induce the resorption of these proteins. In *H. zea*, this coincides with the synthesis of the arylphorin/VHDL storage protein receptor (Haunerland, 1996).

Basic JH-suppressible storage proteins (basic JHSP) are highly homologous to methionine-rich storage proteins (MRSP). Both of these groups of proteins have pIs of 8.0 or greater. The acidic JHSPs have pIs of about 5.8, and share little homology with other storage proteins. The expression of all JHSP and of many MRSP genes has been shown to be suppressed by JH and its analogues (Beintema *et al.*, 1994; Haunerland, 1996; Jones and Sarkari, 1993; Jones *et al.*, 1993). In contrast with arylphorins and VHDLs, the JHSPs and many MRSPs only appear in the final larval stadium (Haunerland, 1996; Jones *et al.*, 1993). Generally, the synthesis of these storage proteins in the final larval stadium begins later than the synthesis of arylphorins or VHDLs (Haunerland, 1996) and increases as the haemolymph JH titre decreases (Jones *et al.*, 1993; Webb and Riddiford, 1988). Resorption of MRSPs and basic JHSPs by the fat body occurs before arylphorin is resorbed (Jones *et al.*, 1996). Both MRSP and basic JHSPs disappear completely from the haemolymph of last-instar larvae and are not found in the pupae or adults (Haunerland, 1996). Whether or not acidic JHSPs are resorbed is not known. The riboflavin-binding proteins (RBP) have only recently been shown to accumulate in the fat body at the end of the last larval instar in Lepidoptera (Pan and Telfer, 1999) in a manner similar to storage proteins. Little is known about how the synthesis and resorption of RBPs are regulated.

6. Effect of parasitization by *Euplectrus* spp. on the haemolymph titre of late larval storage proteins in permissive lepidopteran larvae

Five types of storage proteins have been identified in the permissive host, *Trichoplusia ni*. These include an acidic JH non-suppressible arylphorin (Jones and Sarkari, 1993; Kunkel *et al.*, 1990); an acidic JH-suppressible protein (AJHSP1) (Jones *et al.*, 1988b; Jones and Sarkari, 1993); two basic JH-suppressible proteins (BJHSP1 and BJHSP2) (Jones *et al.*, 1993), and a VHDL-homologous chromolipoprotein (Jones *et al.*, 1988b). Since the effects of parasitization by *Euplectrus* spp. on VHDL titres in the haemolymph of *T. ni* have not been measured, the following discussion will be limited to only arylphorin and JHSPs. The haemolymph titres of arylphorin, BJHSP1 and BJHSP2, and AJHSP1 increase markedly during the final larval stadium of *T. ni* (Jones *et al.*, 1993). However, the temporal patterns of gene expression and of the appearance of protein in the haemolymph differ for each type of storage protein. Gene transcripts (mRNA) for arylphorin appear shortly before gene transcripts for BJHSP1, BJHSP2, and AJHSP1, which appear on day 2 of the final stadium. Arylphorin transcripts decline concurrently with the pupal commitment peak of haemolymph 20E, whereas the transcripts of the JHSPs decline one day later (Jones *et al.*, 1993). The four proteins produced from these genes all appear in the haemolymph of *T. ni* on day 2 of the final larval stadium. Subsequently, BJHSP1 and BJHSP2 disappear from the haemolymph by day 4 of the final larval stadium (i.e. shortly before pupation), AJHSP1 remains in the haemolymph until the pupal stage, and arylphorin persists in the haemolymph until the adult stage (Jones *et al.*, 1993). The disappearance of these proteins is a post-translational process, since gene transcripts for these proteins disappear before the protein titre declines in the haemolymph (Jones *et al.*, 1993).

In contrast to the natural occurrence of storage proteins in the final larval stadium of unparasitized *T. ni*, these four types of storage proteins types appear in the haemolymph of earlier instar larvae when these larvae are parasitized or envenomated by *Euplectrus* or are injected with *Euplectrus* venom (Coudron *et al.*, 1994; T.A. Coudron and S. L. Brandt, unpublished results). Within 2 days of parasitization of 3rd or 4th-instar *T. ni* larvae by *E. comstockii*, a large increase in the haemolymph titres of 70- to 80-kDa proteins was detected (Coudron *et al.*, 1994) (*Figure 7*). An immunoblot of haemolymph from unparasitized and parasitized 4th-instar *T. ni* larvae, probed with polyclonal antibodies to AJHSP1, BJHSP1, and BJHSP2, confirmed that these proteins were absent from the haemolymph of unparasitized 4th-instar *T. ni* larvae, but significantly contributed to the large increase in 70- to 80-kDa haemolymph proteins detected in parasitized larvae (*Figure 8*). The horseshoe-shaped banding pattern of the basic JHSPs in the immunoblot was also indicative of high titres of arylphorin in the haemolymph of parasitized larvae (*Figure 8*) (Coudron *et al.*, 1994). The titre and profile of haemolymph storage proteins induced by parasitism in 4th-instar larvae was

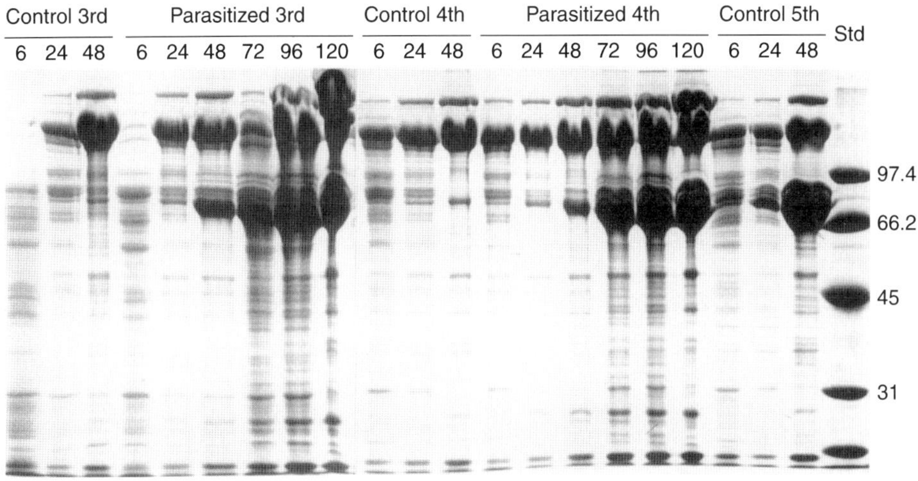

Figure 7. Effect of parasitization by E. plathypenae *on appearance of haemolymph storage proteins in* Trichoplusia ni. *Figure reprinted from Coudron* et al. *(1994)* Arch. Insect Biochem. Physiol. © *1994 with permission of Wiley-Liss, Inc., a subsidiary of John Wiley & Sons, Inc.*

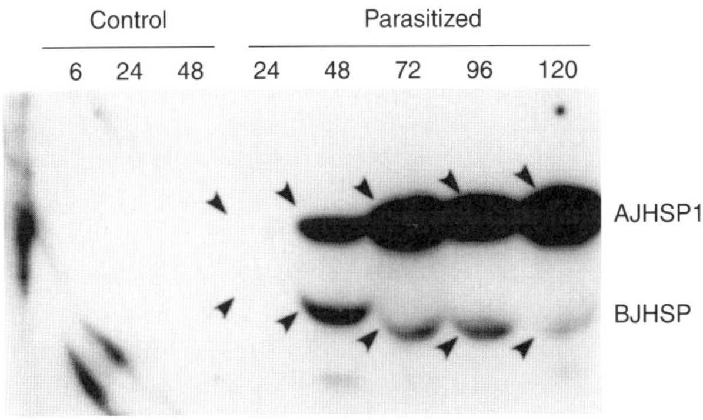

Figure 8. Western blot of haemolymph of parasitized and unparasitized 4th-instar larvae of Trichoplusia ni *probed with antisera to JH-suppressible storage proteins, BJHSP1, BJHSP2, and AJHSP1. Figure reprinted from Coudron* et al. *(1994)* Arch. Insect Biochem. Physiol. © *1994 with permission of Wiley-Liss, Inc., a subsidiary of John Wiley & Sons, Inc.*

similar to the titre and profile of storage proteins naturally appearing in the 5th stadium unparasitized larvae.

However, the expression of transcripts of at least one of these storage proteins (arylphorin) in the parasitoid-envenomated larvae differs from that of the unparasitized larvae. (*Figure 9*). When mRNA isolated from parasitized 4th-instar *T. ni* larvae and from unparasitized 4th and 5th-instar *T. ni* larvae was subsequently probed with an arylphorin cDNA, the arylphorin transcript was not detected in unparasitized 4th-instar larvae. However, a significant quantity of this transcript was detected by 12–24 h post-ecdysis in unparasitized 5th-instar larvae and in parasitized 4th-instar larvae. Additionally, there was an abnormally high level of arylphorin expression in parasitized larvae that was maintained until the end of the sampling period – 4 days later (*Figure 9*). Thus, *Euplectrus* venom appears to up-regulate storage proteins initially by increasing arylphorin gene transcription. However, the possibility that the increase in arylphorin transcription is due to decreased degradation of the transcript cannot be ruled out. Since the pupal commitment peak of ecdysteroid coincides with the disappearance of arylphorin transcripts in unparasitized 5th-instar larvae, arylphorin transcription may remain high in parasitized hosts because haemolymph ecdysteroid titres remain low. Since the presence or absence of intact parasitoid larvae affects the haemolymph ecdysteroid titre in the host, transcription of arylphorin may differ in hosts with intact parasitoid larvae.

Although suppression of ecdysteroid synthesis or release by the venom may be sufficient to up-regulate the level of arylphorin in *T. ni*, this suppression would probably not be sufficient to cause the increase in JHSPs. Rather than acting on specific hormones individually, *Euplectrus* venom may be affecting how these hormones act in concert to induce entire programmes of gene expression. For instance, *Euplectrus* venom may cause precocious expression of a 5th (final) stadium developmental programme, similar to that discovered in *T. ni* parasitized by *Chelonus* sp. near *curvimaculatus* (Farkas *et al.*, 1999; Jones *et al.*, 1996). In both of these parasitoid–host systems, the pattern of expression of storage proteins in 4th-instar parasitized larvae is similar to that found in unparasitized 5th-instar *T. ni* larvae. In addition, an analysis of the pattern of haemolymph ecdysteroid peaks in parasitized 4th-instar *T. ni* larvae, after its induction in the presence of feeding *Euplectrus* larvae, is similar to the pattern of haemolymph ecdysteroid in unparasitized 5th-instar *T. ni* larvae (Grossniklaus-Bürgin and Lanzrein, 1990a). This pattern consists of a small peak of ecdysteroid at about 84 h (reminiscent of the pupal commitment peak), which is followed by a larger ecdysteroid peak at about 100 h (similar to the pre-pupal ecdysteroid peak responsible for inducing

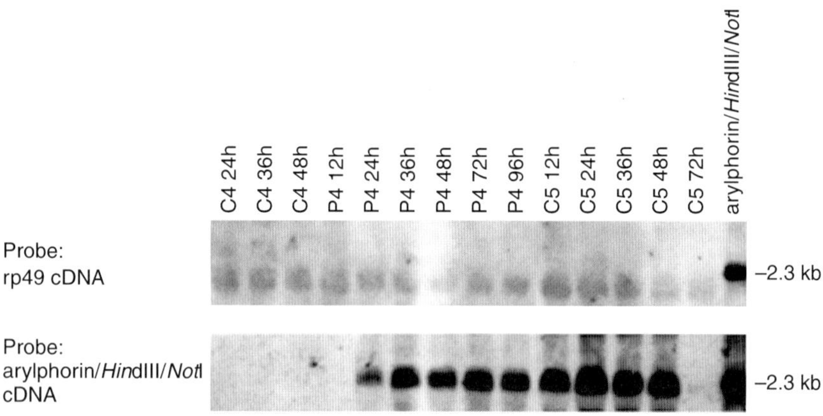

Figure 9. *Northern analysis of the developmental profile of arylphorin transcripts in unparasitized and parasitized* Trichoplusia ni *larvae.* T. ni *larvae were parasitized by* E. comstockii *shortly after ecdysis to the penultimate (4th) stadium. Parasitized and unparasitized penultimate and final (5th)-instar larvae were held at 29°C and 16:8 light/dark photoperiod. Subsets of 3–15 larvae were sampled every 12–24 h, frozen in liquid nitrogen, and held at –80°C. For each time point, whole larvae were pulverized in liquid nitrogen, and total RNA was extracted from 200 mg of tissue (Totally RNA Kit, Ambion). mRNA was extracted from total RNA (Micro Fasttrack Kit, Invitrogen), separated on a formamide 1.2% agarose gel, and transferred to a nylon membrane for Northern analysis. The membrane was probed with a biotinylated control cDNA to Ribosomal Protein 49 (RP49) to ensure even loading of lanes, and was visualized using chemiluminescence of an alkaline phosphatase conjugate (Brightstar Detection Kit, Ambion). The membrane was then stripped of all probes and reprobed with biotinylated arylphorin cDNA/*Not*I/*Hind*III. Visualization of the probe was again via chemiluminescence (Brightstar Detection Kit, Ambion). Unpublished results in collaboration with D. Jones and G. Jones.*

moulting in unparasitized 5th-instar larvae) (see *Figure 4b*). Alternatively, *Euplectrus* spp. may be activating unique programmes of gene expression, not yet understood.

Storage proteins also increase in the haemolymph of the heliothine noctuids *Heliothis virescens* and *Helicoverpa zea* in response to parasitization by *Euplectrus* spp. Parasitization of 3rd and 4th-instar *H. virescens* or *H. zea* larvae by either *E. comstockii* or *E. plathypenae* resulted in a marked increase in the titre of 70- to 80-kDa proteins in the host's haemolymph (Coudron *et al.*, 1997, 1998). Western blots of the haemolymph of parasitized and unparasitized *H. virescens* larvae, sampled over a 120 h period post-parasitism and probed with antisera to the 74-kDa and 76-kDa subunits of *H. virescens* arylphorin and the 82-KDa subunit of the *H. virescens* riboflavin-binding protein (LeClerc and Miller, 1990), confirmed that the increase in 70- to 80-kDa proteins in the haemolymph was due to an increased titre of these particular proteins (Coudron *et al.*, 1997, 1998). The titres and patterns of occurrence of these proteins in the haemolymph of 4th-instar *H. virescens* larvae after parasitization by *E. comstockii* or *E. plathypenae* closely parallel the titres and patterns of occurrence of these proteins in the haemolymph of unparasitized 5th-instar *H. virescens* larvae. A 150–160-kDa protein (which may correspond to the 160-kDa subunit of the VHDL storage protein) was also detected in the haemolymph of both parasitized and unparasitized 4th-instar *H. virescens* larvae. Its titre (as determined qualitatively from Coomassie-blue staining of the same gel) did not increase significantly in parasitized hosts, suggesting that the titre of VHDL in the haemolymph of *H. virescens* is not affected by *E. comstockii* or *E. plathypenae* venoms.

The haemolymph storage protein profile of 4th-instar *H. zea* larvae parasitized by *E. comstockii* was similar to that of parasitized *H. virescens* larvae. An increased amount of 70–80-kDa proteins in the haemolymph was in part due to an increase in arylphorin. Again, the titre of the putative VHDL (about 150 kDa) was not affected by parasitism (Coudron *et al.*, 1998).

In unparasitized 5th-instar *T. ni* larvae, an increase in JHSPs may be mediated by an increase in gene transcription due to declining haemolymph JH titre (Jones *et al.*, 1990b, 1993). *Euplectrus* venom may also be capable of affecting the titre of JH in the haemolymph through action on JH esterase (JHE). When haemolymph JHE activity was measured in venom-injected and 'in-control' final-instar larvae, a lower level of JHE activity was detected in the haemolymph of venom-injected larvae (*Table 4*). The magnitude of the increase in titres of haemolymph storage proteins in parasitoid-envenomated final-instar larvae, however, was lower than the endogenous increase detected in the same stage of unparasitized larvae (Coudron *et al.*, 1997). Why the levels of storage proteins in the haemolymph of parasitoid-envenomated final, 5th-instar *H. virescens* larvae decline is yet to be determined. However, this evidence does suggest that *Euplectrus* venom may be capable of affecting JH titre and/or JHE activity in the host's haemolymph and, thus, may be another mechanism by which the venom could potentially regulate storage proteins in the host. Whether *Euplectrus* venom increases the transcription of JHSP genes in *T. ni* and, if so, whether these changes occur concurrently with alterations in the level of JH or JHE is currently being explored.

Although *Euplectrus* venom arrests moulting in a wide range of lepidopteran species, only hosts in the families Sphingidae, Geometridae, and Noctuidae have been found to support parasitoid larval development (Coudron and Puttler, 1988). Coudron *et al.* (1998) looked at the effect of parasitization by *E. comstockii* on the growth and haemolymph protein profiles of penultimate and final-instar larvae of three non-permissive, pyralid

Table 4. Level of JH esterase activity in haemolymph of 5th-instar Heliothis virescens *larvae at burrowing*

Larval weight (mg)	Venom injected (GE)[1]	JHE activity (nmol JH metabolized/ nmol in 1 μl)
363	0.00 GE (saline)	127 ± 8.5
300	0.05 GE	25.3 ± 9
269	0.10 GE	32.9 ± 11.2

[1]GE, venom gland equivalents
Unpublished results in collaboration with D.Q. Zhang and D.L. Dahlman.

hosts: *Ostrinia nubilalis*, *Diatraea grandiosella*, and *Anagasta kuehniella*. While parasitism arrested development in all of these non-permissive hosts, it resulted in only small increases in the titres of putative storage proteins in the 70–80 kDa range in the haemolymph of penultimate stadium pyralid hosts, and little or no increases in putative storage proteins in the final-instar larvae. The failure of storage proteins to increase (prematurely) to the same levels in non-permissive hosts as they do in permissive hosts is currently the only biochemical difference that has been found between hosts that support *Euplectrus* development and those that do not. On a permissive host, the eggs of *Euplectrus* hatch 72–80 h after parasitization – when the titres of storage proteins in the haemolymph have increased as a result of envenomation. In the non-permissive pyralid hosts, the storage proteins never rise to extremely high levels, and *Euplectrus* eggs (laid on the surface of the integument) fail to hatch. Experiments are currently underway to determine whether a high titre of haemolymph storage protein is a prerequisite for the hatching of *Euplectrus* eggs and for the proper development of *Euplectrus* larvae.

7. Conclusions

Successful parasitism of lepidopteran larvae by *Euplectrus* spp. is dependent upon the induction of at least two significant physiological alterations in the host – the arrestment of moulting and the induction of a high titre of storage proteins in the host haemolymph. Apolysis and ecdysis are regulated by different endocrine events, and yet one substance in the venom is able to arrest both. *Euplectrus* venom is also able to increase the haemolymph titres of different classes of storage proteins, in spite of the fact that the endocrine regulation of these proteins is very different. These physiological alterations are accompanied by numerous changes to related biochemical events in the host. However, the physiological importance of these biochemical changes to the mechanism by which *Euplectrus* spp. mediate the host's physiology and the ecological significance of these biochemical changes to the parasitoid's survival are only partially understood.

Euplectrus spp. arrest apolysis when venom is delivered prior to the release of PTTH from the brain or ecdysteroid from the PG in the host larvae, and arrest only ecdysis when venom is injected later in a larval stadium (after early events in apolysis have been initiated). Suppression of haemolymph ecdysteroid titre, perhaps via a reduction in ecdysteroid synthesis or release from the PG, precedes arrestment of apolysis. This suppression may be due to inhibition of PTTH release from the brain, inhibition of PTTH action on the PG, or direct inhibition of biochemical events involved in ecdysteroid synthesis or release. The venom of a closely related eulophid, *Eulophus pennicornis* (Tribe: Eulophinae) (see *Table 1*) also arrests moulting in lepidopteran larvae (Marris and Edwards, 1995; Weaver *et al.*, 1997; see also Chapter 8). Evidence suggests that *E. pennicornis* arrests moulting by reducing the receptivity of the PG to PTTH (Marris *et al.*, 2001). Whether this mechanism is also involved in the reduced ecdysteroid synthesis detected in *Euplectrus*-parasitized hosts awaits further study.

From a physiological perspective, a reduction in ecdysteroid synthesis may be sufficient to arrest apolysis in hosts parasitized by *Euplectrus* spp. From an ecological perspective, however, a reduction in ecdysteroid synthesis may not be sufficient to arrest apolysis because of the presence of additional peaks of haemolymph ecdysteroid induced in parasitized, permissive host larvae in the presence of viable *Euplectrus* larvae. In these hosts, a change in tissue receptivity to the moulting hormone, inactivation via glycosylation of the active form of ecdysteroid, 20E, and/or an inhibition of gene expression upstream of apolysis may also be required to maintain arrestment.

The only biochemical change that has been measured in hosts in which ecdysis, but not apolysis, has been arrested by *Euplectrus* spp. is a decrease in haemolymph ecdysteroid titre. However, since envenomation by *Euplectrus* still arrests ecdysis after the appearance of the endogenous peak in the host's ecdysteroid titre still arrests ecdysis, a reduction in haemolymph ecdysteroid is, most likely, not the only mechanism through which ecdysis arrestment can be mediated. Changes in the titres of other ecdysis-regulating hormones, changes in the form of these hormones, changes in tissue receptivity to these hormones, and/or other biochemical events upstream of ecdysis may also mediate ecdysis arrestment. Injection of moderate doses of a single 66-kDa protein, purified from crude venom, is sufficient to arrest apolysis, whereas injection of a lower dose (insufficient to arrest apolysis) is sufficient to arrest ecdysis. This suggests that a substance associated with the venom is able to disrupt both apolysis and ecdysis, and is able to do so via a mechanism different from the known action of 20E.

The venom of *Euplectrus* spp. also induces high titres of specific storage proteins in the haemolymph of penultimate and earlier instar larvae of permissive lepidopteran hosts. Haemolymph storage protein titres are not affected, or are only increased slightly, in the haemolymph of parasitized penultimate-instar larvae of non-permissive hosts. In the case of at least one of the storage proteins affected (arylphorin), the increased level of protein in the haemolymph of a permissive host is due to an increase in gene transcription. The mechanism by which *Euplectrus* spp. facilitates this change in hosts' storage proteins is not known. In permissive hosts, the pattern of appearance storage proteins in the haemolymph of parasitized penultimate-instar larvae is similar to the natural pattern detected in the unparasitized final larval stages. A decreased ecdysteroid titre or changes in the titre or activity of JHE in the haemolyph may account for the changes in storage protein production. Therefore, *Euplectrus* venom may be affecting the ways in which hormones act in concert to induce existing endogenous (or possibly unique) programmes of gene expression.

These findings suggest that parasitism by, or treatment with the venom from, *Euplectrus* spp. results in the selective regulation of multiple genes and events in lepidopteran larvae in a pattern different from that observed in normally developing insects. The ability of *Euplectrus* venom to regulate many endocrine events in the host, as well as to regulate important processes in the host independently of these endocrine events, will be valuable to our understanding of the regulation of gene expression and, subsequently, developmental events in insects.

Acknowledgements

We recognize and are grateful to D.L. Dahlman, D.B. Gelman, K. Hiruma, D. Jones, G. Jones, T. J. Kelly (deceased), D.R. O'Reilly, L.M. Riddiford, and D.Q. Zhang, for their collaborative efforts and support in pursuit of this research; and to F. Consoli, J.P. Edwards, D.R. Nelson, Q. Song, R.M. Wagner, and R.J. Weaver for their critical reading of this manuscript.

10

The influence of parasitoids on wing formation in aphids: possible interactions with the endocrine system?

Jim Hardie and Petra Christiansen-Weniger

1. Introduction

One of the reasons for aphids having a high pest profile, particularly in temperate regions, is that they show a high degree of polyphenism (Dixon, 1998; Hille Ris Lambers, 1966; Lees, 1966). That is, up to seven adult forms occur during the annual cycle and these forms, or morphs, are dictated by environmental conditions. The morphs are associated with different modes of reproduction and with population dispersal. Asexual forms predominate during summer while sexual forms tend to be induced by the short days of autumn. This strategy allows for rapid increases in number during the warmer months, particularly as asexual reproduction is associated with viviparity and a form of paedogenesis that leads to the telescoping of generations, while sexual reproduction during autumn leads to the production of diapausing, overwintering eggs (Dixon, 1998; Moran, 1992). Some of the forms are winged and allow populations to disperse.

During summer, as numbers increase, the populations become more crowded and host-plant quality changes. These factors promote the development of winged individuals. Wing induction may be maternal (prenatal), i.e. where crowding the mother affects the next generation and the offspring develop as winged adults, or the environmental factors can act postnatally, directly on the newly-born insect (Müller *et al.*, 2001). Winged forms can also be associated with sexual reproduction. Although sexual females are wingless in host-alternating species of aphid which cycle between herbaceous summer hosts and woody winter hosts, a winged autumn female morph is responsible for effecting the host-plant switch. These forms are often induced by short day length and give birth to sexual forms (Dixon, 1998). In addition, although

Endocrine Interactions of Insect Parasites and Pathogens, edited by J.P. Edwards and R.J. Weaver.

males are also induced by short days, they may be winged or wingless depending upon species or clone. The adult form of the male is genetically determined and unaffected by environmental influences. It is a genetic polymorphism, rather than a polyphenism directly controlled by external conditions.

This review will consider the influence of endophagous, braconid parasitoids on wing development of three distinct aphid forms; the crowding- or nutritionally-induced, long-day winged virginopara, the short-day induced winged autumn female (gynopara) and the short-day-induced male. It will also examine the possible mechanisms involved in the phenotypic responses observed and discuss benefits that may accrue to the parasitoid.

Although Johnson (1959) reported earlier comments that aphids parasitized after the second moult tended to have imperfect wings (Webster and Phillips, 1912) and that aphid 'mummies' tended to be wingless (Veval, 1942), he made the first detailed examination of parasitoid effects on wing development in aphids. Johnson (1959) worked with winged virginoparae of the black legume aphid, *Aphis craccivora*, and the parasitoid, *Aphidius pratensis*. Subsequent to this, Bonnemaison (1970) studied the effects of *Diaeretiella rapae* on winged virginoparae and males of the non-host-alternating cabbage aphid, *Brevicoryne brassicae*, while Liu and Hughes (1984) examined the effects of *Aphidius sonchi* on wing development in virginoparae of the blackcurrant sowthistle aphid, *Hyperomyzus lactucae*. Each of these studies indicated that the earlier parasitism occurred, the greater the effect on wing development. Thus, when parasitoid oviposition occurred during the host's first stadium, there was a complete suppression of wing development, whereas later oviposition interfered with normal wing development, and while wings were present, they were frequently distorted. These authors recognised that the early stages of parasitism must contribute to these effects as they appear 'well before it [the larva] starts any destructive feeding' (Liu and Hughes, 1984). Johnson (1959) proposed that unhatched parasitoid eggs influenced wing development in the host and that the hatched parasitoid larvae had different effects. Bonnemaison (1970) found similar morphological effects after oviposition with either fertile or sterile eggs and proposed that perhaps the mechanics of ovipositor insertion or glandular secretions injected with the egg could be involved. Each of these authors proposed that hormonal changes, due either to influences on the host's endocrine system or by direct parasitoid secretion of endocrine factors into the host haemolymph, were the mechanism by which parasitoids exerted their effects.

More recently, we have published a number of findings on the effect of parasitism by *Aphidius ervi* on wing development in different winged morphs of two aphid species. These studies examined winged virginoparae and males in the non-host-alternating pea aphid, *Acyrthosiphon pisum* (Christiansen-Weniger and Hardie, 2000), and winged virginoparae, gynoparae and males of the blackberry-cereal aphid, *Sitobion fragariae* (Christiansen-Weniger and Hardie, 1998). Given the differences in the environmental induction of the different winged forms, we were interested to see if this was reflected in the morphological changes exerted by parasitism. The morphs and species studied in these and earlier papers are shown in *Table 1*, which also indicates that the winged virginoparae examined were induced by different means. The use of mature or detached leaves by some authors implies a nutritional stimulus, while crowd-induced stimuli were applied with different timings to effect wing induction.

Table 1. *Summary of aphid species and morphs investigated for the effect of parasitism on wing development*

Species	Life cycle	Inhibition of wing formation		
		Winged virginopara	Gynopara	Male
A. pisum[1]	Non-host-alternating	Yes[a]	NA	Yes
A. craccivora[2]	Non-host-alternating	Yes[b]	NA	–
B. brassicae[3]	Non-host-alternating	Yes[c]	NA	Yes
H. lactucae[4]	Host-alternating	Yes[b]	–	–
S. fragariae[5]	Host-alternating	Yes[d]	Yes	Yes

[1]Christiansen-Weniger and Hardie (2000); [2]Johnson (1959); [3]Bonnemaison (1970); [4]Liu and Hughes (1984); [5]Christiansen-Weniger and Hardie (1998).
Wing induction stimulated by: [a]prenatal crowding; [b]nutritional factors; [c]prenatal and postnatal crowding; [d]postnatal crowding. NA, not applicable as these species do not host-alternate.

Wing induction in aphids has been investigated at the physiological level. Part of this review will examine the evidence for endocrine control and attempt to relate these findings to the morphogenetic effects of parasitoids.

2. Morphological effects of parasitism on wing development

When early-stadium aphids are parasitized they become mummified during the fourth nymphal stadium and do not reach the adult stage. These fourth-stadium mummies are usually completely wingless with no evidence of the wingbuds that are present on normal fourths (*Figure 1(a)*). A minority will be winged/wingless intermediates and show wingbud rudiments (*Figure 1(b)*). When later stadia are parasitized, mummy formation occurs during the adult stage of the host and these can again be fully wingless or have an intermediate morphology. These adult intermediate forms can show rudimentary wingbuds (*Figure 1(c)*) or possess wings with a variety of deformities (*Figure 1(d)* to *1(f)*). In other cases the wings are fully formed (*Figure 1(g)*). The species depicted in *Figure 1* is *S. fragariae*, parasitized by *A. ervi*, and mummies with similar morphologies were found in all three winged forms examined, virginoparae, gynoparae and males (Christiansen-Weniger and Hardie, 1998). Indeed, parasitized presumptive winged forms of other species show similar mummy morphology. *Figure 2* shows winged/wingless mummies formed by *A. ervi* in adult *A. pisum*. *Figure 2(a)*, *(b)* and *(d)* correspond to the drawings in *Figure 1(c)*, *(d)* and *(e)*, while *Figure 2(c)* has one crippled wing and one wing partially unfolded.

In some cases, where aphids are exposed to parasitoid attack, mummy formation does not occur, but development can be inhibited beyond the fourth nymphal stadium in males by attack during early host stadia development (Christiansen-Weniger and Hardie, 2000). These forms again tend to be fully wingless and seem to occur when either an egg has not been deposited or when the host's immune system has successfully challenged and destroyed the parasitoid egg. In other cases, normal development occurs, indicating that either attack did not occur or nothing was transferred by the parasitoid.

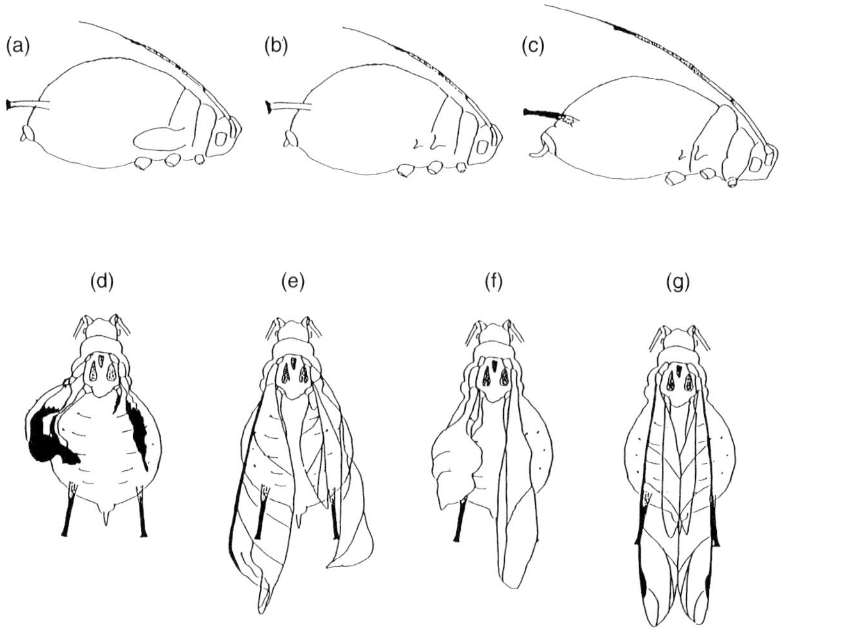

Figure 1. *Control and winged/wingless intermediate forms observed in* Sitobion fragariae *mummies after parasitization by* Aphidius ervi. *(a) Control fourth-stadium nymph with normal wingbuds. (b) Fourth-stadium nymph with rudimentary wingbuds. (c) Adult with rudimentary wingbuds. (d–f) Adults with deformed wings: (d) crippled, (e) partially unfolded and (f) blistered, fluid distended veins. (g) Adult with normally developed wings. From Christiansen-Weniger and Hardie (1998).*

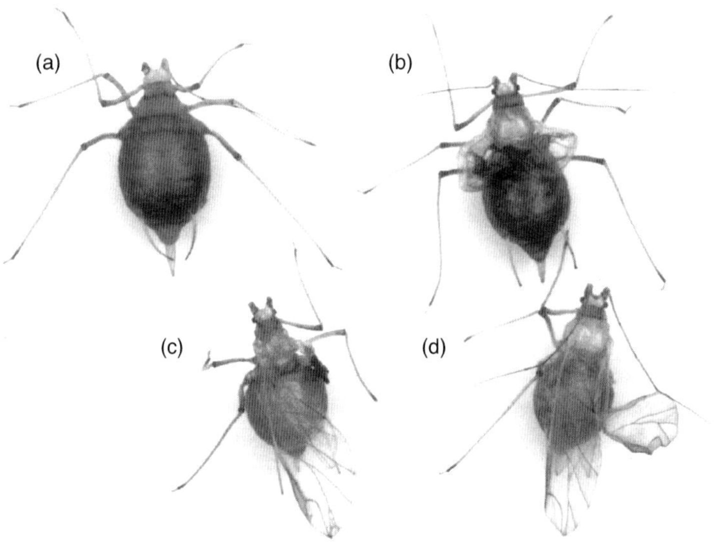

Figure 2. *Adult stadium* Acyrthosiphon pisum *mummies formed in the presumptive winged virginoparae after parasitization by* Aphidius ervi. *(a) Adult with rudimentary wingbuds. (b) Crippled wings. (c) One wing unfolded and the other deformed. (d) Wings partially unfolded.*

3. The effect of timing of parasitoid attack on wing development

Figure 3 shows the effect of parasitism on wing development in three presumptive winged morphs of *S. fragariae*. The final form of the host was assessed as fully winged, winged/wingless intermediate and fully wingless, irrespective of the final host stadium. Data are compared with control aphids that were not parasitized. Controls for the virginoparae (*Figure 3(a)*) show varying proportions of wingless and a few intermediates up to day 9. This is because in younger stadia it is not possible to determine whether the individual is destined to be winged or wingless as an adult.

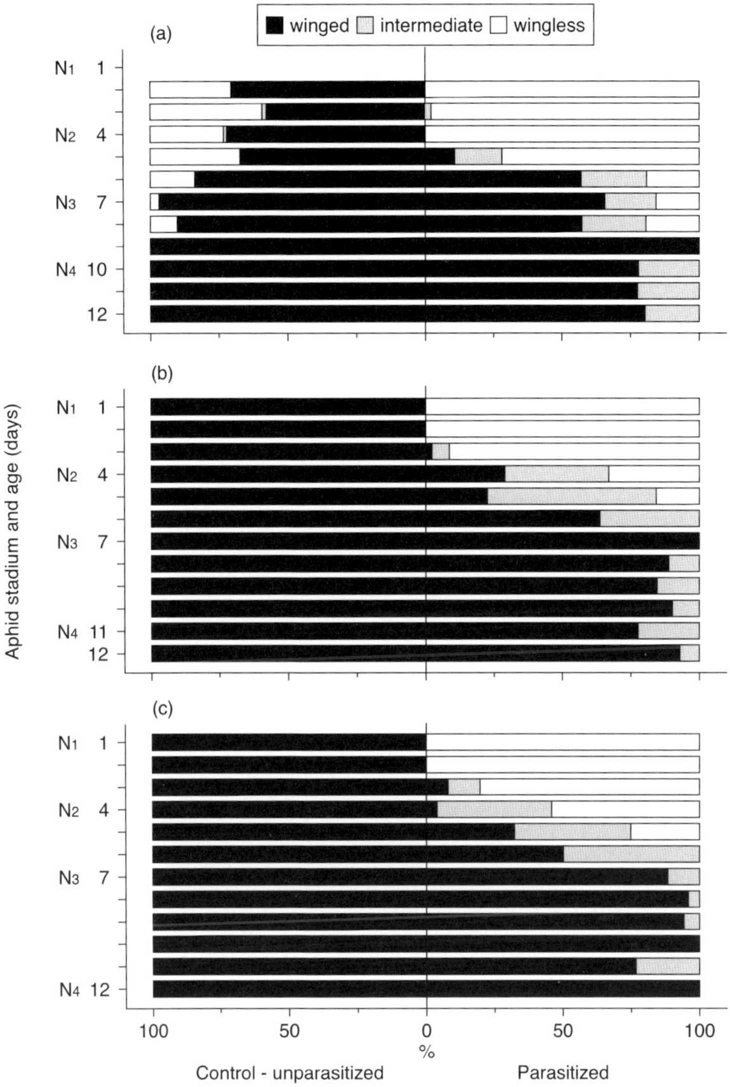

Figure 3. *Percentage of winged, wingless or winged/wingless intermediate forms of* Sitobion fragariae *developing from insects parasitized by* Aphidius ervi *at different times of their development. (a) Virginoparae, (b) gynoparae, and (c) males. From Christiansen-Weniger and Hardie (1998).*

In later stadia, wingbuds become visible and nymphs destined to be winged adults can be specifically selected for control and experimental procedures. The photoperiodic conditions relevant for the induction of gynoparae and males ensure that even the first-stadium insects selected are destined to be winged and so control groups are always 100% winged (*Figure 3(b)* and *(c)*).

The data shown in *Figure 3* confirmed previous reports that the earlier the host is parasitized the greater the effect on wing development with a maximum effect during the host's first stadium. It also emphasises that this observation is similar for all three of the winged aphid morphs despite the different environmental stimuli and whether wing induction is a polyphenic effect (virginoparae and gynoparae) or a genetic polymorphism (males).

Such findings indicate that the effect of the parasitoid on the host occurs at, or soon after, parasitoid oviposition. When oviposition takes place in later stadia, wing development is affected before the egg hatches (5 days at 15°C for *A. ervi* in *S. fragariae*, *Figure 3*). When we examined the efficiency of single attacks by *A. ervi* on individual first-stadia male pea aphids, we found, by dissection, that 73% had resulted in successful oviposition (Christiansen-Weniger and Hardie, 2000). This correlated well with 76% of the control cohort left intact and which developed into wingless fourth-stadium mummies. The remaining intact aphids did not mummify and developed as winged adults (9%), wingless adults (12%) or as arrested fourth-stadium, wingless nymphs (3%). By extrapolation, this result suggested that even when oviposition is not complete and an egg is not deposited, there are developmental effects on the insect host.

4. Physiological control of wing formation

In order to interpret how parasitoids might inhibit wing development in aphids it is necessary to review what we know of the physiological control of wing determination.

Insect juvenile hormones (JH) are produced by the corpora allata (CA) and universally control metamorphosis (Chapman, 1998; see also Chapter 1). When aphids are treated with exogenous hormone, or a hormone mimic, metamorphosis is inhibited and supernumerary nymphs or adult/nymphal intermediates develop in the fifth, normally, adult stadium (Hardie, 1981; Lees, 1980). Wingless forms have been considered neotenous as, superficially, the adults more closely resemble their nymphal forms than do winged adults (Kennedy and Stroyan, 1959). This led to the idea that JH could control both wing development and metamorphosis. Indeed, multiple roles for JH are well known, and in many insects they control not only metamorphosis in the younger stadia, but take on a reproductive role in the adult (Chapman, 1998), including aphids (Hardie, 1987b).

A number of studies have examined the effect of exogenous JH treatments either prenatal, to the mother, or postnatal to the individual test insect in an early stadium. This treatment would be expected to induce wingless forms and produced either supportive (Kohno and Takaoka, 1977; Srivastava and Chhibber, 1980), contradictory (Applebaum *et al.*, 1975; Hardie, 1980; Hardie *et al.*, 1995) or equivocal results (White, 1968; White and Lamb, 1968). All of these experiments were done on crowding-induced, long-day aphids. With the short-day-induced gynoparae of *A. fabae*, not only were nearly all (99.5%) of the experimental insects winged, but postnatal exposure to short days induced apterization (Hardie, 1980, 1981) and this morphogenetic effect could be mimicked by JH treatment. Wing-inhibition (apterization) effects decline with the age at which the insect is treated, and both treatments are ineffective when applied to the early second stadium. Later application of the hormone, particularly during the third nymphal stadium, inhibits metamorphosis. In these experiments, the morphological characteristics of apterization and juvenilization were well defined and separate.

Due to the complexity of environmental factors controlling wing diphenism (see above), it is essential that studies of wing diphenism are critically controlled. The most robust experimental protocols involve species where wing induction is maternally controlled, where direct crowding of the adult female causes her to produce winged progeny, and where wing induction is photoperiodically induced as in the case of gynoparae (see above). Here, treating adult mothers with hormone (prenatal treatment of the progeny) or treating the progeny after birth (postnatal treatment) should reveal any effect, but the timing of the treatment is always crucial. It should also be emphasised that where nymphal and wingless adult characters overlap, these characters cannot be used to assess wing dimorphism, particularly in studies on the effects of JH.

More recently, the precocenes (2-2-dimethylchromene analogues; Bowers, 1985) have been used to investigate aphid wing dimorphism. Some of these compounds, while being species-specific, selectively destroy CA cells and prevent the synthesis of JH (Henrick, 1991; Pratt, 1983). Classically, this promotes early metamorphosis and thus induces precocious adult development where the adult form develops in an earlier than usual stadium (Hales and Mittler, 1981; Hardie, 1986). The first study on wing formation indicated that precocene II (6,7-dimethoxy-2,2-dimethylchromene) induced the appearance of winged progeny when applied to adult pea aphids, *A. pisum* (Mackauer *et al.*, 1981). While the pea aphid data were confirmed, this compound did not induce wing formation in two other aphid species, *Megoura viciae* or *Aphis fabae* (Hardie, 1986). In addition, the crucial evidence that this compound did indeed act by destroying CA cells is lacking, indeed at high doses it can induce an extra moult indicating a JH-agonist effect rather than antagonist-like activity (Hardie, 1986). Detailed investigations of the effects of 2,2-dimethylchromene derivatives on metamorphosis and wing induction in the pea aphid have shown that wing induction is effected by compounds that tend to have JH-agonist properties rather than JH-antagonist properties (Hardie *et al.*, 1995, 1996). In contrast, compounds that have JH-antagonist effects (in particular, precocene III, 6-methoxy-7-ethoxy-2,2-dimethylchromene) inhibit the production of winged progeny after application to crowded mothers. Precocene III also inhibits wing development when applied postnatally to young presumptive winged virginoparae (Hardie *et al.*, 1995, 1996). These results run counter to the idea that JH promotes wing induction in the pea aphid.

If the effects of precocene derivatives are due to interference with CA cells and a reduction in the synthesis of JH it should be possible to effect a 'rescue' with hormone treatment. Such experiments have shown that while precocious adult development induced by precocene III can be rescued by subsequent treatment with JH, precocene effects on wing development cannot (Gao and Hardie, 1996). Thus, neither the induction of wings (precocene II) nor the inhibition of wing development (precocene III) can be countered by subsequent JH application (Gao and Hardie, 1996; Hardie *et al.*, 1995). The precise mode of action of these 2,2-dimethylchromene compounds, with respect to wing induction in aphids, remains unknown but seems unrelated to juvenile hormone titres.

In *Myzus persicae*, precocene III has been shown to induce precocious adults and to induce male production by treated adult virginoparae (Hales and Mittler, 1983). Male production was interpreted as indicating that low JH titres promoted the production of male eggs, but this effect could not be countered by subsequent treatment with JH. Males born after prenatal treatment with precocene III were winged.

5. Evidence for parasitoid interference with hormones

Detailed observations on the morphological changes induced by *A. platensis* in *B. brassicae* made by Johnson (1959) indicated to him that parasitism induced changes in JH titres which affected both metamorphosis and wing formation. However, for many observed characters it was not possible to distinguish these two developmental pathways. For example, ocelli are only present in winged adults and not in nymphs or wingless adults, and this is true also for the olfactory secondary rhinaria on the antennae. Other characters can be considered to be strictly metamorphic. Thus, the shape of the cauda from short, broad and triangular in the nymphs with the major change to the distinctive adult long, conical shape at the final moult. Thus, ambiguity in the interpretation of why certain morphologies occur after parasitism can lead to confusion when attempting to infer physiological control mechanisms. The crucial evidence would come from measurements of JH titres, but this is not possible for individuals or small groups of aphids.

Recently, we have searched for evidence to indicate that parasitism of *A. pisum* by *A. ervi* affects JH titres by looking for morphological characters and changes that would indicate an influence on metamorphosis (Christiansen-Weniger and Hardie, 2000). A distinct and unequivocal indicator of excess JH is an extra moult or attempted extra moult while, as mentioned above, the retention of juvenile characters in what should be the adult stadium is often ambiguous. We could not induce an extra moult in parasitized larvae, but some larvae did not progress further than the fourth stadium, either because they were mummified by the developing parasitoid, or their development to the adult stadium was inhibited. Similar observations were noted by Pennacchio *et al.* (1995) and were deemed to be the result of low ecdysteroid levels that prevented a further moult. There is no implication that metamorphosis had been directly affected as the arrested forms were larval and they showed no adult characters (i.e. there was no precocious adult development indicating low JH titres). These observations were made both in our work and Pennacchio *et al.* (1995).

Counts of secondary rhinaria on the antennae were inconclusive as reduced numbers could indicate development towards the wingless form or an inhibition of metamorphosis (see above). However, adult parasitized males that developed with rudimentary wingbuds had fewer on the third segment than did control winged males, although the numbers on the fifth segment were similar (Christiansen-Weniger and Hardie, 2000). As there is no wing polyphenism in males there is no equivalent of the winged/wingless polyphenism seen in virginoparae, but despite the reduced numbers on the third antennal segment, an average of over 40 secondary rhinaria is not convincing evidence of inhibition of metamorphosis. Completely wingless males only appeared as fourth stadium insects and, in accordance with their nymphal status, no secondary rhinaria were observed.

Precocene III could inhibit wing formation in progeny of treated adult female pea aphids as well as when applied directly to young virginoparae destined to be winged (i.e after pre- and postnatal application) (Gao and Hardie, 1996). These results apply strictly to virginoparae, and precocene III had no effect on wing development in male pea aphids (Christiansen-Weniger and Hardie, 2000). It seems that despite the differences in response of the two winged morphs to precocene III, parasitoids exert identical effects on wing formation in these morphs, indicating a very different mode of action for these wing-inhibiting factors.

6. How do parasitoids exert their effects on aphid wing development?

Not only do four species of parasitoid wasps inhibit wing development in five aphid species examined, but they also have the same effect on different winged morphs, which are induced by different environmental factors and by different physiological mechanisms. Detailed observations on the wing abnormalities promoted by *A. ervi* indicate similar morphological effects on wing differentiation in each morph. There were some indications that the wing rudiment intermediates were more prevalent in parasitized male *A. pisum* than in virginoparae (Christiansen-Weniger and Hardie, 2000), but this was not so in *S. fragariae* (Christiansen-Weniger and Hardie, 1998). The similarities in morphogenetic effect indicate a common parasitoid effector mechanism. These observations, taken together with the different wing induction factors, appear to indicate that the parasitoid effect on wing differentiation might well take place downstream of any interference with the input signal or any endocrine effector system. The full cascade of events from gene to wing differentiation is not known and the possibility of pinpointing a precise mechanism remains elusive. However, recent work on the castration of *A. pisum* virginoparae by venom proteins of *A. ervi* has indicated a direct effect upon the germarial cells (Digilio *et al.*, 2000). Evidence presented above indicates that it is not the developing parasitoid embryo, but rather a factor injected with the egg during oviposition that is responsible for wing inhibition. It is quite possible that similar, or the same, protein factors that effect castration also affect wing development and both may act directly upon the target cells. For the developing parasitoid, the result is the same. Metabolically active tissues are shut down and nutritional suitability for the parasitoid is optimized. Experimental approaches to support/confirm this proposed direct action, rather than an indirect effect via the endocrine system, are now feasible using purified protein preparations, injection into the insect and microscopic examination of wingbud tissue (Digilio *et al.*, 2000). *In vitro* culture of ovarian tissue is also feasible (Hardie, 1987b) and offers further possibilities to study the effect of parasitoid proteins on host germarial cells.

7. Summary

Aphids show a high degree of polymorphism, and wings are developed in at least three distinct adult morphs promoted by particular environmental stimuli. The winged virginoparae are long-day, summer forms, which can be laboratory-induced under crowded conditions. Winged autumn forms (gynoparae) are produced by some host-alternating species under short-day conditions, and in many species there are winged males, again induced by short days, but this is a genetic polymorphism. We have examined the effect of parasitism by *Aphidius ervi* on wing formation in each of these aphid morphs in two host species, the non-host-alternating pea aphid, *Acyrthosiphon pisum*, and the host-alternating blackberry-cereal aphid, *Sitobion fragariae*. *Aphidius ervi* pupates within the cuticle of the dead host, forming an aphid 'mummy'. Mummy formation occurs during the host's fourth (nymphal) stadium or in the adult stadium, depending upon the time of parasitism.

When each of the potentially winged aphid morphs were parasitized by *A. ervi*, wingless and winged/wingless intermediate forms were induced at the mummy stage. Significantly more unparasitized, control aphids developed into winged adults. This developmental effect was most pronounced when test aphids were parasitized during the first stadium – almost all developed completely without wings, i.e. they were apterized. When parasitoid attack was delayed until later host stadia, fewer aphids developed without wings, but more winged/wingless intermediates and aphids with deformed wings were observed. Developmental effects could be seen even when nymphs in the fourth (last) nymphal stadium with well-developed wing buds were parasitized, indicating that the very early stages of parasitism, prior to eclosion of the larval parasitoid, were influential. Indeed, aphids that survived parasitoid attack and did not form mummies (indicating that oviposition had not taken place), developed as wingless individuals, suggesting that reproductive tract fluids from the female parasitoid might be important in wing inhibition.

The endocrine control of wing formation is not clear. A role for JH has been demonstrated in gynoparae, which are induced by short days, but not in the winged virginoparae induced by crowding. There is also evidence that JH may control induction of the male genotype. In the pea aphid, treatment with 2,2-dimethylchrome compounds (precocenes), which classically act as pro-allatocidins and selectively destroy CA cells, thus preventing JH synthesis (Bowers, 1985), have a variety of effects, some of which mimic the wing inhibition seen after parasitoid attack. Data now available indicate that the influence of the parasitoid on the host may not be via the host endocrine system but through a direct effect on wingbud tissue.

11

The parasitoid–host endocrine interaction from a nutritional perspective

S. Bradleigh Vinson, Francesco Pennacchio and Fernando L. Consoli

1. Introduction

Insect parasitoids are carnivores that grow and develop at the expense of their host, which ultimately dies having failed to reproduce. For this review we are going to focus on the entomophagous parasitoid Hymenoptera, but before we examine the parasitoid–host endocrine interaction from a nutritional perspective, we hope it will be useful to first examine some of the different insect carnivore strategies from a food procurement point of view, and to explore how the different strategies have influenced nutrient resource utilization. Further, we will separate the parasitoids on an ecological basis in order to provide a better framework on which to discuss their different endocrine-resource utilization strategies.

2. Entomophagous insect strategies

The prey and hosts of entomophagous insects are containerized, i.e. they represent a resource that is contained within the confines of the cuticle (or the chorion in the case of eggs) of the prey or hosts. The 'container' may be within a second container (e.g. a gall, a mine, a nest cell, or hibernaculum), but regardless, the result is that, in the case of parasitoids, the host represents the only nutritional resource that is available to the developing parasitoid progeny. Thus, a major difference between a predator and a parasitoid is that developing predators require several prey items, while developing parasitoids require only one. However, gaining access to the resource, whether once or many times, presents a number of challenges to the lifestyle of any carnivore. The carnivorous lifestyle is further complicated by several other biological realities. The first challenge to be faced is to overcome the defences of the host or prey. These

Endocrine Interactions of Insect Parasites and Pathogens, edited by J.P. Edwards and R.J. Weaver.

defences can be external (e.g. the physical and behavioural attributes of the prey or host), or they can be internal. For predators and parasitoids alike, these internal defences may include toxins, but for parasitoids in particular, there are the additional challenges of blood clotting and the immune defence system of the host (Strand and Pech, 1995; Vinson, 1993a). In addition, there may be biochemical or developmental incompatibility and competition with the host tissues. While the host immune system is extremely important, and in many cases is manipulated by parasitoids (Salt, 1968; Strand and Pech 1995; Vinson, 1990a), there is little evidence that the immune system is controlled by the endocrine system (Vinson, 1993b). However, it should be noted that the immune system may be indirectly influenced by the endocrine system in a number of ways, an aspect that will not be covered further here.

The other major challenge for entomophagous predators or parasitoids is to secure and manipulate the host as a food resource. If we examine this from the perspective of the different kinds of carnivorous insect, the challenges differ greatly between them (*Table 1*). Relatively large predators, such as preying mantids or adult ladybirds, capture and physically consume their prey. The relatively large size of these carnivores, their restraining grasp and chewing mouthparts allow them to capture and hold the prey, and to consume the resource directly, with little additional processing. The only processing that may be required might include shredding or chewing along with the addition of digestive enzymes and, in some cases, detoxification of the resource. Relatively small predators must subdue their prey before they can process them. For example, fire ants, *Solenopsis invicta*, sting and inject lethal venoms into their prey, which they then cut up and feed to their young, which digest the food resource (Vinson, 1997). Sucking predators, such as the stinkbugs (*Podisus* spp.) must first immobilize their prey. They then secrete powerful enzymes into the prey to digest the resource, and suck out the digested nutrients until only the outer cuticle of the prey remains (Cohen, 1992, 1995).

Parasitoids are organisms that are free living as adults, but their young are dependent on the adult females selecting a single host from which the immature stages must derive all of their nutrition. There is no opportunity for immature parasitoids to chose, or move to, a different resource. The immature parasitoid stages are initially parasitic, but later in their development they often act more like predators (although they only consume one prey item – the host). Parasitoids have been characterized in a number of different ways (see reviews by Askew, 1971 and Godfray, 1994). These include the life stage(s) of the host that they attack (i.e. egg, egg-larval, larval, larval-pupal, pupal, or adult), their location either in or on the host (i.e. endo- or ecto- parasitoids), or whether the female lays one or many eggs per host (i.e. solitary or gregarious parasitoids, respectively). From an ecological perspective, parasitized hosts have been subdivided into koinobionts or idiobionts based on whether the parasitized host can move or not (Haeselbarth, 1979). More specifically, parasitoids that allow their hosts to continue to grow after parasitization are called koinobionts, whilst those termed idiobionts do not permit such growth, and must survive entirely on the nutritional resource provided by the host at the time of parasitization (Askew and Shaw, 1986). This approach is useful when considering the parasitoid–host interaction from a nutritional perspective. The idiobionts (*Table 1*) include egg parasitoids, pupal parasitoids, and ectoparasitoids of concealed larval hosts. The koinobionts (*Table 1*) include exposed ectoparasitoids of larval hosts, larval endoparasitoids, egg-larval parasitoids and endoparasitoids of adult hosts. The koinobionts can be further separated into host regulators (Vinson and Iwantsch, 1980a) or conformers (Lawrence, 1986a), however

Table 1. The different strategies used by carnivorous insects (including parasitoids), to obtain their nutritional needs from their prey or host

Predators	(a) *Large mandibulate predators*: these capture and physically process their prey by chewing/shredding, digestion and, in some cases, detoxification. (b) *Small mandibulate predators*: these must subdue their prey before processing. This often involves the injection of a toxic compound to immobilize their prey. (c) *Sucking predators*: these also must immobilize their prey, but then they must secrete powerful enzymes into their prey to digest the resources.
Parasitoids (idiobionts)	(a) *Egg parasitoids*: in these parasitoids, adult female wasps inject compounds that halt host egg development (a form of host regulation). In addition, enzymes are released that aid in the digestion of the egg contents. The resource is relatively uniform in quantity, but options (iv) and (v) (below) can be important. (b) *Pupal parasitoids*: these inject compounds to halt host development and preserve host tissue. In addition, enzymes are released to aid in the digestion of pupal tissues. The resource is relatively uniform in quantity, but options (iv) and (v) (below) can be important. (c) *Ectoparasitoids of concealed larval hosts*: these avoid being physically removed by injecting venoms that paralyze the host and also prevent host moulting. However, having paralysed the host, the parasitoid must make do with the totality of nutritional resource that the host represents at the time of the attack. In such circumstances there are five options: (i) to locate and attack only large hosts; (ii) to attack hosts of any size, and to adjust the size or number of parasitoid larvae accordingly; (iii) to measure the resource and then lay only male eggs in the smaller resource; (iv) to measure the resource and adjust the number of eggs laid in accordance with host size; (v) to lay many eggs and let the parasitoid larvae adjust their numbers by scramble competition.
Parasitoids (koinobionts)	(a) *Ectoparasitoids of exposed larval hosts*: these parasitoids should not paralyze the host, but must prevent the host moulting, and must obtain their nutrition from the host haemolymph in a way that limits damage to the host. (b) *Endoparasitoids of larval hosts*: these parasitoids must compete with host tissues for the resources in the haemolymph. However, the haemolymph protein levels are low in young hosts, but increase in older hosts as these approach pupation. Depending on the host developmental stage attacked, these parasitoids either adjust their development to be in line with the physiological changes in the host, or they must adjust the host's development to suit the needs of the parasitoid. (c) *Endoparasitoids of adult hosts*: these parasitoids must compete with the reproductive system of the host for available resources. Often, this results in reduced host reproductive capacity (e.g. castration).

most parasitoid species (including the idiobionts) posses qualities of both regulators and conformers and function somewhere between these two extremes.

Egg parasitoids derive their nutrition from a single host egg. While they often prefer younger host eggs (and a few appear to have an absolute requirement for young eggs; Strand, 1986; Vinson, 1998), many can successfully attack older eggs. Often, the ovipositing female injects, along with her egg, compounds that halt host egg development (Strand *et al.*, 1986), a form of host regulation. In addition, enzymes are often released (usually following the hatching of the larval parasitoid) that aid in the digestion of the egg contents releasing the nutritional content for the benefit of the parasitoid (Strand *et al.*, 1985, 1986). The egg parasitoids, particularly after hatching are, in reality, feeding on preserved host tissue (Vinson, 1998).

Pupal parasitoids have not received as much intensive study as either the egg or larval parasitoids, but whether they are ecto- or endoparasitic, they are similar to the egg parasitoids in that they inject venoms that usually stop host development (Beard, 1978), and redirect the flow of resources to the parasitoid (Rivers and Denlinger, 1994b). As the pupal host is relatively quiescent and confined, the need of the parasitoid is probably directed more towards the preservation and utilization of the nutritional resource, than to interacting with the host endocrine system. However, unlike the host of the egg parasitoids, in pupal hosts there is a well-developed endocrine system in place with which the parasitoid may have to contend.

Larval ectoparasitoids have different challenges, depending on how they obtain their nutrition from the host. This, in turn, depends on the host's life style. Ectoparasitoids that attack active larval hosts must reduce the possibility of the juvenile parasitoid stages being physically removed. This is often accomplished by injecting venoms that paralyse the host (Beard, 1952; Shaw, 1981) and which may also preserve the host for relatively long periods (Steiner, 1986). In addition, this paralysis may prevent any further moulting by the host. However, by paralyzing the host or by preventing further moulting, the parasitoid is constrained to completing its development entirely on the finite food resource that the host constitutes at the moment of a successful attack. This is of particular importance as, unlike host eggs or pupae, the various larval stadia of the host can vary greatly in the amount of resource that each represents. There are five options that larval parasitoids have when confronted by this potentially variable resource. These are listed in *Table 1* (idiobionts). *Bracon compressitarsis* is a good example of a species that always selects a host that provides all of the nutrition required by its progeny (*Table 1(i)*). This species responds to cotton buds containing third-instar host larvae, but rarely responds to those containing younger larvae (Rojas, 1993). In contrast, some parasitoids are able successfully to attack hosts of varying size or varying age. This often occurs in gregarious parasitoid species (Godfray, 1994) and, if more or less resource is available, the developing parasitoid larvae adjust their size or number accordingly (*Table 1(ii)*). For example, Schmidt and Smith (1985, 1987) have shown that *Trichogramma minutum* measures its host and lays just enough eggs to utilize the resource contained therein (*Table 1(iv)*). Another strategy is for the adult female parasitoid to determine the host size, and lay a male egg in a small host and a female egg in larger hosts (*Table 1(iii)*). An excellent example of a species that uses this strategy is *Aphitis mealnius* that lays male eggs in smaller hosts (Murdoch *et al.*, 1992). Some parasitoids lay a large number of eggs in a single host, and let scramble competition among the developing parasitoid larvae determine the number and size of parasitoid progeny that the resource can support (*Table 1(v)*). A possible example is *Melitobia digatata* in which a female wasp will lay eggs over a period of days, and sometimes the total number of eggs laid far exceeds the number that can be supported (S.B.Vinson, unpublished results). As the resource is utilized, some of the sisters die and are partially consumed by their siblings. In all these cases, there is no advantage to the parasitoid in manipulating the endocrine system of the host, or in responding to host hormones (other than as nutritional precursors to the parasitoid's own hormones).

The other group of parasitoids, the koinobionts (*Table 1*), all attack hosts that usually vary in size, and these parasitoids do not paralyze their hosts, but allow them to continue to respond to the environment and, in some cases, even to feed. Although, as discussed above, ectoparasitoids of concealed hosts usually paralyze them, the paralysis of exposed hosts by ectoparasitoids may not be a good option. This is because it is usually important for hosts (and therefore, for the parasitoid) to be able to respond to, and to move away from, dangerous situations (e.g. exposure to sunlight, or attack by a predator). Thus, the host must not be killed, paralyzed, or incapacitated by digestive enzymes. Instead, the ectoparasitoid must obtain its nutrition from the host's tissues in a way that limits damage to the host. For the endoparasitoids, the challenges are even more complex. The parasitoid egg must evade the host's immune response, but equally challenging for the developing parasitoid is to obtain all of the nutrition that it needs before it kills the host. For many parasitoids of larval insects these challenges are exacerbated by the pressures on adult female wasps to oviposit in young hosts, because the first egg oviposited appears to have an advantage (Vinson and Hegazi, 1998; Vinson and Mourad, 2000). As a result, at the time of attack the young larval host often does not represent sufficient resource for the development of an adult parasitoid (Mackauer and Sequeira, 1993). Although the options faced by koinobionts are the same as those faced by idiobionts, not killing or paralysing the host has allowed the koinobiont larval endoparasitoids opportunities to evolve towards the utilization of very young hosts that, initially, do not represent sufficient nutritional resource. This is accomplished by allowing small hosts to continue to feed and grow. The ultimate example of this strategy may be seen in the egg-larval parasitoids. In many ways, these resemble larval endoparasitoids. Although the egg-larval parasitoids are constrained to initiate their development in young larvae, whether or not larval endoparasitoid species are restricted to certain larval sizes depends on the biology of the host. Many potential host species (especially those of concealed hosts) are exposed or accessible as first-instar larvae, but then become inaccessible as older larvae, only to become accessible again as the larvae re-emerge to seek pupation sites. As a result, some species, such as *Biosteres (Opius) longicaudatus*, prefer later-stage host larvae (Lathrop and Newton, 1933; Lawrence *et al.*, 1976). In such cases, the need to manipulate or respond to the endocrine system of the host is very different from that of those species that attack very young larvae. Before we examine the role of endocrine manipulation with regard to parasitoid nutrition, we will outline the nutritional needs of the koinobiont parasitoids over time, the nutritional content focusing on the protein changes of the host over time, and the parasitoid's manipulation of the endocrine system of the host and the nutritional consequence of such manipulation.

3. Nutritional needs of the koinobiont parasitoids

House (1977) suggested that parasitoids have essentially the same nutritional requirements as other insects. This view has not changed dramatically (Grenier 1994;

Grenier *et al.*, 1993; Vinson, 1994), although Nettles (1990) suggested that certain chemicals present in the host may have become essential dietary requisites for some parasitoids, as a result of evolutionary changes. While this may be an important consideration for the development of diets for the *in vitro* rearing of parasitoids, it is not a problem for a parasitoid developing in its habitual host. The nutrients available to a developing parasitoid are also determined by a number of factors that include:

- the nutritional history of the host, and its nutritional intake after parasitism (Vinson and Barbosa, 1987);
- the presence of deleterious substances within the food consumed by the host (Barbosa, 1988; Flanders, 1942a);
- the modification, storage, and utilization of the nutrients by the host;
- the endocrine and developmental status of the host (Lawrence, 1986a; Lawrence and Lanzrein, 1993; Vinson and Iwantsch, 1980b).

While certain specific nutrients may occasionally be in short supply in the host, it is the amino acids and proteins that are most likely to be limiting to the growth of the parasitoid. In addition, the specific amino acid composition of the host proteins may be particularly important. In chordates, there is a dietary need for calcium as a major component of the skeleton. By contrast, in insect lines there is a reduced need for calcium (Taylor, 1986), but an increased dietary requirement for components of the cuticle (i.e. chitin and proteins). Of the cuticular components, the proteins constitute the bulk of the dry weight of the insect cuticle (Richards, 1951, 1978). Although the amino acid composition of the cuticle is complex (Hackman, 1976), certain amino acids (e.g. tyrosine, ß-alanine, phenylalanine, and asparagine) may be particularly important and abundant. For example, both tyrosine and ß-alanine appear to play an important role in cuticle sclerotization (Andersen, 1985; Chen, 1985; Sherald, 1980). Furthermore, tyrosine has low solubility, and may be more accessible as a resource when present as a constituent of a protein. In addition, most of these amino acids are locked-up in the cuticle of the host (and therefore unavailable), and may only become available during a moult, or in the latter stages of host development when storage proteins appear in the haemolymph. Thus, the composition of the proteins that are available to the developing parasitoid may change as the host develops, and the nutritional needs of the parasitoid may also change as it develops.

The eggs of most koinobiont parasitoids, and particularly the endoparasitoid species, are small and hydropic (Flanders, 1942b), i.e. they have little or no stored nutrients other than some lipid (Le Ralec, 1995). This decreases both the female parasitoid's investment in the egg, and minimizes the size of the egg – a factor that may have been essential for the evolution of a narrow, piercing ovipositor. In addition, at oviposition, the egg is effectively placed in a nutrient broth (the host) where it can readily increase in size (Schlinger and Hall, 1960; Smith, 1952; Vinson and Hegazi, 1998) by absorbing amino acids and other nutrients from the host haemolymph (Ferkovich and Dillard, 1986).

Although both the eggs and the newly hatched larvae must compete with host tissues for certain resources, their needs and opportunities to fulfil these needs differ between the two stages. Newly oviposited eggs are most likely to absorb fairly simple molecules, and passive diffusion probably plays an important role. However, as reported by Greany *et al.* (1990), certain larger molecules (e.g. proteins) may also be important for egg development. As the parasitoid embryo develops there is a substantial change in the capacity of the embryo to survive in a less than an ideal medium. As shown by Pennacchio *et al.* (1992b) and Hu and Vinson (1997a, b), parasitoid eggs placed in various artificial media for *in vitro* rearing are more likely to complete their development if their embryonic membranes have developed. This change in viability is the direct result of the formation of the embryonic membrane, which facilitates the absorption of nutrients (Vinson and Hegazi, 1998). There is another major change in the nutritional opportunity that occurs as soon as the larvae hatch. From this time onwards oral intake is possible, and the range of nutrients available to the developing larvae increases to include larger, more complex molecules such as complex glyco- and lipoproteins.

Near the completion of larval development, the larvae of most endoparasitoids begin actively to feed on host tissues as well as haemolymph (Liu, 1985; Sequeira and Mackauer, 1992). This has been referred to as 'destructive feeding' (Hemerik and Harvey, 1999). For example, fourth-instar *Campoletis sonorensis* larvae only begin to feed on vital host tissues just prior to egression (Wilson and Ridgway, 1975). By contrast, third-instar *Cardiochiles nigriceps* egress and then re-insert their head into the body of the host and consume all the soft tissues (S.B.Vinson, unpublished results). In effect, many parasitoids only become tissue predators just prior to or just after egression from the host. Thus, over time there is a sequential change in the nutritional opportunities available to the parasitoid, and also in the nutritional needs of the developing parasitoid larva (Vinson and Hegazi, 1998). The parasitoid changes during its development from a passive absorber of simple nutrients, to a consumer of more complex foodstuffs, but only begins to destroy the host (by feeding on vital host tissues) when parasitoid development has been virtually completed. It is during this last phase, tissue predation, that the fat body of both *C. sonorensis* and *C. nigriceps* become enlarged and opaque (S.B. Vinson, unpublished results), presumably due to the storage of nutrients prior to pupation.

In addition to their changing nutritional needs, the egg and newly hatched parasitoid larva have small surface areas in comparison to the total cellular surface area of the host, and all are bathed by the haemolymph of the host. This disparity in surface area is probably more serious when an egg is placed into an early host instar, where the titres of various nutrients in the host haemolymph are low (Chen, 1985; Munn *et al.*, 1967). In addition, the rapid host cellular growth that occurs during the early host instars probably intensifies the competition between the parasitoid and the host tissues for the nutrients that are available. Further, the levels of nutrients in the haemolymph are influenced and regulated by the endocrine system (Wyatt, 1980). Out of this myriad of nutritional challenges and opportunities, the developing parasitoid has several options. One is to wait until the nutritional conditions in the host change – the conformer strategy (Lawrence, 1986a, 1990). The other is to alter the haemolymph environment of the host in favour of the developing parasitoid – the regulation approach (Vinson and Iwantsch, 1980a).

4. Protein changes in the host over time

The eggs of most host insects are anhydropic, that is they contain a lot of stored lipid and protein (Flanders, 1942a), mainly in the form of vitellin (Kunkel and Nordin, 1985). As embryonic development proceeds, these stored nutrients are broken down and utilized to form the tissues of the developing immature insect, resulting in a gradual decrease in the nutritional value of the egg. Thus, the value of an insect egg for parasitoid exploitation by egg parasitoids decreases with age (Strand, 1986).

When insect larvae hatch they have used most, if not all, of the stored resources of the egg. After hatching, they begin to feed, but being small, they are usually constrained to consuming softer, less nutritious tissues. For example, very young *Heliothis* larvae feed on the mesophyll tissue of the leaf, and only later do they begin to feed on whole leaf tissue. Still later in larval development are they able to feed on the more protein-rich, flowering or fruiting structures (Ellsbury *et al.*, 1989; Farrar and Bradley, 1985; Shaver and Parrott, 1970). At each step in this feeding succession, the protein richness of the diet increases. Thus, the haemolymph of young larvae is low in protein and amino acids, partly due to the low quality of the initial food resources. In addition to a diet that is often low in amino acids and protein, the low protein titre in the haemolymph of these younger larvae is accentuated by the high demand of the growing larval tissues for these resources.

As the host larvae grow and develop (and in some cases increase their protein intake) they begin to accumulate some of the excess nutrients, and the haemolymph protein titre begins to increase (Wyatt and Pan, 1978). In general, the protein content of the haemolymph (*Figure 1*, H-PT) increases rapidly during the latter half to one third of larval development (Chen, 1985). In *H. virescens*, this rapid increase occurs following the moult to the last larval instar (Pennacchio *et al.*, 1993). In general, the increase in the concentration of protein in the haemolymph that occurs before pupation is accomplished by the synthesis and release of protein from the fat body to the haemolymph. This occurs as the fat body prepares to switch from protein synthesis to protein storage in preparation for pupation (Levenbook, 1985). These proteins have been referred to as storage proteins (Roberts and Brock, 1981), of which

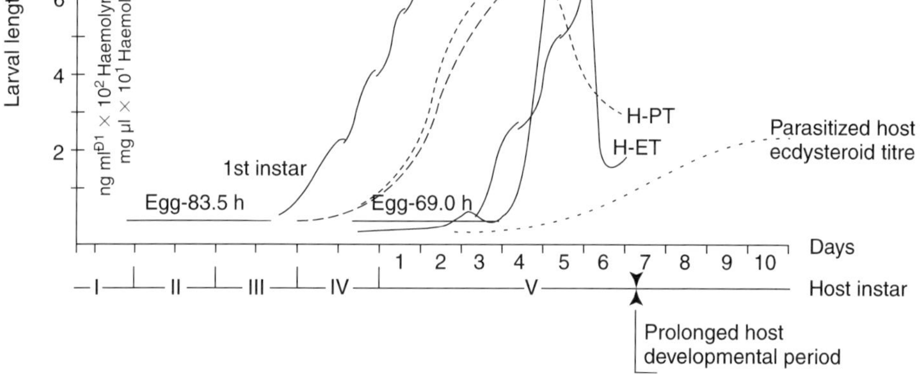

Figure 1. Development of Campoletis sonorensis *larvae when oviposited in a 1st- or 4th- instar host with an overlay of the haemolymph protein and ecdysteroid titres from control and parasitized hosts. H-PT, Heliothis (control) protein titre; H-PE, Heliothis (control) ecdysone titre.*

there are several different groups (Kanost *et al.*, 1990). One group is the arylphorins (Telfer *et al.*, 1983), which are rich in aromatic amino acids (Munn *et al.*, 1967; Ryan *et al.*, 1984). A second group is the methionine-rich storage proteins (Kanost *et al.*, 1990), which in *Manduca sexta* appear in the 5th (final) larval instar (Riddiford and Hice, 1985), and increase in response to elevated juvenile hormone (JH) (Webb and Riddiford, 1988). In *Helicoverpa* (=*Heliothis*) *zea*, the arylphorin consists of hexamers of a single type of subunit (Haunerland and Bowers, 1986). In *H. virescens*, an increase in the storage proteins follows the moult to the last (5th) larval instar, and reaches a peak at around day 4 (*Figure 1*) before dropping again (Pennacchio *et al.*, 1993). A key point is that, at this stage in the development of many holometabolous insects, the accumulated proteins are stored in the haemolymph in a form that is not being taken up by other tissues.

As the developing larva approaches the moult leading to pupation, there is reabsorption of the storage proteins from the haemolymph resulting in a drop in the protein titre of the haemolymph and, at the time of pupation, these protein levels are often less than half their original level (Chen, 1985). The uptake of protein by the fat body, and the corresponding drop in the haemolymph titre of proteins, appears to be initiated by the release of ecdysone and its conversion to 20-hydroxyecdysone (20E) (Tojo *et al.*, 1981, 1985). As in many other holometabolous insects, in *H. virescens* there is a small peak of ecdysteroids (*Figure 1*, H-ET) that appears around the middle of day 3 of the final larval instar (Tanaka and Vinson, 1991a). This is referred to as the wandering or pupal commitment peak (Bollenbacher, 1988; Riddiford, 1985). The subsequent, major release of ecdysone begins on day 4, and peaks on day 5 of the final larval instar, but has dropped to a low level by the time of the moult to the pupal stage (Tanaka and Vinson, 1991a) (*Figure 1*). During the pupal stage the protein titre in the haemolymph is moderately high, but when the insect moults to the adult, there is often a rise in the haemolymph protein titre, which may subsequently rise or fall depending on the sex and reproductive strategy of the adult.

5. The host endocrine system

To help understand the nutritional interactions of a parasitoid with the endocrine system of its host, one must first look at the components of the system, and determine the timing and titres of both hormones and nutrients in the haemolymph. There are many hormones that circulate in the haemolymph of insects (Riddiford and Truman, 1978). These range from those that control moulting and metamorphosis, to those regulating carbohydrates, amino acids, proteins, lipids, and water balance. Some of these potential nutrients, for example, carbohydrates (Dahlman and Vinson, 1975, 1976; Thompson and Dahlman 1998), lipids (Barras *et al.*, 1970; Hayakawa, 1987; Rivers and Yoder, 1997) and nitrogen compounds (Thompson and Cohen, 1984) show changes during host development, and may be very important for parasitoid development, but have, in general, received little study. By contrast, the hormonal systems concerned with moulting and metamorphosis have been explored by a number of authors, and are reviewed in Chapter 1 of this volume. In the following sections we shall concentrate our attention on how these endocrine systems, and their manipulation by parasitoids, might impinge on the nutritional requirements of the developing parasitoid stages.

In addition to the roles and effects of insect hormones outlined in Chapter 1, hormones such as JH and ecdysteroids have additional effects on a number of insect tissues and physiological processes that may affect the nutrition of the developing parasitoid. These include the synthesis and storage of proteins, lipids and carbohydrates. The regulation of these nutrients in the host depends on the timing and titre of hormonally active compounds, and on their interaction with each other (Collins, 1974; Locke, 1980). In addition to endogenous changes in host nutrient levels, it is also clear that 'unusual' protein changes occur in the haemolymph of parasitized host insects (Rolle and Lawrence, 1994; Soldevila and Jones, 1994). Often, these parasitism-specific proteins are highly abundant (Harwood and Beckage, 1994; Richards and Edwards, 1999a; Schepers *et al.*, 1998; Shi *et al.*, 1999; Soldevila and Jones, 1994). Some of these 'new' proteins may be derived from the host in response to parasitism, or may be produced and secreted by the developing parasitoids or its associated teratocytes and/or polydnaviruses (PDV). Whilst there is evidence that some of these proteins may be involved in regulating host endocrinology, some may be more important for achieving optimal nutrition for the parasitoid than for the regulation of host developmental processes.

6. Parasitoid developmental strategies, the endocrine system and the nutritional consequence

The primary challenge for a larval parasitoid developing on or in a host is to extract the needed resources required for its development. But the strategies utilized to gain the nutrients are influenced by the life history strategies of the host. As noted in *Table 1*, there are three major strategies used by koiniobiont parasitoids.

6.1. *Ectoparasitoids of exposed hosts*

As discussed earlier (Section 2), it is important for ectoparasitoids of exposed hosts to prevent moulting, but not to prevent the ability of the host to move away from danger. Although parasitoids with this strategy are not particularly common, the genus *Euplectrus* provides a good example. *Euplectrus comstockii,* which parasitizes several lepidopteran larvae (e.g. *Trichoplusia ni*), prevents host moulting following attack (Coudron and Brandt, 1996) and this result is achieved with the venom injected by the ovipositing female (*Figure 2*). This venom appears to inhibit ecdysone release and ecdysis (Kelly and Coudron, 1990). As a result of this inhibition of ecdysone release, the protein titre of the haemolymph is elevated (see Coudron *et al.*, 1994, and Chapter 9 this volume). This elevation in protein is due to an increase in the haemolymph titre of storage proteins and, interestingly, this increase occurs prematurely even if earlier instars are attacked (Coudron *et.al.*, 1994; Chapter 9 this volume). This premature increase in storage proteins is favourable to the developing parasitoid in that it provides a rich source of protein, high in certain amino acids (e.g. tyrosine), while at the same time eliminating the competition with host tissues for these same resources.

6.2. *Endoparasitoids of larval hosts*

The endoparasitoids of larval hosts (including the egg-larval parasitoid species) have received the most attention with regard to the effects of parasitism on the hosts

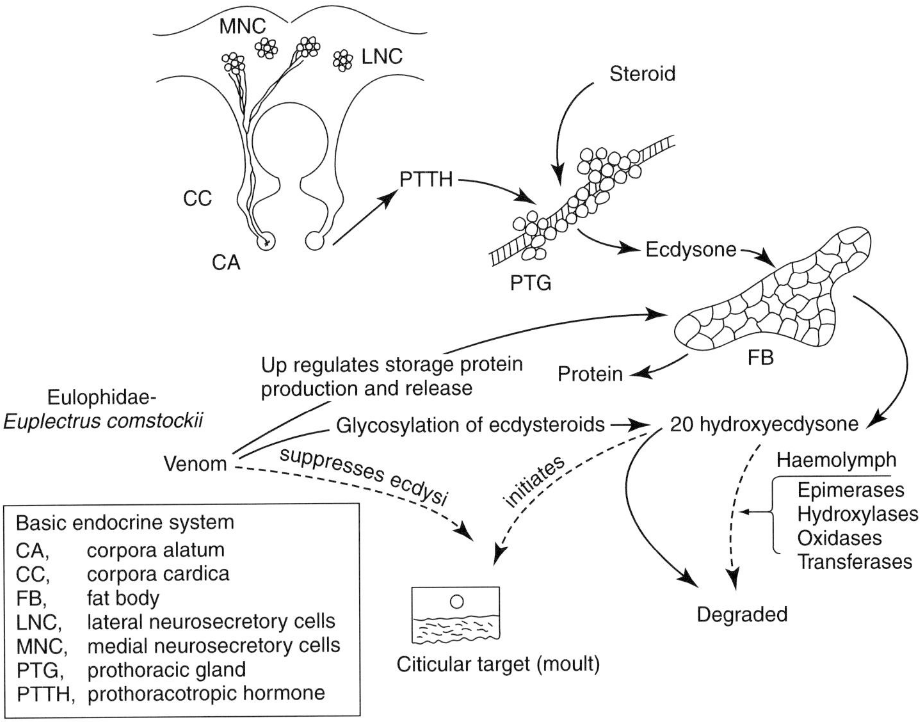

Figure 2. *Diagram showing some of the targets of the venom of* Euplecterus comstockii *(an ectoparasitoid of exposed larval hosts) on the host endrocrine system, which results in increased protein titres in the host haemolymph (circled). Developed from Coudron (1991), Coudron and Brandt (1996), Coudron and Brandt (1996) and Kelly and Coudron (1990). For further details see Chapter 9, this volume.*

endocrine system (Beckage, 1991, 1993a, 1997a; Coudron, 1991; Dahlman *et al.*, 1990a, b; Dover *et al.*, 1995a; Fathpour and Dahlman, 1995; Grossniklaus-Bürgin *et al.*, 1998; Iwantsch and Smilowitz, 1975; Lawrence and Lanzrein, 1993; Pennacchio *et al.*, 1992b, 1993, 1994a, b; Reed and Brown, 1998; Tanaka *et al.*, 1987; Zhang *et al.*, 1992; see also Chapters 3, 4, 5 and 7). Again, however, little attention has been paid to the nutritional implications of such endocrine manipulation.

In the following section, we will focus on three different parasitoid strategies. The first strategy is illustrated by *Diachasmimorpha (Biosteres) longicaudata*, which develops in the dipteran, *Anastrepha suspensa* (the Caribbean fruit fly). This braconid, like others of the genus (Lathrop and Newton, 1933), primarily attacks larval Diptera, particularly the later instars (Lawrence *et al.*, 1976). In these hosts, the 1st-instar parasitoid larva must reach a minimal critical size (Lawrence, 1986b), but will then only moult to a 2nd instar when ecdysone is released by the host to initiate pupariation (Lawrence, 1982, 1988). While the results of these studies provide evidence that ecdysone is necessary to initiate a moult by the 1st-instar parasitoid to a 2nd instar, it is not clear whether this also provides the parasitoid with optimal access to the nutritional resources that the host represents. However, the information suggests that this developing parasitoid delays its own development until the host is ready to initiate pupation, since by this time the host has accumulated the maximum nutritional resource. Only then does the parasitoid proceed to consume the host (which now represents an optimal resource) and by this time the parasitoid has little need to manipulate the host hormones, or even to maintain the host as a living organism.

In contrast to *D. longicaudata*, which generally attacks later host instars, many endoparasitoids of larval hosts attack very young hosts that, at the time of parasitization, do not represent the nutritional resource necessary for the complete development of the parasitoid. In many of these species the host will continue to grow and develop, although if the rate at which the host grows is markedly reduced there is a danger that ultimately the host could represent an insufficient nutritional resource. Some parasitoids have evolved strategies to counteract this potential problem. Thus, if we examine the development of the ichneumonid, *Campoletis sonorensis* and the braconid, *Cardiochiles nigriceps*, before and after the pupal commitment peak of ecdysone and the concomitant increase in protein in the host, we see examples of these different strategies.

Working with *C. sonorensis*, Gunasena *et al.* (1989) reported that the weight of the egressing parasitoid larvae was greater when they developed from an egg that was deposited in a 4th-instar host than from an egg deposited in a 1st or 2nd-instar host. Hu and Vinson (2000), while examining the *in vivo* development of the larvae found an identical result. Vinson and Hegazi (1998) found that the embryonic development of *C. sonorensis* is rather long and that during embryogenesis the egg increases in size from 0.35 × 0.09 mm to 0.5 × 0.17 mm (length x breadth). Such growth is typical of hydropic eggs (Schlinger and Hall, 1960; Smith, 1952). Furthermore, the period of egg development is variable, lasting 83.5 ± 24.0 h when oviposited into late 1st-instar larvae, but was significantly shorter (69 ± 21.8 h) when oviposited in late 4th-instar larvae (Hu and Vinson, 2000) (*Figure 1*). However, the development of the 1st-instar parasitoid larvae in younger hosts was more rapid, and these larvae were smaller than comparable 1st-instar larvae developing in later instar hosts. This difference is maintained throughout larval development, and eventually results in adults that are smaller when they have developed in early host instars than those in later host instars (*Figure 1*). A similar observation has been made with the parasitoid *Venturia canescens* (Harvey *et al.*, 1994, 1995).

These differences in the development of *C. sonorensis* larvae can be related to several physiological and developmental constraints. One of these constraints is that, irrespective of the host instar in which the parasitoid commences development, these parasitoids must reach a critical size and their tissues must be competent to moult – just as appears to be the case in other insects (Nijhout, 1975). This is shown diagrammatically in *Figure 3*, where a 1st-instar larva grows and develops in an ideal environment (*Figure 3*, line L), but must then, within a certain time period, develop to a physiological state where the tissues are capable of responding to the moulting hormone (*Figure 3*, Tx). Additionally, the larva must reach a critical size (*Figure 3*, line C) before it can moult, but this critical size is usually attained sometime later (*Figure 3*, Ty). By contrast, larvae hatching into a nutrient-deficient environment grow more slowly (*Figure 3*, line A) and pass beyond time Ty long before they have reached the normal critical size. However, despite having failed to reach the critical size, some time later these larvae moult (*Figure 3*, Ta) and this results in smaller than normal 2nd-instar larvae (*Figure 3*, Sa). However, when this occurs, it 're-programmes' development by reducing the 'critical size' that the larvae in each subsequent instar must attain before moulting.

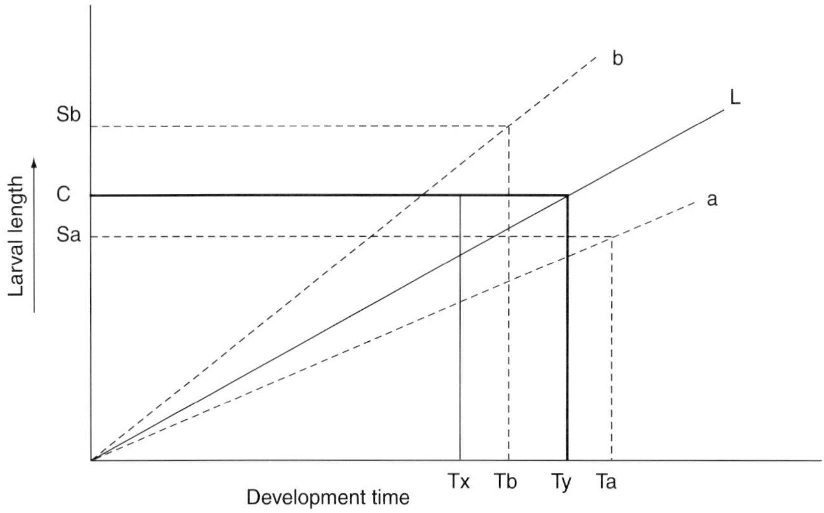

Figure 3. *Diagram showing effect of the nutritional environment on the growth and moult of* Campoletis sonorensis *1st-instar larva, which influences the development of the later instars. Line L, growth of larvae in an ideal environment. Line A, growth of larvae in a sub-optimal environment. Line B, growth of larvae in a rich environment. C, critical size to moult in an ideal environment. Tx, earliest time at which tissues are capable of moulting; Ty, time of moult when larvae reach a critical size in an ideal environment. Time Ta, the time of the moult when larvae reach a critical size in a sub-optimal environment; Tb, the time of moult when larvae reach a critical size in a rich environment; Sa, reset critical size of larvae developing in a sub-optimal environment; Sb, reset critical size of larvae developing in a rich environment.*

Conversely, 1st-instar larvae hatching into a nutrient-rich environment (*Figure 3*, line B) reach their critical size for moulting (*Figure 3*, C) before the tissues are ready to respond to the moulting hormone As a result, the moult is delayed. However, these larvae continue to grow, and eventually reach a stage capable of responding to the moulting hormones. At this point (*Figure 3*, Tb), the moult occurs, albeit somewhat earlier than normal. This results in a larger than normal 2nd-instar larva (*Figure 3*, Sb), and also 're-programmes' development by increasing the 'critical size' that the larvae in each subsequent instar must achieve before moulting. The net result is that, as both of these 'types' of 2nd-instar larvae recommence their development, both are now in a more appropriate nutritional environment. However, the re-programming of their respective critical sizes means that the smaller larvae will always remain smaller than the larger larvae (*Figure 3*) and will, therefore, eventually produce smaller adults.

Harvey (1996) and Hemerik and Harvey (1999), working with *V. canescens* attacking different hosts with differing larval size ranges, have reported a similar phenomenon, but discuss the implications in a slightly different way. The larger host (*Galleria mellonella*) has six larval stadia and reaches a weight of over 300 mg, while the smaller host (*Anagasta kuehniella*) has five larval stadia and only reaches around 40 mg. The development of *V. canescens* depends on the host attaining a critical size for parasitoid development, and this occurs earlier in the larger *G. mellonella* (during the late 4th stadium when the host is around 30 mg) than in the smaller *A. kuehniella* (where critical size is attained during the late 5th stadium when the host is around 10 mg).

Although developmental arrest occurs earlier in *G. mellonella* than *A. kuehniella*, the developing parasitoid has a higher developmental 'critical weight threshold' in *G. mellonella* (Harvey, 1996). As discussed by Harvey (1996), these data indicate that parasitoid development is synchronized with physiological changes in the host that are not detected by the developing parasitoid in *G. mellonella* until the host is larger. In both small and large host species, the development of the parasitoid larvae is delayed allowing the host to grow and reach a 'critical stadium' (a nutritionally suitable stadium) at which time the parasitoid larva begins exponential growth and consumes the host.

We will now consider how the regulation of the host as a nutritional resource is linked to the regulation of the neuroendocrine system. If we examine the development of a *C. sonorensis* larva oviposited in an early host instar (*Figure 1*, late 1st host instar), we see that the parasitoid larva is able to complete much of its development prior to the time of the ecdysone peak in the host (*Figure 1*, day 3 of 5th host instar), which initiates host pupation and the reabsorption of storage proteins. Further, the developing parasitoid larvae complete their last two larval instars within the host's early 5th instar – a time when the storage proteins in the host haemolymph naturally increase. Subsequently, the parasitoid larvae consume all of the soft internal tissues of the host, egress and pupate (Wilson and Ridgway, 1975). In this example, alteration of the neuroendocrine system of the host is not particularly important. In fact, Dover *et al.* (1995a) found that the parasitoid-induced degeneration of the prothoracic glands (PG) was stadium-specific, and did not occur before the 5th (final) instar. However, in other circumstances, the alteration of the host's neuroendocrine system does become important for *C. sonorensis* larvae – especially if they commence their development in later host instars. This is because the parasitoid-induced degeneration of the host's PG (which normally occurs in the 5th instar; Dover *et al.*, 1988, 1995a; Dover and Vinson, 1990) not only prevents host pupation, but also prevents the re-absorption of host haemolymph proteins, which, as a result, remain elevated. Although, the haemolymph protein titre of *Heliothis virescens* larvae parasitized by *C. sonorensis* is about half that of same stage larvae parasitized by *C. nigriceps* (S.B.Vinson, unpublished results), *C. sonorensis* larvae that have commenced their development in the later host instars are larger (*Figure 1*, mid to late 5th host instar), and take longer to develop.

As is the case with the development of *C. sonorensis*, the development of *C. nigriceps* is different in different host instars. However, in *C. nigriceps* there is no difference in the duration of the egg stage, which remains constant regardless of whether or not the egg is deposited in a nutrient-low 1st-instar host or a nutrient-rich 4th-instar host. However, the 1st-instar larva of *C. nigriceps* remains as a 1st instar for a longer period of time (6.5 days) when hatching into a nutrient-low 1st- or 2nd-instar host (*Figure 4*). Furthermore, in *C. nigriceps,* either the critical size that a larva must attain before it is competent to moult is more specific than in *C. sonorensis*, or *C. nigriceps* larvae respond to a rise in protein titre. As a result, *C. nigriceps* larvae that hatch in a 2nd- or 3rd-instar host, only moult to the 2nd instar in early 5th-instar hosts (when the host's haemolymph protein level increases) (Pennacchio *et al.*, 1993). This moult to the 2nd instar occurs at about the same time (*Figure 4,* day 2 of the 5th instar) for all parasitoid larvae if they have hatched in 1st-, 2nd-, or 3rd-instar host. Furthermore, the moult of the 1st instar does not appear to be related to any obvious rise in host ecdysteroid titre, and occurs even in parasitoid larvae hatching in 5th-instar hosts (when the normal peak of ecdysteroid is, in any case, suppressed by parasitism (Pennacchio *et al.*, 1993).

However, in all cases, moulting clearly coincides with an elevation in the titre of the host haemolymph protein (*Figure 4*). The importance of the elevation of protein in stimulating the growth of *C. nigriceps* larvae was underlined when *C. nigriceps* 1st-instar larvae maintained *in vitro* were found to moult to a 2nd instar only when the protein concentration of the rearing medium was increased (Pennacchio *et al.*, 1992b). Thus, while it is clear that ecdysone (acting in its role as a moulting hormone) may be endocrinologically important for the development of larval endoparasitoids (Lawrence, 1986a), the development of some parasitoids is not dependent on the presence of physiologically active titres of ecdysteroid in the haemolymph of the host, although there is a nutritional need for the steroid structure. When *C. sonorensis* larvae commence their development in early host instars, the parasitoid larvae do not moult, yet they are probably exposed to natural changes in host ecdysteroids, since parasitized host larvae continue to moult (Gunasena *et al.*, 1989). However, when *C. sonorensis* larvae commence their development in final (5th)-instar host larvae they are not exposed to ecdysone (because of the parasitoid-induced degeneration of the host PG), yet they still moult three times (Hu and Vinson, 2000). In the case of *C. nigriceps*, the parasitoid larvae moult in the early part of the host's 5th instar, and this occurs regardless of which host instar was originally attacked (Pennacchio *et al.*, 1994b). Moreover, this is a time when the host's ecdysteroid titre is very low (Tanaka and Vinson, 1991a).

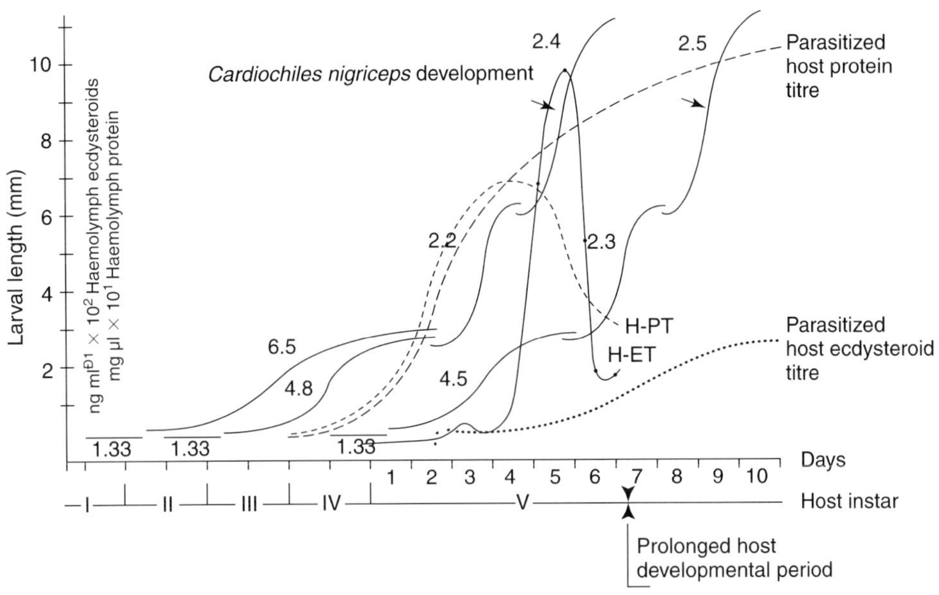

Figure 4. *Development of* Cardiochiles nigriceps *larvae when oviposited in a 1st, 2nd, or 4th-instar host, with an overlay of the haemolymph protein and ecdysteroid titres from control and parasitized hosts. H-PT, Heliothis (control) protein titre; H-PE, Heliothis (control) ecdysone titre. Numbers refer to development time in days (egg stage was consistently 1.33 days).*

In the case of many endoparasitoids of larval hosts, alteration of the host neuroendocrine system leads to an increase in the availability of nutrients. However, this optimization of nutrition is accomplished in different ways in different species. The females of many ichneumonid and braconid wasps inject PDV (Krell, 1991a, b; Stoltz, 1993; Stoltz and Vinson, 1979a; Stoltz *et al.*, 1984; Webb, 1998), together with other venom components (Coudron, 1991; Piek and Spanjer, 1986), all of which may influence the physiology of the host (Stoltz, 1986; Vinson, 1988). Other factors may be released by the parasitoid egg (Fisher, 1971; Mackauer, 1986; Silvers and Nappi, 1986; but see Vinson and Hegazi, 1998), and yet more factors may be produced both by larvae (Vinson *et al.*, 1994) and, in some braconids, by teratocytes. These teratocytes are giant cells that are derived from the serosal membrane of the egg, and are released into the haemocoel of the host when the parasitoid larva hatches. There is evidence that factors released by teratocytes may impact both on the host and on the developing parasitoid (Dahlman and Vinson, 1993). In some systems (see Section 6.3) teratocytes may have more of a direct nutritional function.

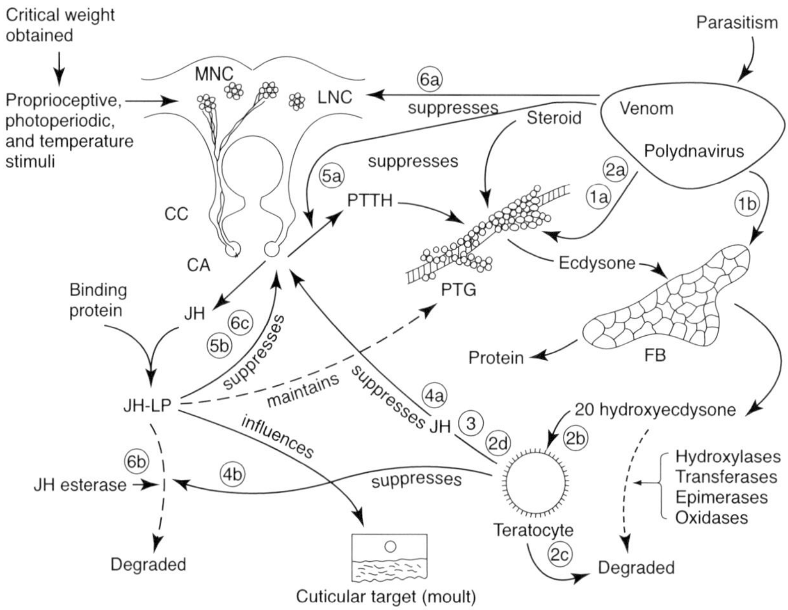

Basic endocrine system
CA, corpora allatum
CC, corpora cardica
FB, fat body
HEDE, hemolymph ecdysone degrading enzymes
JH, juvenile hormone
JHE, juvenile hormone esterase
JH-LP, juvenile hormone lipoprotein complex
LNC, lateral neurosecretory cells
MNC, medial neurosecretory cells
PTG, prothoracic gland
PTTH, prothoracotropic hormone

Parasitoid – host endocrine interactions
1 = *Campoletis sonorensis*
2 = *Cardiochiles nigriceps*
3 = *Biosteres longicaudatus*
4 = *Microplitis croceipes*
5 = *Apanteles kariyai*
6 = *Cotesia congregata*

Figure 5. *Diagram showing some of the targets of the secretions released into the haemolymph by various stages of several larval endoparasitoids (see numbers and key) on the neuroendrocrine system of their host, leading to increased protein titres in the haemolymph. Developed from many authors (see text).*

Although the function of many of the secretions contained in the venom are unknown, the presence of the *C. sonorensis* PDV (*Figure 5*, 1a) results in the premature degeneration of the PG in 5th-instar host larvae (Dover and Vinson, 1990), and results in changes in the regulation of protein synthesis in the fat body (*Figure 5*,1b) (Shelby and Webb, 1997, Chapter 4). In *C. nigriceps* (*Figure 5*, 2a) the PDV suppresses ecdysone synthesis by the PG (Tanaka and Vinson, 1991a; Pennacchio *et al.*, 1993). In addition, the teratocytes (*Figure 5*, 2b, 2c) appear to degrade any remaining ecdysone (Pennacchio *et al.*, 1994a). Tanaka *et al.* (1987) reported that parasitism by *Apanteles kariyai* suppressed prothoracicotropic hormone (PTTH) release in its host *Pseudaletia separata* (*Figure 5*, 5a), while Zitnan *et al.* (1995a, b) reported that parasitism by *Cotesia congregata* suppressed the release of various neurosecretions from the lateral and medial neurosecretory cells of host brains (*Figure 5*, 6a).

Parasitism by endoparasitoids of larval hosts may also lead to changes in host JH titres. Teratocytes (*Figure 5*, 2d) appear to release a molecule with JH activity (Joiner *et al.*, 1973), which may, in turn, suppress the release of PTTH (Pennacchio *et al.*, 1994a, b). The possible release of JH by teratocytes of the parasitoid *B. longicaudatus* has been suggested by Lawrence *et al.* (1990) (*Figure 5*, 3), as has the down-regulation of JH esterase (JHE) by the teratocytes of *Microplitis croceipes* (Dong *et al.*, 1996) (*Figure 5*, 4b). What do all of these alterations have in common? One common feature is a reduction of ecdysteroids that is often accompanied by an elevation of JH (*Figure 5*, 4a, 5b, 6c). As a result of these changes there is either a premature or a sustained rise in the haemolymph titre of the storage proteins from the fat body. This may be very important to the developing parasitoid for several reasons. One is because the parasitoid is bathed in a high titre of proteins. Secondly, the parasitoid is not in competition with host tissues for these proteins, as these proteins are stored in the haemolymph while various tissues are undergoing preparation for metamorphosis.

6.3. *Endoparasitoids of adult insects*

The last group of parasitoids that we will discuss are those that parasitize adult insects. The PG are absent in the adult insect and usually degenerate after pupation (Lockshin, 1985). However, ecdysteroids are still found in the adult insect, in which they are mainly produced by the ovaries and testes (Hagedorn, 1983). In some adult insects, ecdysteroids have been implicated in the regulation of vitellogenin synthesis and egg production (Sahota, 1969). Further, ecdysteroids can be abundant in the eggs of some insects (Tawfik *et al.*, 1999). In addition, JH is still produced by the adult CA (Pratt and Tobe, 1974) and has also been implicated in the regulation of vitellogenin synthesis and uptake by the egg (Engelmann, 1983). However, this is not the case for all insects (Raabe, 1986). Levels of JH in adult insects may be substantially higher than those found in larval stages (see Chapter 1).

In the adult insect, most of the nutritional resource is directed towards the reproductive system, and developing parasitoids must compete with the reproductive system of the host for these resources. Thus, it is not surprising that many parasitoids cause the host's reproductive system to atrophy (Brown and Kainoh, 1992; Kring and Kring, 1988). One of the best examples of this phenomenon involves the venom of the braconid, *Aphidius ervi*, and its effects in the adults of its aphid host, *Acyrthosiphon pisum*. In the unparasitized aphid, the ovary produces ecdysone, which stimulates the production of the vitellogenins that are necessary for egg production (Hardie, 1987a).

The amino acids required for this protein production are synthesized *de novo* by bacterial endosymbionts in the host aphid (Douglas, 1998). The venom of *A. ervi* suppresses the production of ecdysteroids by the host ovaries, and induces the degeneration of the germarium tissue necessary for egg production. However, the symbiotic microorganisms continue to produce amino acids. Once the parasitoid larvae hatch, the associated teratocytes take up the amino acids and release proteins (*Figure* 6), which are, presumably, used by the developing parasitoid (Digilio *et al.*, 1998, 2000; Falabella *et al.*, 2000; Pennacchio *et al.*, 1995).

Another example involves the parasitization of several different adult weevil species by the braconid, *Microtonus aethiopoides*. Parasitism by *M. aethiopoides* results in the increased levels of a 150-kDa protein in some host weevil species, but not in others. As reported by Lovallo *et al.* (2000), this increase in a particular protein was not correlated with whether or not the hosts were permissive or non-permissive, but was up-regulated only in the females of two of the species. Interestingly, a protein with a similar molecular weight was present, regardless of parasitism, in two other species. Such results might suggest that these proteins are important for parasitoid development, being up-regulated only in those species in which they are normally low.

Another example of regulation of host proteins is provided by *Dinocampus (= Perilitus) coccinellae* (Okuda and Kadono-Okuda, 1995). Although interactions of this parasitoid with the endocrine system of its host have not yet been well defined, the

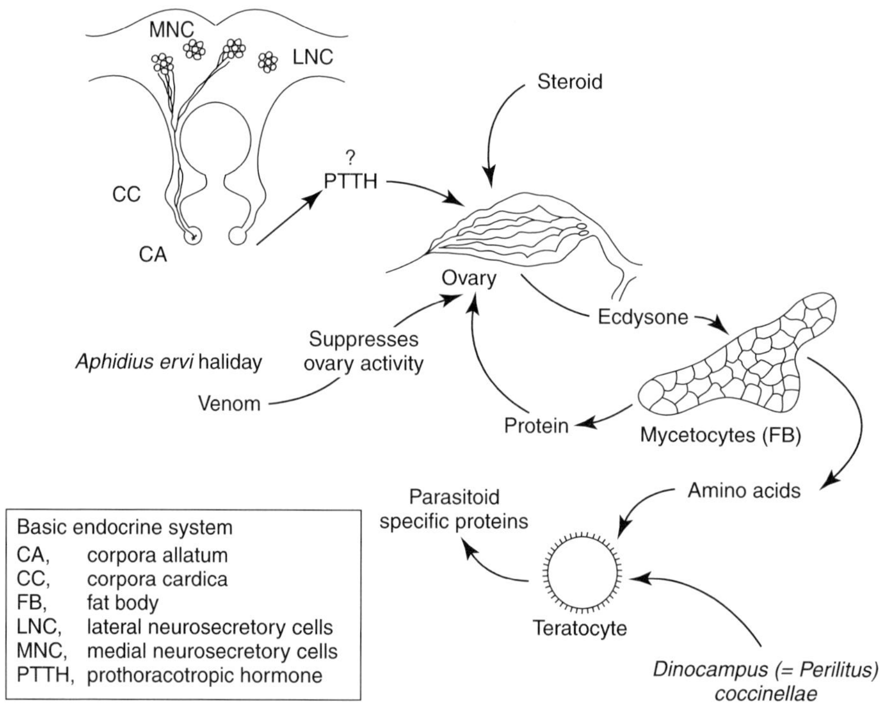

Figure 6. *Diagram showing some of the targets of the secretions released into the haemolymph by two parasitoids of adult insects,* Aphidius ervi *(from Digilio* et al., *1998, 2000; Falabella* et al., *2000) and* Dinocampus coccinellae *(from Kadono-Okuda* et al., *1995, 1998; Okuda and Kadono-Okuda, 1995), on the neuroendrocrine system of their host, leading to increased protein titers in the haemolymph.*

teratocytes produce a protein that accumulates within these cells. The protein is a high mannose glycoprotein with an amino acid composition similar to that of vitellogenin (Kadono-Okuda *et al.*, 1998) (*Figure 6*). The teratocytes become hypertrophied, and are consumed by the developing parasitoid (Kadono-Okuda *et al.*, 1995). In each of these cases, the production of proteins is targeted towards the parasitoid in a way that eliminates or reduces the access of host tissues to these proteins. In the case of the accumulation of proteins in teratocytes, these cells are accessible only to the mandibulate parasitoid larvae. Another strategy is where the teratocytes secrete proteins that the host tissues are unable to assimilate or utilize. These proteins may be considered parasitoid-specific, and it is possible that they are targeted to specific parasitoid receptors for absorption.

7. Summary

Most of the studies concerning the alterations to the host neuroendocrine system as a result of parasitism are largely descriptive (Lawrence and Lanzrein, 1993), although some molecular information is becoming increasingly available. As shown in *Figures 2, 5* and *6*, many different aspects of the endocrine system are influenced, and we have not attempted to document them all. However, there are several aspects that must be considered. One is that there is some pressure for the koinobiont parasitoids to attack their hosts early in host development, but such hosts are nutritionally poor (even though they may be less immunologically competent than older hosts). The other aspect to be considered is that parasitoids must exploit the nutritional resource that the host represents and, in this respect, host proteins may be of particular importance. Furthermore, it is likely that the amino acid composition of these proteins may be especially important for the successful development of the parasitoid. For the koinobiont parasitoids this scenario presents several challenges. The first challenge is not to kill the host until the parasitoid larva is able to make optimum use of the nutritional resource that is represented by the host. The second challenge is for the developing parasitoid to be able to compete with the developing host tissues for the nutritional resources that would normally be available for host development. This competition is probably important for parasitoids in young hosts, but may be a greater problem for parasitoids developing in older hosts. This is because the relatively small surface area of the parasitoid, compared to the much larger surface area of the cells that comprise the tissues of the host, means that the surface area available for the absorption of nutrients is biased in favour of the developing host. This disparity becomes even greater as the tissues of the host grow, so that parasitoids attacking older hosts may be at an even greater disadvantage. Furthermore, the cells of host tissues during the later stages of host development should be at their peak of fitness, having accumulated resources during their development. However, there is one period in the development of host insect tissues when much of the stored nutritional resource is transferred to the haemolymph. This happens when the host insect prepares to undergo metamorphosis. During this period most of the host tissues are entering a phase of change. This results in a reduction in host tissue competition with the parasitoid as the tissues of the host switch from the accumulation of resources to the initiation of metamorphosis. Many endoparasitoids either delay their development until these changes occur in the host, or they induce these changes prematurely.

Parasitoids employing the 'waiting for host metamorphosis option' have some additional time-related challenges, since the mortality of the parasitoid is directly proportional to the mortality of the host. As discussed by Slansky (1986), as exposed hosts age they are increasingly likely to be subjected to predation and disease, and the parasitoids are more likely to be hyperparasitized. This should have resulted in an increasing selection pressure towards the evolution of parasitoids able to develop in earlier host stages and this, in turn, would have selected for the ability of parasitoids to induce the premature availability of host resources. However, this selection pressure would be less for parasitoids that attack hosts that are protected late in their development from these mortality factors (i.e. concealed hosts). Hemerik and Harvey (1999) developed a model based on the idea that in order to survive in a host, developing parasitoids must reach a certain size within a limited time span. When attacking different sized host species or host stadia, parasitoids have to optimize both their size and development time. As discussed (Section 6.2), *V. canescens* completes its development during the last larval stadium of its small host, *A. kuehniella*, but completes its development in the antepenultimate larval stadium when developing in the larger *G. mellonella*. These data support the hypothesis of Slansky (1986), but this would be subject to the developing parasitoid first satisfying its nutritional needs. However, the ultimate challenge for the parasitoid is to optimize its nutritional resources in a way that optimally fulfils its nutritional requirements, but this depends on the host species, the stadium attacked, and the nutritional resources available to the host. However, a favourable nutritional environment is not only dependent on the presence of resources, but the parasitoid must have competitively favourable access to these resources. Depending on the species and stadium attacked, the parasitoid may wait for the host haemolymph to reach a favourable nutritional status, or it may manipulate the physiology of the host in a way that results in the production of favourable nutritional conditions. One way to accomplish this is to manipulate the host endocrine system in a way that promotes the release of stored resources into the host haemolymph. Another way is to up-regulate the production of a type of resource that is more accessible to the parasitoid than to host tissues. For example, teratocytes may synthesize resources that are primarily available only to the developing parasitoid. Moreover, the PDV injected into some hosts may also result in the up-regulation or synthesis of nutritionally important compounds that are preferentially absorbed or consumed by the developing parasitoid. Thus, in the final analysis, it is the preferential availability of the necessary resources that is the key to understanding nutritional aspects of the physiological interactions between parasitoids and their hosts.

12

Parasite regulation of insect reproduction: similar strategies, different mechanisms?

Hilary Hurd

1. Life-history traits of parasites and hosts

All organisms exhibit trade-off positions with respect to the alloction of resources between egg production and metabolic maintenance. Decisions concerning these positions may be crucial determinants of longevity, as evolutionary theory suggests that delayed reproduction is correlated with increased life span (Rose, 1991; Stearns, 1989). As regulators of reproduction and metabolism, hormones are key elements in the modulation of these trade-off decisions (Harshman, 1999) and may play a particularly important role in insects, as these organisms generally invest heavily in reproduction and have short life spans.

Host life span will be a determining factor in the fitness (transmission success) of many parasites. This is particularly so if parasite transmission to the next host is delayed during a period of development prior to transmission, if host longevity increases transmission opportunities, or if a healthy host will act as a good vector to transport the parasite to the next host. Evolutionary pressures are thus likely to select parasites that can manipulate host resource management such that host longevity is favoured at the expense of host reproductive fitness (Forbes, 1993; Hurd, 1998, 2001).

Curtailment of host reproductive potential does indeed occur in many parasite–insect associations (Hurd, 1990, 1993). Parasitic organisms from a wide variety of taxa induce such an effect, and host insects from many different orders are compromised. However, the adaptive nature of such strategies is difficult to demonstrate (Moore and Gotelli, 1990; Poulin, 1995) and, so far, few studies have identified a selective advantage (in terms of increased host longevity) associated with fecundity reduction (see below).

Endocrine Interactions of Insect Parasites and Pathogens, edited by J.P. Edwards and R.J. Weaver.

Despite there being numerous examples of parasite-induced fecundity reduction, it has yet to be demonstrated, unequivocally, that insect reproductive fitness is manipulated by a parasite. Thus, when host reproduction is compromised by parasitism, this may simply be a by-product of pathology (Dawkins, 1990; Hurd and Webb, 1997). One way in which to gain insight into this issue is to study the mechanisms underlying parasite-induced reduction of insect egg production. By understanding the regulatory pathways involved, and how they have been modulated by infection, we hope to gather evidence in support of, or against, the manipulation hypothesis. In particular, the identification of modulator molecules and their source, be it parasite or host, will give clues to the evolutionary history of this interaction.

In this chapter, the evidence provided by two very different parasite–insect associations is reviewed. The first, a cestode, develops rapidly in the haemocoel of its coleopteran host. The parasite undergoes morphological changes and grows, but does not multiply. The second, a protozoan, exhibits both sexual reproduction in the midgut of its mosquito host, and asexual reproduction beneath the midgut basal lamina. Both of these parasites reduce the reproductive success of their insect hosts and, although the end result appears similar, they may use different mechanisms to attain these ends. The different modes of interaction may reflect innate differences in the mode of reproduction and in its hormonal control in these two insect hosts.

2. Parasite–insect models

2.1 *A tapeworm and its intermediate host*

The beetle, *Tenebrio molitor*, along with other tenebrionid beetles such as *Tribolium* spp., will act as intermediate host for *Hymenolepis diminuta* (Cestoda), the rat tapeworm (Arai, 1980). This is a common tapeworm in the UK, and Webster and Macdonald (1995) found that around 22 % of wild rats were infected with the parasite. Embryonated eggs are shed by the adult worm and disseminated in rat faeces. Once ingested by the intermediate host, the hexacanth tapeworm larvae hatch in the beetle gut and migrate through the gut wall into the haemolymph. The tapeworm larvae pass through several stages of development (*Figure 1*) to become mature metacestodes within a period of approximately 10 days, depending upon temperature. The parasite is not able to infect the definitive, vertebrate host during this maturation period. Transmission to vertebrates such as the rat is passive, via the food chain, and thus the longer that the intermediate beetle host remains alive, the more protracted is the period during which it may be predated by a potential vertebrate host. This extended opportunity for parasite transmission, plus the presence of a maturation period, suggests that any strategy that enhances intermediate host longevity will enhance parasite fitness.

The metacestodes of *H. diminuta* have been shown to cause a marked long-term depression of *Tribolium castaneum* and *T. confusum* populations (Keymer, 1980, 1981). Reduction in host fecundity was related to parasite biomass (Keymer, 1980) and was greatly reduced by 14 days post-infection (Maema, 1986). Infection by *H. diminuta* also significantly reduced the fertility of female *Tenebrio molitor* (Hurd and Arme, 1986). In *T. molitor*, mating behaviour is also negatively affected by infection. Sex pheromone, produced by infected females and coated on a glass rod, induced males to make fewer attempts to copulate than pheromone from uninfected females. In

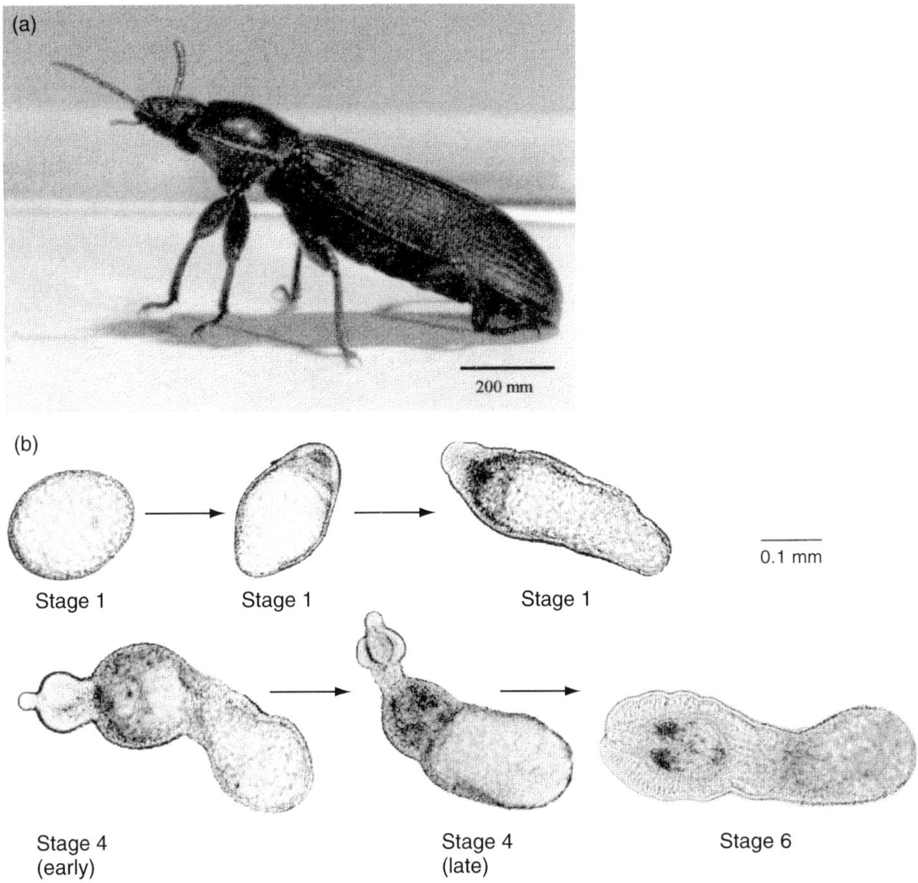

Figure 1. *The rat tapeworm–beetle association: host and parasite. (a) The intermediate host* Tenebrio molitor. *(b) The developmental stages of metacestodes of Hymenolepis diminuta that grow in the haemocoel of* Tenebrio molitor. *Note that at stage 2 a dense cell mass has grown at one pole and the primitive lacuna is clearly visible; by stage 4 a presumptive scolex containing four suckers has developed and the retraction capsule is visible; by stage 6 the scolex has retracted, the suckers are just visible as shadows and the mid-body wall has become fibrous (reproduced from Hurd and Parry, 1991* J. Invert. Pathol., *with permission).*

addition, infected males were less responsive to sex pheromone than uninfected males (Hurd and Parry, 1991). Worden *et al.* (2000) reported that uninfected female *T. molitor* are less attracted to infected males, and that mating uninfected females with very heavily infected males resulted in the production of fewer offspring (although a density of infection below 257 metacestodes per male did not affect the potential reproductive output of the insect). In contrast to the latter report, we have found that uninfected female *T. molitor* mated with infected males produce significantly more eggs, and that the increase in egg production is correlated with the density of infection in the males (H. Hurd, unpublished results). The males used in our recent study were less than 10 days post-emergence when mated, and had lower parasite burdens than the considerably older males used in the study of Worden *et al.* (2000).

Comparison of survivorship of small populations of *H. diminuta*-infected *T. molitor* with that of uninfected beetles did, as predicted, demonstrate an increase of approximately 50% in the longevity of infected females. However, the life span of infected male beetles was only slightly increased by infection (H. Hurd, 2001).

2.2 *Malaria and the mosquito*

Protozoan parasites of the genus *Plasmodium* are transmitted from host to host by vector mosquitoes. Sexual reproduction of the parasite is initiated immediately an infected blood meal enters the mosquito midgut. Factors such as the drop in temperature that occurs on transmission to a poikilotherm and certain mosquito products,

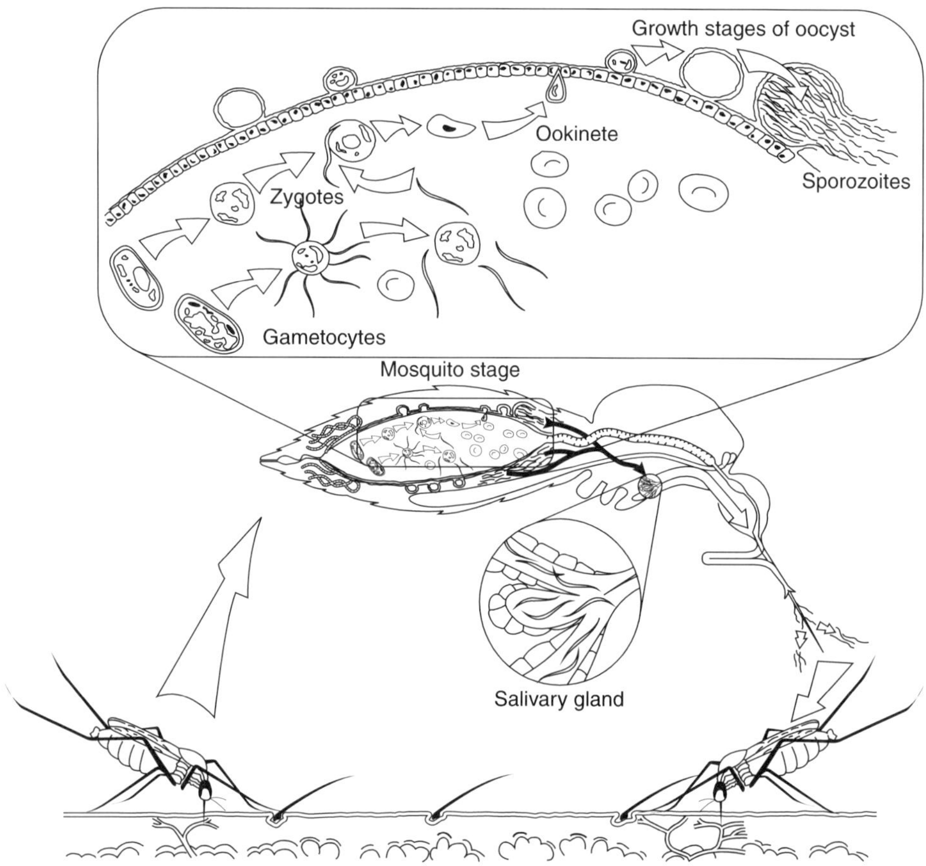

Figure 2. Mosquito stages of the malaria life cycle. Male and female gametes escape from the gametocytes within minutes of the infected blood arriving in the mosquito midgut. The male microgametes fertilize the macrogametes and the resultant zygote develops into an ookinete in approximately 24 h. The ookinete invades the midgut wall and transforms into an oocyst below the midgut basal lamina. Growth, then asexual reproduction take place in the oocyst and thousands of sporozoites are produced, which are released from the mature oocyst and migrate through the mosquito haemolymph to invade the salivary glands (adapted from Hurd and Sinden, 2001, with permission from Cambridge University Press).

notably xanthurenic acid (Billker *et al.*, 1997, 1998), act as cues to stimulate gametogenesis, and fertilization occurs within the first few minutes after feeding (Sinden *et al.*, 1996). The resultant zygote develops into a motile ookinete within 15–30 h and these leave the blood bolus and migrate through the host midgut epithelium. The ookinete transforms into an oocyst below the basal lamina of the midgut. Here sporogony occurs and over 1000 sporozoites are formed, which, once mature, enter the mosquito haemocoel. Many sporozoites invade the salivary glands where they are ready to be transmitted to a new host in the saliva that the mosquito injects during any subsequent blood meal (see *Figure 2*). In the wild, anopheline mosquitoes that transmit human malaria are rarely infected with more than one or two oocysts, but some of the rodent malarias (commonly used as laboratory models), can cause parasite burdens in excess of 100 oocysts per host. The sporogonic cycle of malarial infections is very long, relative to the life of a mosquito, and the period from the initial intake of an infected blood meal to the presence of transmissible infection in the salivary gland can be as long as 20 days. During this time the female mosquito takes several further blood meals (and can mature 2–3 batches of eggs) without transmitting the parasite.

In common with other vector-transmitted protozoans (e.g. *Leishmania*; El Sawaf *et al.*, 1994), *Plasmodium* spp. induce significant fecundity reductions in their insect hosts. The avian malaria parasite, *Plasmodium gallinaceum*, is known to reduce the reproductive success of the yellow fever mosquito, *Aedes aegypti*, although the response varies with the strain of vector (Freier and Friedman, 1976; Hacker, 1971). Hacker and Kilama (1974) reported that low levels of parasitaemia in the chicken caused mosquito fecundity reduction. The anaemia associated with rising parasitaemia may have resulted in a lower blood intake in *A. aegypti* and hence decreased egg production (Freier and Friedman, 1976). This is because, unlike anopheline mosquitoes, *A. aegypti* is unable to concentrate blood meals where the ratio of red blood cells to total blood volume is very low. We have shown that wild-caught *Anopheles gambiae* infected with oocysts of the causative agent of human cerebral malaria, *Plasmodium falciparum*, produced significantly fewer eggs than uninfected females (Hogg and Hurd, 1997). However, the remainder of our studies have been conducted in the laboratory with the rodent malaria, *Plasmodium yoelii nigeriensis*. Two vectors of human malaria, *Anopheles stephensi* and *Anopheles gambiae*, exhibit reduced fertility during the gonotrophic cycle when ookinetes of this parasite are invading the gut and, at a later stage of infection, when oocysts are present (Ahmed *et al.*, 1999b; Hogg and Hurd, 1995; Jahan and Hurd, 1997).

Both of the parasites that we are studying have a negative effect upon the reproductive physiology of their insect hosts. However, the process of egg development in beetles differs in many respects from that in mosquitoes. The following brief description of these processes in each of these insects will provide the background for a review of the various effects that metacestodes and malaria exert on their respective hosts.

3. Insect reproduction and its control

The developing oocytes in the ovaries of insects are provisioned with yolk proteins (vitellins), which are large glycolipophosphoproteins. In the majority of species, vitellin precursors (vitellogenins, Vg) are synthesized exclusively in the fat body. Vitellogenin synthesis has been studied in detail in the yellow fever mosquito, *A. aegypti* (Dhadialla and Raikhel, 1990; Hagedorn and Judson, 1972). A dispersed multigene family consisting of five Vg genes has been identified, and four of these genes are actively transcribed (Gemmill *et al.*, 1986). Vitellogenin genes have also been cloned from *A. gambiae* (Romans *et al.*, 1998) and also form a Vg gene family of four or five genes, three in tandem array. The protein products of the *A. gambiae* Vg genes are very similar to the protein product of the abundantly expressed *A. aegypti* VgA1 gene (Romans *et al.*, 1995). In *A. gambiae*, Vg mRNA synthesis begins to increase within 1–2 h of a blood meal, peaks 24–36 h after the blood meal, and then falls rapidly as the mosquito gonotrophic cycle is completed (Ahmed *et al.*, 2001). Vitellogenin undergoes post-translational modification, and a molecule consisting of 200- and 66-kDa subunits is secreted into the haemolymph and transported to the ovaries (reviewed in Clements, 1992).

The molecular biology of Vg synthesis in *T. molitor* has not been studied, but we do know that the major, female-specific haemolymph protein consists of two groups of subunits, one large and one smaller (Harnish and White, 1982), which we have found to consist of polypeptides of 154 and 145, and 56 and 45 kDa, respectively (Webb and Hurd, 1995a). Synthesis of *T. molitor* Vg is not related to feeding cycles of the beetle, instead the molecule is continuously synthesized and secreted by the fat body.

The paired ovaries of *T. molitor* consist of ovarioles of the merioistic type, characteristic of the Endopterygota. In such ovarioles, nutritive trophocytes are aggregated into the germarium of each ovariole and are connected to the developing oocytes via cytoplasmic strands (Ullmann, 1973). Developing oocytes within each ovariole are surrounded by a follicular epithelium. There is some evidence (Hurd and Arme, 1987) that the uptake of Vg in *T. molitor* occurs in a similar manner to that in the bug *Rhodnius prolixus* and the locust *Locusta migratoria* (Wyatt and Davey, 1996). In these insects, the follicular epithelium controls the access of Vg to the oocyte via intercellular protein diffusion channels, which appear when the cells of vitellogenic follicles change shape (a process called patency) permitting the uptake of Vg by the oocytes (Wyatt and Davey, 1996). Vitellogenin is likely to be taken up by the oocytes by a process of receptor-mediated endocytosis, which has been well studied in mosquitoes (Snigirevskaya *et al.*, 1997), but not in *T. molitor*. In insects generally, the yolk precursor, Vg, is transformed in the oocyte to the yolk protein, vitellin. In *T. molitor*, egg production is asynchronous and continuous, with all of the proximate follicles at different stages of vitellogenesis at any particular time (Ullmann, 1973).

Mosquito ovaries contain a series of approximately 50 ovarioles that radiate from a calyx. Each ovariole consists of a germarium and two follicles or egg chambers. The latter are composed of one oocyte and seven nutritive trophocytes (known as nurse cells) and are enclosed in a follicular epithelium (Clements, 1992; Sokolova, 1994). In contrast to beetles, egg production in mosquitoes is cyclical. A blood meal initiates the maturation of the proximate follicles resulting in the oviposition of a batch of eggs at the end of each gonotrophic cycle. During this process, the follicular epithelial cells become patent and allow passage of Vg to the oolemma, where it is taken up by receptor-mediated endocytosis (Sappington *et al.*, 1995; Snigirevskaya *et al.*, 1997). Monitoring the effects of adenlyate cyclase activators has provided indirect evidence that cAMP is involved in Vg uptake in the mosquito (Sappington *et al.*, 1998), but it is not known whether cAMP affects the follicle cells or the process of endocytosis. Nurse cells synthesize rRNA that is transported to the oocyte cytoplasm via intercellular canals (Clements, 1992). By 48–50 h after the blood meal vitellogenesis is completed, the epithelial cells secrete a chorion (egg shell), and the nurse cells degenerate (see review by Clements, 1992).

3.1 *Endocrine control of reproduction*

The entire process of reproduction in insects is under endocrine control. Two hormones, juvenile hormone (JH) and ecdysone, play major (but different) roles in mosquitoes and beetles. In most insects JH plays a pivotal role as a primer and regulator of both Vg synthesis in the fat body, and its uptake by the ovary (Wyatt and Davey, 1996). Wyatt and Davey (1996) proposed a model for the regulatory action of JH and neuropeptides on the fat body, based largely on research on the locust, *L. migratoria*. They suggest that Vg gene transcription is regulated by the product of an immediate-response gene (which is itself controlled by JH when this hormone is delivered to the nucleus bound to a carrier protein). The synthesis of JH is regulated by brain neuropeptides (allatotropins and allatostatins) that enhance or depress its synthesis in the corpora allata (see Chapter 1, Section 2.4). Wyatt and Davey (1996) further suggest that Vg synthesis may also be controlled at the translational level by adipokinetic hormone (an inhibitor of Vg production in *Locusta migratoria*; Glinka *et al.*, 1995), and by a neuropeptidergic brain factor.

The development of patency in the follicular epithelium of the ovary is also controlled by JH. Juvenile hormones act on the follicle cells via a protein kinase C-dependent cascade to activate JH-sensitive Na^+/K^+ ATPase (Sevala and Davey, 1989). The resultant ionic changes cause the follicle cells to shrink, and allow intracellular spaces to develop – facilitating uptake of Vg into the oocyte. It is likely that activation of a cAMP-dependent protein kinase reverses this process and brings about the termination of Vg uptake when the oocytes are fully developed (Wang and Telfer, 1997). Specific, saturable binding of JH to follicle cell membranes (reviewed by Wyatt and Davey, 1996) has been described in a few insects, and a 35-kDa JH-binding protein has been partially purified from locusts (Sevala *et al.*, 1995). It seems likely that processes similar to those described above control the synthesis and uptake of Vg by the ovaries of insects like *T. molitor*. In mosquitoes, JH is responsible for the separation of follicle cells from the oolemma, creating a perioocytic space (Raikhel and Lea, 1991), but the development of patency in the mosquito has been little studied.

To initiate vitellogenesis, the mosquito fat body requires pre-exposure to JH, which increases the level of polyploidy (Dittman *et al.*, 1989) and the number of ribosomes in fat body cells. However, in these insects, synthesis and secretion of Vg is primarily under the control of 20-hydroxyecdysone (20E), the titre of which rises following a blood meal. Juvenile hormone also has a priming function on previtellogenic mosquito ovaries (Clements, 1992). During this previtellogenic phase, endocytotic machinery such as vitellogenin receptors and clathrin-coated vesicles accumulate. However, following a blood meal, the haemolymph JH titre of *A. aegypti* declines rapidly (Shapiro *et al.*, 1986), and ovarian-produced ecdysteroids take over the control of Vg synthesis.

Titres of 20E follow the progress of the mosquito gonotrophic cycle, rising shortly after a blood meal and falling at the end of the egg maturation cycle. Changes in Vg mRNA levels are closely correlated with changes in 20E titres (Racioppi *et al.*, 1986) and the hormone can stimulate the production of Vg *in vitro*, but its specific role in the regulation of Vg gene expression is not fully understood. It is thought that 20E operates indirectly through a cascade of genes (Deitsch *et al.*, 1995) in a manner similar to the model proposed by Ashburner *et al.* (1974) for the action of 20E in *Drosophila melanogaster*. A homologue of the *Drosophila* E75 early ecdysone-inducible gene has been identified in *A. aegypti* fat body, and is induced by 20E (Pierceall *et al.*, 1999).

In addition, the presence in mosquitoes of AHR3, a homologue of the *Drosophila* HR3 early-late ecdysone-inducible gene, provides further evidence for the existence of a regulatory hierarchy in the fat body of the mosquito that is modulated by 20E (Kapitskaya *et al.*, 2000). However, a recent report now provides evidence that the action of 20E, bound to a heterodimeric ecdysteroid receptor/ultraspiracle in *A. aegypti*, directly activates transcription of a Vg gene (Martin *et al.*, 2001)

The endocrine events that control the production of eggs in discrete batches by haematophagous insects like mosquitoes appear to be more complex than those in insects that produce eggs continuously (e.g. tenebrionid beetles). Thus, in mosquitoes, in addition to JH and 20E, neuropeptides are involved in the initiation and modulation of reproductive processes (see Adams, 1999). In the adult female mosquito, neuronal signals from the midgut, when this becomes distended with blood, together with blood meal digestion products, appear to be involved in initiating the release of an ovarian ecdysteroidogenic hormone, which in turn initiates the production of ecdysone and 20E in the ovaries. A complex web of communication between the brain, ovaries, fat body and gut is emerging (*Figure 3*), but the finer details of these interactions are not yet fully understood.

Many adult insects are able to regulate their reproductive output in relation to prevailing circumstances. One method of controlling reproductive output available to insects is the ability to resorb developing eggs. This process is often associated with nutritional stress, or with lack of mating success (Bownes and Reid, 1990; Clements and Boocock, 1984; Mordue, 1965). Little is known about the mechanism that induces

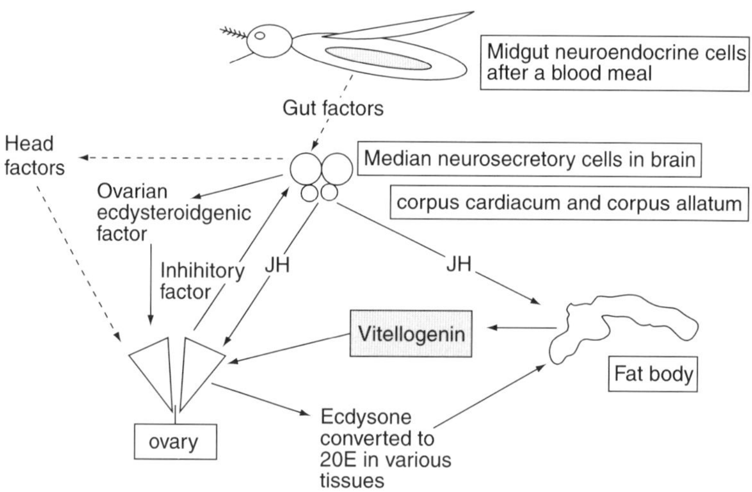

Figure 3. *A schematic diagram of some of the factors involved in the control of egg production in mosquitoes. Solid lines represent established control pathways; dotted lines represent pathways in which the endocrine factors are unknown. JH renders the fat body competent to synthesize vitellogenin, which occurs following stimulation by 20E. In the ovary, juvenile hormone stimulates the development of follicles in the previtellogenic phase and makes them competent to respond to ovarian ecdysteroidogenic hormone, produced after a blood meal. The release of ovarian ecdysteroidogenic factor from the neurohaemal organ of the brain leads to ovarian synthesis of ecdysone and the eventual synthesis of vitellogenin in the fat body.*

oocyte resorption, but in *D. melanogaster* the process has recently been associated with the occurrence of apoptosis (programmed cell death) in the ovarian nurse cells (Soller *et al.*, 1999).

4. Effects of parasite infection on host vitellogenesis

Infections by tapeworm metacestodes, or by malaria oocysts, cause significant reductions in the synthesis of Vg in the fat bodies of their respective hosts. Methods used to detect these infection-associated changes have been constrained, in part, by the experimental tools currently available. Thus, it is not possible to make direct comparison of the mechanisms underlying this inhibition in the different types of insect host. This is because different steps in the process of vitellogenesis have been examined in mosquitoes and in beetles.

In vitro incubation of fat bodies from *H. diminuta*-infected *T. molitor* females demonstrated a marked decrease in the synthesis of Vg, measured in terms of the incorporation of radiolabelled leucine into the secreted protein. No such effect was observed in the fat bodies of infected male beetles (Hurd and Arme, 1986). Immunoprecipitation of secreted proteins from female fat bodies, using polyclonal antibodies against *Tenebrio* vitellin, detected inhibition of Vg synthesis at 3 days post–infection, and at various time points during the development of metacestodes. Synthesis of Vg was also affected when beetles harboured mature parasites, and production of Vg was decreased by a maximum of 75% on day 24 post-infection (Webb and Hurd, 1996). However, when parasites were incubated *in vitro* with fat body from uninfected female *T. molitor*, inhibition of Vg synthesis was found to be stage-specific – the effects being caused by parasites up to stage 2 of development (see *Figure 1*), but not by later stages (Webb and Hurd, 1996, 1999). Vitellogenin synthesis was also inhibited in a density-dependent manner *in vitro* (Webb and Hurd, 1996), although even as few as two metacestodes (grown *in vivo* or axenically) were able to cause a significant reduction in total Vg synthesis by the fat body from an individual female beetle. Inhibition of Vg synthesis increased when 2-10 parasites were present, but any further increase in the parasite number caused no further decrease in Vg synthesis (Webb and Hurd, 1996). This short-term culture system also enabled us to demonstrate that the inhibition of Vg synthesis was caused directly by a factor from the developing metacestodes, and which could be extracted in acetic acid. The active material was pronase-sensitive, heat resistant, and had a molecular mass below 50 kDa, indicating that it was probably a peptide (Webb and Hurd, 1999).

Thus, it appears that tapeworm metacestodes produce a potent inhibitor of Vg synthesis at a time when they are rapidly developing in the beetle haemocoel. This molecule may be long-lived and/or it may have lasting effects upon the fat body, because Vg synthesis appears to be permanently inhibited in infected female beetles. At the present time we have no further information about this molecule. We think it unlikely that the parasite is manipulating host JH titres, since these are not altered *in vivo* (Hurd *et al.*, 1990); neither is JH synthesis by corpora allata of infected beetles affected (Hurd and Weaver, 1987). In addition, *H. diminuta* metacestodes do not contain or produce JH or JH-degrading enzymes (H. Hurd, unpublished results).

With regard to mosquitoes, Vg genes have now been cloned from *A. gambiae* (Romans *et al.*, 1998). These genes are located together at 18A/B on chromosome 2R, and the genes are virtually identical in their coding regions (P. Romans, personal communication). Vitellogenin mRNA abundance has been measured in *Plasmodium y. nigeriensis*-infected and uninfected *A. gambiae*, using a probe derived from the central 3.3-kb Sal 1 fragment common to all *A. gambiae* Vg genes. In common with other mosquitoes, the presence of *A. gambiae* Vg mRNA was detected 1–2 h after the blood meal, peaked 24–36 h after feeding, and disappeared by 48 h (Ahmed *et al.*, 2001). During the first gonotrophic cycle post-infection, when ookinetes are forming and transforming into oocysts below the midgut basal lamina, no differences in the abundance of Vg mRNA were detected late in the gonotrophic cycle. However, during the second gonotrophic cycle, parasite oocysts developing on the midgut induced a significant reduction in Vg mRNA abundance at 2, 8, 12 and 24 h after the blood meal. As predicted, this led to a significant decrease in the titre of circulating yolk protein precursor in the haemolymph of oocyst-infected mosquitoes (Ahmed *et al.*, 2001).

Again, we do not fully understand the exact mechanism by which the malarial parasite affects the fat body of its mosquito host. At this stage we cannot determine whether Vg gene transcription is down-regulated, or if Vg mRNA turnover rate is affected. However, it should be noted that during both gonotrophic cycles in infected mosquitoes a significant number of early-stage oocytes are undergoing resorption (see below). This partial 'degeneration' of the ovary may remove a source of ecdysone and hence affect Vg synthesis in the fat body.

The vitellin content of the ovaries of both metacestode-infected *T. molitor* and of *Plasmodium*-infected mosquitoes is significantly reduced by comparison with uninfected controls (Hogg *et al.*, 1997; Hurd and Arme, 1986). In *T. molitor* this depletion was detected at various stages during metacestode development (3, 6, 9 and 15 days post-infection), and all sizes of follicles were affected (Webb and Hurd, 1995a). It would appear that the majority of follicles in infected females accumulate less vitellin compared to follicles of the same size in uninfected females. This suggests that uptake of Vg occurs at a slower rate in infected females. In addition, a significant number of follicles in the ovaries of mated, infected females were resorbing (Hurd and Arme, 1987).

There is some evidence that the mechanism underlying inhibition of Vg uptake in tapeworm-infected beetles involves the action of JH on the follicular epithelium. The binding of JH to the cytosolic and nuclear fractions of homogenized ovaries was not affected by infection, but binding to follicle membranes was significantly reduced on days 3 and 6 post-infection (Webb and Hurd, 1995b). Binding was restored to normal on day 15 post-infection, suggesting that inhibition of binding is associated with the early stages of parasite development. Analysis of binding kinetics led us to conclude that an infection-related binding inhibitor was present rather than a reduction in JH binding sites (Webb and Hurd, 1995b). According to the model developed by Davey and colleagues (see Wyatt and Davey, 1996), events down-stream of JH binding to the follicular epithelium result in the development of patency (see above). Our findings are in accord with this model, as the parasite was shown to cause a decrease in ovarian microsomal Na^+/K^+ ATPase activity in a stage-specific manner, and activity was restored when parasites were mature (Webb and Hurd, 1995b). Furthermore, patency was significantly reduced in infected females (Hurd and Arme, 1987), but could be rescued by incubating ovaries from infected females with JH III. Consistent with the hypothesis that a competitive inhibitor of JH binding is present in infected females was the observation that patency could also be restored by prolonged washing of the ovaries (Webb and Hurd, 1995b). These results are all suggestive of the presence of an antigonadotrophin in beetles that are infected with early stages of the parasite.

Transfer of haemolymph from infected donors to uninfected recipients resulted in reduced vitellin content in ovaries of the uninfected recipients, but only if the haemolymph came from female beetles (Major *et al.*, 1997). This indicates that the modulator molecule does not come directly from the developing metacestodes, but that the presence of metacestodes in female beetles stimulated the synthesis and/or release of an antigonadotrophin originating from infected female insects (see discussion in Hurd, 1998).

A consequence of the inhibition of Vg uptake due to reduced patency is that this yolk protein precursor accumulates in the haemolymph of infected female *T. molitor*. Increased Vg titre does not, however, benefit the parasite directly, as it is unable to take up or metabolize these proteins. However, amino acids, resulting from the degradation of Vg, may become available to the parasite, and thus enhance the host as a nutritional resource. In addition, there is some evidence from mosquitoes that high titres of Vg in the haemolymph may inhibit synthesis of this protein by the fat body (Van Handel and Lea, 1984).

Infection-induced follicle resorption also occurs in anopheline mosquitoes, and is thought to be the major cause of reduced fecundity in *Plasmodium*-infected females. In uninfected mosquitoes, follicle resorption is usually the result of an inadequate blood meal (Clements and Boocock, 1984), but our studies have shown that resorption occurs even when the size of the infective blood meal was similar to that taken by control (uninfected) mosquitoes (Carwardine and Hurd, 1997). A significant increase in oocyte resorption has been detected as early as 16 h post-infection, and is correlated with the appearance of follicular cells undergoing apoptosis. Moreover, injection of an inhibitor of caspase (a key part of the apoptosis cascade in mammalian cells) into infected mosquitoes inhibits egg resorption (Hopwood, *et al.*, 2001).

Little is known about the process of follicle resorption in insects, but it has recently been shown that 20E induces apoptosis in the nurse cells of *D. melanogaster* ovaries, and that subsequently the follicles are resorbed. Furthermore, it has been suggested that JH may inhibit apoptosis in these nurse cells and stimulate the progression of oocyte development beyond the critical point (stage 9, Soller *et al.*, 1999) at which resorption may occur. The process of oogenesis in mosquitoes does not progress in exactly the same manner as it does in *D. melanogaster*, making it difficult to speculate about any link between infection, endocrine status, and follicle resorption in the former insects.

5. Summary

The use of these very different models has enabled us to make some headway towards an understanding of the process of parasite-induced fecundity reduction in insects. In both cestode-infected beetles and *Plasmodium*-infected mosquitoes, fat body synthesis of Vg is down-regulated, the process of vitellin accumulation in the ovaries is inhibited, follicles are resorbed during oogenesis, and fewer eggs are produced. However, many interesting questions still remain unanswered, and of particular interest is the nature of the mechanisms that initiate and control both parasite-induced changes in Vg synthesis and in oogenesis. Both of the models discussed in this chapter have provided evidence that suggests that we are not observing anything as simple as direct competition for nutrients between the parasite and host tissue (see Hurd and Webb, 1997; Jahan *et al.*, 1999). Rather, it is feasible to speculate that pathways that are part of the normal control process are being usurped by the parasite. This could occur if parasite-produced modulator molecules mimic host hormones. Alternatively, the presence of parasites could be detected by the insect immune defence system, which then produces signals that alter the endocrine system, as is seen in trematode-infected snails (De Jong-Brink *et al.*, 1997). The endocrine system may thus be the key element that links the parasite to the host reproductive system.

The majority of changes in reproductive capacity that we observe in *H. diminuta*-infected T. *molitor* are, at some point, regulated by JH. However, JH itself does not appear to be affected by the parasite. Our observations concerning JH binding to beetle follicular membranes indicate that changes may instead occur at the receptor level and involve an additional tier of regulatory control. It is possible that examination of the effects of agonists and antagonists of JH and ecdysteroids will help to clarify this issue. However, it is equally possible that additional (neuropeptide) regulators are involved.

Once we have identified the exact point(s) at which host reproduction is manipulated by parasites, our investigations will need to focus on earlier stages in the process. Do these manipulators originate from the parasite and are we observing the extended phenotype of the parasite in action (Dawkins, 1990)? In other words, are products of the parasite genome directly regulating host physiology, or are there further layers of complexity? Are we actually observing a facet of the insect's response to stress, initiated by the presence of the parasite? There is evidence, for instance, that the mosquito immune defence system recognises and responds to malaria infection, even if it is not capable of eliminating all of the parasites (Luckhart *et al.*, 1998; Richman *et al.*, 1997). Could there be a connection between the parasite and the host immune defence, endocrine and reproductive systems, as has been demonstrated in schistosome-infected snails (De Jong-Brink *et al.*, 1997)? If so, we may still be far from elucidating the mechanisms underlying parasite-induced fecundity reduction in insects, and the parasites will retain their mysteries for many years to come.

13

Structure of salivary glands of the human body louse, *Pediculus humanus* L., and erythematous and immunogenic properties of its saliva

Jared Bramlett, Susanne Wache and Davy Jones

1. Introduction

Human lice infestations affect an estimated 6–12 million people in the United States (Sokoloff, 1994) and many millions more worldwide (Gratz 1997). The continuing importance of lice to human health is exemplified by recent outbreaks of louse-transmitted relapsing fever, trench fever and epidemic typhus (reviewed by Raoult and Roux, 1999). Even in the USA, head louse infestation (pediculosis) appears in some surveys to be more prevalent among children than all other childhood communicable diseases combined (Donnelly et al., 1991). Likewise, the misperception of infestation has often led to the use of control measures that were in disproportion to the actual situation (Jones and Pollack, 2001).

The worldwide appearance of insecticide resistance in lice is ominous (Gratz, 1997). For this reason, the use of carbaryl against head lice has been restricted in the UK (Boulton, 1995; Scowen, 1995). Malathion resistance has been found in France (Izri and Briere, 1995) and elsewhere. Recently, louse resistance to permethrin has arisen in Israel (Mumcuoglu *et al.*, 1995), the Czech Republic (Rupes *et al.*, 1995), England (Burgess *et al.*, 1995), and now in the United States (Jones and Pollack, 2001).

Apart from their significance as disease vectors, lice have saliva with certain biological activities (Jones, 1998; Jones and Pollack, 2001; Peck *et al.*, 1943). The louse

Endocrine Interactions of Insect Parasites and Pathogens, edited by J.P. Edwards and R.J. Weaver.

feeding process results in certain clinical reactions, such as erythema (reddening of the bitten area, *Figure 1*), and even more severe symptoms may appear in sensitive individuals or when infestations are heavy (Moore, 1918). The human louse, *Pediculus humanus*, has two pairs of salivary glands – 'reniform' and 'U-shaped' – located in the thorax (*Figure 2*). To study adverse human skin reactions, Peck *et al.* (1943) used the head of *P. humanus* as a tissue homogenate for injection, but this preparation would not have included the thoracic salivary glands. In the first study of anticoagulant activity in the saliva of *P. humanus*, Nuttall (1918) combined both pairs of salivary glands together before assay, and thus was unable to identify the exact glandular source of the anticoagulant activity. Mumcuoglu *et al.* (1996a) partially purified anticoagulant activity, but had difficulty separating the two types of salivary gland to determine the specific glandular source of the anticoagulant activity. The biochemical composition of saliva from each pair of glands has not previously been characterized and, despite the vasodilatory and salivary anticoagulant effects, it has not been demonstrated biochemically that any salivary gland molecules are actually injected when *P. humanus* feeds on a human host.

Because of the paucity of information on the salivary gland toxins of human lice, the fact that many millions of people remain infested all over the world, and because there are increasing reports of insecticide resistance in lice, we have investigated the structure of the salivary glands of the human body louse (*P. humanus*) and the nature of the proteins produced in these glands.

2. Materials and methods

2.1 *Insects and tissue collection*

Pediculus humanus (Orlando strain) was provided by Robin Todd (Insect Control Inc., Catonsville, OH, USA). In accordance with an approved IACUC protocol, this strain is maintained on New Zealand white rabbits – a host to which this strain has

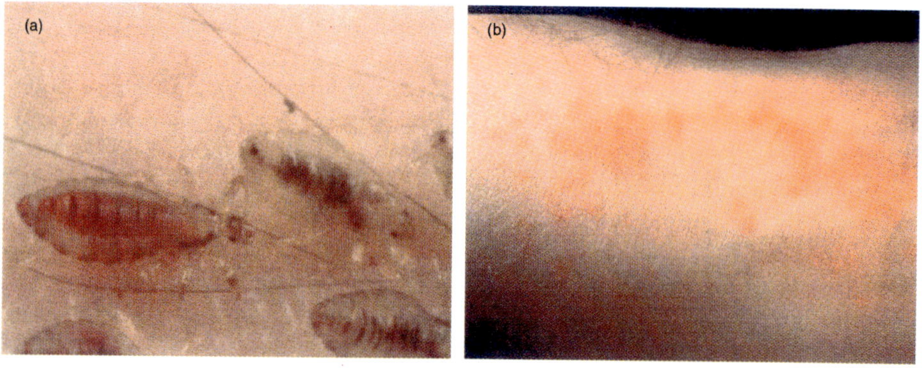

Figure 1. *Erythematous (reddening) vasodilation induced by human body louse saliva injected into human skin during bites. (a) The lice on the right (with brown digestive tracts) have not fed for 24 hours, while the louse on the left (with the red digestive tract) has recently completed engorgement. (b) Vasodilator-induced reddening on the arm of an author (D. Jones), following the feeding of several dozen lice a few minutes earlier.*

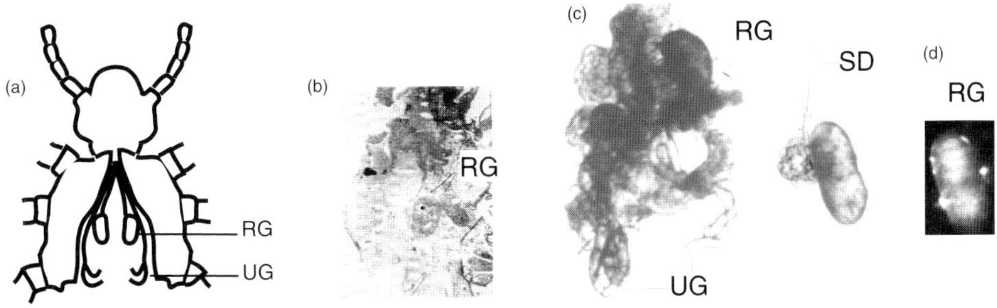

Figure 2. *(a) Line drawing of the internal morphology of the thorax of the human body louse, adapted from Buxton (1947). RG, reniform gland; UG, U-shaped gland. (b) Cross-section of fixed louse thorax (toluidine blue counter-staining), showing the position of the two reniform glands in the thorax. (c) Appearance of the two pairs of louse salivary glands when removed from the body together with supporting tissue. The left reniform gland in this panel has been removed from its position over the supporting tissue and isolated on the right, making its salivary duct (SD) and an attached grape-like mass of cells more easily visible (see text). The other reniform gland and the two U-shaped glands remain in place on supporting tissue. (d) The same isolated reniform gland as shown in (c), but under bright field illumination to show its mottled white appearance.*

been adapted for several decades. In this strain, entry of blood into the louse during feeding can be easily seen through its translucent cuticle using a binocular microscope.

Salivary glands were dissected in isotonic saline, washed further in saline, and then either placed immediately in electrophoresis buffer (see below), or into a tube on dry ice. The U-shaped salivary glands were dissected free of any other tissues. The reniform salivary glands were dissected free of any other tissue where specifically noted; otherwise, a small cluster of cells physically adhering to the glands (see below) was allowed to remain, because analysis (reported below) showed that the protein contribution of these cells was very small compared to the protein content of the salivary glands themselves.

2.2 *Fixed-tissue microscopy*

Intact louse thoracic regions, or individually dissected salivary glands of *P. humanus* adults, were fixed for 1 h or longer in 2.0% formaldehyde, 2.5% glutaraldehyde, and 0.03% picric acid in 0.1 M cacodylate buffer. Samples were then washed in cacodylate buffer and placed in 1% uranyl acetate in the same buffer and then dehydrated in cold ethanol solutions. Propylene oxide was used before embedding in Epon Araldite^{Å} or in a low viscosity Epon\diepoxyoctane (Luft, 1973). Sections were cut at 0.5–1.0 microns for toluidine blue staining for light microscopy, and at 0.08–1.0 microns for electron microscopy. Thin sections on copper grids were stained in saturated uranyl acetate 1:1 with acetone, and then with 0.2% lead citrate, before being examined with a JEOL 1200 EX electron microscope.

2.3 *Electrophoretic analysis*

The various louse tissues were heated for several minutes at 95°C in denaturing loading buffer containing ß-mercaptoethanol, and then subjected to sodium dodecyl sulfate polyacrylamide gel electrophoresis (SDS-PAGE), at the percentage acrylamide indicated in each given experiment. Silver staining was performed by the method of Morrissey (1981).

2.4 *Immunological analysis*

In some experiments, the glandular proteins were fractionated by SDS-PAGE and then transferred to Immobilon® membrane. The blots were then subjected to immunoblotting to detect antigenic proteins as described by Jones *et al.* (1992a) using serum from rabbits serving as hosts to maintain the louse colony, and with ^{125}I-labelled secondary antibody. In other experiments, tissue samples were maintained under native conditions, dot-blotted to the membrane, and subjected to similar immunoblotting detection procedures. Signals were visualized either by autoradiography or phosphorimage scanning.

2.5 *Protein sequencing*

Proteins from the U-shaped salivary glands were subjected to SDS-PAGE, transferred to Immobilon® membrane, and subjected to N-terminal protein sequencing by the Yale University Keck Foundation protein sequencing facility.

2.6 In vivo *video microscopy*

In this technique, a rat arteriolar preparation including second-, third- and fourth-order resistance arterioles (Arden *et al.*, 1994) is viewed with a Zeiss ACD microscope fitted with a microchip camera and high-resolution monochrome monitor. The final tissue-to-screen magnification is 1250x. Images of second-, third- and fourth-order resistance arterioles are made, and recordings taken from the same segment of each arteriole for the duration of the protocol. The arteriolar intraluminal diameters are measured with a calibrated video-image shearer. In addition, the erythrocyte velocity within the selected arteriolar segment and the arteriolar volume flow can both be computed. After the equipment had been set up, each preparation was allowed to equilibrate for at least one hour, at which time the arteriolar diameters, erythrocyte velocities and volume flows were recorded. The preparations are pre-constricted with norepinephrine (noradrenaline), and then treated with test compounds such as calcitonin gene reactive peptide (CGRP) – a positive control for vasodilatory activity – or with the purified putative louse salivary vasodilator. Recordings began 2 minutes after administration of test compounds. At the termination of the experiment, all preparations are maximally dilated with nitroprusside and re-measured. Further details on these methods can be found in Arden *et al.* (1994).

3. Results

3.1 *Morphology and identity of salivary glands of* Pediculus humanus

Classical morphological line drawings of human lice (e.g. Buxton, 1947) depict two independent pairs of salivary glands. There are a pair of reniform glands (RG), and a pair of U-shaped glands (UG). Each RG and UG has its own duct, with all four ducts running into a common salivary duct taking saliva to the proboscis (*Figure 2(a)*). The more anterior location of the RG within the thorax is confirmed in cross-sectioned material examined under a compound microscope (*Figure 2(b)*). The RG, dissected away from thoracic tissues, appears as shown in *Figure 2(c)* when viewed through a binocular microscope. The RG and UG are clearly quite different in overall structure, and are not physically connected (except by the confluent ducts). The isolated RG shown under bright-field illumination in *Figure 2(c)* is also shown under different illumination in *Figure 2(d)*, to illustrate its mottled white colour (see below for significance). When the RGs were physically removed (e.g. the isolated RG in *Figure 2(c)*), the UGs remained in place, indicating the lack of a direct tissue connection between these two salivary gland types. Dissections to isolate individual glands were difficult, as they did not readily extrude from the thorax when the head was pulled off.

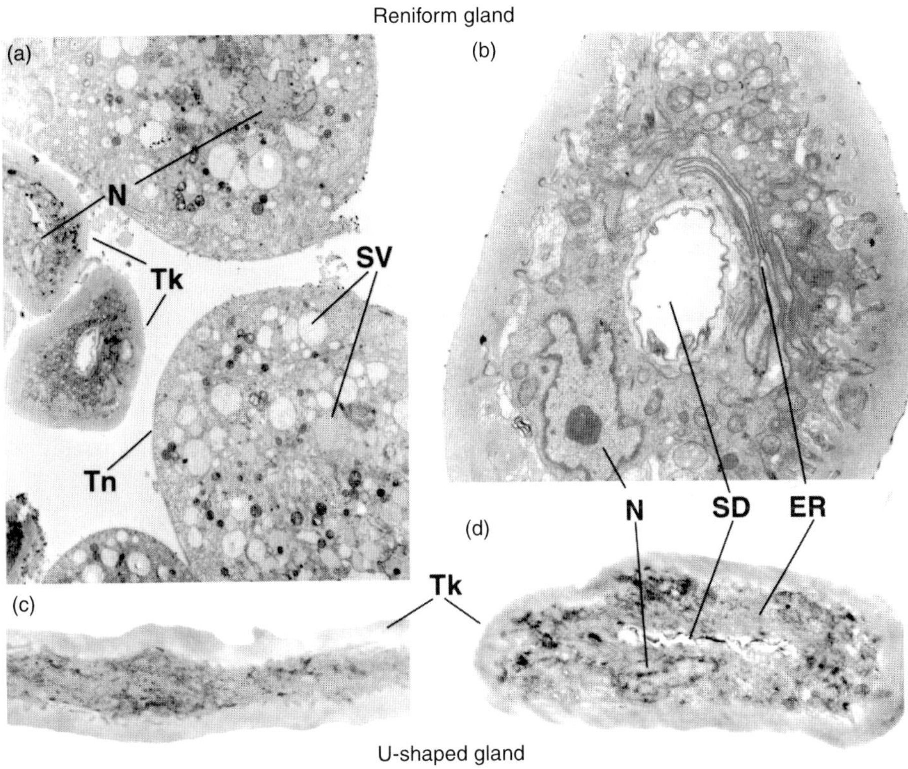

Figure 3. *Electron microscopy of reniform and U-shaped glands of the human body louse. (a) and (b) Sections through the reniform glands. SD, salivary duct; Tk, thick basement-like membrane,; Tn, thin basement-like membrane; SV, secretory vacuoles; ER, endoplasmic reticulum; N, nucleus. (c) and (d) Sections through the U-shaped glands. Magnifications: 1600×, 5700×, 5700× and 6700× in (a), (b), (c) and (d), respectively. Note that while the sections through the U-shaped glands show the entire structure to be enveloped by a thick, flocculent-appearing membrane (Tk), it is only the subset of cells in the reniform gland (i.e. those that immediately surround the salivary duct) that possess such a membrane. The remaining groups of cells in the reniform gland are covered by a typically thin basement-like membrane (Tn). These latter groups of cells also house a large number of secretory vacuoles (SV) that appear to contain granules.*

A grape-like cluster of cells, varying in colour from white to rusty-red, was found attached to each RG near the junction of this gland with the salivary duct (SD, *Figure 2(c)*). Closer examination showed that the rusty-red colour was concentrated in foci in each cell, reminiscent of granules. Silver staining of SDS-10% PAGE fractionated proteins from white glands versus rust-red coloured glands revealed no significant difference in the protein profile (results not shown).

Electron microscopy revealed some unique features, and supported a secretory function for the salivary glands. A section of the RG across the salivary duct revealed an (apparently chitinous) duct lining, lacking microvilli, and surrounded by tissues enclosed by a thick, amorphous basement membrane with much cellular infiltration (*Figure 3(b)*). Tissues in other areas of the RG have a second cell type, with the groups of cells enclosed within a thin basement membrane (*Figure 3(a)*), and do not have the cellular infiltration that is seen in the salivary gland cells of mosquitoes (James, 1994). There is also a difference in the mitochondrial to secretory vacuole ratio of these two types of RG cells. The mitochondria are much more abundant than the secretory vacuoles in the structure containing the duct (*Figure 3(b)*), whereas the secretory vacuoles predominate in that part of the structure that does not contain the duct (*Figure 3(a)*). Sections made through the region of the RG associated with the grape-like cluster of cells revealed no direct connection of these cells to the RG (result not shown).

The U-shaped glands were found to be enveloped completely by a thick amorphous basement membrane with some cellular infiltration (*Figure 3(c)* and *(d)*). The salivary duct that runs down the interior of this gland is easily visible (*Figure 3(c)*), although, once again, no strong evidence of microvilli was observed in the salivary ducts of these glands.

3.2 *Salivary gland proteins*

The RGs were generally found to contain a more complex array of proteins (especially in the range of 50–80 kDa) than the UGs, which had fewer proteins in that size range (*Figure 4(a)* and *(b)*). Attempts were made to correlate any changes in the intensity of particular protein bands that occurred after the lice had fed. However, no consistent qualitative change in SDS-PAGE protein banding pattern after feeding was observed after feeding with either the RG (*Figure 4*) or UG (not shown). The most abundant UG protein apparent by silver staining (arrowed in *Figure 4(b)*) was found (by both SDS-PAGE mobility and N-terminal sequence, *Figure 4(c)*) to resemble closely the glutathione transferases found in some other insects.

3.3 *Vasodilatory activity of louse saliva*

The saliva of *P. humanus* has potent vasodilatory activity, as indicated by the strong, rapid erythematous reaction shown in *Figure 1*. Preliminary observations indicated that the red spots characteristic of this reaction are induced within several minutes after the louse begins feeding, and remain visible for about 2 hours. This erythematous reaction was induced both by male and female adult lice fed earlier in the day, and by those lice that remained unfed for 1, 2 or 3 days previously. The time course of appearance and disappearance of the erythema was the same for one author (D. Jones) who had been exposed to heavy feeding by lice for 6 months, as it was for a different

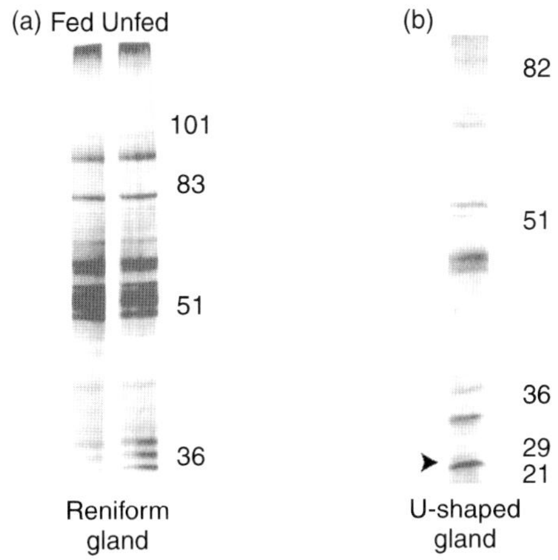

(c)

N-terminal sequence	Species source	Accession #
* ** * * * * **		
PIDIYYLPGSPPXXAVLLTAK	Pediculus humanus (body louse)	
PVDLYYMPLSAPERSVMLTA	Penaeus japonicus (shrimp)	gi 40930
MDFYYLPGSTPYHSVLMTAK	Lucilia cuprina (greenbottle fly)	pir S43851
VDFYYLPGSSPCRSVIMTAK	Drosophila domestica (fruit fly)	pir G4668
MDFYYLPGSAPCRSVLMTAK	Musca domestica (house fly)	sp P28338
MDFYYLPGSAPCRAVQMTAA	Anopheles gambiae (mosquito)	emb z81292

Figure 4. *SDS-12% PAGE analysis of the proteins extracted from (a) reniform glands (20 glands per lane) or (b) U-shaped glands (20 glands) of human body lice. The sharp qualitative differences in the contents of these glands suggest that the molecules of each gland injected via the saliva have different functions or target sites. Values on the right-hand side indicate the relative mobility in kDa equivalents. (a) Proteins from reniform salivary glands taken from lice that had either just fed, or were unfed for two days. No marked qualitative differences were observed in the protein content of the glands in relation to the length of time since last feeding. A similar result was obtained for U-shaped glands (not shown). (c) The N-terminal sequence of the glutathione transferase from U-shaped glands compared to sequences of the same enzyme from other insects and from a crustacean. The enzyme from the glands of the louse is clearly related to the other arthropod glutathione transferases, as indicated by the similarity of the N-terminal sequences, and by its molecular size on SDS-PAGE (arrow next to protein band indicates the glutathione transferase that was sequenced).*

author (J. Bramlett) who had not previously been bitten by human lice. Repeated feedings of several hundred lice on the same region of the arm or leg of the author D. Jones, every few days for 6 months, did not result in noticeable change in the visible intensity or timing of the vasodilatory reaction sequence.

Using lice that had not fed in the previous 24 h, we observed that in all cases the erythema could be observed on the forearm of an author after 11–17 min (i.e. at the completion of the natural bite), and in some cases could be observed by the time during the bite that blood began entering the louse (*Figure 5(a)*, 1–5 minutes). No erythema was observed when the bite was interrupted after just 15 seconds of feeding. During these tests it became apparent that when a louse had not fed for 24 hr, at the first subsequent feeding the time until the appearance of the red spot was longer than it was for subsequent attempts to feed. Thus, it was hypothesized that the full entry of the saliva is delayed in the first feeding of lice, and, therefore, the time until appearance of the red spot is concomitantly delayed. As a test of this hypothesis, lice that had not fed for 24 h were allowed to feed on the forearm of an author (D. Jones) until the time at which blood first entered the louse. At this point (i.e. when blood began to be ingested) feeding was disrupted, and the presence or absence of a red spot assessed, whereupon the lice were moved gently to a new place on the skin. Subsequently, the time until ingestion of blood commenced and the time taken for the red spot to appear at the new site were both recorded. This process was repeated a number of times, and the results are shown in *Figure 5(b)*. There was a clear trend that in the initial feeding

(a) **Time course of induction of erythema after initiation of lice bites**

Time of observation for red spot (after initiation of bite)	Red spot visible
45 minutes	10 out of 10
11–17 minutes [a]	10 out of 10
1–5 minutes [b]	3 out of 10
15 seconds	0 out of 10

[a]moment of completion of bite;
[b]entry of blood into louse begins (unfed previous 24h).

(b) **Induction of erythema by saliva of lice that were repeatedly interrupted in feeding at the moment host blood entered louse body**

	First Replication		Second Replication	
Feeding attempt	Seconds until entry of blood	Presence of red spot*	Time until entry of blood	Presence of red spot*
1	100	–	56	–
2	105	–	73	–
3	157	+	83	–
4	140	+	80	+
5	100	+	170	+
6	110	+	62	–
7	45	+	170	+
8	255	+	85	+
9	(louse ceased feeding)		120	+
10			110	+
11			140	+
12			50	+
			(investigator stopped test)	

*Presence of red spot assessed at time of entry of blood into louse.

(c)

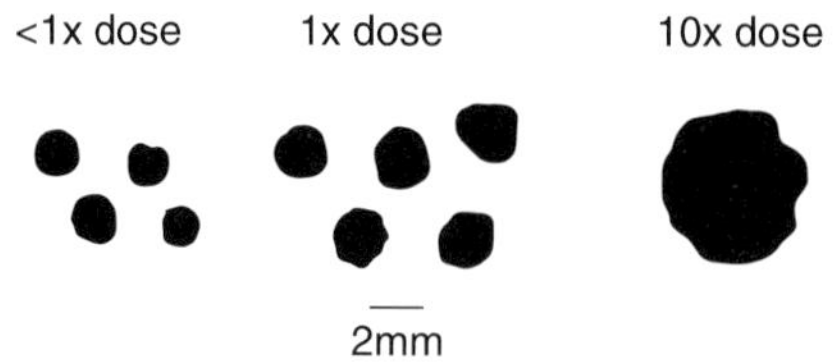

Figure 5. The induction of erythema at bite sites of human body lice. (a) Time-course of induction of erythema after initiation of louse feeding. Detectable vasodilator has begun to enter the bite site 1–5 minutes after initiation of the bite. (b) Tests on the potency of louse bites in inducing erythema, as a function of the number of times biting has been attempted. The results show that erythema is induced more rapidly at bite sites if the louse has previously initiated bites at other sites. (c) Dose-dependence of the size of the erythema versus the number of bites at the bite site. Bar indicates 2 mm.

attempts, insufficient vasodilator had been injected to cause the presence of any erythema before the time that blood began to be ingested. However, in later feeding attempts, it was seen that sufficient vasodilator had been injected to cause erythema at the moment of ingestion of blood (generally 1–2 minutes after initiation of the bite). These data further confirm that the red spot assay is rapid, and that the effects of injected vasodilator can be assessed within 1–2 minutes.

That the size of the red spot is dose-dependent was indicated by additional tests. Lice were allowed to feed on an author's forearm until the moment of entry of blood into each louse, at which time each louse was removed. In this way, lice were allowed to inject saliva in a quantity that we have called a 1× dose. In another treatment, 10 lice were placed in rapid succession to feed at a single site, each being allowed to feed until the moment of ingestion of blood by the given louse. Thus, this site received a 10× dose of putative vasodilator. Finally, another set of lice were allowed to feed for only 30 seconds before being removed, and this was long before the entry of any blood into the insects. These latter feeding sites, therefore, received somewhat less than a 1× dose. The eventual size of the red spots that developed after each of these treatments was traced, and the tracings scanned so that areas could be calculated. As shown in *Figure 5(c)*, there is clearly an increase in the size of the red spot that is linked to increasing dose of the putative vasodilator from louse saliva. Moreover, we have also observed that the size of the red spots induced by the smaller, nymphal lice were smaller than those induced by larger, adult lice. Collectively, these data confirm that the vasodilator acts in a dose-dependent manner. Confirmation that an excess stock of vasodilator exists in the glands was also demonstrated by the experimental results shown in *Figure 5(b)*. Individual lice were moved from one feeding site to the next before feeding had been completed, for an average of 10 times. Each of the red spots that appeared at each bite site attained approximately the same final size, and these final sizes were almost the same as those appearing at the sites of fully completed bites. Thus, an adult louse appears to contain at least 10× the dose needed to induce a strong, easily scoreable red spot.

The biochemical properties of the vasodilatory components of the saliva were also investigated. The possibility that the causative agent might be a peroxidase (as has been reported for *Anopheles* spp.) was investigated in such a way that the results could be compared with those obtained in *A. albimanus* (Ribeiro and Nussenzsveig, 1993) and in *A. gambiae* (Ribeiro *et al.*, 1994). First, each of the two *P. humanus* gland types was individually tested for peroxidase activity against *o*-dianisidine. During a time course of 3 min, 0.1 gland equivalents per/100 μl of reaction volume of either gland type (RG or UG) gave no detectable oxidization of the substrate, as measured at 450 nm. In contrast, strong oxidizing activity was detected under these conditions with 600 pg of horseradish peroxidase, which was used as a positive control (Ribeiro and Nuessenzveig, 1993; Ribeiro *et al.*, 1994).

As an alternative possibility, the catechol oxidase activity of the glands was tested, using both noradrenaline and dihydroxyphenylalanine as substrates. As shown in *Table 1*, neither type of gland exhibited significant activity compared with that of the blank control, and both glands showed only very weak activity compared to that shown by 12 ng of tyrosinase. In contrast, a comparable salivary gland equivalent of *A. albimanus*, exhibited about 100-fold increase in the oxidation of noradrenaline during the same period (Ribeiro and Nussenzveig, 1993), and that of *A. gambiae* exhibited a 20-fold increase in oxidase activity (Ribeiro *et al.*, 1994). Thus, it does not appear that the vasodilator of the louse saliva contains either oxidase or peroxidase activity that could be

***Table 1**. Test for catechol oxidase activity (OD at 450 nm) in the reniform glands and U-shaped glands of human body lice*

			Reniform gland		U-shaped gland	
Substrate	Blank	Tyrosinase[a]	rep. 1	rep. 2	rep. 1	rep. 2
DOPA (300 μM)	0.02	0.59	0.02	-	0.02	0.02
	0.02	0.63	0.01	0.02	0.02	0.02
Noradrenaline (1 mM)	0.01	1.92	0.03	0.04	0.04	0.04
	0.01	2.09	0.04	0.04	0.04	0.04

[a]Tyrosinase, 12 ng, was used as a positive control.
Each preparation of potential activity was tested in duplicate against each substrate. Two independent preparations (replications) of each gland type (0.1 gland equivalent per 100 μl reaction) were made and tested. Reactions were allowed to proceed for 2 h and the OD_{450} was measured.

responsible for destroying biogenic amines at the bite site (although this does seem to be the case with the saliva of some anopheline mosquitoes).

The hypothesis that the causative agent inducing vasodilation could be a haeme protein was also investigated. Such salivary proteins carry nitric oxide in other blood-sucking insects (e.g. *Rhodnius prolixus*, and *Cimex* spp.). One indication that this might not be the case in *P. humanus* salivary glands, is that the colour of these glands is either white or translucent (*Figure 2(c)* and *(d)*). This is in marked contrast to the orange or cherry-red colour of the glands in the other blood-sucking insects that do utilize haeme proteins to alter host haemostasis. Nevertheless, homogenates of each of the *P. humanus* glands (50 glands/100 μl) were analyzed biochemically for evidence of a haeme protein, using UV spectrophotometry and scanning from 350 to 550 nm. No diagnostic maximum Soret band was visible at 413 nm (result not shown), whereas such a band is distinctly visible in homogenates of salivary glands from *R. prolixus* (Champagne *et al.*, 1995; Ribeiro *et al.*, 1995), and in homogenates of *Anopheles* glands that contain a haeme peroxidase (Ribeiro and Nussenzveig, 1993).

Our initial experiments using *in vivo* video microscopy to qualitatively assess the combined extracts of the two types of glands on rat resistance arterioles did not produce evidence of a strong or consistent vasodilatory effect. Similarly, intradermal injection of rats with the extracts did not result in evidence of vasodilatory activity in the form of red spots. In addition, close inspection of the rabbits being used as laboratory hosts for the lice did not reveal the presence of red spots at the bite sites, although these are clearly visible at bite sites on human hosts (see *Figure 1*).

3.4 *Immunogenic properties of the louse salivary proteins*

After hundreds of lice had fed on D. Jones several times per week for over 6 months, the subject did not develop a significant antibody titre against continuous epitopes of reniform salivary gland proteins (SDS-PAGE/immunoblot data, 1:40 dilution, results not shown). However, immunoblot results with serum from host rabbits (1:40 dilution) indicated that the rabbits had developed a moderately strong immune

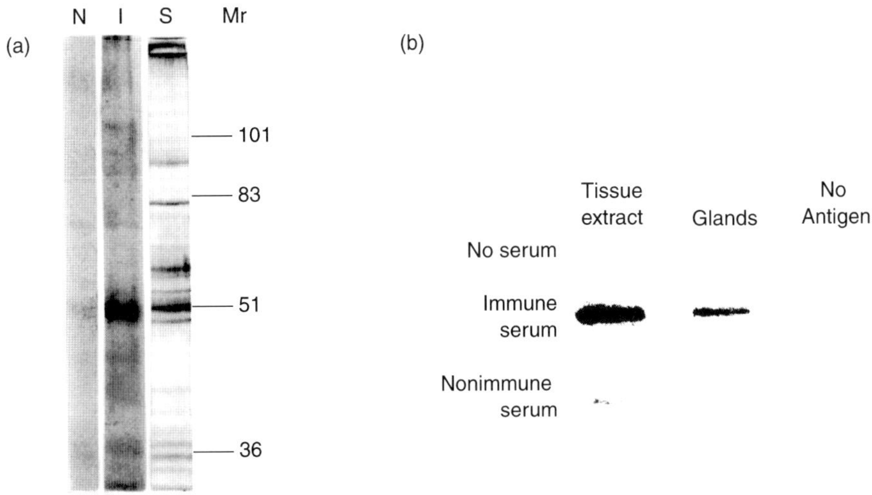

Figure 6. (a) Immunoblot analysis of host antibody titre towards louse salivary gland proteins separated by SDS-PAGE. Each lane contains the protein from 100 reniform glands. S, silver staining. The non-immune serum control (N) showed no specific signal, while the sera from rabbits serving as lice hosts (I) gave a single signal for non-immune serum. (b) Immunoslot-blot analysis of native proteins extracted either from louse thoracic tissues containing salivary glands, or from salivary glands dissected free from other thoracic tissues. Each slot received the equivalent of six of each of the two types of salivary gland. The same antisera used for the tests in (a) were also used in these tests, but at 1:40 final dilution. The strong specific signal for the immune serum slots demonstrates that salivary gland proteins are injected into hosts, and a component of these salivary gland extracts is antigenic. Comparison of the results shown in (b) with those shown in (a), suggests that the most antigenic epitope of the proteins in the louse saliva is a native, discontinuous epitope.

reaction (*Figure 6(a)*). When the same dilution of rabbit (host) antisera were tested under native slot blot conditions for reactivity against antigens, a distinct reaction with the antigens was seen (*Figure 6(b)*). The fact that the antisera appeared to bind to the proteins more strongly in the native slot blot, and more weakly in the SDS-PAGE immunoblot, indicated that the antibody response in rabbits is directed mainly toward discontinuous epitopes of the salivary proteins. Thus, the data suggest that the primary sequence of the antigenic salivary proteins did not elicit as strong an antigenic response in the rabbit host by comparison with the response to the discontinuous epitopes formed by the native antigen structure.

4. Discussion

4.1 *Morphology and identity of* P. humanus *salivary glands*

The presence of two pairs of salivary gland (as opposed to one pair) distinguishes human lice and several haematophagous bugs from most other blood-sucking insects. Both types of *P. humanus* glands have been shown to possess a salivary duct structure that is evidence for the functional contribution of both gland types to the louse saliva. That different regions of the salivary glands of some mosquitoes are known to produce different compounds (James *et al.* 1991; Ribeiro, 1992), suggests that the two louse glands may produce different compounds, each with different putative target sites.

Indeed, the different morphological structure of the two types of louse glands suggests different gland functions. The reniform glands possess two distinct types of cell, each apparently associated with the different functions of production and transport of salivary components, whereas the U-shaped glands had only one of these two cell types.

4.2 *The nature of salivary gland proteins*

Although there were marked qualitative differences in the protein contents of the two types of glands, no specific changes were detected in relative protein abundance in either gland after feeding. Nor did the relative abundance of any particular protein change visibly during periods of fasting. Perhaps the proteins actually injected during feeding are not sufficiently abundant to be visible after SDS-PAGE fractionation/silver staining of gland extracts. Alternatively, only a small portion of the protein content of each gland may be injected during louse feeding, although this contrasts with the clear changes in protein abundance detected after feeding in some other blood-sucking insects. For example, in a single bite *A. albimanus* can inject 40% of its total salivary protein (Ribeiro and Nussenzveig, 1993), and most of the vasodilatory activity contained in the salivary glands of *Culicoides varipennis* is injected at a single bite (Perez de Leon *et al.*, 1997). Recently, it has been shown that approximately 80% of the adenosine deaminase activity in the salivary glands of the sand fly is lost after a single bite (Charlab *et al.*, 2000). Interestingly, in the louse *P. humanus*, one of the most abundant proteins in the saliva of the U-shaped glands is apparently a glutathione transferase. One of the reasons why this enzyme might be in the saliva may be that it is an enzyme that is normally associated with the terminal steps in the synthesis of peptidoleukotrienes (Penrose, 1999). In humans, intradermal injection of the peptidoleukotriene LTD4 induces vasodilation of the dermal capillaries, superficial venules and arterioles (Bisgaard, 1987; Larkin *et al.*, 1995; Soter *et al.*, 1983). Intradermal injection of LTD4 can also cause increased microvascular leakage of plasma (Woodward *et al.*, 1989).

4.3 *Vasodilatory activity in louse saliva*

Vasodilators that cause reddening (erythema) at the bite site have been isolated from a wide variety of blood-sucking insects. These include a maxadilan in the sand fly (Lerner and Shoemaker, 1992; Ribeiro *et al.*, 1989), a vasodilatory peptide from the black fly (Cupp *et al.*, 1994), a tachykinin from *Aedes aegypti* (Champagne and Ribeiro, 1994), and catechol oxidase/peroxidase activity found in *Anopheles albimanus* (Ribeiro and Neussenzveig, 1993). As noted by Champagne and Ribeiro (1994), it seems that different blood-sucking insects have each utilized a different 'domestic' protein, originally performing a different function, for the specialized needs of the blood-sucking habit. In this study, the absence in the louse salivary glands of a diagnostic colour for haeme proteins, such as the cherry-red colour in *Rhodnius prolixus* (Champagne *et al.*, 1995) or the brownish-red colour in *Anopheles albimanus* (Ribeiro and Nussenzveig, 1993), coupled with the fact that we did not detect a diagnostic maximum Soret band, suggests that lice do not utilize the haeme-type vasodilator found in other blood-sucking insects. We were also unable to demonstrate that the louse vasodilator had either catechol oxidase or peroxidase activity. Our studies here also suggest that the human body louse (which in nature is restricted to human hosts), possesses a salivary gland vasodilator that is unusually specific for its target(s) in human skin microvasculature. The rat does not show vasodilation in response to intradermal injection of combined louse gland extracts, nor does the rabbit show vasodilation at bite sites. In contrast, both sandfly and blackfly vasodilators cause erythema in rabbit skin (Cupp *et al.*, 1994; Lerner *et al.*, 1991). The *Aedes aegypti* tachykinin, the *Rhodnius prolixus* NO_2-releasing vasodilator, and the *Anopheles albimanus* catechol oxidase/peroxidase-based vasodilatory activity, are all able to relax rabbit aorta (Ribeiro *et al.*, 1990; Ribeiro, 1992). The *A. aegypti* tachykinin can also affect guinea pig ileum (Champagne and Ribeiro, 1994). To date, the most host-specific haematophagous arthropod in which the salivary vasodilator has been identified is the cattle tick, *Boophilus microplus*, which secretes prostaglandin F2α (PGF2α) in its saliva, and this molecule is the most active tick prostaglandin on cattle skin (Kemp *et al.*, 1983). The apparent biochemical differences between louse vasodilators and those of other insects, combined with the high specificity of the louse vasodilator, suggest that its structure may offer a new lead for the design of clinical agents that control blood flow (Ribeiro, 1987).

4.4 *Immunological properties of louse saliva*

The immunoblotting results presented here are the first biochemical evidence that proteins in human louse saliva actually enter the host tissues during feeding. The antibody response in rabbits to louse salivary proteins appears to be primarily directed against their discontinuous epitopes. However, the rabbit antibody raised in response to louse salivary components is either not targeted against, or is ineffective against, the louse salivary gland vasodilator. Balbi (1933) was not able to demonstrate a detectable level of antibody response to louse saliva in the serum of persons infested with large numbers of lice. However, he was able to transfer the rapid hypersensitive reaction to louse bites from heavily louse-infested persons to louse-free persons, by injecting the louse-free persons with serum from louse-infested persons. These results, together with other observations reported by Balbi (1933), were interpreted by Nelson *et al.* (1977) to implicate IgE as the basis of the immediate hypersensitive reaction to lice. Abdel Fattah *et al.* (1994) reported a marked correlation between the high serum levels of IgE and swollen cervical lymph nodes in louse-infested children. There is also a recent report of abnormal distribution of histocompatibility antigens in louse-infested patients (Morsy *et al.*, 1996). These authors found a significant increase in the histocompatibility antigens HLA-A11 and HLA-B5, in persons infested with lice.

Whether the specific components of the louse saliva that elicit the antigenic response and hypersensitivity are the same components that cause vasodilation or other feeding-associated alterations in the host is not known. In fact, it may be that the primary antigens causing hypersensitive responses are not salivary in origin, but are actually present in the faeces of the lice (Mumcuoglu *et al.*, 1996b; Peck *et al.*, 1943). Should it transpire that the louse salivary proteins are not, in fact, antigenic, this may be a consequence of the fact that, unlike many other well-studied blood-sucking insects, the body louse is an obligate parasite of humans. The possible evolutionary selection for the reduced antigenicity of louse salivary toxins in humans may offer a fruitful approach to identification of structures of therapeutic value with minimal allergenicity.

In summary, although it is estimated that up to 6–12 million Americans, and millions of others worldwide, are injected each year with salivary toxins by human body lice, there is little histological or biochemical information about the secretory structure of the glands or of the proteins they produce. We report here the structure and histology of both the reniform-type and U-shaped salivary glands of these parasites. The reniform glands were found to have two major types of cellular organization. Those adjacent to the salivary duct possess an unusually thick and flocculent basement-like membrane, whereas the other type of cells (not surrounding the ductwork) are larger with a thin basement membrane. Both types of cell in the reniform glands possess numerous secretory vacuoles apparently containing salivary protein granules. The U-shaped glands were observed to contain the type of cells surrounded by a thick flocculent membrane. A central salivary duct runs the length of each salivary gland. Vasodilatory erythema of human host skin was caused by the saliva of *P. humanus*. The intensity and persistence of the wheal did not appear to vary significantly between naïve and experienced host individuals. Lice contained a stock of vasodilator, enabling them to rapidly reinitiate new feeding attempts after successively interrupted bites. The vasodilator exerts its action in a rapid, dose-dependent manner, although lice that had not fed for several days took slightly longer to inject the vasodilator than those that had fed more recently. The vasodilator appears not to be a haem-binding protein, and does not exhibit peroxidase or catechol peroxidase activity at gland equivalent concentrations where such activity was detected in other blood-sucking insects. In addition, the vasodilator appears highly specific for human tissue target(s), since it was not detected to exert activity on rat resistance arterioles (by *in vivo* videomicroscopy), nor in intradermal injection in rats, nor via natural bites on rabbits. Immunoblot analysis of antibodies produced by hosts (rabbit and human) that have been subjected to heavy feeding of lice over several months showed only very weak reactivity against denatured salivary proteins, while a stronger reaction was detected against native proteins. Electrophoretic analysis of the proteins contained in each type of gland revealed that the U-shaped glands have a less complex protein content. The N-terminal sequence of one of the dominant proteins in the U-shaped glands closely matched insect glutathione transferases, not known to be a major salivary gland enzyme in other blood-sucking insects, but which is known to be a terminal enzyme in the vertebrate synthesis of peptidoleukotrienes that induce vasodilation in skin.

Acknowledgements

The studies reported here were supported in part by the University of Kentucky program in Research and Graduate Studies. Much of the research was conducted at the laboratory of Dr. Charles Shoemaker, who graciously hosted the senior author during his sabbatical visit to the Harvard School of Public Health, Boston, MA, USA. Appreciation is profusely expressed to Dr. Susumu Ito (Anatomy Department) and Louise Trakimas (Pathology Department), Harvard Medical School, for their invaluable assistance in the microscopy studies reported here. Dr. Patrick Skelly provided additional helpful advice and assistance. Dr. Tony James, University of California, provided important interpretations of electron micrographs. Dr. Darlyne Murawski took the pictures of lice feeding.

14

Host manipulation by insect pathogens: the effect of the baculovirus *egt* gene on the host–virus interaction

Jenny S. Cory, Katharine R. Wilson, Rosemary S. Hails and David R. O'Reilly

1. Introduction

Of all natural enemies, pathogens have the most intimate relationship with their hosts. This means that in order to survive, reproduce and compete they need to overcome numerous barriers within their host. While the interaction of parasitoid wasps and their insect hosts has been the subject of considerable research (Beckage, 1985, 1997c; Lawrence, 1986a), we know very little about the way in which insect pathogens manipulate their host environment to increase their likelihood of success. Insects are infected by a wide range of pathogens, including several families of virus. The most intensively studied group is the baculoviruses, a family of double-stranded DNA viruses that occur exclusively in arthropods, especially insects (Miller, 1997). Baculoviruses are most common in Lepidoptera and only infect the larval stage. The virus needs to be ingested to initiate infection. The viral occlusion bodies (OB) are broken down in the midgut and from there the (non-occluded) virus particles spread throughout the body. The majority of baculovirus infections involve a wide range of tissues, including the fat body, tracheoles and epidermis. The widespread nature of the infection means that a large proportion of the insect's body is converted into viral OB. The virus yield from a single, final instar noctuid larva can be in excess of 10^9 OB. These viral OB are not released until the insect dies, when they are then available to infect other susceptible hosts. This is thought to be the major route of transmission (Cory *et al.*, 1997). The structure of the OB provides stability outside the host; the virus particles are embedded in a protein matrix, which protects the virus from environmental degradation. If the OBs are protected from UV irradiation, they can persist in the environment for years (e.g. Carruthers *et al.*, 1988; Thompson *et al.*, 1981).

Endocrine Interactions of Insect Parasites and Pathogens, edited by J.P. Edwards and R.J. Weaver.

Passage of baculoviruses in cell culture has shown that they contain genes that are not essential for their survival *in vitro*, implying that such genes may only act at the level of the host organism. Several baculovirus genes have been shown to be non-essential in cell culture. One of the best-characterized genes of this type is the ecdysteroid UDP-glucosyltransferase, or *egt* gene (O'Reilly and Miller, 1989). One particularly interesting feature of this gene is that it appeared to manipulate the survival and development of the insect host, demonstrating hitherto unimagined subtleties in the relationship between an insect virus and its host. In this chapter we describe what is currently known about the baculovirus *egt* gene at the molecular and biochemical levels, and the ways in which it alters the interaction between the baculovirus and its host at the individual and population levels.

2. The ecdysteroid UDP-glucosyltransferase *(egt)* gene

The ecdysteroid UDP-glucosyltransferase (*egt*) gene was first identified in *Autographa californica* nucleopolyhedrovirus (AcMNPV) during a study of regions of the genome known to be non-essential for viral replication (O'Reilly and Miller, 1989). Studies by Kumar and Miller (1987) have shown that a region within the PstI G fragment of the AcMNPV genome is not required for replication in cell culture. Characterization of this region revealed that two open reading frames (ORF) were disrupted by these deletions, one of which was the *egt* gene (O'Reilly and Miller, 1990; O'Reilly *et al.*, 1990). The AcMNPV *egt* gene was found to encode a protein of 506 amino acids with homology to several mammalian UDP-glucuronosyltransferases (UDPGT) (O'Reilly and Miller, 1989). Since its identification in AcMNPV, homologues of *egt* have been identified in a large number of baculovirus genomes, including representatives of both the nucleopolyhedroviruses (NPV) and the granuloviruses (GV). The *egt* gene is moderately well conserved between different baculoviruses (Clarke *et al.*, 1996; Hu *et al.*, 1997; O'Reilly, 1995). For example, *Plutella xylostella* GV *egt* is between 34% and 38% identical to various NPV *egt* genes (Hashimoto *et al.*, 2000). Phylogenetic studies show that the *egt* gene phylogeny parallels accepted baculovirus phylogeny, suggesting that *egt* represents an ancestral gene that was present in the most recent common ancestor of baculoviruses (Clarke *et al.*, 1996; Hu *et al.*, 1997; Smith and Goodale, 1998). An *egt* gene is notably absent from the *Xestia c-nigrum* GV (XcGV) genome (Hayakawa *et al.*, 1999). The most parsimonious explanation for this is that *egt* was lost during the evolution of XcGV (Herniou *et al.*, 2001). This is the only baculovirus that is definitively known not to possess an *egt* gene.

The AcMNPV *egt* is expressed early in infection as two 5′ coterminal transcripts of 1.8 and 3 kb, initiating 43 nucleotides upstream of the translational start site (O'Reilly and Miller, 1990). Transcript levels decline late in infection. Based on transcription in the presence of cycloheximide, *egt* was initially classified as an immediate early gene. However, more recent work has shown that cotransfection of a viral transcriptional activator such as IE-1 is required for detection of *egt* expression in transient expression assays (Clarke *et al.*, 1996; Rodrigues *et al.*, 2001; Toister-Achituv and Faktor, 1997). Our current interpretation of these data is that the AcMNPV *egt* promoter is active at a low level, independent of any other viral gene, but is significantly activated by IE-1. Most other baculovirus *egt* genes are also thought to be expressed as early genes. The notable exception to this is the *egt* gene of *Lacanobia oleracea* GV (LoGV), which is expressed predominantly as a late gene (Smith and Goodale, 1998). The significance of this late expression for the biology of the LoGV is not clear.

3. The baculovirus *egt* gene encodes an ecdysteroid glycosyltransferase enzyme (EGT)

The AcMNPV *egt* expressed protein (EGT) is secreted from infected cells, and N-terminal sequence analysis of EGT has shown that a signal sequence of 18 amino acids is cleaved during export (O'Reilly *et al.*, 1992). Examination of other baculovirus EGT proteins shows that, although they do not share high levels of sequence identity in this leader region, they all have a high proportion of hydrophobic residues, suggestive of a signal sequence. In addition, the homology between the viral EGT proteins becomes high precisely at the position corresponding to the first amino acid of the mature AcMNPV EGT. The AcMNPV EGT is known to lack a signal motif at its C-terminus for retention of the protein in the endoplasmic reticulum (O'Reilly and Miller, 1989). The related mammalian UDPGTs are found in the lumen of the endoplasmic reticulum, and each possesses a short region at the extreme C-terminus that is rich in lysine residues, and is thought to anchor the protein in the endoplasmic reticulum. This motif appears to be absent from all the baculovirus EGT proteins (O'Reilly, 1995). Thus, it seems likely that all baculovirus EGT enzymes are secreted.

The mature AcMNPV EGT protein has an apparent mass of 56 kDa on denaturing SDS-PAGE (Evans and O'Reilly, 1999). However, EGT enzyme activity always elutes from gel-filtration columns as a single peak corresponding to an approximately 260-kDa protein (Evans and O'Reilly, 1999). Thus, the native enzyme is likely to be an oligomer, possibly a tetramer. Immunoblot analysis demonstrated that the protein is produced rapidly after infection and secreted efficiently from infected cells, in agreement with the transcription studies discussed above. There is a good correspondence between EGT protein levels and EGT activity in time-course experiments, suggesting there is little post-translational control of EGT activity (Evans and O'Reilly, 1999). The AcMNPV EGT has been shown to be N-glycosylated, with approximately 11 kDa of sugar residues on each protein molecule (Evans and O'Reilly, 1999). There are seven potential N-linked glycosylation sites in the AcMNPV EGT amino acid sequence, although removal of the sugar residues by glycosidase treatment has shown that N-linked glycosylation is not essential for enzyme activity. Several other baculovirus EGTs have potential N-glycosylation sites, but it is not known whether any are actually glycosylated.

4. Substrate specificity of the AcMNPV EGT enzyme

The enzyme activity of EGT was first elucidated as a result of the protein's homology to several mammalian UDPGT structures (O'Reilly and Miller, 1989). These enzymes catalyse the conjugation of a variety of exogenous and endogenous lipophilic substrates with glucuronic acid (Burchell and Coughtrie, 1989). This led to the speculation that EGT catalyzed a similar conjugation reaction. Assay of various substrates led to the initial discovery that EGT catalysed the conjugation of ecdysteroids with UDP-glucose (O'Reilly and Miller, 1989). The ecdysone-glucoside formed by the AcMNPV EGT-catalyzed conjugation of ecdysone and UDP-glucose has been purified and its structure determined (O'Reilly *et al.*, 1991). The glucose is linked to the hydroxyl group at position C-22 of the ecdysone molecule, yielding the conjugate ecdysone 22-*O*-β-D-glucopyranoside (see *Figure 1*). In addition to ecdysone, 20-hydroxyecdysone (20E), 26-hydroxyecdysone, 3-dehydroecdysone and makisterone A

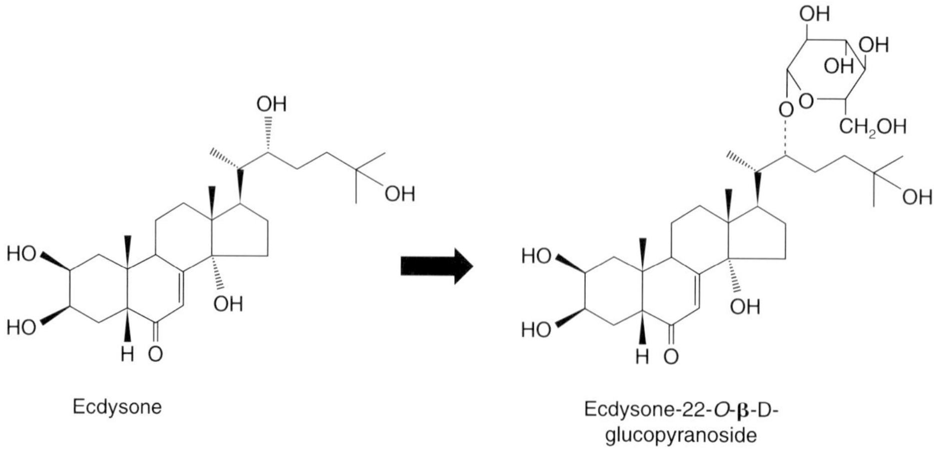

Figure 1. *Conversion of ecdysone to ecdysone 22-O-β-D-glucopyranoside catalysed by the enzyme ecdysteroid UDP-glycosyltransferase (EGT), the protein product of the baculovirus* egt *gene.*

can be conjugated with glucose by AcMNPV EGT (Evans and O'Reilly, 1998; O'Reilly *et al.*, 1991). All of these ecdysteroids possess a hydroxyl group at C-22. In contrast, any ecdysteroid lacking a hydroxyl group at C-22 is not conjugated by EGT. These data strongly suggest that the presence of a hydroxyl moiety at position C-22 of the ecdysteroid is a primary requirement for EGT-catalysed conjugation. A wide range of non-ecdysteroid steroids has also been tested as potential substrates for AcMNPV EGT (O'Reilly and Miller, 1989). These include cholesterol (a precursor of ecdysteroids in insects), selected mammalian steroid hormones, and a variety of other typical substrates of the mammalian UDPGT enzymes. No conjugation to any of these substrates was observed, supporting the idea that EGT is specific for ecdysteroids. The fact that EGT can conjugate ecdysone, 3-dehydroecdysone and 20E is of interest, because these molecules form the central axis of ecdysteroid metabolism in Lepidoptera. In most species, ecdysteroids are produced and secreted by the prothoracic glands in the form of 3-dehydroecdysone. This hormone is rapidly converted by the circulating enzyme 3-dehydroecdysone 3ß-reductase into ecdysone. Finally, ecdysone is converted into 20E, usually the most abundant ecdysteroid in the haemolymph, by intracellular enzymes (ecdysone 20-mono-oxygenases) found in a range of different tissues. The ecdysteroid 20E is the most active form of the hormone *in vivo*. AcMNPV EGT has been purified to near-homogeneity, and its relative ability to conjugate the three ecdysteroids mentioned above determined by kinetic analysis (Evans and O'Reilly, 1998). Ecdysone is the preferred substrate and is conjugated approximately 8-fold more efficiently than 3-dehydroecdysone (k_{cat}/K_m ecdysone = 4229.91; k_{cat}/K_m 3-dehydroecdysone = 528.34). This difference is due almost exclusively to the different apparent K_m for the two substrates (K_m ecdysone = 0.97 μM; K_m 3-dehydroecdysone = 6.27 μM). The V_{max} is similar for both substrates (V_{max} ecdysone = 1.11 pmol min^{-1}; V_{max} 3-dehydroecdysone = 0.89 pmol min^{-1}). Thus, EGT binds less efficiently to 3-dehydroecdysone, but once bound it can conjugate the hormone equally rapidly. The active hormone (20E) is a very poor substrate for the enzyme, being conjugated approximately 200-fold and 25-fold less efficiently than ecdysone and 3-dehydroecdysone, respectively (k_{cat}/K_m = 21.82). This is due both to poor binding (K_m = 8.22 μM) and inefficient conjugation once bound (V_{max} = 0.05 pmol min^{-1}) (Evans and O'Reilly, 1998). It is thought that the hydroxyl group present at C-20 in 20E interferes in some way, possibly by steric hindrance, with the transfer of a sugar to the hydroxyl group at C-22.

Analysis of the AcMNPV EGT kinetic data revealed that the enzyme mechanism involves the formation of an ecdysteroid-UDP-sugar ternary complex (i.e. both the ecdysteroid and the UDP-sugar are bound by the enzyme at the same time; Evans and O'Reilly, 1998). The equivalent mammalian enzymes (UDPGT) are thought to broadly comprise two domains, the N-terminal half of the protein being involved in aglycone binding, and the C-terminal half binding the UDP-sugar (Mackenzie, 1990). In agreement with this, the C-termini of baculovirus EGT show greater sequence similarities to the mammalian UDPGT C-termini, than do their N-termini (O'Reilly, 1997).

The specificity of baculovirus EGT for the UDP-sugar used as donor has also been investigated. In addition to UDP-glucose, UDP-galactose was found to function as a sugar donor for AcMNPV EGT *in vitro* (O'Reilly *et al.*, 1992). However, no other UDP-sugars were utilized. The *Lymantria dispar* and *Mamestra brassicae* NPV (LdMNPV and MbMNPV) EGT can also use both UDP-glucose and UDP-galactose as substrates *in vitro* (Clarke *et al.*, 1996; Kelly *et al.*, 1995). However, characterization of the conjugation reaction in AcMNPV-infected *Spodoptera frugiperda* haemolymph demonstrated that only ecdysteroid-galactosides are formed (O'Reilly *et al.*, 1992). In contrast, the ecdysteroid conjugates formed in the haemolymph of *L. dispar* infected with LdMNPV are exclusively glucosides (Kelly *et al.*, 1995). Kinetic studies of purified AcMNPV EGT have shown that it can use UDP-glucose and UDP-galactose with similar efficiencies (Evans and O'Reilly, 1998). Thus, the differences observed *in vivo* presumably reflect the availability of the UDP-sugars in the different species. It is not known if there are any functional differences between ecdysteroid glucosides and ecdysteroid galactosides. Ecdysteroid galactosides have not been detected in insects that are not infected with virus; however, ecdysteroid 22-glucosides have been identified in *Manduca sexta* embryos and neonate larvae (Thompson *et al.*, 1987; Warren *et al.*, 1986) and are thought to be involved in the arrestment of moulting in the host (*Trichoplusia ni*), which is induced by the ectoparasitic chalcid wasps, *Euplectrus plathypenae* and *Euplectrus comstockii* (see Chapter 9).

5. The effects of the EGT enzyme on host physiology and development

The effects of baculoviral EGT on host ecdysteroid levels following infection are not consistent. Early experiments suggested that ecdysteroid levels never rose above basal levels in final-instar *S. frugiperda* larvae infected with AcMNPV (O'Reilly *et al.*, 1992), suggesting that conjugated ecdysteroids are cleared very rapidly from the haemolymph of the infected insect. However, other experiments with the same species detected high levels of ecdysteroids in the infected insects. These ecdysteroids were found to be conjugated (O'Reilly *et al.*, 1995). In *L. dispar*, the developmental stage of the insect at the time of infection may be important. Park *et al.* (1993) reported that ecdysteroids are undetectable in the haemolymph of final-instar *L. dispar* infected by LdMNPV. They also found that ecdysteroid levels are higher in LdMNPV-infected 4th-instar *L. dispar* larvae than in uninfected larvae of the same stage (Park *et al.*, 1996). Again, these ecdysteroids were all conjugated (Park *et al.*, 1993). Linked to the changes in the endocrine physiology of the host are alterations in host development, which are brought about by EGT. It was clear from the initial studies, using *S. frugiperda* and AcMNPV, that the presence of the *egt* gene altered host development. In infected hosts, larval to larval moulting was inhibited by wild-type virus, whereas those larvae infected with the mutant virus in which the *egt* gene was deleted (*egt*-minus), underwent a normal moult (O'Reilly and Miller, 1989). Similarly, in the final stadium, insects infected with the wild-type AcMNPV showed no signs of pupation, whereas those infected with the corresponding *egt*-minus AcMNPV displayed the same behavioural changes as the untreated controls (O'Reilly and Miller, 1989). Infection in this study was initiated by injecting the insects with high doses of viruses. It has since been shown that other factors moderate the effects of the virus on host development. For example, a lower dose of virus produces less complete developmental arrest (Bianchi *et al.*, 2000; O'Reilly *et al.*, 1998) and timing of infection within the stadium can also affect the outcome. In our recent studies using AcMNPV wild-type virus, final-instar *Heliothis virescens* larvae infected early in the stadium showed complete developmental arrest, and had only basal levels of ecdysteroids. On the other hand, insects infected later in the final stadium showed only a partial developmental arrest, and had extremely high levels of ecdysteroids in their haemolymph (O'Reilly *et al.*, 1998). How baculovirus EGT enzyme activity influences these developmental changes and differences in host response is not well understood. Preliminary data in *Trichoplusia ni* indicate that infection early in the final stadium with an *egt*-minus virus resulted in normal pre-pupal spinning behaviour (K.R. Wilson, unpublished results). However, when insects were infected in the latter half of the final stadium the differences between wild-type and *egt*-minus viruses disappeared, and all virus-treated insects showed spinning behaviour (K.R. Wilson, unpublished results).

6. The significance of the *egt* gene

6.1 *What benefit does a baculovirus gain from the* egt *gene?*

Although the detailed studies described above tell us a great deal about the nature of EGT, its substrate specificity, and its effect on host endocrinology, perhaps the most important question is, how does the *egt* gene affect virus fitness? The first issue is therefore to identify those components that are likely to contribute to baculovirus fitness – for example, infectivity, speed of kill, productivity and environmental persistence (e.g. Anderson and May, 1981; Cory *et al.*, 1997, see also below). One of the advantages of studying baculoviruses is that it is possible to delete specific genes. Thus, it is possible to estimate these components by directly comparing the biological properties of baculoviruses with or without the *egt* gene. This is the approach that has been taken in the studies described below.

6.2 *Effects of the* egt *gene on infectivity and speed of kill*

The effect of *egt* has now been investigated in five baculoviruses: AcMNPV, LdMNPV, *Helicoverpa zea* SNPV (HzSNPV), *Helicoverpa armigera* SNPV (HaSNPV) and *Cydia pomonella* granulovirus (CpGV). Even with such a small number of systems, the *egt* story is not clear-cut. Deletion of the *egt* gene does not appear to alter the infectivity of the virus (Bianchi *et al.*, 2000; Chen *et al.*, 2000; O'Reilly and Miller, 1991; Popham *et al.*, 1997; Slavicek *et al.*, 1999; Treacy *et al.*, 1997). However, when a recombinant *egt*-minus AcMNPV was constructed to express a lepidopteran (*Bombyx mori*) prothoracicotropic hormone (PTTH), infectivity was decreased by almost two orders of magnitude (O'Reilly *et al.*, 1995). Since PTTH stimulates the production of ecdysone, and as injection of 20E into NPV-infected *Heliothis* larvae causes a delay in the onset of virus-induced pathology, as well as a decrease in mortality of the infected insects (Keeley and Vinson, 1975), it is possible that some insect ecdysteroids can interfere with viral replication. Why a similar reduction in infectivity is not seen with *egt*-minus viruses is not clear.

Deletion of the *egt* gene can increase baculovirus speed of action; however, this effect is not universal (*Table 1*). The only host stage in which the *egt*-minus virus consistently produced a reduction in the time from infection to host death (time-to-death) was the 5th instar. There is some evidence that *egt* exerts an effect across all instars, but this seems to be system-specific. The AcMNPV is the only virus that has been bioassayed in different species. In *T. ni* and *S. frugiperda*, deletion of the *egt* gene has always produced an increased speed of action. In another host species, *Spodoptera exigua*, the results have been more variable. Flipsen *et al.* (1995) showed that the time taken for the virus to kill 50% of exposed larvae (LT_{50}) of an AcMNPV *egt*-minus virus (expressing a β-galactosidase gene), was reduced by 33% compared to a wild-type (also containing the β-galactosidase gene) in 2nd-instar larvae. However, recent studies by the same group, using an AcMNPV *egt*-minus mutant, found no alteration in speed of kill in 2nd and 4th-instar *S. exigua* larvae (Bianchi *et al.*, 2000). So the results to date on baculovirus speed of action and how it is influenced by expression of the *egt* gene are inconsistent. This could be the result of genuine differences in the response of different instars or the action of particular virus isolates in specific host systems. Alternatively, the inconsistencies may be a consequence of more subtle differences in experimental details, such as differing times of infection within an instar combined with a high sensitivity to within-instar changes in hormone levels (see below).

Table 1. *Effect of deletion of the* egt *gene on baculovirus speed of action*[a]

Host	Virus	% reduction in time-to-death	Larval instar
S. frugiperda[1,2]	AcMNPV	22, 17	1st
S. exigua[3]	AcMNPV	0	2nd and 4th
T. ni[4]	AcMNPV	11	2nd and 4th
	AcMNPV[5]	23	5th
H. zea[6]	HzSNPV	0	1st
H. armigera[7]	HaSNPV	27	2nd
L. dispar[8]	LdMPNV	0	4th
	LdMNPV	33	5th
C. pomonella[9]	CpGV	0	1st
	CpGV	12	5th

[a]Bioassays in which the *egt* gene was replaced by another gene, such as *lacZ*, have been omitted as it would not be possible to separate the effect of the additional gene from the gene deletion.
[1]O'Reilly and Miller, 1991; [2]Eldridge *et al.*, 1992; [3]Bianchi *et al.*, 2000; [4]Wilson *et al.*, 2000; [5]Cory *et al.*, unpublished data; [6]Popham *et al.*, 1997; [7]Chen *et al.*, 2000; [8]Slavicek *et al.*, 1999; [9]D. Winstanley, personal communication.

6.3 *The effect of the* egt *gene on virus yield*

Most of the studies on *egt* deletion mutants have been carried out from the point of view of developing a faster-acting bioinsecticide, and thus the bioassays have largely concentrated on measuring the speed of kill. From an ecological point of view, the most important consequence of *egt* expression is likely to be an increase in virus yield. Most of the body of the host insect is converted into virus OB, and baculoviruses essentially have a 'sit-and-wait' strategy for infecting new hosts. The amount of progeny virus produced in the host is therefore likely to be very important for optimizing baculovirus survival. However, not all of the studies cited above have gone on to measure yield. Recent work has shown that there is a clear trade-off between baculovirus yield and speed of kill, because larvae that die later are likely to produce more virus for future infections. This principle has been illustrated both by adding genes (e.g. those encoding insecticidal toxins) to increase speed of kill (Burden *et al.*, 2000; Hernández-Crespo *et al.*, 2001), and deleting the *egt* gene (Wilson *et al.*, 2000). A reduction in yield as a result of deleting the *egt* gene is consistently found in the small number of studies carried out so far. Relatively small decreases in the time-to-death can produce quite large reductions in the resulting virus yield (*Table 2*). In *T. ni*, a decrease in time-to-death by 11% produced a reduction in virus yield of 50% in 4th-instar larvae (Wilson *et al.*, 2000). However, yield differences were not detected in 2nd-instar larvae, despite finding a similar reduction in time-to-death (Wilson *et al.*, 2000). It is possible that any yield differences in the 2nd-instar larvae were not detected, as the counting methods were near the limit of accuracy. There also appear to be differences in the responses of males and females. For example, female 4th-instar *L. dispar* larvae treated with an *egt*-minus LdMNPV always produced fewer OB than those treated with the wild-type virus; however, a significant difference was not found in male larvae (Slavicek *et al.*, 1999). Interestingly, although significant yield differences were found in 4th-instar *L. dispar* larvae, there was no difference in time-to-death between insects treated with wild-type and *egt*-minus virus (Slavicek *et al.*, 1999). The general picture that emerges from these studies is that when deletion of the *egt* gene results in a reduced time-to-death, this is usually accompanied by a reduction in virus yield. The data on 4th-instar *L. dispar* larvae infected with LdMNPV, indicate that this might not always be the case. This difference may be a feature of the different ecdysteroid levels found in different host stadia (see above), or the interaction between virus dose, timing and developmental arrest. However, the results certainly indicate that more work is needed on these factors in relation to *egt* expression.

6.4 *The influence of host developmental stage on virus yield*

As noted above, high doses of wild-type virus are more likely to lead to developmental arrest. In addition, the timing of infection within a particular host stadium can also influence the capacity of the virus to block host moulting. Studies on wild-type AcMNPV in *H. virescens* clearly show that the later within the host stadium that the insect is infected, the less likely it is for complete developmental arrest to occur, and this is associated with a concomitant reduction in yield (O'Reilly *et al.*, 1998). Therefore, we would expect the maximum differences in yield between wild-type and *egt*-minus viruses to occur when a larva is infected early in any particular stadium. Thus, the degree of developmental arrest may be more critical to changes in yield than

Table 2. *The effect of deletion of the* egt *gene on virus productivity*

Host	Virus	% reduction in yield	Instar
S. frugiperda[1]	AcMNPV	23	5th
T. ni[2]	AcMNPV	0	2nd
	AcMNPV	50	4th
L. dispar[3]	LdMNPV	60	4th - female
	LdMNPV	0[a]	4th – male
	LdMNPV	90	5th – female
	LdMNPV	80	5th – male
C. pomonella[4]	CpGV	25	5th

[1]O'Reilly and Miller, 1991; [2]Wilson *et al.*, 2000; [3]Slavicek *et al.*, 1999 (approximate values only); [4]D. Winstanley, unpublished data.
[a]Less virus was produced in insects infected with the *egt* minus-virus, but the differences were not statistically significant.

to a reduction in speed of kill *per se*. Few studies have measured time-to-death and virus yield, as well as recording any changes in host development, therefore it is difficult to come to a firm conclusion on this issue. However, it is interesting to note that Flipsen *et al.* (1995), investigating the interaction between AcMNPV and *S. exigua*, found that although there was a significant reduction in speed of kill with the *egt*-minus mutant, this was not accompanied by any developmental changes in the host. By contrast, in a later study, using the same larval stadium, there was no difference in speed of kill, but moulting was arrested at higher virus doses (Bianchi *et al.*, 2000). The reason why the insect actually dies earlier when the *egt* gene is deleted is not clear, although one theory is that it could be related to the early degeneration of the Malpighian tubules in larvae infected with the *egt*-minus virus (Flipsen *et al.*, 1995).

6.5 *Effects of the* egt *gene on larval growth and feeding behaviour*

Two studies have investigated the way in which deletion of the *egt* gene affects growth and feeding of the host after infection. In *L. dispar* there was a distinct difference between the response of male and female larvae. Infection of 4th and 5th-instar female larvae with an *egt*-minus LdMNPV resulted in a reduction in cumulative weight gain, whereas no difference was measured in similarly infected male larvae (Slavicek *et al.*, 1999). In *T. ni*, infection with an *egt*-minus mutant in the 4th larval stadium resulted in a significant reduction in the production of faecal matter, although production peaked earlier in insects infected with the AcMNPV *egt*-minus mutant than in uninfected controls or in insects infected with the wild-type virus, suggesting early enhanced feeding in insects infected with the *egt*-minus virus (Wilson *et al.*, 2000). An analysis of virus yield per unit body weight of the infected host showed no differences in viral productivity between the two types of virus (wild-type and *egt*-minus). Cadaver weights of the insects infected with the *egt*-minus mutant were significantly lower, suggesting that the reduction in virus yield in *egt*-minus infected insects is mainly due to a reduction in host feeding and growth (Wilson *et al.*, 2000).

6.6 *Effects of the* egt *gene on virus transmission and persistence*

Most research on the role of the *egt* gene has focussed on its effect on speed of kill and virus yield. However, it is also possible that the *egt* gene could influence other components of virus fitness, such as the transmission and persistence of the virus. For example, the transmission process in the field is dependent not only on the susceptibility of the insect host and the pathogenicity of the virus, but also on the behaviour of the host and the impact of this behaviour on virus acquisition. Baculoviruses are thought to manipulate host behaviour so that the infected larvae tend to climb towards the top of the plant before dying (Vasconcelos *et al.*, 1996). Presumably, this virus-linked change in host behaviour is a means of enhancinges the likelihood of virus transmission or dispersal. Final instar lepidopteran larvae usually go through a wandering phase, and often burrow into the soil to pupate. This behaviour would remove the virus from the vicinity of the next generation of susceptible larvae (as these are likely to be feeding on plant material above the soil surface), as well as altering the environmental persistence of the virus. The developmental arrest induced in the host by viral *egt* expression might also alter both the distribution of virus after an epizootic, as well as modifying its rate of release from the cadaver. Both of these factors are likely to influence virus transmission and persistence in the field. Recent field studies indicate that transmission in the field is enhanced in wild-type AcMNPV when compared to an *egt*-minus virus (K.R. Wilson, unpublished results). However, further field-based studies are crucial if we are to understand the full impact of the *egt* gene on virus dynamics.

7. Mathematical models as tools to investigate the invasiveness and persistence of *egt* deletion mutants

The use of mathematical modelling to study the comparative ecology of *egt* deletion mutants and their wild-type counterparts can provide a useful tool with which to investigate their potential behaviour. Is it possible to extrapolate from the laboratory and small-scale field experiments described above, in order to predict the likelihood of invasion and persistence of *egt*-minus or wild-type genotypes? Furthermore, how will the different virus genotypes influence the population dynamics of the host insect?

As discussed above, a central question is how the addition or deletion of a gene will affect the fitness of the virus. While fitness has a very precise ecological meaning, its definition in mathematical terms will depend upon the theoretical framework used. Models phrased in continuous time use R_0 (the basic reproductive rate) as the measure of fitness. This is defined as the number of new infections resulting from each primary infection. Briggs and Godfray (1995) developed a theoretical framework that involved delay differential equations. In comparison with earlier models (e.g. Anderson and May, 1981) this new framework more explicitly modelled the delay between infection and death, and took account of the presence of different insect life stages, which vary in their susceptibility to viral infection (i.e. a stage-structured context). However, the expression for R_0 incorporates the same basic life history parameters of the virus as the earlier models, that is:

$$R_0 = L\frac{\upsilon \sigma_I \Lambda}{\mu}$$

where L = density of host larvae (the susceptible stage), υ = virus transmission rate, σ_I = the probability of survival of an infected individual (life stage), Λ = the yield of virus from a single cadaver, and μ = the rate of decay of the OB in the environment. A number of trade-offs between the different virus life-history parameters limit the size of R_0. For example, viruses that lyse the cadaver will enhance virus transmissibility, but may also increase their rate of degradation. The evidence presented earlier (see Section 6.3) shows that the *egt* gene affects another of these trade-offs (that between the speed with which the host is killed and the resulting yield of OB from cadavers). Does deletion of the *egt* gene result in reduced fitness, or would we predict the coexistence of both wild-type and *egt*-minus genotypes in the field?

Mathematical approaches to this problem are most useful, even in the absence of comprehensive data, as they allow a number of potential scenarios to be explored. Current data suggest that the *egt*-minus mutants suffer principally from a reduction in the eventual yield of viral OB from an infected host. Considering continuous time models and pure infections, one would predict that the *egt*-minus virus would have a lower R_0, and would therefore be out-competed by the wild-type virus. However, the lower yields of the *egt*-minus viruses are usually accompanied by an increase in the speed of kill and, therefore, the potential to have a greater number of cycles of infection within one generation (or season) of host larvae. Consequently, alternative measures of fitness include the number of new hosts infected in any current generation per infected host of the previous generation (Dwyer *et al.*, 1999).

Dushoff and Dwyer (2001) have examined the possibility of coexistence of baculovirus genotypes with differing 'speed of kill/transmission trade-offs' (yield being a prime component of transmission). They found that there are certain conditions in which coexistence of genotypes could occur, or in which a genotype with reduced transmission and enhanced speed of kill (such as an *egt*-minus virus) could even become dominant. However, the slope of the trade-off must be shallow (i.e. there must be little loss of yield associated with enhanced speed of kill). Dushoff and Dwyer (2001) also compared the observed transmission and speed of kill parameters of a LdMNPV *egt*-minus virus with the predictions of the model and found that the *egt*-minus genotype was well within the parameter space that predicted dominance of the wild-type virus. However, bootstrapped 95% confidence intervals did just overlap with the area of parameter space in which the *egt*-minus virus could coexist or dominate. Most interestingly, in their models (based on *L. dispar* population dynamics) they discovered that the time to extinction of the virus could be long, even if the competitive disadvantage was substantial. With repeated introductions of an engineered virus (as may happen in a pest management context), the modified virus could effectively be present for many host generations. A similar approach in a stage-structured context has also found conditions in which the wild-type and *egt*-minus viruses can coexist (M. Bonsall and R. S. Hails personal communication). The latter study shows that the likelihood of coexistence of the two genotypes depends upon the relative survival of infected hosts during the viral incubation period, and on any yield advantage of the wild-type virus.

Some general conclusions can be drawn from these theoretical studies. Coexistence of the two genotypes can occur for many generations if the *egt*-minus virus is able to gain a significant advantage in transmission and/or speed of kill over the wild-type virus. Secondly, there are a number of important knowledge gaps. Dushoff and Dwyer (2001) illustrated that the rate of decay of the virus during the winter could be a crucial parameter in determining the relative competitive ability of the two viruses at the beginning of the next season, as it determines the relative size of the inoculum able to initiate the first round of infection. If pathogen densities are low at the beginning of the season, a faster-killing recombinant (e.g. *egt*-minus) can exploit the population of susceptible hosts more quickly. By contrast, if pathogen densities are high, then the wild-type virus has an advantage due to its higher yields. However, as mentioned above, it is also possible that the relative winter survival rates of the wild-type and *egt*-minus viruses may differ as a result of the behavioural and developmental changes induced in the host by viral *egt* expression.

8. Conclusions

The information available to date strongly supports the notion that the advantage that a baculovirus gains from possessing the *egt* gene is that it increases virus yield and, thus, the likelihood of further virus transmission. However, the data are certainly not consistent and before we can fully describe the process we need more information on the importance of age structure (i.e. the relative effect of the wild-type and *egt*-minus viruses on different host stadia), and on the time of infection within the host stadium in different species. Conventional wisdom would suggest that by deleting the *egt* gene from a baculovirus, its fitness would be reduced. However, because there might be a cost to the virus of generating extra yield (such as a reduction in secondary cycling as a result of a prolonged duration of infection), both empirical and theoretical studies suggest that the actual consequences of deleting or adding a gene may not be so easy to predict. More data are needed on the behaviour of the wild-type and *egt*-minus viruses in the field, particularly in relation to their longer-term persistence. We also need to know about the behaviour of mixed wild-type and *egt*-minus infections within an individual host. What proportion of a mixed infection of the two virus genotypes must be wild-type (*egt*-plus) for the infection to behave as if it was exclusively wild-type? Will the *egt*–minus viruses have any replication advantage? As NPVs exist as a collection of hundreds of virus particles within an OB, it is also possible that *egt*-minus viruses could persist as 'parasites' within a predominately wild-type virus OB. Most importantly, perhaps, we need to establish whether wild-type and *egt*-minus viruses can coexist in natural populations. Answering these questions will not only allow us to more fully understand the role of the *egt* gene in baculovirus–host interactions, but also help us to elucidate those factors that are particularly important for baculovirus survival.

Acknowledgements

We would like to thank Mike Bonsall, Imperial College at Silwood Park, for his modelling expertise. We would also like to thank Doreen Winstanley, Greg Dwyer and Jonathan Dushoff for access to unpublished data. This review was supported by NERC grants GR3/8967 to David O'Reilly and Jenny Cory and GR3/A8967 to Jenny Cory, David O'Reilly and Rosemary Hails.

15

Interactions with the host of recombinant baculoviruses expressing components of the insect endocrine system – towards environmentally benign biological control

Jian-Hua Chen, Steven M. Sait, Hajime Takeuchi, David R. O'Reilly, Philip C. Turner and Huw H. Rees

1. Introduction

Our laboratory has been involved in investigation of the regulation of moulting hormone (ecdysteroid) titres during insect development, in particular the key enzymes catalysing hormone transformations. Recently, we have purified, cloned and characterized some of these enzymes. Construction of recombinant baculoviruses over-expressing such enzymes has not only allowed production of protein *in vitro* for further study, but has also permitted evaluation of the effects *in vivo* of such enzyme over-expression on host insects. Expression of these enzymes at the incorrect time may derange normal development and, thus, provide a potential approach to increase the efficiency of the baculovirus for biological control of insect pests.

Although the use of baculoviruses as biological insect control agents offers considerable advantages over chemical insecticides, particularly due to their high degree of host specificity, their practical use is limited by a number of factors, particularly their slow speed of action (for review see Moscardi, 1999). Over the last decade, there has been much interest in enhancing the speed of kill of insects by baculoviruses. This has been investigated using recombinant viruses expressing a variety of peptides/proteins, including components of the insect endocrine system (for reviews see Bonning and

Endocrine Interactions of Insect Parasites and Pathogens, edited by J.P. Edwards and R.J. Weaver.

Hammock, 1996; Miller, 1995; O'Reilly, 1995; Possee *et al.*, 1997). In this article, we outline the relevant aspects of the insect endocrine system, summarize studies on the effects of recombinant baculoviruses over-expressing components of this system on host insects, and discuss our initial studies on the effects of over-expression of 3-dehydroecdysone 3β-reductase on the cabbage looper, *Trichoplusia ni.*

2. Aspects of the insect endocrine system

Recent reviews are available on the insect endocrine system, including ecdysteroid biosynthesis and inactivation (Gade *et al.*, 1997; Grieneisen, 1994; Henrich *et al.*, 1999; Rees, 1995). The basic components of the insect endocrine system and the physiological roles of the hormones involved are reviewed in Chapter 1. In immature stages of insects, in response to various stimuli, neurosecretory cells in the brain secrete a peptidic prothoracicotropic hormone (PTTH), which stimulates the prothoracic glands (PGs) to produce ecdysteroids. In most species, this ecdysteroid is ecdysone (*Figure 1*), which primarily, but not entirely, serves as a precursor of the generally much more active hormone, 20-hydroxyecdysone (20E). The conversion of ecdysone into 20E (*Figure 2*) is catalysed by the enzyme ecdysone 20-monooxygenase, and occurs in certain peripheral tissues. It has been demonstrated recently that in several lepidopteran species, 3-dehydroecdysone (3DE), accompanied by varying amounts of ecdysone, is a major product of the PG (*Figure 1*). The 3DE is then reduced to ecdysone by a haemolymph NAD(P)H-dependent 3DE 3ß-reductase enzyme (Chen *et al.*, 1996; Kiriishi *et al.*, 1990; Warren *et al.*, 1988).

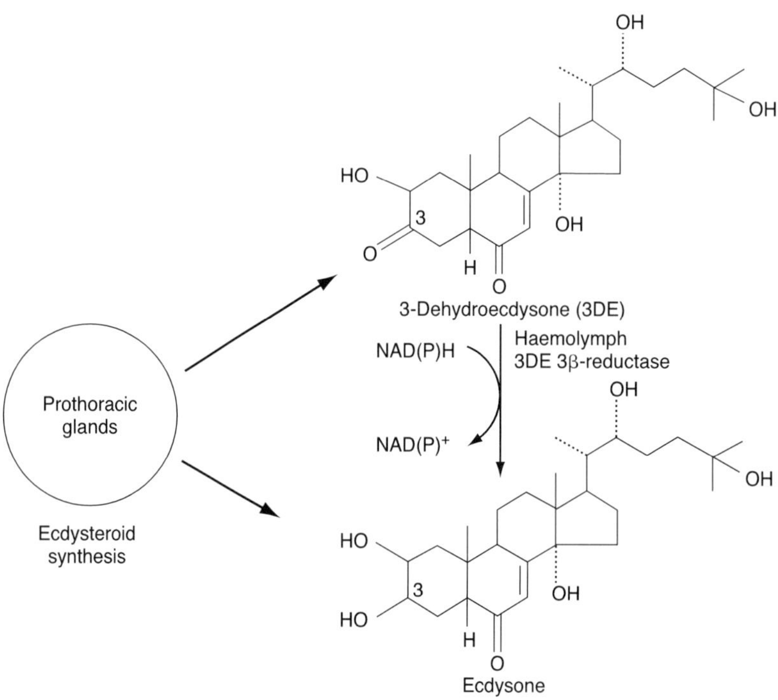

Figure 1. *Formation of ecdysone in Lepidoptera.*

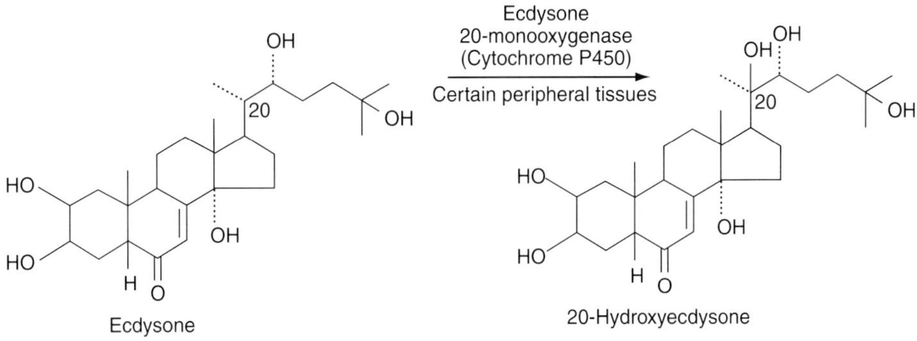

Figure 2. *Conversion of ecdysone into the major active hormone, 20-hydroxyecdysone.*

Ecdysteroids not only function in immature stages of insects, but also in adults. In adult females, ecdysteroids are produced in ovarian follicle cells and may function in regulation of meiotic reinitiation in oocytes or in vitellogenin synthesis in the fat body (Rees, 1995). Furthermore, ovarian ecdysteroids may be passed to the oocytes (frequently as conjugates) and can function in the regulation of membrane/cuticular events in early embryogenesis (Hoffmann and Lagueux, 1985). There is also evidence that ecdysteroids are produced in testis, where they function in the control of spermatogenesis (Loeb *et al.*, 1996).

Ecdysteroid inactivation may occur via several routes (Lafont and Connat, 1989; Rees, 1995; Weirich, 1989). These include:

- i 26-hydroxylation followed by conversion into the ecdysteroid 26-ioc acid;
- ii formation of conjugates at certain positions, primarily phosphates, although others (e.g. glucosides and acetates) occur;
- iii ecdysone oxidase-catalysed oxidation to yield the 3-dehydro derivative, which is then irreversibly reduced by an NAD(P)H-dependent 3-dehydroecdysone 3α-reductase enzyme to produce 3-epiecdysteroid (see *Figure 3*).

Analogous reactions occur with both ecdysone and 20E. It appears that this latter cytosolic pathway is particularly prominent in Lepidoptera, especially in midgut tissue (Rees, 1995). However, in many tissues of several species investigated, the 3-dehydroecdysteroid can undergo NAD(P)H-dependent 3-dehydroecdysone 3β-reductase-catalysed reduction back to the 3β-hydroxyecdysteroid (Lafont and Connat, 1989). This occurs even in tissues incapable of producing 3-epiecdysteroids. However, the physiological significance of this apparently futile cycle is unclear.

In immature stages of insects, whereas moulting is triggered by ecdysteroids, the nature of the moult is determined by the juvenile hormone(s) produced by the corpora allata (see Riddiford, 1994). Juvenile hormone (JH) maintains the juvenile character of larvae. Decrease in JH titre occurs through a decrease in biosynthesis caused, at least in part, by neurohormone(s) and neuromodulators (see Chapter 1). As JH titre decreases, JH esterase (JHE) increases. This enzyme catalyzes hydrolysis of the methyl ester of JH, yielding JH acid, and is regarded as important in JH inactivation (Hammock, 1985). Although JH acid has been viewed as being biologically inactive, there is now evidence that it is a hormone in its own right in certain biological systems

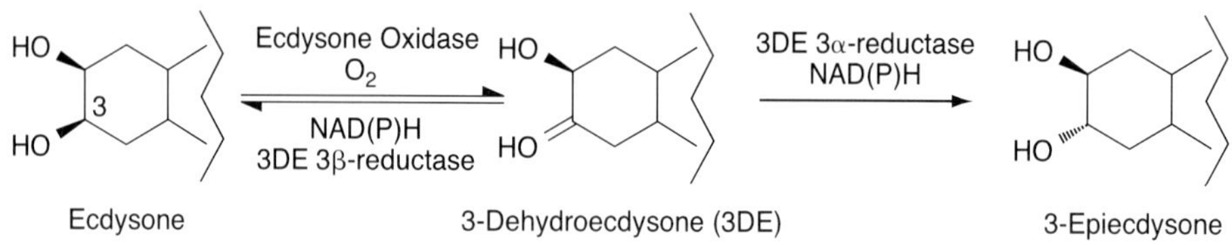

Figure 3. *Ecdysteroid interconversions.*

(for review see Gilbert *et al.*, 2000). In adult females, the corpora allata again synthesize JH, and the hormone may be involved in regulation of synthesis of vitellogenin and its uptake by the ovaries.

Eclosion hormone (EH) is a neuropeptide having multiple functions (for review see Truman, 1985). These include release of motor programmes required for ecdysis (Copenhaver and Truman, 1982), initiation of stage-specific behaviour after moulting (Levine and Truman, 1983), release of the insect tanning hormone (Reynolds *et al.*, 1979), and initiation of intersegmental muscle death following ecdysis (Schwartz and Truman, 1984).

3. Effects of genetically-modified baculoviruses expressing components of the insect endocrine system

There has been much interest in modifying baculoviruses to express insect-selective toxins or components of the insect endocrine system, in an endeavour to enhance the speed of kill of the viruses. However, there are certain problems associated with this approach. For example, a recombinant virus leading to a more rapid kill of the host may be subject to strong negative selection pressure and, thus, may not be able to compete with its wild-type counterpart (but see Chapter 14, this volume). Therefore, recombinant baculoviruses for insect control would have to be used as biological insecticides that are repeatedly applied, rather than as true biological control agents that are able to persist in the environment (see Section 6). This overview will be limited to a consideration of the over-expression of components of the insect endocrine system resulting from viral infection.

The *egt* gene of *Autographa californica* nucleopolyhedrovirus (AcMNPV) encodes the enzyme, ecdysteroid UDP-glucosyltransferase (EGT), which catalyses the conjugation of ecdysteroids with glucose or galactose (for review see O'Reilly, 1995). Comparison of insects infected by an *egt*-minus mutant of AcMNPV, with those infected by wild-type virus demonstrated that EGT expression causes a suppression of host moulting and pupation (O'Reilly and Miller, 1989, 1991). Wild-type AcMNPV-infected insects do not experience the feeding arrest normally associated with ecdysis and, therefore, become larger and yield more virus than those infected with *egt*-minus viruses (O'Reilly and Miller, 1989, 1991). In fact, deletion of the *egt* gene results in a 40% reduction in feeding damage, a 10–20% reduction in the time from infection to host death (time-to-death), and a 30% reduction in the yield of virus progeny. A further discussion of the effects of the *egt* gene on host physiology is given in Chapter 14.

A cDNA encoding the eclosion hormone (EH) from *Manduca sexta* (tobacco hornworm) has been expressed in both wild-type and *egt*-minus AcMNPV virus (Eldridge *et al.*, 1992a). The recombinant hormone was shown to be biologically active, but the ability of the recombinant virus either to limit insect feeding, or to hasten host death, was not affected compared to the wild-type virus.

Similarly, PTTH from the silkworm, *Bombyx mori,* has been over-expressed in AcMNPV (O'Reilly *et al.*, 1995). The recombinant viruses produced large amounts of PTTH, but no reduction in time-to-death was observed following infection of larvae of the fall armyworm, *Spodoptera frugiperda,* with such virus. However, most interestingly, expression of PTTH markedly reduced the infectivity of *egt*-minus AcMNPV.

There has been considerable interest in the development of anti-JH agents as novel insecticides against crop pests, since induction of precocious development would lead to a reduced feeding and fewer reproducing adults (Schooley and Edwards, 1996; Staal, 1986). Although over-expression of JHE would be expected to have an analogous effect to an anti-JH agent, significant improvement in the insecticidal efficacy of various engineered AcMNPVs, which efficiently expressed the JHE, has not generally been observed (Bonning *et al.*, 1992; Eldridge *et al.*, 1992b; Hammock *et al.*, 1990, 1993). However, infection of larvae with a recombinant baculovirus expressing a doubly modified JHE, in which Lys-29 and Lys-524 (within the putative lysosome-targeting sequence of JHE) were replaced with arginines, resulted in a significant reduction in time-to-death, with a 36–50% reduction in feeding damage, compared with larvae infected with the wild-type virus (Bonning *et al.*, 1997). The time-to-death resulting from the doubly modified virus increased when it was used in conjunction with JH analogues or JHE inhibitors, suggesting that JHE activity is required for the enhanced speed of kill. However, when the doubly modified virus was bioassayed in conjunction with anti-JH agents, the time-to-death decreased, suggesting that the effect results, at least partly, from a depletion of JH (Bonning and Hammock, 1996). The improved insecticidal activity was attributed to an apparent failure of lysosomal targeting of JHE in the pericardial cells. On the other hand, in other studies directed at altering the catalytic mechanism of JHE expressed by a modified baculovirus, enzymic activity was completely removed by the replacement, in an engineered AcMNPV, of the active site Ser-201 by glycine (Bonning *et al.*, 1995). Nonetheless, this AcJHE-SG was highly insecticidal, with a 20–30% reduction in the time-to-death and 55–66% reduction in feeding damage, compared to larvae infected with the wild-type virus. In this case, the time-to-death shown by infection with the AcJHE-SG was increased when bioassayed with anti-JH agents, suggesting that the lethal effect of the virus was not related to the degradation or sequestration of JH.

The foregoing studies indicate that specific genetic changes to enhance the stability of an expressed protein can be used to advantage in improving the insecticidal efficacy of recombinant viruses. However, such modification of expressed proteins may increase virus effectiveness without necessarily achieving this outcome in the expected manner.

4. Cloning of 3DE 3β-reductase cDNA and production of a baculovirus construct

4.1 *Haemolymph 3DE 3β-reductase in the cotton leafworm,* Spodoptera littoralis

In agreement with other reports that PGs in lepidopteran species mainly synthesize and release 3DE, we observed that the PGs of last-instar *Spodoptera littoralis* larvae primarily secreted 3DE (~82%) with a lesser amount of ecdysone (~18%) (Chen *et al.*, 1996). As alluded to in Section 2, the conversion of 3DE to ecdysone occurs after its secretion into the haemolymph, and is catalysed by the haemolymph 3DE 3β-reductase. Ecdysone undergoes C-20 hydroxylation in certain peripheral tissues yielding 20E, which is the moulting hormone in most insects. Thus, 3DE 3β-reductase-catalysed conversion of 3DE to ecdysone is an important step in the production of the moulting hormone in lepidopteran species. The enzyme activity of 3β-reductase in *S. littoralis* haemolymph shows developmental changes during the last larval stadium. The highest activity was detected in the late stadium larvae, and coincided with the major ecdysteroid peak, suggesting that this enzyme contributes to the rise in ecdysteroid titre.

The 3β-reductase from haemolymph of *S. littoralis* was purified to homogeneity using a combination of polyethylene glycol 6000 precipitation and sequential fractionation on Mono-Q, Phenyl Superose and hydroxyapatite columns. The purified protein showed an apparent molecular mass of 36 kDa (Chen *et al.*, 1996). Limited N-terminal sequences for the intact protein, as well as for two internal fragments generated by trypsin digestion, were obtained from the purified enzyme.

At first, degenerate primers were designed based on the partial amino acid sequences: forward primers were made on the basis of the limited N-terminal sequences of the intact protein and reverse primers were made on the basis of the two internal fragment sequences. These primers were used in RT-PCR in which the cDNA templates were synthesized from the poly $(A)^+$ RNA isolated from total insect tissues (with the cuticle, head and gut contents removed, but including haemolymph). This allowed us to obtain a cDNA fragment corresponding to a region between the N-terminal area and an internal fragment initially sequenced from the purified protein. Gene-specific primers derived from this cDNA fragment were subsequently synthesized and used for 5′- and 3′-RACE (rapid amplification of cDNA ends), which, respectively, allowed us to obtain the 5′- and 3′-ends of the clone. Taken together, the overlapping cDNAs spanned a total of 1425 bp that encode a protein of 345 amino acids, including a predicted signal peptide of 17 amino acids (Chen *et al.*, 1999). This protein contained all the partial amino acid sequences initially obtained from the purified enzyme. This protein sequence was further confirmed by sequencing a cDNA clone that was obtained by PCR using gene-specific primers, which covered the open reading frame and the 3′-non-coding region. The predicted molecular weight of the mature protein was 37 689, which is similar to the apparent M_r of the purified protein as analysed by SDS-PAGE.

Northern analysis revealed that the mRNA for 3β-reductase is widely distributed, but is differentially expressed in some tissues. Of the tissues examined, Malpighian tubules had the highest expression level, followed by midgut and fat body, while the lowest expression levels were found in haemocytes. It is not clear at present whether all of these tissues are responsible for the production of the enzyme in the haemolymph. Northern analysis also revealed that the mRNA for the enzyme is only expressed in the second half of the last larval stadium. The fluctuation of the expression during the stadium mirrored the change in the enzyme activity, suggesting that the change in activity is largely regulated by gene expression at the transcriptional level (Chen *et al.*, 1999).

4.2 *Construction of a recombinant baculovirus expressing a functional recombinant 3DE 3β-reductase*

A 3β-reductase cDNA containing the entire coding sequence was amplified by PCR using two gene-specific primers in which restriction sites for *Eco*RI and *Pst*I were respectively incorporated. Such amplified product was purified and digested with *Eco*RI and *Pst*I and inserted into a transfer plasmid pSynXIV VI$^+$X3 (Wang *et al.*, 1991), which had been digested with the same restriction enzymes. This was cotransfected into insect cells (Sf21) by the calcium phosphate coprecipitation method with a parent virus vSynVI$^-$gal in which the polyhedrin gene had been replaced by the *lac*Z gene (Wang *et al.*, 1991). Recombination between viral sequences flanking the 3β-reductase cDNA in the transfer vector and the homologous sequences in the viral genome resulted in the replacement of the *lac*Z gene with the entire 3β-reductase cDNA and the polyhedrin gene in the vector. The resultant recombinant viruses express 3β-reductase cDNA under the control of a strong polyhedrin promoter. These viruses were screened by successive plaque purifications based on their occlusion-positive, *lac*Z-negative plaque phenotype, as detailed in O'Reilly *et al.* (1994). The structure of four selected clones was confirmed by restriction digestion and Southern blot analysis.

The expression of recombinant 3β-reductase by the above clones was analysed on Western blots using antibodies raised against the purified protein. The recombinant protein could be detected in both the culture medium and in the cell lysate fraction. The recombinant protein detected in the culture medium showed the same molecular mass as that of the purified natural enzyme, suggesting that the recombinant protein was expressed, secreted and cleaved correctly. Indeed, the secreted recombinant protein had the same N-terminal sequence as that of the natural 3β-reductase, as revealed by protein sequencing. The expression could be detected by Western blotting at 24 h post-infection, and the level of expression increased with time post-infection up to 3 days. At this stage, one out of four clones that showed the highest level of expression was chosen for further use. Enzyme assay, followed by HPLC analysis, indicated that the recombinant protein was functional in reducing 3DE to ecdysone in the presence of a cofactor, NADPH.

Occlusion bodies (OB) of both wild-type and recombinant viruses were obtained 3 days post-infection from infected Sf21 cells. A multiplicity of infection (MOI) of 10 to 20 plaque-forming units (pfu/cell) was routinely used to ensure a synchronous infection of the cells, as well as to drive viral replication at its fastest rate. The infected Sf21 cells were harvested by centrifugation. The pellet was resuspended in 0.5% SDS solution to release the OB, which were sedimented by centrifugation. This OB pellet was then resuspended in 0.5M NaCl solution, subjected to further centrifugation, and the final pellet was resuspended in a small volume of water. The concentration of OB was determined by counting the number of OB in a measured aliquot under a microscope.

5. Effects of wild-type (egt$^+$) baculovirus (AcMNPV) and virus (egt$^+$) expressing 3β-reductase on larvae of *Trichoplusia*

During lethal infection of the host insect, as the virus proliferates and spreads throughout the tissues of the infected insect, the recombinant virus over-expresses 3DE 3β-reductase. At times of substrate (3DE) availability, this should give rise to a steady and progressive increase in the titre of ecdysone. This virus-induced pattern of hormone production would be in stark contrast to the normal pattern of ecdysteroid production, and would be likely to lead to a disruption of host development, to the detriment of the insect. Thus, the recombinant virus may possess qualities that improve its potential as a biological insecticide.

Three key parameters that determine the efficacy of a baculovirus are pathogenicity, time-to-death and virus yield (when the infected insect dies). These three factors are closely linked with one another, and dictate, in part, the probability of the virus being transmitted to other susceptible hosts. For example, a highly pathogenic virus will kill the host quickly, but this has a potential cost, in that increased speed of kill is likely to lead to reduced virus yield. The likelihood of virus transmission will also be reduced if fewer virus particles have been produced when the host dies (see Burden *et al.*, 2000; Cory *et al.*, 1997). For a fuller discussion of the implications of genetic modification on baculovirus ecology see Chapter 14.

5.1 *Bioassay procedure*

It is possible to quantify all of the above biological characteristics in standard laboratory assays. In a preliminary study, we have measured and compared mortality and time-to-death caused by infections with wild-type AcMNPV and the recombinant virus expressing 3β-reductase.

Bioassays were carried out using newly moulted 2nd-instar larvae of the cabbage looper *T. ni* (Lepidoptera; Noctuidae), a species that is particularly sensitive to AcMNPV (Cory *et al.*, 1994). Larvae were starved overnight, selected randomly from mass culture, and dosed with virus using a modified droplet feeding technique (Hughes *et al.*, 1986). This method reduces variation in virus uptake by the treated insects. Virus suspensions were made up in sterile distilled water containing 4% (v/v) green food colouring. A maximum of 45 larvae per dose were inoculated with increasing doses of virus ranging from approximately 3×10^2 to 2×10^5 OB ml^{-1}. Twelve different virus concentrations (wild-type or recombinant) were used for each dose range. A control group of larvae was dosed with distilled water containing only food colouring. After treatment, larvae were transferred individually to 12 ml plastic containers and were provided with artificial diet (Hunter *et al.*, 1984). The larvae were checked after 24 h, and any that had died as a result of handling were discarded. Subsequently, the larvae were inspected every 8 h (until control insects pupated) to record virus induced mortality and time of death. Insects that did not exhibit obvious virus symptoms were checked for the presence of virus polyhedra by examining Giemsa-stained preparations of cadavers under oil immersion, phase contrast microscopy (400× magnification). It was interesting to note that none of the larvae that were killed by the recombinant virus displayed the typical symptoms of NPV infection, hence it was important to determine the likely cause of death in each case (see discussion, Section 6).

5.2 *Mortality and time to death*

There was no difference in the virus-induced host mortality between the wild-type and recombinant viruses (*Figure 4*), as determined by the overlap of 95% confidence limits (CL). The LC_{50} value for the wild-type virus was 6.45×10^4 OB ml^{-1} (lower and upper 95% CL 4.43×10^4 and 1.06×10^5 OB ml^{-1}, respectively) and for the recombinant virus 4.0×10^4 OB ml^{-1} (lower and upper 95% CL 2.91×10^4 and 5.92×10^4 OB ml^{-1}, respectively).

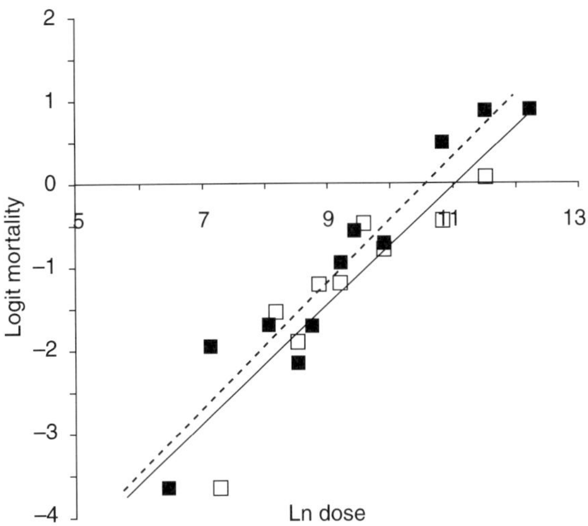

Figure 4. *Dose-mortality response curves for 2nd-instar* Trichoplusia ni *infected with AcMNPV wild-type (open squares) and AcMNPV modified to express 3DE 3β-reductase (closed squares). The lines correspond to the fitted values for the wild-type (solid line, logit mortality = 0.703(ln dose)–7.781, SE of slope = 0.079) and recombinant viruses (dotted line, logit mortality = 0.760(ln dose)–8.052, SE of slope = 0.079). In both cases the* X^2 *goodness-of-fit test was not significant* (P>0.05) *so no heterogeneity factor was used in calculating the 95% confidence limits (see results). Mortality data were logit transformed [ln(p/1–p)], where p is proportionate virus mortality.*

Over the upper range of the doses of recombinant virus used (5×10^3, 1×10^4, 2×10^4, 5×10^4, 1×10^5 and 2×10^5 OB ml^{-1}), there was no evidence of a dose-dependent decrease in speed of kill. However, the median times to death were significantly less for the recombinant virus compared with the wild-type virus (117, 121, 121, 113, 113 and 113 h for the recombinant, and 161, 137, 161, 161, 149 and 137 h for the wild-type, respectively; Mann-Whitney U test, $U_{0.05,(2),6,6} = 31$, $P < 0.005$).

6. Discussion

The insertion of the 3DE 3β-reductase cDNA into the genome of AcMNPV had no effect on its infectivity towards larvae of *T. ni*. This result is typical of many recombinant viruses (reviewed by Bonning and Hammock, 1996; Moscardi, 1999; Richards, *et al.*, 1998), and is to be expected unless the transgene itself plays a role in the actual initiation of the infection process, such as the passage of the virus through the gut wall of the host (Volkman, 1997). This lack of effect on virus infectivity has important implications in terms of the risk assessment of using recombinant baculoviruses as biological insecticides, but highlights the paradox that exists between the desirable biological characteristics of the recombinant virus and, on the other hand, the perceived environmental risks of using them. One of the main concerns in using recombinant viruses as insecticides is that increased mortality in the target pest species would result in the greater likelihood of transmission to non-target, but susceptible, species. Greater mortality results in increased virus productivity, and hence higher contact rates with other target and non-target hosts, thereby increasing the rate of virus transmission (Cory *et al.*, 1997). Similarly, increased virus levels, as a result of a greater number of insects dying, would lead to more virus entering refuges, such as the soil, in which the virus is protected (D'Amico and Elkinton, 1995). The rate at which virus leaves the refuge, and is again available to infect other susceptible hosts, will affect the long-term persistence of the virus in the environment, again posing a threat to non-target species (Cory *et al.*, 1997). While, in terms of pest control, it might be desirable for transformed baculoviruses to have greater pathogenicity than wild-type viruses, the risks they might pose to susceptible non-target insect species may be considered unacceptable (because more virus is in the environment for longer periods). Similarly, greater pathogenicity may itself increase the probability of infection in non-target species that are not normally susceptible to the wild-type virus.

A more widespread use of wild-type baculoviruses has been somewhat limited by the time it takes for them to kill the host in a typical infection. Whilst this aspect of virus pathology is advantageous in terms of increased virus yield, and hence a greater likelihood of transmission (Cory *et al.*, 1997), it may lead to unacceptable levels of damage in crops with low economic thresholds (Moscardi, 1999). Thus, perhaps the most important effect in larvae infected with the 3DE 3β-reductase virus, may be the significant reduction in time-to-death (up to two days) in comparison with its wild-type counterpart. Such a reduction in time-to-death, caused by a toxin-expressing recombinant baculovirus, has been shown to significantly reduce crop damage in controlled field trials (Cory *et al.*, 1994).

Again, the issues of pest control and risk assessment must be considered together, but depending on the intended use for the recombinant virus, these issues need not be conflicting. An increase in the speed of kill of the virus would lead to a reduction in virus yield when the host dies, which would serve to reduce the probability of other hosts becoming infected (Cory *et al.*, 1997). This would pose a reduced threat to non-target insects and, in this scenario, the recombinant baculovirus could be applied repeatedly in order to achieve control of target pests, in a manner analogous to conventional chemical pesticides. However, for classical biological control, sustained pest management is brought about by the long-term maintenance of the pathogen in the host population. A reduction in the transmission rate of the virus would be detrimental to this aim but, arguably, the establishment and persistence of a recombinant baculovirus may lead to a greater risk to non-target species.

At this stage, the precise mechanism leading to the reduction in time to death by the 3DE 3β-reductase virus is unknown, although the pathology of the modified virus did appear to be compromised. The lack of visible symptoms normally associated with NPV infection suggests that infected insects die before the virus is able to induce the normal range of pathological effects. In addition, larvae infected with the recombinant virus do not become flaccid and lyse at death, in stark contrast to larvae infected with the wild-type virus. Under normal conditions, the characteristic liquefaction of the host (the endpoint of many wild-type virus infections) would lead to the release of vast numbers of virus particles onto surrounding foliage where they would be available to infect other susceptible hosts (the normal cycle of events in horizontal transmission). The absence of lysis in larvae infected with the modified virus may have been due simply to the premature death of the host, which would have precluded the production of two virus-expressed enzymes (chitinase and a cathepsin protease) that are primarily responsible for breaking down the host's cuticle (Volkman, 1997). Alternatively, there is the intriguing possibility that the over expression of 3β-reductase, and potential hormonal effects, may have inhibited the replication cycle of the virus or the expression of these two crucial enzymes.

The possibility that a protein expressed by a modified virus might impinge on virus as well as host physiology raises the interesting question of how the virus itself (which is replicating throughout the tissues of the infected host) is affected by the products of the foreign gene it is expressing (see O'Reilly *et al.*, 1995). The expression of the gene in this study might have been expected to lead to an increase in the titre of moulting hormone. However, it should be remembered that this particular recombinant retains the *egt* gene normally present in the wild-type virus, the expression of which would lead to a decrease in the titre of moulting hormone in the host (see Chapter 14). Thus, the activity of the *egt* gene product (retained in the modified baculovirus) may well have counteracted the anticipated effect of insertion of the 3DE 3β-reductase cDNA.

Acknowledgments

Work in the authors' laboratories was supported by the Leverhulme Trust, the BBSRC and the NERC.

References

Abdel Fattah, S.M., el Sedfy, H.H., el Sayed, H.L., Abdel Aziz, H., Morsy, T.A., Salama, M.M. and Hamdi, K.N. (1994) Seropositivity against pediculosis in children with cervical lymphadenopathy. *J. Egypt. Soc. Parasitol.* 24: 59–67.

Ables, R. and Vinson, S.B. (1981) Regulation of host larval development by the egg-larval endoparasitoid *Chelonus insularis* (Hym.: Braconidae). *Entomophaga* **26:** 453–456.

Adamo, S.A. (1997) How parasites alter the behavior of their insect hosts. In: *Parasites and Pathogens: Effects on Host Hormones and Behavior* (ed Beckage, N.E.), pp. 231–245. Chapman & Hall, New York.

Adamo, S.A. (1998) Feeding suppression in the tobacco hornworm, *Manduca sexta*: costs and benefits to the parasitic wasp *Cotesia congregata*. *Can. J. Zool.* **76:** 1634–1640.

Adamo, S.A., Linn, C.E. and Beckage, N.E. (1997) Correlation between changes in host behavior and octopamine levels in the tobacco hornworm, *Manduca sexta*, parasitized by the gregarious braconid parasitoid wasp *Cotesia congregata*. *J. Exp. Biol.* **200:** 117–127.

Adams, T.S. (1999) Hematophagy and hormone release. *Ann. Entomol. Soc. Am.* **92:** 1–13.

Ahmed, A., Martin, D., Manetti, A.G., Han, S.J., Lee, W.J., Mathiopoulos, K.D., Muller, H.M., Kafatos, F.C., Raikhel, A. and Brey, P.T. (1999a) Genomic structure and ecdysone regulation of the prophenoloxidase 1 gene in the malaria vector *Anopheles gambiae*. *Proc. Natl Acad. Sci. USA* **96:** 14795–14800.

Ahmed, A.M., Taylor, P., Maingon, R. and Hurd, H. (1999b) The effect of *Plasmodium yoelii nigeriensis* on the reproductive fitness of *Anopheles gambiae*. *Invert. Reprod. Devel.* **36:** 217–222.

Ahmed, A.M., Maingon, R., Romans, P. and Hurd, H. (2001) Effects of malaria infection on vitellogenesis in *Anopheles gambiae*. *Insect Mol. Biol* (in press).

Albrecht, U., Wyler, T., Pfister-Wilhelm, R., Gruber, A., Stettler, P., Heiniger, P., Kurt, E., Schümperli, D. and Lanzrein, B. (1994) Polydnavirus of the parasitic wasp *Chelonus inanitus* (Braconidae): characterization, genome organization and time point of replication. *J. Gen. Virol.* **75:** 3353–3363.

Alleyne, M. and Beckage, N.E. (1997) Parasitism-induced effects on host growth and metabolic efficiency in tobacco hornworm larvae parasitized by *Cotesia congregata*. *J. Insect Physiol.* **43:** 407–424.

Alleyne, M., Chappell, M.A., Gelman, G.B. and Beckage, N.E. (1997) Effects of parasitism by the braconid wasp *Cotesia congregata* on metabolic rate in host larvae of the tobacco hornworm, *Manduca sexta*. *J. Insect Physiol.* **43:** 143–154.

Altschul, S.F., Madden, T.L., Schaffer, A.A., Zhang, J., Zhang, Z., Miller, W. and Lipman, D.J. (1997) Gapped BLAST and PSI-BLAST: a new generation of protein database search programs. *Nucleic Acid Res.* **25:** 3389–3402.

Andersen, S.O. (1985) Sclerotization and tanning of the cuticle. In: *Comprehensive Insect Physiology Biochemistry and Pharmacology*, Vol. 3. (eds Kerkut, G.A. and Gilbert, L.I.), pp. 59–74. Pergamon Press, New York.

Anderson, R.M. and May, R.M. (1981) The population dynamics of microparasites and their invertebrate hosts. *Phil. Trans. R. Soc.* B **291:** 451–524.

Applebaum, S.W., Raccah, B. and Leiserwitz, R. (1975) Effect of juvenile hormone and β-ecdysone on wing determination in the aphid, *Myzus persicae*. *J. Insect Physiol.* **21:** 1279–1281.

Arai, H.P. (1980) *Biology of the tapeworm* Hymenolepis diminuta. Academic Press, New York.

Arden, W.A., Fiscus, R.R., Beihn, L.D., Derbin, M., Oremus, R. and Gross, D.R. (1994) Skeletal muscle microcirculatory response to rat alpha-calcitonin gene-related peptide. *Neuropeptides 27*: 39–51.

Asgari, S. and Schmidt, O. (1994) Passive protection of eggs from the parasitoid *Cotesia rubecula*, in the host, *Pieris rapae*. *J. Insect Physiol.* **40:** 789–795.

Asgari, S., Hellers, M. and Schmidt, O. (1996) Host haemocyte inactivation by an insect parasitoid: transient expression of a polydnavirus gene. *J. Gen. Virol.* **77:** 2653–2662.

Asgari, S., Schmidt, O. and Theopold, U. (1997) A poldnavirus-encoded protein of an endoparasitoid wasp is an immune suppressor. *J. Gen. Virol.* **78:** 3061–3070.

Asgari, S., Theopold, U., Wellby, C. and Schmidt, O. (1998) A protein with protective properties against the cellular defense reactions in insects. *Proc. Natl Acad. Sci. USA* **95:** 3690–3695.

Ashburner, M., Chihara, C., Meltzer, P. and Richards, G. (1974) Temporal control of puffing activity in polytene chromosomes. *Cold Spring Harbor Symp. Quant. Biol.* **38:** 655–662.

Askew, R.R. (1971) Parasitic Hymenoptera. In: *Parasitic Insects* (ed Askew, R.R.), pp. 113–185. Heinemann Educational Books, London.

Askew, R.R. and Shaw, M.R. (1986) Parasitoid communities: their size, structure and development. In: *Insect Parasitoids. 13th Symposium of the Royal Entomological Society of London* (eds Waage, J.K. and Greathead, D.J.), pp. 225–264. Academic Press, London.

Audsley, N., Weaver, R.J. and Edwards, J.P. (1998) Enzyme linked immunosorbent assay of *Manduca sexta* allatostatin (Mas-AS), isolation and measurement of Mas-AS immunoreactive peptide in *Lacanobia oleracea*. *Insect Biochem. Mol. Biol.* **28:** 775–784.

Audsley, N., Weaver, R.J. and Edwards, J.P. (1999) Juvenile hormone synthesis by corpora allata of tomato moth, *Lacanobia oleracea* (Lepidoptera: Noctuidae), and the effects of allatostatins and allatotropin *in vitro*. *Eur. J. Entomol.* **96:** 287–293.

Audsley, N., Weaver, R.J. and Edwards, J.P. (2000) Juvenile hormone biosynthesis by corpora allata of larval tomato moth, *Lacanobia oleracea*, and regulation by *Manduca sexta* allatostatin and allatotropin. *Insect Biochem. Mol. Biol.* **30:** 681–689.

Böcking, D. and Hiruma, K. (1998) Functional significance of ecdysteroid diversity in arthropods. *Ann. N. Y. Acad. Sci.* **839:** 147–151.

Baehr, J.C., Pradelles, P., Lebreux, V., Cassier, P. and Dray, F. (1976) A simple and sensitive radioimmunoassay of juvenile hormones using an iodinated tracer. *FEBS Letts* **69:** 123–128.

Baker, F.C., Jamieson, G.C., Morallo-Rejesus, B. and Schooley, D.A. (1985) Identification of the juvenile hormones from adult *Attacus atlas*. *Insect Biochem.* **15:** 321–324.

Baker, F.C., Tsai, L.W., Reuter, C.C. and Schooley, D.A. (1987) *In vivo* fluctuations of JH, JH acid and ecdysteroid titer, and JH esterase activity, during development of fifth stadium *Manduca sexta*. *Insect Biochem.* **17:** 989–996.

Balbi, E. (1933) Studi allergici nella pediculosi. *Soc. Med. Chir. Padova atti*. **11:** 858–881.

Balgopal, M.M., Dover, B.A., Goodman, W.G. and Strand, M.R. (1996) Parasitism by *Microplitis demolitor* induces alterations in the juvenile hormone titers and juvenile hormone esterase activity of its host, *Pseudaletia includens*. *J. Insect Physiol.* **42:** 337–345.

Baltensweiler, W. and Moreau, J.P. (1957) Ein Beitrag biologisch-systematischer Art zur Kenntnis der Gattung *Phytodietus* (Hymenoptera). *Z. Ang. Entomol.* **41:** 272–276.

Bangs, P. and White, K. (2000) Regulation and execution of apoptosis during *Drosophila* development. *Dev. Dynamics* **218:** 68–79.

Banks, B.E.C. and Schipolini, R.A. (1986) Chemistry and pharmacology of honey-bee venom. In: *Venoms of the Hymenoptera* (ed Piek, T.), pp. 329–416. Academic Press, London.

Barbosa, P. (1988) Natural enemies of herbivore-plant interactions: influence of plant allelochemicals and host specificity. In: *Novel Aspects of Insect-Plant Interactions*, Vol. 1. (eds Barbosa, P. and Letourneau, D.), pp. 201–210. John Wiley & Sons, Inc., New York.

Barras, D. J., Joiner, R. L. and Vinson, S. B. (1970) Neutral lipid composition of the tobacco budworm, *Heliothis virescens* (Fab.), as affected by its habitual parasite *Cardiochiles nigriceps* Viereck. *Comp. Biochem. Physiol.* **36:** 775–783.

Beard, R.L. (1952) The toxicology of *Habrobracon* venom: a study of a natural insecticide. *Conn. Agric. Exp. Stn. Bull.* **562:** 1–27.

Beard, R.L. (1964) Pathogenic stinging of housefly pupae by *Nasonia vitripennis* (Walker). *J. Invert. Pathol.* **6:** 1–7.

Beard, R.L. (1978) Venoms of Braconidae In: *Handbuch der Experimentellen Pharmacologie, Vol 48. Arthropod Venoms* (ed Bettini, S.), pp. 773–800. Springer-Verlag, Berlin.

Beck, M., Theopold, U. and Schmidt, O. (2000) Evidence for serine protease inhibitor activity in the ovarian calyx fluid of the endoparasitoid *Venturia canescens*. *J. Insect Physiol.* **46:** 1275–1283.

Beckage, N.E. (1985) Endocrine interactions between endoparasitic insects and their hosts. *Annu. Rev. Entomol.* **30:** 371–413.

Beckage, N.E. (1991) Host–parasite hormonal relationships: a common theme? *Exp. Parasitol.* **72:** 332–338.

Beckage, N.E. (1993a) Games parasites play: the dynamic roles of proteins and peptides in the relationship between parasite and host. In: *Parasites and pathogens of insects,* Vol. 1 (eds Beckage, N.E., Thompson, S.N. and Federici, B.A.), pp. 25–57. Academic Press, San Diego.

Beckage, N.E. (1993b) Endocrine and neuroendocrine host–parasite relationships. *Receptor* **3:** 233–245.

Beckage, N.E. (1997a) New insights: how parasites and pathogens alter the endocrine physiology and development of insect hosts. In: *Parasites and Pathogens: Effects on Host Hormones and Behavior* (ed Beckage, N.E.), pp. 3–36. Chapman & Hall, New York.

Beckage, N.E. (1997b) The parasitic wasp's secret weapon. *Sci. Am.* **277:** 50–55.

Beckage, N.E. (1997c) *Parasites and Pathogens: Effects on Host Hormones and Behaviour.* Chapman and Hall, New York.

Beckage, N.E. (1998a) Modulation of immune responses to parasitoids by polydnaviruses. *Parasitology* **116:** S57–S64.

Beckage, N.E. (1998b) Insect growth regulators. In: *Biological and Biotechnological Control of Insect Pests* (eds Rechcigl, J.E. and Rechcigl, N.A.), pp. 123–137. CRC Publishers, Boca Raton.

Beckage, N.E. and de Buron, I. (1993) Lack of prothoracic gland degeneration in developmentally arrested host larvae of *Manduca sexta* parasitized by the braconid wasp *Cotesia congregata*. *J. Invert. Pathol.* **61:** 103–106.

Beckage, N.E. and Riddiford, L.M. (1978) Developmental interactions between the tobacco hornworm, *Manduca sexta,* and its braconid parasite *Apanteles congregatus*. *Entomol. Expl. Appl.* **23:** 139–151.

Beckage, N.E. and Riddiford, L.M. (1982a) Effects of parasitism by *Apanteles congregatus* on the endocrine physiology of the tobacco hornworm, *Manduca sexta*. *Gen. Comp. Endocrinol.* **47:** 308–322.

Beckage, N.E. and Riddiford, L.M. (1982b) Effects of methoprene and juvenile hormone on larval ecdysis, emergence, and metamorphosis of the endoparasitic wasp *Apanteles congregatus*. *J. Insect Physiol.* **28:** 329–334.

Beckage, N.E. and Riddiford, L.M. (1983a) Growth and development of the endoparasitic wasp *Apanteles congregatus*: dependence on host nutritional status and parasite load. *Physiol. Entomol.* **8:** 231–241.

Beckage, N.E. and Riddiford, L.M. (1983b) Lepidopteran anti-juvenile hormones: effects on development of *Apanteles congregatus* in *Manduca sexta*. *J. Insect Physiol.* **29:** 633–647.

Beckage, N.E. and Templeton, T.J. (1985) Temporal synchronization of emergence of *Hyposoter exiguae* and *H. fugitivus* with apolysis preceding larval molting in *Manduca sexta*. *Ann. Entomol. Soc. Am.* **78:** 775–782.

Beckage, N.E. and Templeton, T.J. (1986) Physiological effects of parasitism by *Apanteles congregatus* in terminal-stage tobacco hornworm larvae. *J. Insect Physiol.* **32:** 299–314.

Beckage, N.E., Templeton, T.J., Nielsen, B.D., Cook, D.I. and Stoltz, D.B. (1987a) Parasitism-induced haemolymph polypeptides in *Manduca sexta* (L.) larvae parasitized by the braconid wasp *Cotesia congregata* (Say). *Insect Biochem.* **17:** 439–455.

Beckage, N.E., Templeton, T.J., Stirling, B.A. and Nielsen, B.D. (1987b) Disruptive developmental effects of benzyl-1,3-benzodioxole derivatives on unparasitized and parasitized *Manduca sexta* larvae. *J. Insect Physiol.* **33:** 603–611.

Beckage, N.E., Metcalf, J.S., Nielsen, B.D. and Nesbit, D.J. (1988) Disruptive effects of azadirachtin on development of *Cotesia congregata* in host tobacco hornworm larvae. *Arch. Insect Biochem. Physiol.* **9:** 47–65.

Beckage, N.E., Metcalf, J.S., Nesbit, D.J., Schleifer, K.W., Zetlan, S.R. and de Buron, I. (1990) Host hemolymph monophenoloxidase activity in parasitized *Manduca sexta* larvae and evidence for inhibition by wasp polydnavirus. *Insect Biochem.* **20:** 285–294.

Beckage, N.E., Tan, F.F., Schleifer, K.W., Lane, R.D. and Cherubin, L.L. (1994) Characterization and biological effects of *Cotesia congregata* polydnavirus on host larvae of the tobacco hornworm, *Manduca sexta. Arch. Insect Biochem. Physiol.* **26:** 165–195.

Bedwin, O. (1979) An insect glycoprotein: a study of the particles responsible for the resistance of a parasitoid's eggs to the defense reactions of its insect host. *Proc. R. Soc. Lond. B.* **205:** 271–286.

Beeson, C.F.C. and Chatterjee, S.N. (1935) Biology of the Braconidae. *Indian Forest Rec. New Ser. Entomol.* **1:** 105–138.

Beintema, J.J., Stam, W.T., Hazes, B. and Smidt, M.P. (1994) Evolution of arthropod hemocyanins and insect storage proteins (hexamerins). *Mol. Biol. Evol.* **11:** 493–503.

Belinskaya, V.Z. (1973) Permeability of first stage larvae of *Hypoderma bovis* De Geer integument for amino acids tagged with radioactive isotopes. *Parazitologiya* **7:** 116–122.

Béliveau, C., Laforge, M., Cusson, M. and Bellemare, G. (2000) Expression of a *Tranosema rostrale* polydnavirus gene in the spruce budworm, *Choristoneura fumiferana*. *J. Gen. Virol.* **81:** 1871–1880.

Bennett, A.W. and Sullivan, D.J. (1978) Defensive behaviour against tertiary parasitism by the larva of *Dendrocerus carpenteri,* an aphid hyperparasitoid. *J. N. Y. Entomol. Soc.* **86:** 153–160.

Bergot, B.J., Jamieson, G.C., Ratcliff, M.A. and Schooley, D.A. (1980) JH Zero: new naturally-occurring insect juvenile hormone from developing embryos of the tobacco hornworm. *Science* **210:** 336–338.

Bergot, B.J., Ratcliff, M. and Schooley, D.A. (1981a) Method for the quantitative determination of the four known juvenile hormones in insect tissue using gas chromatography-mass spectroscopy. *J. Chromatogr.* **204:** 231–244.

Bergot, B.J., Baker, F.C., Cerf, D.C., Jamieson, G. and Schooley, D.A. (1981b) Qualitative and quantitative aspects of juvenile hormone titres in developing embryos of several insect species: discovery of a new JH-like substance extracted from eggs of *Manduca sexta*. In: *Juvenile Hormone Biochemistry* (eds Pratt, G.E. and Brooks, G.T.), pp. 33–45. Elsevier/North-Holland Biomedical Press, Amsterdam.

Beydon, P., Permana, A-D., Colardeau, J., Moriniere, M. and Lafont, R. (1989) Ecdysteroids from developing eggs of *Pieris brassicae*. *Arch. Insect Biochem. Physiol.* **11:** 1–11.

Bhaskaran, G., Dahm, K.H., Barrera, P., Pacheco, J.L., Peck, K.E. and Muszynska-Pytel, M. (1990) Allatinhibin, a neurohormonal inhibitor of juvenile hormone biosynthesis in *Manduca sexta*. *Gen. Comp. Endocrinol.* **78:** 123–136.

Bianchi, F.J., Snoeijing, I., van der Werf, W., Mans, R.M.W., Smits, P.H. and Vlak, J.M. (2000) Biological activity of SeMNPV, AcMNPV, and three AcMNPV deletion mutants against *Spodoptera exigua* larvae (Lepidoptera: Noctuidae). *J. Invert. Pathol.* **75:** 28–35.

Biliotti, E. and Daumal, J. (1969) Biologie de *Phanerotoma flavitestaca* (Hymenoptera). Mise au point d'un élevage permanent en vue de la lutte biologique contre *Ectomyelois ceratoniae. Ann. Zool. Ecol. Anim.* **1:** 379–394.

Billker, O., Shaw, M.K., Margos, G. and Sinden, R.E. (1997) The roles of temperature, pH and mosquito factors as triggers of male and female gametogenesis of *Plasmodium berghei in vitro*. *Parasitology* **114:** 1–7.

Billker, O., Lindo, V., Panico, M., Etienne, A.E., Paxton, T., Dell, A., Rogers, M., Sinden, R.E. and Morris, H.R. (1998) Identification of xanthurenic acid as the putative inducer of malaria development in the mosquito. *Nature* **392:** 289–292.

Bischoff, H. (1929) Zur biologie des *Euplectrus bicolor* Swed. *Z. wiss. Insekt. Biol.* **24:** 78–82.

Bisgaard, H. (1987) Vascular effects of leukotriene D4 in human skin. *J. Invest. Dermatol.* **88:** 109–114.

Blackford, M.J.P. and Dinan, L. (1997a) The tomato moth *Lacanobia oleracea* (Lepidoptera: Noctuidae) detoxifies ingested 20HE, but is susceptible to the ecdysteroid agonists RH-5849 and RH-5992. *Insect Biochem. Molec. Biol.* **27:** 167–177.

Blackford, M.J.P. and Dinan, L. (1997b) The effects of ingested 20HE on the larvae of *Aglais urticae*, *Inachis io*, *Cynthia cardui* (Lepidoptera: Nymphalidae) and *Tyria jacobaeae* (Lepidoptera: Arctiidae). *J. Insect Physiol.* **43:** 315–327.

Blackford, M. and Dinan, L. (1997c) The effects of ingested ecdysteroid agonists (20-hydroxyecdysone, RH5849 and RH5992) and an ecdysteroid antagonist (cucurbitacin B) on larval development of two polyphagous lepidopterans (*Acherontia atropos* and *Lacanobia oleracea*). *Entomol. Exp. Appl.* **83:** 263–276.

Blackford, M.J.P., Clarke, B. and Dinan, L. (1996) Tolerance of the Egyptian cotton leafworm *Spodoptera littoralis* (Lepidoptera: Noctuidae) to ingested phytoecdysteroids. *J. Insect Physiol.* **42:** 931–936.

Blissard, G.W. (1982) Cytology of fat body development in fifth instar larvae of the cabbage looper, *Trichoplusia ni*. M.Sc. thesis. University of California, Riverside.

Blissard, G.W., Fleming, J.G.W., Vinson, S.B. and Summers, M.D. (1986) *Campoletis sonorensis* virus: expression in *Heliothis virescens* and identification of expressed sequences. *J. Insect Physiol.* **32:** 351–359.

Blissard, G.W., Smith, O.P. and Summers, M.D. (1987) Two related viral genes are located on a single superhelical DNA segment of the multipartite *Campoletis sonorensis* virus genome. *Virology* **160:**120–134.

Blissard, G.W., Theilmann, A. and Summers, M.D. (1989) Segment W of the *Campoletis sonorensis* genome: expression, gene products, and organization. *Virology* **169:** 78–79.

Bloch, G., Borst, D.W., Huang, Z-Y., Robinson, G.E., Cnaani, J. and Hefetz, A. (2000a) Juvenile hormone titers, juvenile hormone biosynthesis, ovarian development and social environment in *Bombus terrestris*. *J. Insect Physiol.* **46:** 47–57.

Bloch, G., Hefetz, A. and Hartfelder, K. (2000b) Ecdysteroid titer, ovary status, and dominance in adult worker and queen bumble bees (*Bombus terrestris*). *J. Insect Physiol.* **46:** 1033–1040.

Bocchino, F.J. and Sullivan, D.J. (1981) Effects of venoms from two aphid hyperparasitoids, *Asaphes lucens* and *Dendrocerus carpenteri* (Hymenoptera: Pteromalidae and Megaspilidae), on larvae of *Aphidius smithi* (Hymenoptera: Aphidiidae). *Can. Ent.* **113:** 887–889.

Boenisch, A. and Jürgens, M. (1994) Enhanced reproduction of an aphid in the presence of its hyperparasitoid. *Norw. J. Agric. Sci. Suppl.* **16:** 123–129.

Boenisch, A., Petersen, G. and Wyss, U. (1997) Influence of the hyperparasitoid *Dendrocerus carpenteri* on the reproduction of the grain aphid *Sitobion avenae*. *Ecological Entomol.* **22:** 1–6.

Bogus, M. and Scheller, K. (1996) Allatotropin released by the brain controls larval moulting in *Galleria mellonella* by affecting juvenile hormone synthesis. *Int. J. Devel. Biol.* **40:** 205–210.

Bollenbacher, W.E. (1988) The endocrine regulation of larval-pupal development in the tobacco hornworm, *Manduca sexta*: a model. *J. Insect Physiol.* **34:** 941–947.

Bollenbacher, W.E. and Gilbert, L.I. (1981) Neuroendocrine control of post-embryonic development in insects: The prothoracicotropic hormone. In: *Neurosecretion: Molecules, Cells, Systems* (eds Farner, D.S. and Lederis, K.), pp. 361–370. Plenum Press, New York.

Bollenbacher, W.E. and Granger, N.A. (1985) Endocrinology of the prothoracicotropic hormone. In: *Comprehensive Insect Physiology, Biochemistry and Pharmacology*, Vol. 7, (eds Kerkut, G.A. and Gilbert, L.I.) pp. 109–151. Pergamon Press, Oxford.

Bollenbacher, W.E., Vedeckis, W.V., Gilbert, L.I. and O'Connor, J.D. (1975) Ecdysone titers and prothoracic gland activity during the larval-pupal development of *Manduca sexta*. *Devel. Biol.* **44:** 46–53.

Bollenbacher, W.E., Zvenko, H., Kumaran, A.K. and Gilbert, L.I. (1978) Changes in ecdysone content during postembryonic development of the Wax moth *Galleria mellonella*: the role of the ovary. *Gen. Comp. Endocrinol.* **34:** 169–179.

Bollenbacher, W.E., Smith, S.L., Goodman, W. and Gilbert, L.I. (1981) Ecdysteroid titer during larval-pupal-adult development of the Tobacco hornworm, *Manduca sexta. Gen. Comp. Endocrinol.* **44:** 302–306.

Bollenbacher, W.E., Granger, N.A. and Gilbert, L.I. (1987) Developmental endocrinology of larval moulting in the tobacco hornworm, *Manduca sexta. J. Exp. Biol.* **128:** 175–192.

Bonnemaison, L. (1970) Action de *Diaretiella rapae* sur le dévelopment et la morphologie de *Brevicoryne brassicae*. *Ann. Soc. Ent. Fr.* **6:** 167–191.

Bonning, B.C. and Hammock, B.D. (1996) Development of recombinant baculoviruses for insect control. *Annu. Rev. Entomol.* **41:** 191–210.

Bonning, B.C., Hirst, M., Possee, R.D. and Hammock, B.D. (1992) Further development of a recombinant baculovirus insecticide expressing the enzyme juvenile hormone esterase from *Heliothis virescens*. *Insect Biochem. Mol. Biol.* **22:** 453–458.

Bonning, B.C., Hoover, K., Booth, T.F., Duffey, S. and Hammock, B.D. (1995) Development of a recombinant baculovirus expressing a modified juvenile hormone esterase with potential for insect control. *Arch. Insect Biochem. Physiol.* **30:** 177–194.

Bonning, B.C., Ward, V.K., van Meer, M.M.M., Booth, T.F. and Hammock, B.D. (1997) Disruption of lysosomal targeting is associated with insecticidal potency of juvenile hormone esterase. *Proc. Natl Acad. Sci. USA* **94:** 6007–6012.

Borst, D.W. and O'Connor, J.D. (1972) Arthropod molting hormone: Radioimmune assay. *Science* **178:** 418–419.

Bottrell, D.G. (1968) Notes on parasites attacking the yellow-striped armyworm in Oklahoma. *Ann. Entomol. Soc. Am.* **62:** 250–252.

Bouèek, Z. (1959) A study of central European Eulophidae. *Acta Entomol. Mus. Nat. Prague* **33:** 163–164.

Bouèek, Z. (1988) *Australasian Chalcidoidea (Hymenoptera)*. CAB International, Wallingford, UK.

Bouèek, Z. and Askew, R.R. (1968) Hymenoptera Chalcidoidea: Palearctic Eulophidae. In: *Index of Entomophagous Insects* (eds Delucchi, V. and Reaudiere, G.), pp. 1–254. Le Francois, Paris.

Boulton, A. (1995) Britain restricts lice treatment. *Brit. Med. J.* **311:** 1322.

Bowers, W.S. (1983) Endocrine strategies for insect control. In: *Pesticide Chemistry: Human Welfare and the Environment*, Vol. 2 (eds Takahashi, N., Yoshioka, H., Misato, T. and Matsunaka, S.), pp. 29–36. Pergamon Press, Oxford.

Bowers, W.S. (1985) Antihormones. In: *Comprehensive Insect Physiology, Biochemistry and Pharmacology*, Vol. 8, (eds Kerkut, G.A. and Gilbert, L.I.) pp. 551–563. Pergamon Press, Oxford.

Bownes, M. and Reid, G. (1990) The role of the ovary and nutritional signals in the regulation of fat body yolk protein gene expression in *Drosophila melanogaster. J. Insect Physiol.* **36:** 471–479.

Briggs, C.J. and Godfray, H.C.J. (1995) The dynamics of insect-pathogen interactions in stage-structured populations. *Amer. Nat.* **145:** 855–887.

Broodryk, S.W. and Doutt, R.L. (1966) Studies of two parasites of olive scale, *Parlatoria oleae* (Colvee). II. The biology of *Coccophagoides utilis* Doutt (Hymenoptera: Aphelinidae). *Hilgardia* **37:** 233–254.

Brothers, D.J. (1972) Biology and immature stages of *Pseudomethoca f. frigida*, with notes on other species (Hymenoptera: Mutillidae). *Univ. Kansas Sci. Bull.* **50:** 1–38.

Brothers, D.J. (1978) Biology and immature stages of *Myrmosula parvula* (Hymenoptera: Mutillidae). *J. Kansas Entomol. Soc.* **51:** 698–710.

Brown, J.J. and Kainoh, Y. (1992) Host castration by *Ascogaster* spp. (Hymenoptera: Braconidae). *Ann. Entomol. Soc. Amer.* **85:** 67–71.

Brown, J.J. and Reed-Larsen, D. (1991) Ecdysteroids and insect host/parasitoid interactions. *Biol. Control* **1:** 136–143.

Brown, J.J., Kiuchi, M., Kainoh, Y. and Takeda, S. (1993) *In vitro* release of ecdysteroids by an endoparasitoid, *Ascogaster reticulatus* Watanabe. *J. Insect Physiol.* **39:** 229–234.

Buhler, A., Hanzlik, T.N. and Hammock, B.D. (1985) Effects of parasitization of *Trichoplusia ni* by *Chelonus* sp. *Physiol. Entomol.* **10:** 383–394.

Burchell, B. and Coughtrie, M. (1989) UDP-glucuronosyltransferases. *Pharmac. Ther.* **43:** 261–289.

Burden, J.P., Hails, R.S., Windass, J.D., Suner, M.M. and Cory, J. S. (2000) Pathogenicity, virulence and productivity of a baculovirus expressing the itch mite toxin Txp-1 in second and fourth instar *Trichoplusia ni*. *J. Invert. Pathol.* **75:** 226–236.

Burgess, I.F., Brown, C.M., Peock, S. and Kaufman, J. (1995) Head lice resistant to pyrethroid insecticides in Britain. *Brit. Med. J.* **311:** 752.

Burmester, T., Massey, H.C., Zakharkin, S.O. and Beneš, H. (1998) The evolution of hexamerins and the phylogeny of insects. *J. Mol. Evol.* **47:** 93–108.

Buxton, P.A. (1947) *The Louse: An Account of the Lice which Infest Man, their Medical Importance and Control.* Edward Arnold & Co., London.

Calvez, B., Hirn, M. and De Reggi, M. (1976) Ecdysone changes in the haemolymph of two silkworms (*Bombyx mori* and *Philosamia cynthia*) during larval and pupal development. *FEBS Letts* **71:** 57–61

Campbell, A. and Mackauer, M. (1975) The effect of parasitism by *Aphidius smithi* (Hymenoptera: Aphidiidae) on reproduction and population growth of the pea aphid, *Acyrthosiphon pisum* (Homoptera: Aphididae). *Can. Ent.* **107:** 910–1026.

Carruthers, W.R., Cory, J.S. and Entwistle, P.F. (1988) Recovery of pine beauty moth (*Panolis flammea*) nuclear polyhedrosis virus from pine foliage. *J. Invert. Pathol.* **52:** 27–32.

Carwardine, S. L. and Hurd, H. (1997) Effect of *Plasmodium yoelii nigeriensis* infection on *Anopheles stephensi* egg development and resorption. *Med. Vet. Entomol.* **11:** 265–269.

Champagne, D.E. and Ribeiro, J.M.C. (1994) Sialokinin I and II: Vasodilatory tachykinins from the yellow fever mosquito *Aedes aegypti*. *Proc. Natl Acad. Sci. USA* **91:** 138–142.

Champagne, D.E., Nussenzveig, R.H. and Ribeiro, J.M.C. (1995) Purification, partial characterization, and cloning of nitric oxide-carrying heme proteins (Nitrophorins) from salivary glands of the blood-sucking insect, *Rhodnius prolixus*. *J. Biol. Chem.* **270:** 8691–8695.

Chang, E.S. and O'Connor, J.D. (1979) Arthropod moulting hormones. In *Methods of Hormone Radioimmunoassay* (eds Jaffe, B.M. and Behrman, H.R.), pp. 797–813. Academic Press, New York.

Chapman, R.F. (1998) *The Insects: Structure and Function,* 4th edn. Cambridge University Press, Cambridge.

Charlab, R., Rowton, E.D. and Ribeiro, J.M. (2000) The salivary adenosine deaminase from the sand fly *Lutzomyia longipalpis*. *Exp. Parasitol.* **95:** 45–53.

Chatterjee, P.N. (1945) On the biology and morphology of *Euplectrus parvulus* Ferr. (Hymenoptera: Eulophidae). *Indian J. Entomol.* **6:** 95–101.

Chelliah, J. and Jones, D. (1990) Biochemical and immunological studies of proteins from polydnavirus *Chelonus* sp. near *curvimaculatus*. *J. Gen. Virol.* **71:** 2353–2359.

Chen, J.-H., Tracey, J.W., Powls, R. and Rees, H.H. (1996) Purification and characterisation of haemolymph 3-dehydroecdysone 3β-reductase in relation to ecdysteroid biosynthesis in the cotton leafworm *Spodoptera littoralis*. *Eur. J. Biochem.* **242:** 394–401.

Chen, J.-H., Turner, P.C. and Rees, H.H. (1999) Molecular cloning and characterization of hemolymph 3-dehydroecdysone 3β-reductase from the cotton leafworm, *Spodoptera littoralis*: a new member of the third superfamily of oxidoreductases. *J. Biol. Chem.* **274:** 10551–10556.

Chen, P.S. (1985) Amino acid and protein metabolism. In: *Comprehensive Insect Physiology Biochemistry and Pharmacology,* Vol. 10 (eds Kerkut, G.A. and Gilbert, L.I.), pp. 177–217. Pergamon Press, Oxford.

Chen, X., Sun, X., Hu, Z., Li, M., O'Reilly, D.R., Zuidema, D. and Vlak, J.M. (2000) Genetic engineering of *Helicoverpa armigera* single-nucleocapsid nucleo-polyhedrovirus as an improved pesticide. *J. Invert. Pathol.* **76:** 140–146.

Cherbas, P. and Cherbas, L. (1996) Molecular aspects of ecdysteroid hormone action. In: *Metamorphosis – Postembryonic Reprogramming of Gene Expression in Amphibian and Insect Cells* (eds Gilbert, L.I., Tata, J.R. and Atkinson, B.G.), pp. 175–221. Academic Press, San Diego.

Chow, A. and Mackauer, M. (1999) Host handling and specificity of the hyperparasitoid wasp *Dendrocerus carpenteri* (Curtis) (Hym.: Megaspilidae): importance of host age and species. *J. Appl. Entomol.* **123:** 83–91.

Christiansen-Weniger, P. and Hardie, J. (1998) Wing development in parasitized male and female *Sitobion fragariae. Physiol. Entomol.* **23:** 208–213.

Christiansen-Weniger, P. and Hardie, J. (2000) The influence of parasitism on wing development in the male and female pea aphid, *Acyrthosiphon pisum. J. Insect Physiol.* **46:** 861–867.

Claret, J., Porcheron, P. and Dray, F. (1978) La teneur en circulantes au cours du dernier stade larvaire de l'Hyménoptère endoparasite *Pimpla instigator*, et l'entrée en diapause. *C. R. Acad. Sci. Paris* **286D:** 639–641.

Clark, K., Pech, L.L. and Strand, M.R. (1997) Isolation and identification of a plasmatocyte spreading peptide from hemolymph of the lepidopteran insect *Pseudoplusia includens. J. Biol. Chem.* **272:** 23440–23447.

Clarke, E.E., Tristem, M., Cory, J.S. and O'Reilly, D.R. (1996) Characterization of the ecdysteroid UDP-glucosyltransferase gene from *Mamestra brassicae* nucleopolyhedrosis virus. *J. Gen. Virol.* **77:** 2865–2871.

Clausen, C.P. (1940) *Entomophagous Insects*. Hafner Publishing Co., New York.

Clements, A.N. (1992) *The Biology of Mosquitoes,* Vol. 1. Chapman and Hall, London.

Clements, A.N. and Boocock, M.R. (1984) Ovarian development in mosquitoes: stages of growth and arrest, and follicular resorption. *Physiol. Entomol.* **9:** 1–8.

Cohen, A.C. (1992) Using a systematic approach to develop artificial diets for predators. In: *Advances in Insect Rearing for Research and Pest Management* (eds Anderson, T.E. and Leppla, N.C.), pp. 77–91. Westview Press, Boulder, CO.

Cohen, A.C. (1995) Extra-oral digestion in predaceous terrestrial arthropoda. *Annu. Rev. Entomol.* **40:** 85–103.

Collins, J.V. (1974) Hormonal control of protein sequestration in the fat body of *Calpodes ethliujs* Stoll. *Can. J. Zool.* **52:** 639–642.

Cook, D.I., Stoltz, D.B. and Vinson, S.B. (1984) Induction of a new hemolymph glycoprotein in larvae of permissive hosts parasitized by *Campoletis sonorensis. Insect Biochem.* **14:** 45–50.

Copenhaver, P.F. and Truman, J.W. (1982) The role of eclosion hormone in the larval ecdysis of *Manduca sexta. J. Insect Physiol.* **28:** 695–701.

Corbitt, T.S., Bryning, G., Olieff, S. and Edwards, J.P. (1996) Reproductive, developmental and nutritional biology of the tomato moth, *Lacanobia oleracea* (Lepidoptera: Noctuidae) reared on artificial diet. *Bull. Entomol. Res.* **86:** 647–657.

Cory, J.S., Hirst, M.L., Williams, T., Hails, R.S., Goulson, D., Green, B.M., Carty, T.M., Possee, R.D., Cayley, P.J. and Bishop, D.H.L. (1994) Field trial of a genetically improved baculovirus insecticide. *Nature* **370:** 138–140.

Cory, J.S., Hails, R.S. and Sait, S.M. (1997) Baculovirus ecology. In: *The Baculoviruses* (ed Miller, L.K.), pp. 301–339. Plenum Press, New York.

Coudron, T.A. (1991) Host-regulating factors associated with parasitic Hymenoptera. In: *Naturally Occurring Pest Bioregulators* (ed Hedin, P.A.), pp. 41–65. ACS Symposium Series No. 449. American Chemical Society.

Coudron, T.A. and Brandt, S.L. (1996) Characteristics of a developmental arrestant in the venom of the ectoparasitoid wasp *Euplectrus comstockii. Toxicon* **34:** 1431–1441.

Coudron, T.A. and Kelly, T.J. (1985) *Euplectrus plathypenae* parasitization of *Trichoplusia ni.* Effect on weight gain, ecdysteroid titer and moulting. In: *New Concepts and Trends in Pesticide Chemistry* (ed Hedin, P.), p. 319. ACS Symposium Series No. 276. American Chemical Society.

Coudron, T.A. and Puttler, B. (1988) Response of natural and factitious hosts to the ectoparasite *Euplectrus plathypenae* (Hymenoptera: Eulophidae). *Ann. Ent. Soc. Am.* **81:** 931–937.

Coudron, T.A., Kelly, T.J. and Puttler, B. (1990) Developmental responses of *Trichoplusia ni* (Lepidoptera: Noctuidae) to parasitism by the ectoparasite *Euplectrus plathypenae* (Hymenoptera: Eulophidae). *Arch. Insect Biochem. Physiol.* **13:** 83–94.

Coudron, T.A., Schauff, M.E., Steiner, W.W., Dillwith, J.W., Bergman, D.K., Puttler, B. and Sacks, J.M. (1993) Morphological and chemical differences between the ectoparasitoids *Euplectrus comstockii* and *E. plathypenae* (Hymenoptera: Eulophidae). *Ann. Entomol. Soc. Am.* **86:** 551–559.

Coudron, T.A., Jones, D. and Jones, G. (1994) Premature production of late larval storage proteins in larvae of *Trichoplusia ni* parasitized by *Euplectrus comstockii. Arch. Insect Biochem. Physiol.* **26:** 97–109.

Coudron, T.A., Rice, W.C., Ellersieck, M.R. and Pinnell, R.E. (1995) Mediated pathogenicity of the baculovirus HzSNPV by the venom from *Euplectrus comstockii* (Hymenoptera: Eulophidae). *Biol. Control* **5:** 92–98.

Coudron, T.A., Brandt, S.L. and Raqib, A. (1997) Comparison of the response of *Heliothis virescens* to parasitism by *Euplectrus comstockii* and *Euplectrus plathypenae. Comp. Biochem. Physiol.* **116B:** 197–202.

Coudron, T.A., Raqib, A., Brandt, S.L. and Knop-Wright, M. (1998) Comparison of the hemolymph proteins in permissive and non-permissive hosts of *Euplectrus comstockii. Comp. Biochem. Physiol.* **120B:** 349–357.

Coudron, T.A., Iqbal, M., Rice, W.C., Ellersieck, M.R. and Pinnell, R.E. (1999) Mediated pathogenicity of the baculovirus AcMNPV by the venom from *Euplectrus comstockii* Howard (Hymenoptera: Eulophidae). *Comp. Biochem. Physiol.* **124B:** 231–240.

Cui, L. and Webb, B.A. (1996) Isolation and characterization of a member of the cysteine-rich gene family from *Campoletis sonorensis* polydnavirus. *J. Gen. Virol.* **77:** 797–809.

Cui, L. and Webb, B.A. (1997) *Campoletis sonorensis* polydnavirus W segment family: integration in the wasp genome and segment nesting. *J. Virol.* **71:** 8504–8513.

Cui, L., Soldevila, A.S. and Webb, B.A. (2000) Relationships between polydnavirus gene expression and host range of the parasitoid wasp, *Campoletis sonorensis. J. Insect. Physiol.* **46:**1397–1407.

Cupp, M.S., Ribeiro, J.M. and Cupp, E.W. (1994) Vasodilative activity in black fly salivary glands. *Am. J. Trop. Med. Hyg.* **50:** 241–246.

Cushman, R.A. (1926) Some types of parasitism among the Ichneumonidae. *Proc. Entomol. Soc. Wash.* **28:** 29–51.

Cusson, M., Lucarotti, C., Stoltz, D., Krell, P. and Doucet, D. (1998) A polydnavirus from the spruce budworm parasitoid, *Tranosema rostrale* (Ichneumonidae). *J. Invert. Pathol.* **72:** 50–56.

Cusson, M., Laforge, M., Miller, D., Cloutier, C. and Stoltz, D. (2000) Functional significance of parasitism-induced suppression of juvenile hormone esterase activity in developmentally delayed *Choristoneura fumiferana* larvae. *Gen. Comp. Endocrinol.* **117:** 343–354.

Cymborowski, B., Bogus, M., Beckage, N.E., Williams, C.M. and Riddiford, L.M. (1982) Juvenile hormone titres and metabolism during starvation-induced supernumerary larval moulting of the tobacco hornworm, *Manduca sexta* L. *J. Insect Physiol.* **28:** 129–135.

Dahlman, D.L. (1990) Evaluation of teratocyte functions: an overview. *Arch. Insect Biochem. Physiol.* **13:**159–166.

Dahlman, D.L. (1991) Teratocytes and host/parasitoid interactions. *Biol. Control* **1:** 118–126.

Dahlman, D.L. and Vinson, S.B. (1975) Trehalose and glucose levels in the hemolymph of *Heliothis virescens* parasitized by *Microplitis croceipes* and *Cardiochiles nigriceps*. *Comp. Biochem. Physiol.* **52:** 465–468.

Dahlman, D.L. and Vinson, S.B. (1976) Trehalose level in the hemolymph of *Heliothis virescens* parasitized by *Campoletis sonorensis*. *Ann. Entomol. Soc. Am.* **69:** 523–524.

Dahlman, D.L. and Vinson, S.B. (1993) Teratocytes: developmental and biochemical characteristics. In: *Parasites and Pathogens of Insects*, Vol. 1 (eds Beckage, N.E., Thompson, S.N. and Federici, B.A.), pp. 145–165. Academic Press, New York.

Dahlman, D.L., Coar, D.L., Koller, C.N. and Neary, T.J. (1990a) Contributing factors to reduced ecdysteroid titres in *Heliothis virescens* parasitized by *Microplitis croceipes*. *Arch. Insect Biochem. Physiol.* **13:** 29–39.

Dahlman, D.L., Zhang, D. and Neary, T.J. (1990b) Alteration of juvenile hormone esterase activity by parasitoid-derived factors. [Abstract]. In: *Molecular Insect Science* (eds Hagedorn, H.H., Hildebrand, J.G., Kidwell, M.G. and Law, J.H.), p. 295. Plenum Press, New York.

D'Amico, V. and Elkinton, J.S. (1995) Rainfall effects on the transmission of gypsy moth (Lepidoptera: Lymantriidae) nuclear polyhedrosis virus. *Environ. Entomol.* **24:** 1144–1149.

Darling, D.C. and Werren, J.H. (1990) Biosystematics of *Nasonia* (Hymenoptera: Pteromalidae): Two new species reared from bird's nests in North America. *Ann. Entomol. Soc. Am.* **83:** 352–370.

Davis, A.C. (1928) A note on the parasitism of *Hippodamia*. *Pan-Pacific Ent.* **4:** 184.

Dawei, H. and Dingxi, L. (1987) Studies on the behavior of *Nasonia vitripennis* (Walker). *Sinozoologia* **5:** 177–182.

Dawkins, R. (1990) Parasites, desiderata lists and the paradox of the organism. *Parasitology* **100:** S63–S75.

de Buron, I. and Beckage, N.E. (1992) Characterization of a polydnavirus (PDV) and virus-like filamentous particles (VLFP) in the braconid wasp *Cotesia congregata* (Hymenoptera: Braconidae). *J. Invertebr. Pathol.* **59:** 315–327.

de Buron, I. and Beckage, N.E. (1997) Developmental changes in teratocytes of the braconid wasp *Cotesia congregata* in larvae of the tobacco hornworm, *Manduca sexta*. *J. Insect Physiol.* **43:** 915–930.

Deitsch, W.K., Chen, J.-S. and Raikhel, A.S. (1995). Indirect control of yolk protein genes by 20-hydroxyecdysone in the fat body of the mosquito, *Aedes aegypti*. *Insect Biochem. Mol. Biol.* **25:** 449–454.

De Jong-Brink, M., Hoek, R.M., Lageweg, W. and Smit, A.B. (1997) Schistosome parasites induce physiological changes in their snail hosts by interfering with two regulatory systems, the internal defense system and the neuroendocrine system. In: *Parasites and Pathogens: Effects of Host Hormones and Behaviour* (ed Beckage, N.E.), pp. 57–76. Chapman and Hall, New York.

De Kort, C.A. and Granger, N.A. (1996) Regulation of JH titers: the relevance of degradative enzymes and binding proteins. *Arch. Insect Biochem. Physiol.* **33:** 1–26.

De Loof, A. and Vanden Broeck, J. (1995) Hormones and the cytoskeleton: an overview. *Netherlands J. Zool.* **45:** 219–223.

Deng, L. and Webb, B. (1999) Cloning and expression of a gene encoding a *Campoletis sonorensis* polydnavirus structural protein. *Arch. Insect Biochem. Physiol.* **40:** 30–40.

Denlinger, D.L. (1981) The physiology of pupal diapause in flesh flies. In: *Current Topics in Insect Endocrinology and Nutrition* (eds Bhaskaran, G., Friedman, S. and Rodriguez, J.G.), pp. 131–160. Plenum Press, New York.

Dhadialla, T.S. and Raikhel, A.S. (1990) Biosynthesis of mosquito vitellogenin. *J. Biol. Chem.* **265:** 9924–9933.

Dhadialla, T.S., Carlson, G.R. and Le, D.P. (1998) New insecticides with ecdysteroidal and juvenile hormone activity. *Annu. Rev. Entomol.* **43:** 545–569.

Dib-Hajj, S.D., Webb, B.A. and Summers, M.D. (1993) Structure and evolutionary implications of a 'cysteine-rich' *Campoletis sonorensis* polydnavirus gene family. *Proc. Natl Acad. Sci. USA* **90:** 3765–3769.

Digilio, M.C., Pennacchio, F. and Tremblay, E. (1998) Host regulation effects of ovary fluid and venom by *Aphidius ervi* (Hymenoptera: Braconidae). *J. Insect Physiol.* **44:** 779–784.

Digilio, M.C., Isidoro, N., Tremblay, E. and Pennacchio, F. (2000) Host castration by *Aphidius ervi* venom proteins. *J. Insect Physiol.* **46:** 1041–1050.

Dittman, F., Kogan, P.H. and Hagedorn, H.H. (1989) Ploidy levels and DNA synthesis in the fat body cells of the adult mosquito, *Aedes aegypti* the role of juvenile hormone. *Arch. Insect Biochem. Physiol.* **12:** 133–143.

Dixon, A.F.G. (1998) *Aphid Ecology*, 2nd Edn. Chapman & Hall, London.

Dong, K., Zhang, D. and Dahlman, D.L. (1996) Down-regulation of juvenile hormone esterase and arylphorin production in *Heliothis virescens* larvae parasitized by *Microplitis croceipes*. *Arch. Insect Biochem. Physiol.* **32:** 237–248.

Donnelly, E., Lipkin, J., Clore, E.R. and Altuschuler, D.Z. (1991) Pediculosis prevention and control strategies of community health and school nurses: a descriptive study. *J. Comm. Health Nurs.* **8:** 85–95.

Doucet, D. and Cusson, M. (1996a) Alteration of developmental rate and growth of *Choristoneura fumiferana* parasitized by *Tranosema rostrale:* role of the calyx fluid. *Entomol. Exp. Appl.* **81:** 21–30.

Doucet, D. and Cusson, M. (1996b) Role of calyx fluid in alterations of immunity in *Choristoneura fumiferana* larvae parasitized by *Tranosema rostrale*. *Comp. Biochem. Physiol.* **114A:** 311–317.

Douglas, A.E. (1998) Nutritional interactions in insect–microbial symbioses: aphids and their symbiotic bacteria. *Buchnera*. *Annu. Rev. Entomol.* **43:** 17–37.

Doury, G. and Rojas-Rousse, D. (1994) Reproductive potential in the parasitoid *Eupelmus orientalis* (Hymenoptera: Eupelmidae). *Bull. Ent. Res.* **84:** 199–206.

Doury, G., Rojas-Rousse, D. and Periquet, G. (1995) Ability of *Eupelmus orientalis* ectoparasitoid larvae to develop on an unparalyzed host in the absence of female stinging behaviour. *J. Insect Physiol.* **41:** 287–296.

Doury, G., Bigot, Y. and Periquet, G. (1997) Physiological and biochemical analysis of factors in the female venom gland and larval salivary secretions of the ectoparasitoid wasp *Eupelmus orientalis*. *J. Insect Physiol.* **43:** 69–81.

Dover, B.A. and Vinson, S.B. (1990) Stage-specific effects of *Campoletis sonorensis* parasitism on *Heliothis virescens* development and prothoracic glands. *Physiol. Entomol.* **15:** 405–414.

Dover, B.A., Davies, D.H., Strand, M.R., Gray, R.S., Keeley, L.L. and Vinson, S.B. (1987) Ecdysteroid-titre reduction and developmental arrest of last-instar *Heliothis virescens* larvae by calyx fluid from the parasitoid *Campoletis sonorensis*. *J. Insect Physiol.* **33:** 333–338.

Dover, B.A., Davies, D.H. and Vinson, S.B. (1988a) Degeneration of last instar *Heliothis virescens* prothoracic glands by *Campoletis sonorensis* polydnavirus. *J. Invert. Pathol.* **51:** 80–91.

Dover, B.A., Davies, D.H. and Vinson, S.B. (1988b) Dose-dependent influence of *Campoletis sonorensis* polydnavirus on the development and ecdysteroid titers of last-instar *Heliothis virescens* larvae. *Arch. Insect Biochem. Physiol.* **8:** 113–126.

Dover, B.A., Strand, M.R., Davies, D.H. and Vinson, S.B. (1989) Ultrastructure of host tissues exposed to the calyx fluid of the parasitoid *Campolitis sonorensis*. *Int. J. Insect Morphol. Embryol.* **18:** 47–57.

Dover, B.A., Tanaka, T. and Vinson, S.B. (1995a) Stadium-specific degeneration of host prothoracic glands by *Campoletis sonorensis* calyx fluid and its association with host ecdysteroid titers. *J. Insect Physiol.* **41:** 947–955.

Dover, B.A., Menon, A., Brown, R.C. and Strand, M.R. (1995b) Suppression of juvenile hormone esterase in *Heliothis virescens* by *Microplitis demolitor* calyx fluid. *J. Insect Physiol.* **41:** 809–817.

Drenth, D. (1974) Susceptibility of different species of insects to an extract of the venom gland of the wasp *Microbracon hebetor* (Say). *Toxicon* **12:** 189–192.

Dushay, M.S. and Beckage, N.E. (1993) Dose-dependent separation of *Cotesia congregata*-associated polydnavirus effects on *Manduca sexta* larval development and immunity. *J. Insect Physiol.* **39:** 1029–1040.

Dushoff, J. and Dwyer, G. (2001) Evaluating the risks of engineered viruses: modelling pathogen competition. *Ecol. Appl.* (in press).

Dwyer, G., Dushoff, J., Elkinton, J.S., Burand, J.P. and Levin, S.A. (1999) Host heterogeneity in susceptibility: lessons from an insect virus. In: *Virulence Management: The Adaptive Dynamics of Pathogen–Host Interactions* (eds Dieckmann, U., Metz, H., Sabelis, M. and Sigmund, K.), Cambridge University Press, Cambridge.

Eberhard, W.G. (1990) Early stages of orb construction by *Philoponella vicina*, *Leucauge mariana* and *Nephila clavipes* (Araneae, Uloboridae and Tetragnathidae), and their phylogenetic implications. *J. Arachnol.* **18:** 205–234.

Eberhard, W.G. (2000) A parasitic wasp forces its host to weave a special web for its own ends. *Nature* **406:** 255–256.

Edson, K.M. and Vinson, S.B. (1976) The function of the anal vesicle in respiration and excretion in the braconid wasp, *Microplitis croceipes*. *J. Insect Physiol.* **22:** 1037–1043.

Edson, K.M. and Vinson, S.B. (1977) Nutrient absorption by the anal vesicle of the braconid wasp *Microplitis croceipes*. *J. Insect Physiol.* **23:** 5–8.

Edson, K.M. and Vinson, S.B. (1979) A comparative morphology of the venom apparatus of female braconids (Hymenoptera: Braconidae). *Can. Ent.* **111:** 1013–1024.

Edson, K.M., Vinson, S.B., Stoltz, D.B. and Summers, M.D. (1981) Virus in a parasitoid wasp: suppression of the cellular immune response in the parasitoid's host. *Science* **211:** 582–583.

Edson, K.M., Barlin, M.R. and Vinson, S.B. (1982) Venom apparatus of braconid wasps: comparative ultrastructure of reservoirs and gland filaments. *Toxicon* **20:** 553–562.

Edwards, J.P. (1987) Activity of optical and geometrical isomers of the juvenile hormone analogue hydroprene on endogenous juvenile hormone levels in larval *Manduca sexta*. *Pestic. Sci.* **21:** 203–210.

Edwards, J.P. and Menn, J.J. (1981) The use of juvenoids in insect pest management. In: *Chemie der Pflanzenschcutz-und Schadlingsbekampfungsmittel*, Vol. 6 (ed Wegler, R.), pp. 185–214. Springer-Verlag, Berlin.

Edwards, J.P., Bergot, B.J. and Staal, G.B. (1983) Effects of three compounds with anti juvenile hormone activity and a juvenile hormone analogue on endogenous juvenile hormone levels in the Tobacco hornworm, *Manduca sexta*. *J. Insect Physiol.* **29:** 83–89.

Edwards, J.P., Cerf, D.C., and Staal, G.B. (1985) Inhibition of ootheca production in *Periplaneta americana* (L.) with the anti-juvenile hormone fluoromevalonate. *J. Insect Physiol.* **31:** 723–728.

Edwards, J.P., Corbitt, T.S., McArdle, H.F., Short, J.E. and Weaver, R.J. (1995) Endogenous levels of insect juvenile hormones in larval, pupal and adult stages of the Tomato moth, *Lacanobia oleracea*. *J. Insect Physiol.* **41:** 641–651.

Edwards, J.P., Audsley, N., Marris, G.C., Cusson, M. and Weaver, R.J. (2001) The role of allatostatic and allatotropic neuropeptides in the regulation of juvenile hormone biosynthesis in *Lacanobia oleracea* (Lepidoptera: Noctuidae). *Peptides* **22:** 255–261.

Efremova, Z.A. (1997) Trophic links of Chalcids of the subfamily Eulophinae (Hymenoptera, Eulophidae) and their evolution. *Ent. Rev.* **77:** 309–317.

Eldridge, R., O'Reilly, D.R. and Miller, L.K. (1992a) Efficacy of a baculovirus pesticide expressing an eclosion hormone gene. *Biol. Cont.* **2:** 104–110.

Eldridge, R., O'Reilly, D.R., Hammock, B.D. and Miller, L.K. (1992b) Insecticidal properties of genetically engineered baculoviruses expressing an insect juvenile hormone esterase gene. *Appl. Environ. Microbiol.* **58:** 1583–1591.

Ellsbury, M.M., Burkett, G.A. and Davis, F.M. (1989) Development and feeding behavior of *Heliothis zea* (Lepidoptera: Noctuidae) on leaves and flowers of crimson clover. *Environ. Entomol.* **18:** 323–327.

El Sawaf, B.M., El Sattar, S.A., Shehata, M.G., Lane, R.P. and Morsy, T.A. (1994) Reduced longevity and fecundity in *Leishmania*-infected sandflies. *Am. J. Trop. Med. Hyg.* **5:** 767–770.

Engelmann, F. (1983) Vitellogenesis controlled by juvenile hormone. In: *Endocrinology of Insects* (eds Laufer, H. and Downer, R.), pp. 259–270. A.R. Liss, New York.

Evans, A.C. (1933) Comparative observations on the morphology and biology of some Hymenopterous parasites of carrion-infesting Diptera. *Bull. Ent. Res.* **24:** 385–405.

Evans, O.P. and O'Reilly, D.R. (1998) Purification and kinetic analysis of a baculovirus ecdysteroid UDP-glucosyltransferase. *Biochem. J.* **330:** 1265–1270.

Evans, O.P. and O'Reilly, D.R. (1999) Expression and structural characterisation of a baculovirus ecdysteroid UDP-glucosyltransferase. *J. Gen. Virol.* **80:** 485–492.

Ewer, J., Gammie, S.C. and Truman, J.W. (1997) Control of insect ecdysis by a positive-feedback endocrine system: roles of eclosion hormone and ecdysis triggering hormone. *J. Exp. Biol.* **200:** 869–881.

Fain, M.J. and Riddiford, L.M. (1975) Juvenile hormone titers in the hemolymph during late larval development of the tobacco hornworm, *Manduca sexta* (L.). *Biol. Bull.* **149:** 506–521.

Falabella, P., Tremblay, E. and Pennacchio, F. (2000) Host regulation by the aphid parasitoid *Aphidius ervi*: the role of teratocytes. *Entomol. Exp. Appl.* **97:** 1–9.

Farkas, R., Wache, S. and Jones, D. (1999) Uncoupling of sequential heteromorphic developmental programs. *Arch. Insect Biochem. Physiol.* **40:** 1–16.

Farrar, R.R. and Bradley, J.R. (1985) Effects of within-plant distribution of *Heliothis* (Boddie) (Lepidoptera: Noctuidae) eggs and larvae on larval development and survival on cotton. *J. Econ. Entomol.* **78:** 1233–1237.

Fathpour, H. and Dahlman, D.L. (1995) Polydnavirus of *Microplitis croceipes* prolongs the larval period and changes hemolymph protein content of the host, *Heliothis virescens*. *Arch. Insect Biochem. Physiol.* **28:** 33–48.

Feddersen, I., Sander, K. and Schmidt, O. (1986) Virus-like particles with host protein-like antigenic determinants protect an insect parasitoid from encapsulation. *Experientia* **42:** 1278–1281.

Feldlaufer, M.F., Herbert, E.W. Jr., Svoboda, J.A., Thompson, M.J. and Lusby, W.R. (1985) Makisterone A: the major ecdysteroid from the pupa of the honey bee *Apis mellifera*. *Insect Biochem.* **15:** 597-600.

Feldlaufer, M.F., Herbert, E.W. Jr., Svoboda, J.A. and Thompson, M.J. (1986) Biosynthesis of makisterone A and 20-hydroxyecdysone from labelled sterols in the honey bee *Apis mellifera*. *Arch. Insect Biochem. Physiol.* **3:** 415–421.

Ferkovich, S.M. and Dillard, C.R. (1986) A study of uptake of radiolabeled host proteins and protein synthesis during development of eggs of the endoparasitoid, *Microplitis croceipes* (Cresson) (Braconidae). *Insect Biochem.* **16:** 337–345.

Ferran, M.A. and Daumal, J. (1973) Consequences sur les larves d' *Anagasta kuehniella* Zeller (Lep. Phyticidae) de l'elimination precoce par le froid de son endoparasite *Phanerotoma flavitestacea* Fischer (Hym Braconidae). *C.R. Acad. Sci. Paris* **277D:** 869–872.

Ferran, M.A. and Laforge, J.P. (1974) Mise en evidence d'une action toxique des secretions ovariennes des femelles d'un insecte entomophage [*Phanerotoma flavitestacea* Fischer (Hym. Braconidae)] sur les larves de son hote, *Anagasta kuehniella* Zeller (Lep. Phyticidae). *C.R. Acad. Sci. Paris* **279D:** 1341–1344.

Ferrière, C. (1941) New species of Euplectrini (Hym. Chalcidoidea) from Europe, Africa, and Asia. *Entomol. Res.* **32:** 17–48.

Fincke, O., Higgins, L. and Rojas, E. (1990) Parasitism of *Nephila clavipes* (Araneae, Tetragnathidae) by an ichneumonid (Hymenoptera, Polysphinctini) in Panama. *J. Arachnol.* **18:** 321–329.

Fisher, R.C. (1971) Aspects of the physiology of endoparasitic Hymenoptera. *Biol. Rev.* **46:** 243–278.

Fitton, M.G., Shaw, M.R. and Gauld, I.D. (1988) Pimpline ichneumon-flies. In: *Handbooks for the Identification of British Insects,* Vol. 7, Part 1 (eds Barnard, P.C. and Askew, R.R.). Royal Entomological Society of London, London.

Flanders, S.E. (1942a) Abortive development in parasitic Hymenoptera, induced by the food plant of the insect host. *J. Econ. Entomol.* **35:** 834–835.

Flanders, S.E. (1942b) Oösorption and ovulation in relation to oviposition in the parasitic Hymenoptera. *Ann. Entomol. Soc. Amer.* **35:** 251–266.

Fleming, J.G.W. and Krell, P.J. (1993) Polydnavirus genome organization. In: *Parasites and Pathogens of Insects,* Vol. 1 (eds Beckage, N.E., Thompson, S.N. and Federici, B.A.), pp. 189–225. Academic Press, New York.

Fleming, J.G.W., Blissard, G.W., Summers, M.D. and Vinson, S.B. (1983) Expression of *Campoletis sonorensis* virus in the parasitized host, *Heliothis virescens. J. Virol.* **48:** 74–78.

Flipsen, J.T.M., Mans, R.M.W., Kleefsman, A.W.F., Knebel-Mordsdorf, D. and Vlak, J.M. (1995) Deletion of the baculovirus ecdysteroid UDP-glucosyltransferase gene induced early degeneration of Malpighian tubules in infected insects. *J. Virol.* **69:** 4529–4532.

Forbes, M.L.R. (1993) Parasitism and host reproductive effort. *Oikos* **67:** 444–450.

Freier, J.E. and Friedman, S. (1976) Effect of host infection with *Plasmodium gallinaceum* on the reproductive capacity of *Aedes aegypti. J. Invert. Pathol.* **28:** 161–166.

Fritz, R.S. (1982) Selection for host modification by insect parasitoids. *Evolution* **36:** 283–288.

Führer, E. (1973) Sekretion on Mucopolysacchariden im weiblichen Geschlechtsapparat von *Pimpla turionellae* L. (Hym., Ichneumonidae). *Z. Parasitenk.* **41:** 207–213.

Führer, E. and Willers, D. (1986) The anal secretion of the endoparasitic larva *Pimpla turionellae*: sites of production and effects. *J. Insect Physiol.* **32:** 361–367.

Fukushima, J., Kuwahara, Y., Yamada, A. and Suzuki, T. (1990) New non-cyclic homoditerpene from the sting glands of *Bracon hebetor* Say (Hymenoptera: Braconidae). *Agric. Biol. Chem.* **54:** 809–810.

Furuya, K., Hackett, M., Cirelli, M.A., Schegg, K.M., Wang, H., Shabanowitz, J., Hunt, D.F. and Schooley, D.A. (1999) A cardioactive peptide from the southern armyworm *Spodoptera eridania. Peptides* **20:** 53–61.

Gade, G., Hoffman, K.-H. and Spring, J.H. (1997) Hormonal regulation in insects: facts, gaps, and future directions. *Physiol. Rev.* **77:** 963–1032.

Gao, N. and Hardie, J. (1996) Pre- and post-natal effects of precocenes on aphid morphogenesis and differential rescue. *Arch. Insect Biochem. Physiol.* **32:** 503–510.

Gauld, I.D. (1988) Evolutionary patterns of host utilization by ichneumonid parasitoids (Hymenoptera: Ichneumonidae and Braconidae). *Biol. J. Linn. Soc.* **35:** 351–377.

Gauld, I.D. (1995) Stephanidae. In: *The Hymenoptera of Costa Rica.* (eds Hanson, P.E. and Gauld, I.D.), pp. 181–184. Oxford University Press, Oxford.

Gauld, I.D. and Bolton, B. (1988) *The Hymenoptera.* Oxford University Press, Oxford.

Gauthier, N. and Monge, J.P. (1999) Could the egg itself be the source of the oviposition deterrent marker in the ectoparasitoid *Dinarmus basilis*? *J. Insect Physiol.* **45:** 393–400.

Gauthier, N., LaSalle, J., Quicke, D.L.J. and Godfray, H.C.J. (2000) Phylogeny of Eulophidae (Hymenoptera: Chalcidoidea), with a reclassification of Eulophinae and the recognition that Elasmidae are derived eulophids. *System. Entomol.* **25:** 521–539.

Gelman, D.B. and Beckage, N.E. (1995) Low molecular weight ecdysiotropins in proctodaea of fifth instars of the tobacco hornworm, *Manduca sexta* (Lepidoptera: Sphingidae), and hosts parasitized by the braconid wasp *Cotesia congregata* (Hymenoptera: Braconidae). *Eur. J. Entomol.* **92:** 123–129.

Gelman, D.B. and Woods, C.W. (1983) Haemolymph ecdysteroid titers of diapause- and nondiapause-bound fifth instars and pupae of the European corn borer *Ostrinia nubilalis* (Hubner). *Comp. Biochem. Physiol.* **76A:** 367–375.

Gelman, D.B., Thyagaraja, B.S., Kelly, T.J., Masler, E.P., Bell, R.A. and Borkovec, A.B. (1991) The insect gut: a new source of ecdysiotropic peptides. *Experientia* **47:** 78–80.

Gelman, D.B., Kelly, T.J. and Coudron, T.A. (1997) Mode of action of the venom of the ectoparasitic wasp, *Euplectrus comstockii*, in causing developmental arrest in the European corn borer, *Ostrinia nubilalis. Invertebr. Neurosci.* **3:** 231–238.

Gelman, D.B., Reed, D.A. and Beckage, N.E. (1998) Manipulation of fifth-instar host (*Manduca sexta*) ecdysteroid levels by the parasitoid wasp *Cotesia congregata. J. Insect Physiol.* **44:** 833–843.

Gelman, D.B., Kelly, T.J., Reed, D.A. and Beckage, N.E. (1999) Synthesis and release of ecdysteroids by *Cotesia congregata*, a parasitoid wasp of the tobacco hornworm, *Manduca sexta. Arch. Insect Biochem. Physiol.* **40:** 17–29.

Gelman, D.B., Carpenter, J.E. and Greany, P.D. (2000) Ecdysteroid levels/profiles of the parasitoid wasp, *Diapetimorpha introita*, reared on its host, *Spodoptera frugiperda* and on an artificial diet. *J. Insect Physiol.* **46:** 457–465.

Gemmill, R.M., Hamblin, M., Glaser, R.L., Racioppi, J.V., Marx, J.L., White, B.N. Calvo, J.M., Wolfner, M. and Hagedorn, H.H. (1986) Isolation of mosquito vitellogenin genes and induction of expression by 20-hydroxyecdysone. *Insect Biochem.* **16:** 761–774.

Gerling, D. (1971) Occurrence, abundance, and efficiency of some local parasitoids attacking *Spodoptera littoralis* in selected cotton fields in Israel. *Ann. Entomol. Soc. Am.* **64:** 492–499.

Gerling, D. and Limon, S. (1976) A biological review of the genus *Euplectrus* (Hym.: Eulophidae) with special emphasis on *E. laphygmae* as a parasite of *Spodoptera littoralis* (Lep.: Noctuidae). *Entomophaga* **21:** 179–187.

Gharib, B. and de Reggi, M. (1983) Changes in ecdysteroid and juvenile hormone levels in developing eggs of *Bombyx mori*. *J. Insect Physiol.* **29:** 871–876.

Gharib, B., de Reggi, M., Connat, J.-L. and Chaix, J.-C. (1983) Ecdysteroid and juvenile hormone changes in *Bombyx mori* eggs, related to the initiation of diapause. *FEBS Letts* **160:** 119–123.

Gibson, G.A.P., Heraty, J.M. and Woolley, J.B. (1999) Phylogenetics and classification of Chalcidoidea and Mymarommatoidea – a review of current concepts (Hymenoptera, Apocrita). *Zool. Scripta* **28:** 87–124.

Gilbert, L.I. and Schneiderman, H.A. (1959) Prothoracic gland stimulation by juvenile hormone of insects. *Nature* **184:** 171–173.

Gilbert, L.I. and Schneiderman, H.A. (1961) The content of juvenile hormone and lipid in Lepidoptera: sexual differences and developmental changes. *Gen. Comp. Endocrinol.* **1:** 453–472.

Gilbert, L.I., Bollenbacher, W.E. and Granger, N.A. (1980a) Insect endocrinology: Regulation of endocrine glands, hormone titer and hormone metabolism. *Annu. Rev. Physiol.* **42:** 493–510.

Gilbert, L.I., Bollenbacher, W.E., Goodman, W., Smith, S.L., Agui, N., Granger, N.A. and Sedlak, B. (1980b) Hormones controlling insect metamorphosis. *Recent Prog. Horm. Res.* **36:** 401–402.

Gilbert, L.I., Combest, W.L., Smith, W.A., Meller, V.H. and Rountree, D.B. (1988) Neuropeptides, second messengers and insect molting. *BioEsssays* **8:** 153–157.

Gilbert, L.I., Rybczynski, R. and Tobe, S.S. (1996) Endocrine cascade in insect metamorphosis. In: *Metamorphosis – Postembryonic Reprogramming of Gene Expression in Amphibian and Insect Cells* (eds Gilbert, L.I., Tata, J.R. and Atkinson, B.G.), pp. 59–107. Academic Press, San Diego.

Gilbert, L.I., Granger, N.A. and Roe, R.M. (2000) The juvenile hormones: historical facts and speculations on future research directions. *Insect Biochem. Mol. Biol.* **30:** 617–644.

Glinka, A.V., Kleiman, A.M. and Wyatt, G.R. (1995) Roles of juvenile hormone, a brain factor and adipokinetic hormone in regulation of vitellogenesis biosynthesis in *Locusta migratoria. Biochem. Molec. Biol. Internat.* **35:** 323–328.

Godfray, H.C.J. (1994) *Parasitoids: Behavioral and Evolutionary Ecology*, Princeton University Press, Princeton, NJ.

Goodman, W., Coy, D.C., Baker, F.C., Xu, L. and Toong, Y.C. (1990) Development and application of a radioimmunoassay for the juvenile hormones. *Insect Biochem.* **20:** 357–364.

Granger, N.A. and Goodman, W.G. (1983) Juvenile hormone radioimmunoassays – theory and practice. *Insect Biochem.* **13:** 333–340.

Gratz, N.G. (1997) Human lice. Their prevalence, control and resistance to insecticides. A review 1985–1997. World Health Organization unpublished document WHO/CTD/WHOPE/97.8, 61 pp.

Greany, P.D., Clark, W.R., Ferkovich, S.M., Law, J.H. and Ryan, R.O. (1990) Isolation and characterization of a host hemolymph protein required for development of the eggs of the endoparasite, *Microplitis croceipes.* [Abstract]. In: *Molecular Insect Science* (eds Hagedorn, H.H., Hildebrand, J. G., Kidwell M.G. and Law, J.H.), p. 306. Plenum Press, New York,

Grenier, S. (1994) Rearing of *Trichogramma* and other egg parasitoids on artificial diets. In: *Biological control with egg parasitoids* (eds Wajnberg, E. and Hassan, S.A.), pp. 73–92. CAB International, Wallingford.

Grenier, S., Greany, P.D. and Cohen, A.C. (1993) Potential for mass release of insect parasitoids and predators through development of artificial culture techniques. In: *Pest Management in the Subtropics. Biological Control - a Florida Perspective* (eds Rosen, D., Bennett, F.D. and Capinera, J.L.), pp. 181–205. Intercept Ltd., Andover, UK.

Grieneisen, M.L. (1994) Recent advances in our knowledge of ecdysteroid biosynthesis in insects and crustaceans. *Insect Biochem. Mol. Biol.* **24:** 115–132.

Grossniklaus-Bürgin, C. and Lanzrein, B. (1990a) Qualitative and quantitative analyses of juvenile hormone and ecdysteroids from the egg to the pupal moult in *Trichoplusia ni. Arch. Insect Biochem. Physiol.* **14:** 13–30.

Grossniklaus-Bürgin, C. and Lanzrein, B. (1990b) Endocrine interrelationship between the parasitoid *Chelonus* sp. and its host *Trichoplusia ni. Arch. Insect Biochem. Physiol.* **14:** 201–206.

Grossniklaus-Bürgin, C., Lanz, F. and Lanzrein, B. (1988) Influence of the parasitoid *Chelonus* sp. on the endocrinology of its host, *Trichoplusia ni.* In: *Endocrinological Frontiers in Physiological Insect Ecology* (eds Sehnal, F., Zabza, A. and Denlinger, D.L.), pp. 437–441. Wroclaw Technical University Press, Wroclaw, Poland.

Grossniklaus-Bürgin, C., Connat, J.L. and Lanzrein, B. (1989) Ecdysone metabolism in the host-parasitoid-system *Trichoplusia ni/Chelonus* sp. *Arch. Insect Biochem. Physiol.* **11:** 79–92.

Grossniklaus-Bürgin, C., Wyler, T., Pfister-Wilhelm, R. and Lanzrein, B. (1994) Biology and morphology of the parasitoid *Chelonus inanitus* (Braconidae, Hymenoptera) and effects on the development of its host *Spodoptera littoralis* (Noctuidae, Lepidoptera). *Invertebr. Rep. Dev.* **25:** 143–158.

Grossniklaus-Bürgin, C., Pfister-Wilhelm, R., Meyer, V., Treiblmayr, K. and Lanzrein, B. (1998) Physiological and endocrine changes associated with polydnavirus/venom in the parasitoid-host system *Chelonus inanitus-Spodoptera littoralis. J. Insect Physiol.* **44:** 305–321.

Gruber, A., Stettler, P., Heiniger, P., Schümperli, D. and Lanzrein, B. (1996) Polydnavirus DNA of the braconid wasp *Chelonus inanitus* is integrated in the wasp's genome and excised only in later pupal and adult stages of the female. *J. Gen. Virol.* **77:** 2873–2879.

Gruetzmacher, M.C., Gilbert, L.I., Granger, N.A., Goodman, W. and Bollenbacher, W.E. (1984) The effect of juvenile hormone on prothoracic gland function during the larval-pupal development of *Manduca sexta*: an *in situ* and *in vitro* analysis. *J. Insect Physiol.* **30:** 331–340.

Guillot, F.S. and Vinson, S.B. (1972) The role of the calyx and poison gland of *Cardiochiles nigriceps* in the host-parasitoid relationship. *J. Insect Physiol.* **18:** 1315–1321.

Gunasena, G.H., Vinson, S.B. and Williams, H.J. (1989) Interrelationships between growth of *Heliothis virescens* (Lepidoptera: Noctuidae) and that of its parasitoid, *Campoletis sonorensis* (Hymenoptera: Ichneumonidae). *Ann. Entomol. Soc. Am.* **82:** 187–191.

Gupta, A.P. (1991) Insect immunocytes and other hemocytes: roles in cellular and humoral immunity. In: *Immunology of Insects and Other Arthropods* (ed Gupta, A.P.), pp 19–118. CRC Press Inc., Boca Raton, FL.

Gupta, P. and Ferkovich, S.M. (1998) Interaction of calyx fluid and venom from *Microplitis croceipes* (Braconidae) on developmental disruption of the natural host, *Heliocoverpa zea,* and two atypical hosts, *Galleria mellonella* and *Spodoptera exigua*. *J. Insect Physiol.* **44:** 713–719.

Guzo, D. and Stoltz, D.B. (1985) Obligatory multiparasitism in the tussock moth, *Orgyia leucostigma*. *Parasitol.* **90:** 1–10.

Hacker, C.S. (1971) The differential effect of *Plasmodium gallinaceum* on the fecundity of several strains of *Aedes aegypti*. *J. Invertebr. Pathol.* **18:** 373–377.

Hacker, C.S. and Kilama, W.L. (1974) The relationship between *Plasmodium gallinaceum* density and the fecundity of *Aedes aegypti*. *J. Invertebr. Pathol.* **23:** 101–105.

Hackman, R.H. (1976) The interactions of cuticular proteins and some comments on their adaptation to function. In: *The Insect Integument* (ed Hepburn, H.R.), pp. 107–120. Elsevier, Amsterdam.

Haeselbarth, E. (1979) Zur Parasitierung der Puppen von Forleuhe *(Panolis flammea)* (Schiff), Kiefernspanner *Bupalus piniarius* (L.) und Heidelbeerspanner *Boarmia bistortana* (Goezel) in bayerischen Kiefernwaldern. *Z. Angew. Entomol.* **87:** 186–202.

Hagedorn, H.H. (1983) The role of ecdysteroids in the adult insect. In: *Endocrinology of Insects,* (eds Downer, R.G. and Laufer, H.), pp. 271–304. A.R. Liss, New York.

Hagedorn, H.H. and Judson, C.L. (1972) Purification and site of synthesis of *Aedes aegypti* yolk protein. *Exp. Zool.* **182:** 367–377.

Hagenguth, H. and Rembold, H. (1978) Identification of juvenile hormone 3 as the only JH homologue in all developmental stages of the honey bee. *Z. Naturforsch.* **33C:** 847–850.

Hales, D. and Mittler, T.E. (1981) Precocious metamorphosis of the aphid *Myzus persicae* induced by the precocene analogue 6-methoxy-7-ethoxy-2,2-dimethylchromene. *J. Insect Physiol.* **27:** 333–337.

Hales, D. and Mittler, T.E. (1983) Precocene causes male determination in the aphid *Myzus persicae*. *J. Insect Physiol.* **29:** 819–823.

Hammock, B.D. (1985) Regulation of juvenile hormone titer: Degradation. In: *Comprehensive Insect Physiology, Biochemistry and Pharmacology,* Vol. 7 (eds Kerkut, G.A. and Gilbert, L.I.), pp. 431–472. Pergamon Press, Oxford.

Hammock, B.D. and Quistad, G.B. (1981) Metabolism and mode of action of juvenile hormone, juvenoids and other insect growth regulators. In: *Progress in Pesticide Biochemistry* (eds Hutson, D.H. and Roberts, T.R.), pp. 1–83. John Wiley & Sons, New York.

Hammock, D.B. and Sparks, T.C. (1977) A rapid assay for insect juvenile hormone esterase activity. *Analyt. Biochem.* **82:** 573–579.

Hammock, B.D., Bonning, B.C., Possee, R.D., Hanzlik, T.N. and Maeda, S. (1990) Expression and effects of the juvenile hormone esterase in a baculovirus vector. *Nature* **344:** 458–461.

Hammock, B.D., McCutchen, B.F., Beetham, J. *et al.* (1993) Development of recombinant viral insecticides by expression of an insect-specific toxin and insect-specific enzyme in nuclear polyhedrosis viruses. *Arch. Insect Biochem. Physiol.* **22:** 315–344.

Hanaoka, K. and Ohnishi, E. (1974) Changes in ecdysone titre during pupal-adult development in the silkworm, *Bombyx mori*. *J. Insect Physiol.* **20:** 2375–2384.

Hänel, H. and Koeniger, N. (1986) Possible regulation of the reproduction of the honey bee mite *Varroa jacobsoni* (Merostigmata: Acari) by a host's hormone: juvenile hormone III. *J. Insect Physiol.* **32:** 791–798.

Hardie, J. (1980) Juvenile hormone mimics photoperiodic apterization of the alate gynopara of the aphid, *Aphis fabae*. *Nature* **286:** 602–604.

Hardie, J. (1981) Juvenile hormone and photoperiodically controlled polymorphism in *Aphis fabae*: postnatal effects on presumptive gynoparae. *J. Insect Physiol.* **27:** 347–355.

Hardie, J. (1986) The morphogenetic effects of precocenes on three aphid species. *J. Insect Physiol.* **32:** 813–818.

Hardie, J. (1987a) Neurosecretory and endocrine systems. In: *Aphids. Their Biology, Natural Enemies and Control,* Vol. 2B (eds Minks, A.K. and Harrewijn, P.), pp. 139–152. Elsevier, Amsterdam.

Hardie, J. (1987b) Juvenile hormone stimulation of oocyte development and embryogenesis in the parthenogenetic ovaries of an aphid, *Aphis fabae*. *Int. J. Invert. Reprod. Dev.* **11:** 189–202.

Hardie, J., Honda, K., Timar, T. and Varjas, L. (1995) The effects of 2,2-dimethylchromene derivatives on wing determination and metamorphosis in the pea aphid, *Acyrthosiphon pisum*. *Arch. Insect Biochem. Physiol.* **30:** 25–40.

Hardie, J., Gao, N., Timar, T., Sebok, P. and Honda, K. (1996) Precocene derivatives and aphid morphogenesis. *Arch. Insect Biochem. Physiol.* **32:** 493–501.

Harnish, D.G. and White, B.N. (1982) Insect vitellins: identification, purification and characterisation from eight orders. *J. Exp. Zool.* **220:** 1–10.

Harshman, K.G. (1999) Investigation of the endocrine system in extended longevity lines of *Drosophila melanogaster*. *Exp. Gerontol.* **34:** 997–1006.

Harvey, J.A. (1996) *Venturia canescens* parasitizing *Galleria mellonella* and *Anagasta kuehniella*: is the parasitoid a conformer or regulator? *J. Insect Physiol.* **42:** 1017–1025.

Harvey, J.A., Harvey, I.F. and Thompson, D.J. (1994) Flexible larval growth allows use of a range of host sizes by a parasitoid wasp. *Ecology*, **75:** 1420–1428.

Harvey, J.A., Harvey, I.F. and Thompson, D.J. (1995) The effect of host nutrition on development of the solitary wasp, *Venturia canescens*. *Entomol. Exp. Appl.* **75:** 213–220.

Harwood, S.H. and Beckage, N.E. (1994) Purification and characterization of an early-expressed polydnavirus-induced protein from the hemolymph of *Manduca sexta* larvae parasitized by *Cotesia congregata*. *Insect Biochem. Mol. Biol.* **24:** 685–698.

Harwood, S.H., Grosovsky, A.J., Cowles, E.A., Davis, J.W. and Beckage, N.E. (1994) An abundantly expressed host hemolymph glycoprotein isolated from newly parasitized *Manduca sexta* larvae is a polydnavirus gene product. *Virology* **205:** 381–392.

Hasegawa, K. and Ata, A.M. (1972) Penetration of phytoecdysones through the pupal cuticle of the silkworm, *Bombyx mori*. *J. Insect Physiol.* **18:** 959–971.

Hashimoto, Y., Hayakawa, T., Ueno, Y., Fujita, T., Sano, Y. and Matsumoto, T. (2000) Sequence analysis of the *Plutella xylostella* granulovirus genome. *Virology* **275:** 358–372.

Haunerland, N.H. (1996) Insect storage proteins: gene families and receptors. *Insect Biochem. Mol. Biol.* **26:** 755–765.

Haunerland, N.H. and Bowers, W.S. (1986) Arylphorin from the corn earworm, *Heliothis zea*. *Insect Biochem.* **16:** 617–625.

Haunerland, N.H. and Shirk, P.D. (1995) Regional and functional differentiation in the insect fat body. *Annu. Rev. Entomol.* **40:** 121–145.

Hayakawa, T., Ko, R., Okano, K., Seong, S.-I., Goto, C. and Maeda, S. (1999) Sequence analysis of the *Xestia c-nigrum* granulovirus genome. *Virology* **262:** 277–297.

Hayakawa, Y. (1987) Inhibition of lipid transport in insects by a parasitic factor. *Comp. Biochem. Physiol.* **87:** 279–283.

Hayakawa, Y. (1990) Juvenile hormone esterase activity repressive factor in the plasma of parasitized insect larvae. *J. Biol. Chem.* **265:** 10813–10816.

Hayakawa, Y. (1991) Structure of a growth-blocking peptide present in parasitized insect hemolymph. *J. Biol. Chem.* **266:** 7982–7984.

Hayakawa, Y. (1992) A putative new juvenile peptide hormone in lepidopteran insects. *Biochem. Biophys. Res. Commun.* **185:** 1141–1147.

Hayakawa, Y. (1995) Growth-blocking peptide: an insect biogenic peptide that prevents the onset of metamorphosis. *J. Insect Physiol.* **41:** 1–6.

Hayakawa, Y. and Yashuhara, Y. (1993) Growth-blocking peptide or polydnavirus effects on the last instar larvae of some insect species. *Insect Biochem. Mol. Biol.* **23:** 225–231.

Hayakawa, Y., Ohnishi, A., Yamanaka, A., Izumi, S. and Tomino, S. (1995) Molecular cloning and characterization of cDNA for insect biogenic peptide, growth-blocking peptide. *FEBS Letts* **376:** 185–189.

Hayakawa, Y., Ohnishi, A. and Endo, Y. (1998) Mechanism of parasitism-induced elevation of haemolymph growth-blocking peptide levels in host insect larvae (*Pseudaletia separata*). *J. Insect Physiol.* **44:** 859–866.

Heitmans, W.R.B. (1990) Which information is used to assess host size and sex of the progeny? Experiments with a solitary aculeate ectoparasitoid of cockroaches. *Proceedings of the section Experimental and Applied Entomology of the Netherlands Entomological Soc. Ned. Entomol. Ver.* **1:** 13–18.

Hellers, M., Beck, M., Theopold, U., Kamei, M. and Schmidt, O. (1996) Multiple alleles encoding a virus-like particle protein in the ichneumonid endoparasitoid *Venturia canescens. Insect Mol. Biol.* **5:** 239–249.

Hemerik, L. and Harvey, J.A. (1999) Flexible larval development and the timing of destructive feeding by a solitary endoparasitoid: an optimal foraging problem in evolutionary perspective. *Ecol. Entomol.* **24:** 308–315.

Henrick, C.A. (1991) Juvenoids and anti-juvenile hormone agents: past and present. In: *Insect Chemical Ecology,* (ed Hrdy, I.), pp. 429–452. Academic Praha, Prague.

Henrich, V., Rybczynski, R. and Gilbert, L.I. (1999) Peptide hormones, steroid hormones and puffs: mechanisms and models in insect development. In: *Vitamins and Hormones,* Vol. 55 (ed. Litwack, G.), pp. 73–125. Academic Press, San Diego.

Herman, W.S. and Gilbert, L.I. (1966) The neuroendocrine system of *Hyalophora cecropia* (L.) (Lepidoptera: Saturniidae) I. The anatomy and histology of the ecdysial glands. *Gen. Comp. Endocrinol.* **7:** 275–291.

Hernández-Crespo, P., Sait, S.M., Hails, R.S. and Cory, J.S. (2001) Behaviour of a recombinant baculovirus in lepidopteran hosts of different susceptibilities. *Appl. Env. Micro.* **67:** 1140–1146.

Herniou, E.A., Luque, T., Chen, X., Vlak, J.M., Winstanley, D., Cory, J.S. and O'Reilly, D.R. (2001) The use of whole genome sequence data to infer baculovirus phylogeny. *J. Virol.* **75:** 8117–8126.

Hesterlee, S. and Morton, S.B. (1996) Insect physiology: the emerging story of ecdysis. *Curr. Biol.* 6: 648–650.

Hille Ris Lambers, D. (1966) Polymorphism in Aphididae. *Annu. Rev. Entomol.* **11:** 47–78.

Hiruma, K., Shimada, H. and Yagi, S. (1978) Activation of the prothoracic gland by juvenile hormone and prothoracicotropic hormone in *Mamestra brassicae. J. Insect Physiol.* **24:** 215–220.

Hiruma, K., Hardie, J. and Riddiford, L.M. (1991) Hormonal regulation of epidermal metamorphosis in vitro: control of expression of a larval-specific cuticle gene. *Dev. Biol.* **144:** 369–378.

Hochuli, A., Pfister-Wilhelm, R. and Lanzrein, B. (1999) Analysis of endoparasitoid-released proteins and their effects on host development in the system *Chelonus inanitus* (Braconidae)-*Spodoptera littoralis* (Noctuidae). *J. Insect Physiol.* **45:** 823–833.

Hoffmann, J.A. and Lagueux, M. (1985) Endocrine aspects of embryonic development in insects. In: *Comprehensive Insect Physiology, Biochemistry and Pharmacology,* Vol. 1 (eds Kerkut, G.A. and Gilbert, L.I.), pp. 435–460. Pergamon, Oxford.

Hogg, J.C. and Hurd, H. (1995) *Plasmodium yoelii nigeriensis*: the effect of high and low intensity of infection upon the egg production and blood meal size of *Anopheles stephensi* during three gonotrophic cycles. *Parasitology* **111:** 555–562.

Hogg, J.C. and Hurd, H. (1997) The effects of natural *Plasmodium falciparum* infection on the fecundity and mortality of *Anopheles gambiae* s. l. in north-east Tanzania. *Parasitology* **114:** 325–331.

Hogg, J.C., Carwardine, S.L. and Hurd, H. (1997) The effect of *Plasmodium yoelii nigeriensis* infection on ovarian protein accumulation by *Anopheles stephensi*. *Parasitol. Res.* **83:** 374–379.

Höller, C., Bargen, H., Vinson, S.B. and Braune, H.J. (1993a) Sources of the marking pheromones used for host discrimination in the hyperparasitoid *Dendrocerus carpenteri*. *J. Insect Physiol.* **39:** 649–656.

Höller, C., Borgemeister, C., Haardt, H. and Powell, W. (1993b) The relationship between primary parasitoids and hyperparasitoids of cereal aphids: an analysis of field data. *J. Anim. Ecol.* **62:** 12–21.

Höller, C., Bargen, H., Vinson, S.B. and Witt, D. (1994) Evidence for the external use of juvenile hormone for host marking and regulation in a parasitic wasp, *Dendrocerus carpenteri*. *J. Insect Physiol.* **40:** 317–322.

Holman, G.M. and Meola, R.W. (1978) A high performance liquid chromatography method for the purification and analysis of insect ecdysones: application to measurement of ecdysone titers during pupal-adult development of *Heliothis zea*. *Insect Biochem.* **8:** 275–278.

Hopkins, T.L. and Kramer, K.J. (1992) Insect cuticle sclerotization. *Annu. Rev. Entomol.* **37:** 273–302.

Hopkins, T.L., Starkey, S.R., and Beckage, N.E. (1998) Tyrosine and catecholamine levels in the hemolymph of tobacco hornworm larvae, *Manduca sexta*, parasitized by the braconid wasp, *Cotesia congregata*, and in the developing parasitoids. *Arch. Insect Biochem. Physiol.* **38:** 193–201.

Hopwood, J.A., Ahmed, A.M., Polwart, A., Williams, G.T. and Hurd, H. (2001) Malaria-induced apoptosis in mosquito ovaries: a mechanism to control vector egg production. *J. Exp. Biol.*

Horodyski, F.M. (1996) Neuroendocrine control of insect ecdysis by eclosion hormone. *J. Insect Physiol.* **42:** 917–924.

House, H.L. (1977) Nutrition of natural enemies. In: *Biological Control by Augmentation of Natural Enemies* (eds Ridgway, R.L. and Vinson, S.B.), pp. 105–182. Plenum Press, New York.

Hrdy, I. (1991) *Insect Chemical Ecology*. Academia, Prague.

Hsiao, T.H. and Hsiao, C. (1977) Simultaneous determination of moulting and juvenile hormone titers of the greater wax moth. *J. Insect Physiol.* **23:** 89–93.

Hu, J.S. and Vinson, S.B. (1997a) *In vitro* rearing of *Campoletis sonorensis*, a larval endoparasitoid of *Heliothis virescens* from egg to third instar in an artificial medium devoid of insect source. *Entomol. Exp. Appl.* **85:** 263–273.

Hu, J.S. and Vinson, S.B. (1997b) *In vitro* development of *Campoletis sonorensis (Hym.: Ichneumonidae)*, a larval endoparasitoid of *Heliothis virescens* (*Lep.: Noctuidae*) in an artificial medium with insect sources from egg to third larval instar. *Entomophaga* **42:** 405–415.

Hu, J.S. and Vinson, S.B. (2000) Interaction between the larval endoparasitoid *Campoletis sonorensis* (Hymenoptera: Ichneumonidae) and its host the Tobacco budworm (Lepidoptera: Noctuidae). *Ann. Entomol. Soc. Am.* **93:** 220–224.

Hu, Z.H., Broer, R., Westerlaken, J., Martens, J.W.M., Jin, F., Wang, L.M. and Vlak, J.M. (1997) Characterization of the ecdysteroid UDP-glucosyltransferase gene of a single nucleocapsid nucleopolyhedrosis of *Buzura suppressaria*. *Virus Res.* **47:** 91–97.

Hughes, P.R., van Beek, N.A.M. and Wood, H.A. (1986) A modified droplet feeding method for rapid assay of *Bacillus thuringiensis* and baculoviruses in noctuid larvae. *J. Invertebr. Pathol.* **69:** 112–118.

Hunter, F.R., Crook, N.E. and Entwistle, P.F. (1984) Viruses as pathogens for the control of insects. In: *Microbiological Methods for Environmental Biotechnology* (ed Grainger, J.M.L.), pp. 323–347. Academic Press, New York.

Hurd, H. (1990) Physiological and behavioural interactions between parasites and invertebrate-hosts. *Adv. Parasitol.* **29:** 271–317.

Hurd, H. (1993) Reproductive disturbances induced by parasites and pathogens of insects. In: *Parasites and Pathogens of Insects,* Vol. 1 (eds Beckage, N.E. and Thompson, S.N. and Federici, B.A.), pp. 87–105. Academic Press, San Diego.

Hurd, H. (1998) Parasite manipulation of insect reproduction: who benefits? *Parasitology* **116:** S13–S21.

Hurd, H. (1999) Manipulative parasites. *Biologist* **46:** 162–166.

Hurd, H. (2001) Host fecundity reduction: a damage limitation strategy? *Trends Parasitol.* **17:** 363–368.

Hurd, H. and Arme, C. (1986) *Hymenolepis diminuta*: influence of metacestodes upon synthesis and secretion of fat body protein and its ovarian sequestration in the intermediate host, *Tenebrio molitor. Parasitology* **93:** 111–120.

Hurd, H. and Arme, C. (1987) *Hymenolepis diminuta*: effect of infection upon the patency of the follicular epithelium in the intermediate host, *Tenebrio molitor. J. Invert. Pathol.* **49:** 227–234.

Hurd, H. and Burns, I. (1994) Observations on the axenic culture of metacestodes of *Hymenolepis diminuta. Can. J. Zool.* **72:** 618–623.

Hurd, H. and Parry, G. (1991) Metacestode-induced depression of the production and response to, sex pheromone in the intermediate host *Tenebrio molitor. J. Invert. Pathol.* **58:** 82–87.

Hurd, H. and Sinden, R.E. (2001) Malaria: an example of a vector transmitted parasite. In: *Practical Exercises in Parasitology* (eds Halton, D.W., Behnke, J.M. and Marshall, I.) pp. 89–97. Cambridge University Press, Cambridge.

Hurd, H. and Weaver, R.J. (1987) Evidence against the hypothesis that metacestodes of *Hymenolepis diminuta* inhibit corpora allata functioning in the intermediate host, *Tenebrio molitor. Parasitology* **95:** 93–97.

Hurd, H. and Webb, T.J. (1997) The role of endocrinologically active substances in mediating changes in insect hosts and insect vectors. In: *Parasites and Pathogens: Effects on Host Hormones and Behaviour* (ed Beckage, N.E.), pp. 179–197. Chapman & Hall, New York.

Hurd, H., Strambi, C. and Beckage, N.E. (1990) *Hymenolepis diminuta.* An investigation of juvenile hormone titre, degradation and supplementation in the intermediate host, *Tenebrio molitor. Parasitology* **100:** 445–452.

Ismail, S.M., Satyanarayana, K., Bradfield, J.Y., Dahm, K.H. and Bhaskaran, G. (1998) Juvenile hormone acid: evidence for a hormonal function in induction of vitellogenin in larvae of *Manduca sexta. Arch. Insect Biochem. Physiol.* **37:** 305–314.

Ismail, S.M., Goin, C., Muthumani, K., Kim, M., Dahm, K.H. and Bhaskaran, G. (2000) Juvenile hormone acid and ecdysteroid together induce competence for metamorphosis of the Verson's gland in *Manduca sexta. J. Insect Physiol.* **46:** 59–68.

Iwantsch, G.F. and Smilowitz, Z. (1975) Relationship between the parasitoid *Hyposoter exiguae* and the cabbage looper, *Trichoplusia ni.*: prevention of host pupation at the endocrine level. *J. Insect Physiol.* **21:** 1151–1157.

Iwantsch, G.F. and Smilowitz, Z. (1976) Relationships between the parasitoid *Hyposoter exiguae* and *Trichoplusia ni*: parasitoid-induced histo- and cytopathology. *J. Invert. Pathol.* **27:** 149–160.

Izri, M.A. and Briere, C. (1995) First cases of resistance of *Pediculus capitis* Linne 1758 to malathion in France. *Presse Med.* **24:** 1444.

Jackson, C.G., Neemann, E.G. and Patana, R. (1979) Parasitization of 6 lepidopteran cotton pests by *Chelonus blackburni* [Hym.: Braconidae]. *Entomophaga* **24:** 99–105.

Jacques, R. (1955) Vergleichende Fermentuntersuchungen an tierischen Giften (Cholinesterase, 'Lecithinase', Hyaluronidase). *Helv. Physiol. Pharmacol. Acta* **13:** 113–120.

Jahan, N. and Hurd, H. (1997) The effect of infection with *Plasmodium yoelii nigeriensis* on the reproductive fitness of *Anopheles stephensi*. *Ann. Trop. Med. Parasitol.* **91:** 365–369.

Jahan, N., Docherty, P.T., Billingsley, P.F. and Hurd, H. (1999) Blood digestion in the mosquito, *Anopheles stephensi* the effects of *Plasmodium yoelii nigeriensis* on midgut enzyme activities. *Parasitology* **119:** 535–541.

James, A.A. (1994) Molecular and biochemical analyses of the salivary glands of vector mosquitoes. *Bull. Inst. Pasteur* **92:** 133–150.

James, A.A., Balckmer, K., Marinotti, O., Goshn, C.R. and Racioppi, J.V. (1991) Isolation and characterization of the gene expressing the major salivary gland protein of the female mosquito, *Aedes aegypti*. *Mol. Biochem. Parasitol.* **44:** 245–254.

Jansons, I., Cusson, M., McNeil, J.N., Tobe, S.S. and Bendena, W.G. (1996) Molecular characterization of a cDNA from *Pseudaletia unipuncta* encoding the *Manduca sexta* allatostatin peptide (Mas-AST). *Insect Biochem. Mol. Biol.* **26:** 767–773.

Johner, A., Stettler, P., Gruber, A. and Lanzrein, B. (1999) Presence of polydnavirus transcripts in an egg-larval parasitoid and its lepidopterous host. *J. Gen. Virol.* **80:** 1847–1854.

Johnson, B. (1959) Effect of parasitization by *Aphidius platensis* Bréthes on the developmental physiology of its host, *Aphis craccivora* Koch. *Entomol. Exp. Appl.* **2:** 82–89.

Johnson, J.H., Kral, R.M. Jr. and Krapcho, K. (1996) International Patent – WO 96/25429

Johnson, J.H., Kral, R.M. Jr. and Krapcho, K. (1999) United States Patent No. 5,874,298

Joiner, R.L., Vinson, S.B. and Benskin, J.B. (1973) Teratocytes as a source of juvenile hormone activity in a parasitoid - host Relationship. *Nature New Biology* **246:** 120–121.

Jones, D. (1985a) Parasite regulation of host insect metamorphosis: A new form of regulation of pseudoparasitized larvae of *Trichoplusia ni*. *J. Comp. Physiol.* B **155:** 583–590.

Jones, D. (1985b) The endocrine basis for developmentally stationary prepupae in larvae of *Trichoplusia ni* pseudoparasitized by *Chelonus insularis*. *J. Comp. Physiol. B* **155:** 235–240.

Jones, D. (1986) *Chelonus* sp.: suppression of host ecdysteroids and developmentally stationary pseudoparasitized prepupae. *Expl. Parasitol.* **61:** 471–475.

Jones, D. (1987) Material from adult female *Chelonus* sp. directs expression of altered developmental programme of host Lepidoptera. *J. Insect Physiol.* **33:** 129–134.

Jones, D. (1996) Biochemical interaction between chelonine wasps and their lepidopteran hosts: after a decade of research – the parasite is in control. *Insect Biochem. Mol. Biol.* **26:** 981–996.

Jones, D. (1998) The neglected saliva: medically important toxins in the saliva of human lice. *Parasitology* **116:** S73–S81.

Jones, D. and Coudron, T.A. (1993) Venoms of parasitic hymenoptera as investigatory tools. In: *Parasites and Pathogens*, Vol.1 (eds Beckage, N.E., Thompson, S.N. and Federici, B.A.), pp. 227–244. Academic Press, New York.

Jones, D. and Leluk, J. (1990) Venom proteins of the endoparasitic wasp *Chelonus* sp. near *curvimaculatus*: characterization of the major components. *Arch. Insect Biochem. Physiol.* **6:** 387–394.

Jones, D. and Pollack, R. (2001) Human lice. In: *Encyclopedia of Pest Management* (ed. Pimentel, D.). Marcel Dekker, New York.

Jones, D. and Wache, S. (1998) Preultimate 4th/5th instar *Trichoplusia ni* naturally-injected with venom/calyx fluid from *Chelonus curvimaculatus* precociously metamorphose, rather than obey the metamorphic size threshold that would normally compel molting to a 5th/6th instar. *J. Insect Physiol.* **44:** 755–765.

Jones, D., Jones, G. and Hammock, B.D. (1981a) Developmental and behavioral responses of larval *Trichoplusia ni* to parasitization by an imported braconid parasite *Chelonus* sp. *Physiol. Ent.* **6**: 387–394.

Jones, D., Jones, G. and Hammock, B.D. (1981b) Growth parameters associated with endocrine events in larval *Trichoplusia ni* (Hubner) and timing of these events with developmental markers. *J. Insect Physiol.* **27**: 779–788.

Jones, D., Jones, G., Rudnicka, M., Click, A., Reck-Malleczewen, V. and Iwaya, M. (1986a) Pseudoparasitism of host *Trichoplusia ni* by *Chelonus* spp. as a new model system for parasite regulation of host physiology. *J. Insect Physiol.* **32**: 315–328.

Jones, D., Sreekrishna, S., Iwaya, M., Yang, J.-N. and Eberely, M. (1986b) Comparison of viral ultrastructure and DNA banding patterns from the reproductive tracts of eastern and western hemisphere *Chelonus* spp. (Braconidae: Hymenoptera). *J. Invert. Pathol.* **47**: 105–115.

Jones, D., Taylor, T., Farkas, R., Chelliah, J., Haene, B., Brown, J. and Reed-Larsen, D. (1990a) Intercession of parasitic wasps (Cheloninae) in host developmental and biochemical pathways. In: *Advances in Invertebrate Reproduction 5* (eds Hoshni, M. and Yamashita, O.), pp. 157–162. Elsevier Science Publishers, Amsterdam.

Jones, D., Sawicki, G. and Wozniak, M. (1992a) Sequence, structure, and expression of a wasp venom protein with a negatively charged signal peptide and a novel repeating internal structure. *J. Biol. Chem.* **267**: 14871–14878.

Jones, D., Gelman, D. and Loeb, M. (1992b) Hemolymph concentrations of host ecdysteroids are strongly suppressed in precocious prepupae of *Trichoplusia ni* parasitized and pseudoparasitized by *Chelonus* near *curvimaculatus*. *Arch. Insect Biochem. Physiol.* **21**: 155–165.

Jones, D., Krishnan, A., Sarkari, N. and Wozniak, M. (1994) Isomeric and quaternary properties of homogenous 33 kDa protein from the venom of *Chelonus* near *curvimaculatus*. *Arch. Insect Biochem. Physiol.* **26**: 83–95.

Jones, D., Turner, H. and Chalkier, V. (1996) Altered hexamerin regulation in pre-pupal *Trichoplusia ni* pseudoparasitized by *Chelonus* sp. near *curvimaculatus*. *Arch. Insect Biochem. Physiol.* **32**: 537–548.

Jones, G. and Sarkari, N. (1993) Basic isoforms of hemolymph storage proteins expressed during larval metamorphosis. *Arch. Insect Biochem. Physiol.* **24**: 55–64.

Jones, G., Hiremath, S.T., Hellmann, G.M. and Rhoads, R.E. (1988a) Juvenile hormone regulation of mRNA levels for a highly abundant hemolymph protein in larval *Trichoplusia ni*. *J. Biol. Chem.* **263**: 1089–1092.

Jones, G., Ryan, R.O., Haunerland, N.H. and Law, J.H. (1988b) Purification and characterization of a very high density chromolipoprotein from the hemolymph of *Trichoplusia ni* (Hubner). Arch. Insect Biochem Physiol. **7**: 1–11.

Jones, G., Hanzlik, T., Hammock, B.D., Schooley, D.A., Miller, C.A., Tsai, L.W. and Baker, F.C. (1990b) The juvenile hormone titre during the penultimate and ultimate larval stadia of *Trichoplusia ni*. *J. Insect Physiol.* **36**: 77–83.

Jones, G., Manczak, M. and Horn, M. (1993) Hormonal regulation and properties of a new group of basic hemolymph proteins expressed during insect metamorphosis. *J. Biol. Chem.* **268**: 1284–1291.

Jones, G.L. and Yin, C.-M. (1989) Juvenile hormone biosynthesis by corpus cardiacum-corpus allatum complexes of larval *Lymantria dispar*. *Comp. Biochem. Physiol.* **92A**: 9–14.

Jones, P. and Sands, D.P.A. (1999) *Euplectrus melanocephalus* Girault (Hymenoptera: Eulophidae), an ectoparasitoid of larvae of fruit-piercing moths (Lepidoptera: Noctuidae: Catocalinae) from northern Queensland. *Australian J. Entomol.* **38**: 377–381.

Jones, R.L. and Lewis, W.J. (1971) Physiology of the host-parasite relationship between *Heliothis zea* and *Microplitis croceipes*. *J. Insect Physiol.* **17**: 921–927.

Junnikkala, E. (1985) Testis development in *Pieris brassicae* parasitized by *Apanteles glomeratus*. *Entomol. Expl. Appl.* **37**: 283–288.

Kaatz, H. (1994) Regulation of juvenile hormone biosynthesis in the honey bee. *Apidologie* **25:** 469–470.

Kadono-Okuda, K., Sakurai, H., Takeda, S. and Okuda, T. (1995) Synchronous growth of a parasitoid, *Perilitus coccinellae* and the teratocytes with the development of the host *Coccinella septempunctata*. *Entomol. Exp. Appl.* **75:** 145–149.

Kadono-Okuda, K., Weyda, F. and Okuda, T. (1998) *Dinocampus* (=*Perilitus*) *coccinellae* teratocyte-specific polypeptide: its accumulative property, localization and characterization. *J. Insect Physiol.* **44:** 1073–1080.

Kallenborn, H.G. and Mosbacher, G.C. (1983) The ecdysteroid titres during the last larval instar of *Ephestia kuehniella* Z. (Lepidoptera: Pyralidae). *J. Insect Physiol.* **29:** 749–753.

Kanost, M.R., Kawooya, J.K., Law, J.H., Ryan, R.O., Van Heusden, M.C. and Ziegler, R. (1990) Insect haemolymph proteins. *Adv. Insect Physiol.* **22:** 299–396.

Kapitskya, M.Z., Li., C., Miura, K., Segraves, W. and Raikhel, A.S. (2000) Expression of the early-late gene encoding the nuclear receptor HR3 suggests its involvement in regulating the vitellogenic response to ecdysone in the adult mosquito. *Mol. Cell. Endocrinol.* **160:** 25–37.

Kasparyan, D.R. (1981) *Fauna of the USSR. Hymenoptera 3(1). Ichneumonidae (Subfamily Tryphoninae) Tribe Tryphonini.* Oxonian Press, New Dehli.

Kataoka, H., Toschi, A., Li, J.P., Carney, R.L., Schooley, D.A. and Kramer, S.J. (1989) Identification of an allatotropin from adult *Manduca sexta. Science* **243:** 1481–1483.

Kataoka, H., Nagasawa, H., Isogai, A., Ishizaki, H. and Suzuki, A. (1991) Prothoracicotropic hormone of the Silkworm, *Bombyx mori*: amino acid sequence and dimeric structure. *Agric. Biol. Chem.* **55:** 73–86.

Keeley, L.L. and Vinson, S.B. (1975) Beta-ecdysone effects on development of nuclearpolyhedrosis virus in *Heliothis* spp. *J. Invert. Pathol.* **26:** 121–123.

Keightley, D.K., Lou, K.J. and Smith, W.A. (1990) Involvement of translation and transcription in insect steroidogenesis. *Mol. Cell Endocrinol.* **74:** 229–237.

Kelly, T.J. and Coudron, T.A. (1990) Total and specific ecdysteroids in the haemolymph of *Trichoplusia ni* (Lepidoptera Noctuidae) and its parasite, *Euplectrus plathypenae* (Hymenoptera: Eulophidae). *J. Insect Physiol.* **36:** 463–470.

Kelly, T.J. and Coudron, T.A. (1996) Ecdysteroid glycosylation discovered in wasp-parasitized (*Euplectrus* spp.) Cabbage looper *Trichoplusia ni* larvae: similarity to moult inhibition in baculovirus-infected insects. XII Ecdysone workshop abstracts, Barcelona.

Kelly, T.J., Park, E.J., Masler, C.A. and Burand, J.P. (1995) Characterization of the glycosylated ecdysteroids in the hemolymph of baculovirus-infected gypsy moth larvae and cells in culture. *Eur. J. Entomol.* **92:** 51–61.

Kelly, T.J., Kingan, T.G., Masler, C.A. and Robinson, C.H. (1996) Analysis of the ecdysiotropic activity in larval brains of the tobacco hornworm, *Manduca sexta. J. Insect Physiol.* **42:** 873–880.

Kelly, T.J., Gelman, D.B., Reed, D.A. and Beckage, N.E. (1998) Effects of parasitization by *Cotesia congregata* on the brain-prothoracic gland axis of its host, *Manduca sexta. J. Insect Physiol.* **44:** 323–332.

Kemeny, D.M., Dalton, N., Lawrence, A.J., Pearce, F.L. and Vernon, C.A. (1984) The purification and characterisation of hyaluronidase from the venom of the honey bee. *Eur. J. Biochem.* **139:** 217–223.

Kemp, D.H., Hales, J.R., Schleger, A.V. and Fawcett, A.A. (1983) Comparison of cutaneous hyperemia in cattle elicited by larvae of *Boophilus microplus* and by prostaglandins and other mediators. *Experientia* **39:** 725–727.

Kennedy, J.S. and Stroyan, H. (1959) Biology of aphids. *Annu. Rev. Ent.* **4:** 139–160.

Keymer, A.E. (1980) The influence of *Hymenolepis diminuta* on the survival and fecundity of the intermediate host, *Tribolium confusum. Parasitology* **81:** 405–421.

Keymer, A.E. (1981) Population dynamics of *Hymenolepis diminuta* in the intermediate host. *J. Anim. Ecol.* **50:** 941–950.

Kilincer, N. (1975) Untersuchungen uber die hamacytare Abwerhrreaktion der Puppe von *Galleria mellonella* L. (Lepidoptera) und uber ihre Hemmung durch den Puppenparasiten *Pimpla turionellae* L. *Z. Angew. Ent.* **78:** 340–370.

Kimura, S. (1974) Relationship between hormone titres and RNA and protein synthesis when the change to the pupal programme occurs in the silkworm *Bombyx mori. J. Insect Physiol.* **20:** 887–895.

Kingan, T.G., Gray, S.T., Zitnan, D. and Adams, M.E. (1997a) Regulation of ecdysis-triggering hormone release by eclosion hormone. *J. Exp. Biol.* **200:** 3245–3256.

Kingan, T.G., Zitnan, D., Jaffe, H., and Beckage, N.E. (1997b) Identification of neuropeptides in the midgut of parasitized insects: FLRFamides as candidate paracrines. *Mol. Cell. Endocrinol.* **133:**19–32.

Kiriishi, S., Rountree, D.B., Sakurai, S. and Gilbert, L.I. (1990) Prothoracic gland synthesis of 3-dehydroecdysone and its haemolymph 3β-reductase mediated conversion to ecdysone in representative insects. *Experientia* **46:** 716–721.

Kitano, H. (1982) Effect of the venom of the gregarious parasitoid, *Apanteles glomeratus* on its hemocytic encapsulation by the host, *Pieris. J. Invert. Pathol.* **40:** 61–67.

Kitano, H. (1986) The role of *Apanteles glomeratus* venom in the defensive response of its host, *Pieris rapae crucivora. J. Insect Physiol.* **32:** 369–375.

Kitano, H., Wago, H. and Arakawa, T. (1990) Possible role of teratocytes of the gregarious parasitoid, *Cotesia* (=*Apanteles*) *glomerata* in the suppression of phenoloxidase activity in the larval host, *Pieris rapae crucivora. Arch. Insect Biochem. Physiol.* **13:** 177–185.

Koch, P.B. (1996) Preadult changes of ecdysteroid and juvenile hormone titers in relation to diapause and pigmental variations in two lepidopteran species, *Cerura vinula* and *Araschnia levana* (Lepidoptera: Notodontidae/ Nymphalidae). *Entomol. Gen.* **20:** 143–155.

Kohno, M. and Takaoka, I. (1977) Effects of synthetic juvenile hormones through oral administration on *Myzus persicae* (Sulzer). *Kontyu Tokyo* **45:** 132–136.

Kramer, K.J., Copruz, L., Choi, H.K. and Muthukrishnan, S. (1993) Sequence of a cDNA and expression of the gene encoding epidermal and gut chitinases of *Manduca sexta. Insect Biochem. Mol. Biol.* **23:** 691–701.

Kramer, S.J., Toschi, A., Miller, C. A., Kataoka, H., Quistad, G.B., Li, J.P., Carney, R.L. and Schooley, D.A. (1991) Identification of an allatostatin from the Tobacco hornworm, *Manduca sexta. Proc. Natl Acad. Sci. USA* **88:** 9458–9462.

Krell, P.J. (1991a) The polydnaviruses: multipartite DNA viruses from parasitic Hymenoptera. In: *Viruses of Invertebrates* (ed. Kurstak, E.), pp. 141–177. Marcel Dekker, Inc., New York.

Krell, P.J. (1991b) Polynaviridae. In: *Atlas of Invertebrate Viruses* (eds Adams, J.R and Bonami, J.R.), pp. 321–338. CRC Press, Boca Raton, FL.

Kring, T.J. and Kring, J.B. (1988) Aphid fecundity, reproductive longevity and parasite development in the *Schizaphis graminum* (Rondani) (Homoptera: Aphidiidae) – *Lysiphlebus testaceipes* (Cresson) (Hymenoptera: Braconidae) system. *Can. Entomol.* **120:** 1079–1083.

Krishnan, A., Nair, P.N. and Jones, D. (1994) Isolation, cloning, and characterisation of new chitinase stored in active form in chitin-lined venom reservoir. *J. Biol. Chem.* **269:** 20971–20976.

Krombein, K.V., Hurd, P.D. Jr., Smith, D.R. and Burks, B.D. (1979) *Catalogue of Hymenoptera in America North of Mexico*, Vol. I. Smithsonian Institution Press, Washington D.C.

Kubo, L. and Klocke, J.A. (1982) Azadirachtin, insect ecdysis inhibitor. *Agric. Biol. Chem.* **46:** 1951–1953.

Kubo, I., Klocke, J.A. and Asano, S. (1983) Effects of ingested phytoecdysteroids on the growth and development of two lepidopterous larvae. *J. Insect Physiol.* **29:** 307–316.

Kumar, S. and Miller, L.K. (1987) Effects of serial passage of *Autographa californica* nuclear polyhedrosis virus in cell culture. *Virus Res.* **7:** 335–350.

Kumaran, A.K., Ray, A., Tertakian, J.A. and Memmel, N.A. (1987) Effects of juvenile hormone, ecdysteroids and nutrition on larval hemolymph protein expression in *Galleria mellonella. Insect Biochem.* **17:** 1053–1058.

Kunkel, J.G. and Nordin, J.H. (1985) Yolk protein. In: *Comprehensive Insect Physiology Biochemistry and Pharmacology*, Vol. 1, (eds Kerkut, G.A. and Gilbert, L.I.), pp. 83–112. Pergamon Press, New York.

Kunkel, J.G., Grossniklaus-Bürgin, C., Karpells, S.T. and Lanzrein, B. (1990) Arylphorin of *Trichoplusia ni*: characterization and parasite-induced precocious increase in titer. *Arch. Insect Biochem. Physiol.* **13:** 117–125.

Lafont, R. and Connat, J.-L. (1989) Pathways of ecdysone metabolism. In: *Ecdysone: from Chemistry to Mode of Action* (ed. Koolman, J.), pp. 167–173. Georg-Thieme Verlag, Stuttgart.

Lafont, R., Mauchamp, B., Pennetier, J.L., Tarroux, P., De Hys, J. and Delbeque, P. (1975) α and β-ecdysone levels in insect haemolymph: correlation with developmental events. *Experientia* **31:** 1241–1242.

Lafont, R., Mauchamp, B., Blais, C. and Pennetier, J.-L. (1977) Ecdysones and imaginal disc development during the last larval instar of *Pieris brassicae*. *J. Insect Physiol.* **23:** 277–283.

Lagueux, M., Perron, J.-M. and Hoffmann, J.A. (1976) Ecdysone metabolism and endogenous moulting hormone titre during larval-pupal development in *Choristoneura fumiferana*. *J. Insect Physiol.*, **22:** 57–62.

Lanzrein, B. and Hammock, B. (1995) Degradation of juvenile hormone III *in vitro* by nonparasitized and parasitized *Spodoptera exigua* (Noctuidae) and by the endoparasitoid *Chelonus inanitus* (Braconidae). *J. Insect Physiol.* **41:** 993–1000.

Larkin, S.W., Fraser, L., Showell, H.J., Williams, T.J. and Warren, J.B. (1995) Prolonged microvascular vasodilation induced by leukotriene B4 in human skin is cyclooxygenase independent. *Pharmacol. Exp. Ther.* **272:** 392–398.

Lathrop, F.H. and Newton, R.C. (1933) The biology of *Opius melleus* Gahan. A parasite of the blueberry maggot. *J. Agric. Res.* **46:** 143–160.

Lavine, M.D. and Beckage, N.E. (1995) Polydnaviruses: potent mediators of host insect immune dysfunction. *Parasitol. Today* **11:** 368–378.

Lavine, M.D. and Beckage, N.E. (1996) Temporal pattern of parasitism-induced immunosuppression in *Manduca sexta* larvae parasitized by *Cotesia congregata*. *J. Insect Physiol.* **42:** 41–51.

Lawrence, P.O. (1982) *Biosteres longicaudatus*: developmental dependence on host *Anastrephas suspensa* physiology. *Exp. Parasitol.* **53:** 396–405.

Lawrence, P.O. (1986a) Host-parasite hormonal interactions: an overview. *J. Insect Physiol.* **32:** 295–298.

Lawrence, P.O. (1986b) The role of 20-hydroxyecdysone in the moulting of *Biosteres longicaudatus*, a parasite of the Caribbean fruit fly, *Anastrepha suspensa*. *J. Insect Physiol.* **32:** 329–337.

Lawrence, P.O. (1988) *In vivo* and *in vitro* development of first instars of the parasitic wasp, *Biosteres longicaudatus* (Hymenoptera: Braconidae). *Adv. Parasitic Hymen. Res.* **1:** 351–366.

Lawrence, P.O. (1990) The biochemical and physiological effects of insect hosts on the development and ecology of their insect parasites: an overview. *Arch. Insect Biochem. Physiol.* **13:** 217–228.

Lawrence, P.O. and Hagedorn, H.H. (1986) Relationship between the ecdysteroid titres of a host and those of its parasite. *Insect Biochem.* **16:** 163-167.

Lawrence, P.O. and Lanzrein, B. (1993) Hormonal interactions between insect endoparasites and their host insects. In: *Parasites and Pathogens of Insects, Vol. 1* (eds Beckage, N.E., Thompson, S.N. and Federici, B.A.), pp. 59–86. Academic Press, New York.

Lawrence, P.O., Baranowski, R.M. and Greany, P.D. (1976) Effect of host age on development of *Biosteres (=Opius) longicaudatus)*, a parasitoid of the Caribbean fruit fly, *Anastrepha suspensa*. *Fla. Entomol.* **59:** 33–39.

Lawrence, P.O., Baker, F.C., Tsai, L.W., Miller, C.A., Schooley, D.A. and Geddes, L.G. (1990) JH III levels in larvae and pharate pupae of *Anastrepha suspensa* (Diptera: Tephritidae) and in larvae of the parasitic wasp *Biosteres longicaudatus* (Hymenoptera: Braconidae). *Arch. Insect. Biochem. Physiol.* **13:** 53–62.

Leclerc, R.F. and Miller, S.G. (1990) Identification and molecular analysis of storage proteins from *Heliothis virescens*. *Arch. Insect Biochem. Physiol.* **14:** 131–150.

Lees, A.D. (1966) The control of polymorphism in aphids. *Adv. Insect Physiol.* **3:** 207–277.

Lees, A.D. (1980) The development of juvenile hormone sensitivity in alatae of the aphid *Megoura viciae*. *J. Insect Physiol.* **26:** 143–151.

Leluk, J. and Jones, D. (1989) *Chelonus* sp. near *curvimaculatus* venom proteins: Analysis of their potential role and processing during development of host *Trichoplusia ni*. *Arch. Insect Biochem. Physiol.* **10:** 1–12.

Leluk, J., Schmidt, J. and Jones, D. (1988) Characterization and functions of the proteins of Hymenopteran venoms. In: *Endocrinological Frontiers in Physiological Insect Ecology* (eds Sehnal, F., Zabza, A. and Denlinger, D.L), pp. 457–460. Wroclaw Technical University Press, Wroclaw, Poland.

Leluk, J., Schmidt, J. and Jones, D. (1989) Comparative studies on the protein composition of hymenopteran venom reservoirs. *Toxicon* **27:** 105–114.

Le Ralec, A. (1995) Egg contents in relation to host-feeding in some parasitic Hymenoptera. *Entomophaga* **40:** 87–93.

Lerner, E.A. and Shoemaker, C. (1992) Maxadilan, cloning and functional expression of the gene encoding this potent vasodilator. *J. Biol. Chem.* **267:** 1062–1066.

Lerner, E.A., Ribeiro, J.M.C., Nelson, R.J. and Lerner, M.R. (1991) Isolation of maxadilan, a potent vasodilatory peptide from the salivary glands of the sandfly *Lutzomyia longipalpis*. *J. Biol. Chem.* **266:** 11234–11236.

Levenbook, L. (1985) Insect Storage Proteins. In: *Comprehensive Insect Physiology Biochemistry and Pharmacology*, Vol. 10 (eds Kerkut, G.A. and Gilbert, L.I.), pp. 307–346. Pergamon Press, Oxford.

Levine, R.B. and Truman, J.W. (1983) Peptide activation of a simple neural circuit. *Brain Res.* **279:** 335–338.

Lewis, W.J. and Brazzel, J.R. (1966) Biological relationships between *Cardiochiles nigriceps* and the *Heliothis* complex. *J. Econ. Entomol.* **59:** 820–823.

Lewis, W.J. and Vinson, S.B. (1968) Egg and larval development of *Cardiochiles nigriceps*. *Ann. Entomol. Soc. Am.* **61:** 561–565.

Li, X. and Webb, B.A. (1994) Apparent functional role for a cysteine-rich polydnavirus protein in suppression of the insect cellular immune response. *J. Virol.* **68:** 7482–7489.

Liu, S.S. (1985) Development, adult size and fecundity of *Aphidius sonchi* reared in two instars of its aphid host, *Hyperomyzus latucae*. *Entomol. Exp. Appl.* **37:** 41–48.

Liu, S.-S. and Hughes, R.D. (1984) Effect of host age at parasitization by *Aphidius sonchi* on the development, survival and reproduction of the sowthistle aphid, *Hyperomyzus lactucae*. *Entomol. Exp. Appl.* **36:** 238–246.

Locke, M. (1980) The cell biology of fat body development. In: *Insect Biology in the Future* (eds Locke, M. and Smith, D.S.), pp. 227–252. Academic Press, New York.

Locke, M. (1985) A structural analysis of postembryonic development. In: *Comprehensive Insect Physiology, Biochemistry and Pharmacology*, Vol. 2 (eds Kerkut, G.A. and Gilbert, L.I.), pp. 87–149. Pergamon Press, Oxford.

Lockshin, R.A. (1985) Programed cell death. In: *Comprehensive Insect Physiology Biochemistry and Pharmacology*, Vol. 2 (eds Kerkut, G.A. and Gilbert, L.I.), pp. 301–317. Pergamon Press, Oxford.

Loeb, M.J., Bell, R.A., Gelman, D.B., Kochansky, J.P., Lusby, W.R. and Wagner, R.M. (1996) Action cascade of an insect gonadotropin, testis ecdysiotropin, in male Lepidoptera. *Invert. Reprod. Devel.* **30:** 181–190.

Lonard, D.M., Bhaskaran, G. and Dahm, K.H. (1996) Control of prothoracic gland activity by juvenile hormone in fourth instar *Manduca sexta* larvae. *J. Insect Physiol.* **42:** 205–213.

Loughton, B.G. (1965) The effect of parasitization by *Pimpla turionellae* (L.) on the emergence from the pupal case of its host, *Galleria mellonella* (L.). *Can. Ent.* **97:** 314–316.

Lovallo, N., Barratt, B.I.P., Legge, M. and Cox-Foster, D.L. (2000) Effects of *Microtonus aethiopoides* parasitism on hemolymph protein composition across alternative hosts. *Entomol. Exp. Appl.* **95:** 213–216.

Luckhart, S., Vodovotz, Y., Cui, L. and Rosenberg, R. (1998) The mosquito *Anopheles stephensi* limits malaria parasite development with inducible synthesis of nitric oxide. *Proc. Natl Acad. Sci. USA* **95:** 5700–5705.

Luft, J.H. (1973) Embedding media old and new. In: *Advanced Techniques in Biological Electron Microscopy* Vol. 1 (ed. Koehler, J.), pp. 1–34. Springer Verlag, Berlin.

Mackauer, M. (1986) Growth and developmental interactions in some aphids and their hymenopteran parasites. *J. Insect Physiol.* **32:** 275–280.

Mackauer, M., Nair, K.K. and Unnithan, G.C. (1981) Effects of precocene II on alate production in the pea aphid, *Acyrthosiphon pisum*. *Can. J. Zool.* **57:** 856–859.

Mackauer, M. and Sequeira, R. (1993) Patterns of development in insect parasites. In: *Parasites and Pathogens of Insects,* Vol. 1 (eds Beckage, N.E., Thompson, S.N. and Federici, B.A.), pp. 1–23. Academic Press, New York.

Mackenzie, P. (1990) Expression of chimeric cDNAs in cell culture defines a region of UDP glucuronosyltransferase involved in substrate selection. *J. Biol. Chem.* **265:** 3432–3435.

Maema, M. (1986) Experimental infection of *Tribolium confusum* (Coleoptera) by *Hymenolepis diminuta* (Cestoda): Host fecundity during infection. *Parasitology* **92:** 405–412.

Major, M., Webb, T.J. and Hurd, H. (1997) Haemolymph from female beetles infected with *Hymenolepis diminuta* metacestodes retards the development of ovarian follicles in recipient *Tenebrio molitor* (Coleoptera). *Parasitology* **114:** 175–181.

Marris, G.C. and Edwards, J.P. (1995) The biology of the ectoparasitoid wasp *Eulophus pennicornis* (Hymenoptera: Eulophidae) on host larvae of the tomato moth, *Lacanobia oleracea* (Lepidoptera: Noctuidae). *Bull. Ent. Res.* **85:** 507–513.

Marris, G.C., Weaver, R.J., Olieff, S.M. and Mosson, H.J. (1996) Ectoparasitoid venom as a regulator of lepidopteran host development and moulting. *Proc. Brighton Conf. Crop Protection (Pests and Diseases 1996)* **3:** 1021–1028.

Marris, G.C., Weaver, R.J. and Edwards, J.P. (1999) The role of parasitoid venom in the disruption of noctuid larval development. *Proc. Crop Protection in Northern Britain* **10:** 163–168.

Marris, G.C., Bell, H.A., Naylor J. and Edwards, J.P. (2000) The role of *Pimpla hypochondriaca's* venom in the incidence of microbial infection in a pupal noctuid host. *Entomol. Exp. Appl.* **93:** 291–298.

Marris, G.C., Weaver, R.J., Bell, J. and Edwards, J.P. (2001) Venom from the ectoparasitic wasp *Eulophus pennicornis* disrupts host ecdysteroid production by regulating host prothoracic gland activity. *Physiol. Entomol* (in press).

Martin, D., Wang, S. and Raikhel, A.S. (2001) The vitellogenin gene of the mosquito *Aedes aegypti* is a direct target of ecdysteroid receptor. *Mol. Cell. Endocrinol.* **173:** 75–86.

Martínez, T., Burmester, T., Veenstra, J.A. and Wheeler, D. (2000) Sequence and evolution of a hexamerin from the ant *Camponotus festinatus*. *Insect Mol. Biol.* **9:** 427–431.

Masler, E.P. and Kovaleva, E.S. (1999) Inhibition of larval growth in the gypsy moth (Lepidoptera: Lymantriidae) by venom from the parasitic wasp *Microbracon hebetor* (Hymenoptera: Braconidae). *J. Entomol. Sci.* **34:** 435–444.

Mason, W.R.M. (1967) Specialization in the egg structure of *Exentrus* (Hymenoptera: Ichneumonidae) in relation to distribution and abundance. *Can. Ent.* **99:** 383–393.

Mauchamp, B., Lafont, R. and Jourdain, D. (1979) Mass fragmentographic analysis of juvenile hormone I levels during the last larval instar of *Pieris brassicae*. *J. Insect Physiol.* **25:** 545–550.

Melin, B.E. (1978) The developmental biology of *Euplectrus plathypenae* Howard (Hymenoptera: Eulophidae), a parasitoid of *Pseudaletia unipuncta* (Haw.) (Lepidoptera: Noctuidae), and the effects of a microsporidian disease of the host on the parasitoid. *Dissertation Abstracts International* **39B(5):** 2124.

Mertins, J.W. (1985) Hyperparasitoids from pea aphid mummies, *Acyrthosiphon pisum* (Homoptera: Aphididae). *Ann. Entomol. Soc. Am.* **78:** 186–197.

Meyer-Fernandes, J.R., Gondim, K.C. and Wells, M.A. (2000) Developmental changes in the response of larval *Manduca sexta* fat body glycogen phosphorylase to starvation, stress, and octopamine. *Insect Biochem. Mol. Biol.* **30:** 415–422.

Miles, C.I. and Booker, R. (2000) Octopamine mimics the effects of parasitism on the foregut of the tobacco hornworm *Manduca sexta. J. Exp. Biol.* **203:** 1689–1700.

Miller, L.K. (1995) Genetically engineered insect virus pesticides: present and future. *J. Invert. Pathol.* **65:** 211–216.

Miller, L.K. (1997) *The Baculoviruses.* Plenum Press, New York.

Moore, J. (1984) Parasites that change the behavior of their host. *Sci. Am.* **250:** 108–115.

Moore, J. (1993) Parasites and the behavior of biting flies. *J. Parasitol.* **79:** 1–26.

Moore, J. (1995) The behavior of parasitized animals. *Bioscience* **45:** 89–96.

Moore, J. and Gotelli, N.J. (1990) A phylogenetic perspective on the evolution of altered host behaviour: a critical look at the manipulation hypothesis. In: *Parasitism and Host Behaviour* (eds Barnard, C.J. and Behnke, J.M.), pp. 193–233. Taylor and Francis, London.

Moore, W. (1918) An interesting reaction to louse bites. *J. Amer. Med. Assoc.* **71:** 1481–1482.

Morales-Ramos, J.A. and Cate, J.R (1992) Laboratory determination of age-dependent fecundity, development, and rate of increase of *Catolaccus grandis* (Burks) (Hymenoptera: Pteromalidae). *Ann. Entomol. Soc. Am.* **85:** 469–476.

Morales-Ramos, J.A., Rojas, M.G. and King, E.G. (1995) Venom of *Catolaccus grandis* (Hymenoptera: Pteromalidae) and its role in parasitoid development and host regulation. *Ann. Entomol. Soc. Am.* **88:** 800–808.

Moran, N. (1992) The evolution of aphid life cycles. *Annu. Rev. Ent.* **37:** 321–348.

Mordue, W. (1965) Studies on oocyte production and associated histological changes in the neuro-endocrine system of *Tenebrio molitor. J. Insect Physiol.* **11:** 493–503

Moreno, J. and Jimenez, R. (1993) Parasitization of *Ephestia kuehniella* Zeller (Lep: Pyralidae) by *Phanerotoma* (*Phanerotoma*) *ocularis* Kohl (Hym: Braconidae) – Parasitism, superparasitism and emergency rates. *J. Appl. Entomol.* **115:** 273–276.

Morrissey, J. H. (1981) Silver stain for proteins in polyacrylamide gels: a modified procedure with enhanced uniform sensitivity. *Analyt. Biochem.* **117:** 307–310.

Morsy, T.A., Alalfy, M.S., Sabry, Ah. H., Fikry, A.A. and El Sharkawy, I.M. (1996) Abnormal distribution of the histocompatibility (HLA) in lousy patients. *J. Egypt. Soc. Parasitol.* **26:** 227–235.

Moscardi, F. (1999) Assessment of the application of baculoviruses for control of Lepidoptera. *Annu. Rev. Entomol.* **44:** 257–289.

Muesebeck, C.F.W., Krombein, K.V and Townes, H.K. (1951) *Hymenoptera of America North of Mexico. Synoptic Catalog.* USDA Agriculture Monograph, No. 2, 1420 pp.

Müller, C.B., Williams, I.S. and Hardie, J. (2001) The role of nutrition, crowding and interspecific interactions in wing development in aphids. *Ecol. Ent.* **26:** 330–340.

Mumcuoglu, K.Y., Hemingway, J., Miller, J., Ioffe-Uspensky, I., Klaus, S., Ben-Ishai, F. and Galun, R. (1995) Permethrin resistance in the head louse *Pediculus capitatus* from Israel. *Med. Vet. Entomol.* **9:** 427–432.

Mumcuoglu, K.Y., Galun, R., Kaminchik, Y., Panet, A. and Levanon, A. (1996a) Antihemostatic activity in salivary gands of the human body louse *Pediculus humanus* (Anoplura: Pediculidae). *J. Insect Physiol.* **42:** 1083–1087.

Mumcuoglu, K.Y., Ben-Yakir, E., Gunzberg, S. and Galun, R. (1996b) Immunogenic proteins in body and fecal material of the human body louse, *Pediculus humanus*, and their homology to antigens of other lice species. *Med. Vet. Entomol.* **10:** 105–107.

Munn, E.A., Feinstein, A. and Greville, G.D. (1967) A major protein constituent of pupae of the blowfly *Calliphora erythrocephala* (Diptera). *Biochem. J.* **102:** 5–6.

Murdoch, W.W., Nisbet, R.M., Luck, R.F., Godfray, H.C.J. and Gurney, W.S.C. (1992) Size-selective sex-allocation and host feeding in a parasitoid host model. *J. Anim. Ecol.* **61:** 533–541.

Nafus, D. (1991) Biological control of *Penicillaria jocosatrix* (Lepidoptera: Noctuidae) on mango on Guam with notes on the biology of its parasitoids. *Environ. Entomol.* **20:** 1725–1731.

Nakahara, Y., Hiraoka, T. and Iwabuchi, K. (2000) Growth-promoting effects of ecdysteroids and juvenile hormone on *in vitro* development of the larval endoparasitoid, *Venturia canescens* (Hymenoptera: Ichneumonidae). *J. Insect Physiol.* **46:** 467–476.

Nakajima, T. (1986) Pharmacological biochemistry of vespid venoms. In: *Venoms of the Hymenoptera* (ed Piek, T.), pp. 309–327. Academic Press, London.

Neckemeyer, W.S. (1996) Multiple roles for dopamine in *Drosophila* development. *Develop. Biol.* **176:** 209–219

Nelson, W.A., Bell, J.F., Clifford, C.M. and Keirans, J.E. (1977) Interaction of ectoparasites and their hosts. *J. Med. Entomol.* **13:** 389–428.

Neser, S. (1973) Biology and behaviour of *Euplectrus* species near *laphygmae* Ferrière (Hymenoptera: Eulophidae). *Entomology Memoir, Department of Technical Services,* No. 32. Government of the Republic of South Africa, Pretoria.

Nettles, W.C. Jr. (1990) *In vitro* rearing of parasitoids – role of host factors in nutrition. *Arch. Insect Biochem. Physiol.* **13:** 167–175.

Niimi, S. and Sakurai, S. (1997) Developmental changes in juvenile hormone and juvenile hormone acid titres in the hemolymph and *in vitro* juvenile hormone synthesis by corpora allata of the silkworm *Bombyx mori*. *J. Insect Physiol.* **43:** 875–884.

Nijhout, H.F. (1975) A threshold size for metamorphosis in the tobacco hornworm, *Manduca sexta* (L.). *Biol. Bull.* **149:** 214–225.

Nijhout, H.F. (1994) *Insect Hormones*. Princeton University Press, Princeton, NJ.

Nijhout, H.F. and Wheeler, D.E. (1982) Juvenile hormone and the physiological basis of insect polymorphisms. *Quart. Rev. Biol.* **57:** 109–133.

Nijhout, H.F. and Williams, C.M. (1974a) Control of moulting and metamorphosis in the tobacco hornworm, *Manduca sexta* (L.): growth of the last instar larva and decision to pupate. *J. Exp. Biol.* **61:** 481–491.

Nijhout, H.F. and Williams, C.M. (1974b) Control of moulting and metamorphosis in the tobacco hornworm, *Manduca sexta* (L.): cessation of juvenile hormone secretion as a trigger for pupation. *J. Exp. Biol.* **61**: 493-501.

Noble, N.S. (1936) *Euplectrus agaristae* Crawf., a parasite of the grape-vine moth, *Phalaenoides glycine* Lew. *J. Aust. Inst. Agric. Sci.* **2:** 165–168.

Noble, N.S. (1938) *Euplectrus agaristae* Crawf., a parasite of the grape-vine moth. *Dept. Agric. New South Wales,* **63B:** 5–27.

Noguchi, H., Hayakawa, Y. and Downer, R.G.H. (1995) Elevation of dopamine levels in parasitized insect larvae. *Insect Biochem. Mol. Biol.* **25:** 197–201.

Noguti, T., Adachi-Yamada, T., Katagiri, T., Kawakami, A., Iwami, M., Ishibashi, J., Kataoka, H., Suzuki, A., Go, M. and Ishizaki, H. (1995) Insect prothoracicotropic hormone: a new member of the vertebrate growth factor superfamily. *FEBS Letts* **376:** 251–256.

Nuttall, G.H.F. (1918) Biology of *Pediculus humanus*. *Parasitology* **10:** 1–185.

Oeh, U., Lorenz, M.W., Dyker, H., Lösel, P. and Hoffmann, K.H. (2000) Interaction between *Manduca sexta* allatotropin and *Manduca sexta* allatostatin in the fall armyworm *Spodoptera frugiperda*. *Insect Biochem. Mol. Biol.* **30:** 719–727.

Ohnishi, A., Hayakawa, Y., Matsuda, Y., Kwon, K.-W., Takahashi, T.A. and Sekiguchi, S. (1995) Growth-blocking peptide titer during larval development of parasitized and cold-stressed armyworm. *Insect Biochem. Mol. Biol.* **25:** 1121–1127.

Ohno, K., Yamaguchi, D., Maryana, N., Takesaki, K. and Takemoto, H. (1999) Reproductive efficiency of eulophid parasitoids (Hymenoptera: Eulophidae) attacking the larvae of *Liriomyza trifolii* (Diptera: Agromyzidae). *Japanese J. Entomol.* **2:** 1–9.

Okajima, A. and Watanabe, M. (1989) Electrophysiological identification of neuronal pathway to the prothoracic gland and the change in electrical activities of the prothoracic gland innervating neurons during larval development of a moth, *Mamestra brassicae*. *Zool. Sci.* **6:** 459–468.

Okuda, T. and Kadono-Okuda, K. (1995) *Perilitus coccinellae* teratocyte polypeptide: evidence for production of a teratocyte-specific 540 kDa protein. *J. Insect Physiol.* **41:** 819–825.

O'Reilly, D.R. (1995) Baculovirus-encoded ecdysteroid UDP-glucosyltransferases. *Insect Biochem. Mol. Biol.* **25:** 541–550.

O'Reilly, D.R. (1997) Auxiliary genes of baculoviruses. In: *The Baculoviruses* (ed Miller, L.K.), pp. 267–300. Plenum Press, New York.

O'Reilly, D.R. and Miller, L.K. (1989) A baculovirus blocks insect molting by producing ecdysteroid UDP-glucosyltransferase. *Science* **245:** 1110–1112.

O'Reilly, D.R. and Miller, L.K. (1990) Regulation of expression of a baculovirus ecdysteroid UDP-glucosyltransferase gene. *J. Virol.* **64:** 1321–1328.

O'Reilly, D.R. and Miller, L.K. (1991) Improvement of a baculovirus pesticide by deletion of the *egt* gene. *Biotechnology* **9:** 1086–1089.

O'Reilly, D.R., Passarelli, A.L., Goldman, I.F. and Miller, L.K. (1990) Characterization of the DA26 gene in a hypervariable region of the *Autographa californica* nuclear polyhedrosis virus genome. *J. Gen. Virol.* **71:** 1029–1037.

O'Reilly, D.R., Howarth, O.W., Rees, H.H. and Miller, L.K. (1991) Structure of the ecdysone glucoside formed by a baculovirus ecdysteroid UDP-glucosyltransferase. *Insect Biochem.* **21:** 795–801.

O'Reilly, D.R., Brown, M.R. and Miller, L.K. (1992) Alteration of ecdysteroid metabolism due to baculovirus infection of the fall armyworm, *Spodoptera frugiperda*: host ecdysteroids are conjugated with galactose. *Insect Biochem. Mol. Biol.* **22:** 313–320.

O'Reilly, D.R., Miller, L.K. and Luckow, V.A. (1994) *Baculovirus Expression Vectors: a laboratory manual*. Oxford University Press, Oxford.

O'Reilly, D.R., Kelly, T.J., Masler, E.P., Thyagaraja, B.S., Robson, R.M., Shaw, T.C. and Miller, L.K. (1995) Overexpression of *Bombyx mori* prothoracicotropic hormone using baculovirus vectors. *Insect Biochem. Mol. Biol.* **25:** 475–485.

O'Reilly, D.R., Hails, R.S. and Kelly, T.J. (1998) The impact of host developmental status on baculovirus replication. *J. Invert. Pathol.* **72:** 269–275.

Orme-Johnson, N.R. (1990) Distinctive properties of adrenal cortex mitochondria. *Biochim. Biophys. Acta* **1020:** 213–231.

Osman, S.E. and Führer, E. (1979) Histochemical analysis of accessory genital gland secretions in female *Pimpla turionellae* L. (Hymenoptera: Ichneumonidae). *Int. J. Invert. Reprod.* **1:** 323–332.

Pak, G.A. (1986) Behavioural variations among strains of *Trichogramma* spp.: A review of the literature on host-age selection. *J. Appl. Entomol.* **101:** 55–64.

Palli, S.R., Ladd, T.R., Tomkins, W.L., Shu, S., Ramaswamy, S.B., Tanaka, Y., Arif, B. and Retnakaran, A. (2000) *Choristoneura fumiferana* entomopoxvirus prevents metamorphosis and modulates juvenile hormone and ecdysteroid titres. *Insect Biochem. Mol. Biol.* **30:** 869–876.

Palumbo, J.C., Mullis, C.H. Jr. and Reyes, F.J. (1994) Composition, seasonal abundance, and parasitism of *Liriomyza* (Diptera: Agromyzidae) species on lettuce in Arizona. *J. Econ. Entomol.* **87:** 1070–1077.

Pan, M.L. and Telfer, W.H. (1999) Equivalence of riboflavin-binding hexamerin and arylphorin as reserves for adult development in two saturniid moths. *Arch. Insect Biochem. Physiol.* **42:** 138–146.

Park, E.J., Burand, J.P. and Yin, C.-M. (1993) The effect of baculovirus infection on ecdysteroid titer in gypsy moth larvae. *J. Insect Physiol.* **39:** 791–796.

Park, E.J., Yin, C.-M. and Burand, J.P. (1996) Baculovirus replication alters hormone-regulated host development. *J. Gen. Virol.* **77:** 547–554.

Parkinson, N.M. and Weaver, R.J. (1999) Noxious components of venom from the pupa-specific parasitoid *Pimpla hypochondriaca*. *J. Invert. Pathol.* **73:** 74–83.

Parkinson, N., Smith, I., Weaver, R. and Edwards, J.P. (2001) A new form of arthropod phenoloxidase is abundant in venom of the parasitoid wasp *Pimpla hypochondriaca*. *Insect Biochem. Mol. Biol.* **31:** 57–63.

Parkman, P. (1982) Searching ability and host selection by *Euplectrus plathypenae* Howard (Hymenoptera: Eulophidae) biological control agent for plant pests, particularly *Spodoptera* species. *J. Georgia Entomol. Soc.* **17:** 150–156.

Pawson, B.M., Gold, R.E., Wildey, K. B. and Robinson, W.H. (1993) Evaluating different release strategies for the control of American cockroaches using the egg parasitoid *Aprostocetus hagenowii*. In: *Proceedings of the 1st International Conference on Insect Pests in theUrban Environment* (eds Wildey, K.B. and Robinson, W.H.), pp. 407–413. International Conference on Insect Pests in the Urban Environment, Cambridge.

Peck, O. (1963) Catalogue of the Nearctic Chalcidoidea (Insecta: Hymenoptera). *Can. Entomol.* **30:** 1092.

Peck, S.M., Wright, W.H. and Gant, J.Q. (1943) Cutaneous reactions due to the body louse (*Pediculus humanus*). *J. Amer. Med. Assoc.* **123:** 821–825.

Pennacchio, F., Vinson, S.B. and Tremblay, E. (1992a) Host regulation effects of *Heliothis virescens* (F.) larvae induced by teratocytes of *Cardiochiles nigriceps* Viereck (Lepidoptera, Noctuidae-Hymenoptera, Braconidae). *Arch. Insect Biochem. Physiol.* **19:** 177–192.

Pennacchio, F., Vinson, S.B. and Tremblay, E. (1992b) Preliminary results on *in vitro* rearing of the endoparasitoid *Cardiochiles nigriceps* from egg to second instar. *Entomol. Exp. Appl.* **64:** 209–216.

Pennacchio, F., Vinson, S.B. and Tremblay, E. (1993) Growth and development of *Cardiochiles nigriceps* Viereck (Hymenoptera, Braconidae) larvae and their synchronization with some changes of the hemolymph composition of their host, *Heliothis virescens* (F.) (Lepidoptera, Noctuidae). *Arch. Insect Biochem. Physiol.* **24:** 65–67.

Pennacchio, F., Vinson, S.B., Tremblay, E. and Ostuni, A. (1994a) Alteration of ecdysone metabolism in *Heliothis virescens* (F.) (Lepidoptera, Noctuidae) larvae induced by *Cardiochiles nigriceps* Viereck (Hymenoptera, Braconidae) teratocytes. *Insect Biochem. Mol. Biol.* **24:** 383–394.

Pennacchio, F., Vinson, S.B., Tremblay, E. and Tanaka, T. (1994b) Biochemical and developmental alterations of *Heliothis virescens* (F.) (Lepidoptera, Noctuidae) larvae induced by the endophagous parasitoid *Cardiochiles nigriceps* Viereck (Hymenoptera, Braconidae). *Arch. Insect Biochem. Physiol.* **26:** 211–233.

Pennacchio, F., Vinson, S.B. and Tremblay, E. (1994c) Morphology and ultrastructure of the serosal cells (teratocytes) in *Cardiochiles nigriceps* Viereck (Hymenoptera: Braconidae) embryos. *Int. J. Insect Morphol. Embryol.* **23:** 93–104.

Pennacchio, F., Vinson, S.B. and Tremblay, E. (1994d) Regulation of *Heliothis virescens* (F.) development and hormonal metabolism by the endophagous parasitoid *Cardiochiles nigriceps* Viereck: the role of teratocytes. *Norw. J. Agric. Sci.* **16 (Suppl.):** 293–300.

Pennacchio, F., Digilio, M.C. and Tremblay, E. (1995) Biochemical and metabolic alterations in *Acyrthosiphon pisum* parasitized by *Aphidius ervi*. *Arch. Insect Biochem. Physiol.* **30:** 351–367.

Pennacchio, F., Sordetti, R., Falabella, P. and Vinson, S.B. (1997) Biochemical and ultrastructural alterations in prothoracic glands of *Heliothis virescens* (F.) (Lepidoptera: Noctuidae) last instar larvae parasitized by *Cardiochiles nigiceps* Viereck (Hymenoptera: Braconidae). *Insect Biochem. Mol. Biol.* **27:** 439–450.

Pennacchio, F., Falabella, P. and Vinson, S.B. (1998a) Regulation of *Heliothis virescens* prothoracic glands by *Cardiochiles nigriceps* polydnavirus. *Arch. Insect Biochem. Physiol.* **38:** 1–10.

Pennacchio, F., Falabella, P., Sordetti, R., Varricchio, P., Malva, C. and Vinson, S.B. (1998b) Prothoracic gland inactivation in *Heliothis virescens* (F.) (Lepidoptera: Noctuidae) larvae parasitized by *Cardiochiles nigriceps* Viereck (Hymenoptera: Braconidae). *J. Insect Physiol.* **44:** 845–857.

Pennacchio, F., Fanti, P., Falabella, P., Digilio, M.C., Bisaccia, F. and Tremblay, E. (1999) Development and nutrition of the braconid wasp *Aphidius ervi* in aposymbiotic host aphids. *Arch. Insect Biochem. Physiol.* **40:** 53–63.

Pennacchio, F., Falabella, P., Varricchio, P., Sordetti, R., Vinson, S.B., Graziani, F. and Malva, C. (2000) Molecular basis of developmental alteration in *Heliothis virescens* larvae parasitized by *Cardiochiles nigriceps*. In: *The Hymenoptera: Evolution, Biodiversity and Biological Control* (ed Austin, A.) pp. 17–22. CSIRO Publishing, Canberra.

Penrose, J.F. (1999) LTC4 synthase. Enzymology, biochemistry, and molecular characterization. *Clin Rev. Allergy Immunol.* **17:** 133–152.

Perez de Leon, A.A., Ribeiro, J.M., Tabachnick, W.J. and Valenzuela, J.G. (1997) Identification of a salivary vasodilator in the primary North American vector of bluetongue viruses, *Culicoides variipennis. Am. J. Trop. Med. Hyg.* **57:** 375–381.

Pfister-Wilhelm, R. and Lanzrein, B. (1996) Precocious induction of metamorphosis in *Spodoptera littoralis* (Noctuidae) by the parasitic wasp *Chelonus inanitus* (Braconidae): Identification of the parasitoid larva as the key regulatory element and the host corpora allata as the main targets. *Arch. Insect Biochem. Phyiol.* **32:** 511–525.

Piek, T. (1985) Insect venoms and toxins. In: *Comprehensive Insect Physiology, Biochemistry and Pharmacology*, Vol. 11 (ed. Kerkut, G.A. and Gilbert, L.I.), pp. 595–633. Pergamon Press, Oxford.

Piek, T. (1986) *Venoms of the Hymenoptera*. Academic Press, London.

Piek, T. (1990) Neurotoxins of the Hymenoptera – twenty-five years of research in Amsterdam. *Comp. Biochem. Physiol.* **96C:** 223–233.

Piek, T. (1993) New neurotoxins from venoms of Aculeate Hymenoptera: A contribution to the knowledge of stinging behaviour. *J. Hym. Res.* **2:** 101–105.

Piek, T. and Mantel, P. (1970) The effect of the venom of *Microbracon hebetor* (Say) on the hyperpolarizing potentials in a skeletal muscle of *Philosamia cynthia* Hübn. *Comp. Gen. Pharmacol.* **1:** 87–92.

Piek, T. and Spanjer, W. (1978) Effects and chemical characterization of some paralysing venoms of solitary wasps. In: *Pesticide and Venom Neurotoxicity* (eds Shankland, D.L., Hollingworth, R.M. and Smyth, T. Jr), pp. 211–226. Plenum Press, New York.

Piek, T. and Simon Thomas, T.T. (1969) Paralysing venoms of solitary wasps. Review article. *Comp. Biochem. Physiol.* **30:** 13–31.

Piek, T. and Spanjer, W. (1986) Chemistry and pharmacology of solitary wasp venoms. In *Venoms of the Hymenoptera. Biochemical, Pharmacological and Behavioural Aspects* (ed Piek, T.), pp. 161–307. Academic Press, London.

Piek, T., Spanjer, W., Njio, K.D., Veenendaal, R.L. and Mantel, P. (1974) Paralysis caused by the venom of the wasp, *Microbracon gelechiae. J. Insect Physiol.* **20:** 2307–2319.

Piek, T., Visser, J.H. and Veenendaal, R.L. (1984) Change in behaviour of the cockroach *Periplaneta americana* after being stung by the sphecid wasp *Ampulex compressa*. *Entomol. Exp. Appl.* **35:** 195–203.

Piek, T., Hue, B., Lind, A., Mantel, P., Van Marle, J. and Visser, J.H. (1988) The venom of *Ampulex compressa* changes the behaviour of the cockroach. *Pestic. Sci.* **24:** 268–270.

Piek, T., Hue, B., Lind, A., Mantel, P., Van Marle, J. and Visser, J.H. (1989) The wasp *Ampulex compressa* – effects on behaviour and synaptic transmission of cockroaches. *Comp. Biochem. Physiol.* **92C:** 175–183.

Pierceall, W.E., Li, C., Biran, A., Miura, K., Raikhel, A.S. and Segraves, W.A. (1999) E75 expression in mosquito ovary and fat body suggests reiterative use of ecdysone-regulated hierarchies in development and reproduction. *Mol. Cell. Endocrinol.* **150:** 73–89.

Pitcairn, M.J. and Gutierrez, A.P. (1992) Influence of adult size and age on the fecundity and longevity of *Tetrastichus incertus* (Hymenoptera: Eulophidae). *Ann. Entomol. Soc. Am.* **85:** 53–57.

Plantevin, G. and Grenier, S. (1990) Ecophysiology of host-parasitoid relationships: synchronisation of cycles of development. *Bull. Soc. Entomol. Fr.* **95:** 49–61.

Plantevin, G., De Reggi, M. and Nardon, C. (1984) Changes in ecdysteroid and juvenile hormone titers in the hemolymph of *Galleria mellonella* larvae and pupae. *Gen. Comp. Endocrinol.* **56:** 218–230.

Plantevin, G., Bosquet, G., Calvez, B. and Nardon, C. (1987) Relationships between juvenile hormone levels and synthesis of major haemolymph proteins in *Bombyx mori* larvae. *Comp. Biochem. Physiol.* **86B:** 501–507.

Polaszek, A. (1986) The effects of two species of hymenopterous parasitoid on the reproductive system of the pea aphid, *Acyrthosiphon pisum. Entomol. Exp. Appl.* **40:** 285–292.

Popham, H.J.R., Li, Y. and Miller, L.K. (1997) Genetic improvement of *Helicoverpa zea* nuclear polyhedrosis virus as a biopesticide. *Biol. Contr.* **10:** 83–91.

Possee, R.D., Barnett, A.L., Hawtin, R.E. and King, L.A. (1997) Engineered baculoviruses for pest control. *Pestic. Sci.* **51:** 462–470.

Poulin, R. (1995) 'Adaptive changes' in the behaviour of parasitised animals: a critical review. *Int. J. Parasitol.* **25:** 1371–1383.

Pratt, G.E. (1983) The mode of action of pro-allatocidins. In: *Natural Products for Innovative Pest Management* (eds Whitehead, D.L. and Bowers, W.S.), pp. 323–351. Pergamon Press, Oxford.

Pratt, G.E. and Tobe, S.S. (1974) Juvenile hormones radiobiosynthesized by corpora allata of adult female locusts *in vitro*. *Life Sci.* **14:** 575–586.

Pratt, G.E., Farnsworth, D.E., Siegel, N.R., Fox, K.F. and Feyereisen, R. (1989) Identification of an allatostatin from adult *Diploptera punctata*. *Biochem. Biophys. Res. Commun.* **163:**1243–1247.

Puttler, B., Gordh, G. and Long, S.H. (1980) Bionomics of *Euplectrus puttleri*, new species, an introduced parasite of the velvetbean caterpillar, *Anticarsia gemmatalis*, from South America. *Ann. Entomol. Soc. Am.* **73:** 28–35.

Quicke, D.L.J. (1997) *Parasitic Wasps*. Chapman & Hall, London.

Quicke, D.L.J. and van Achterberg, C. (1990) Phylogeny of the subfamilies of the family Braconidae (Hymenoptera: Ichneumonoidea). *Zool. Verh.* **258:** 1–95.

Quistad, G.B. and Leisy, D.J. (1996) Insecticidal toxins from the parasitic wasp, *Bracon hebetor*. US Patent No. 5,554,592

Quistad, G.B., Cerf, D.C., Schooley, D.A. and Staal, G.B. (1981) Fluoromevalonate acts as an inhibitor of juvenile hormone biosynthesis. *Nature* **289:** 176.

Quistad, G.B., Nduyen, Q., Bernasconi, P. and Leisy, D.J. (1994) Purification and characterization of insecticidal toxins from venom glands of the parasitic wasp, *Bracon hebetor. Insect Biochem. Mol. Biol.* **24:** 955–961.

Raabe, M. (1986) Insect reproduction: regulation of successive steps. *Adv. Physiol.* **19:** 29–154.

Rachinsky, A. and Feldlaufer, M.F. (2000) Responsiveness of honey bee (*Apis mellifera* L.) corpora allata to allatoregulatory peptides from four insect species. *J. Insect Physiol.* **46:** 41–46.

Rachinsky, A., Tobe, S.S. and Feldlaufer, M.F. (2000) Terminal steps in JH biosynthesis in the honey bee (*Apis mellifera* L.): developmental changes in sensitivity to JH precursor and allatotropin. *Insect Biochem. Mol. Biol.* **30:** 729–737.

Racioppi, J.V., Gemmill, R.M., Kogan, P.H., Calvo, J.M. and Hagedorn, H.H. (1986) Expression and regulation of vitellogenin messenger RNA in the mosquito, *Aedes aegypti. Insect Biochem.* **16:** 255–262.

Rafai, J. and King, P.E. (1972) Observations on the structure and possible function of the calyx region of the female reproductive system in *Nasonia vitripennis* (Walker) (Hymenoptera, Pteromalidae). *Pakistan J. Zool.* **4:** 1–11.

Raikhel, A.S. and Lea, A.O. (1991) Control of follicular epithelium development and vitelline envelope formation in the mosquito; role of juvenile hormone and 20-hydroxyecdysone. *Tissue and Cell* **23:** 577–591.

Raoult, D. and Roux, V. (1999) The body louse as a vector of reemerging human diseases. *Clin. Infect. Dis.* **29:** 888–911.

Ratcliffe, N.A. and King, P.E. (1967) The "venom" system of *Nasonia vitripennis* (Walker) (Hymenoptera: Pteromalidae). *Proc. R. Ent. Soc. Lond.* A **42:** 49–61.

Rathmayer, W. (1978) Venoms of Sphecidae, Pompilidae, Mutillidae and Bethylidae. In: *Handbuch der Experimentellen Pharmacologie,* Vol. 48 *(Arthropod Venoms)* (ed. Bettini, S.), pp. 661–690. Springer-Verlag, Berlin.

Ray, A., Memmel, N.A. and Kumaran, A.K. (1987) Developmental regulation of the larval hemolymph protein genes in *Galleria mellonella. Wilhelm Roux Arch. Dev. Biol.,* **196:** 414–420.

Reed, D.A. and Beckage, N.E. (1997) Inhibition of testicular growth and development in *Manduca sexta* larvae parasitized by the braconid wasp *Cotesia congregata. J. Insect Physiol.* **43:** 29–38.

Reed, D.A. and Brown, J.J. (1998) Host/parasitoid interactions: critical timing of parasitoid-derived products. *J. Insect Physiol.* **44:** 721–732.

Reed, D.A., Loeb, M.J. and Beckage, N.E. (1997) Inhibitory effects of parasitism by the gregarious endoparasitoid *Cotesia congregata* on host testicular development. *Arch. Insect Biochem. Physiol.* **36:** 95–114.

Reed-Larsen, D.A. and Brown, J.J. (1990) Embryonic castration of the codling moth, *Cydia pomonella* by an endoparasitoid *Ascogaster quadridentata. J. Insect Physiol.* **36:** 111–118.

Rees, H.H. (1995) Ecdysteroid biosynthesis and inactivation in relation to function. *Eur. J. Entomol.* **92:** 9–39.

Rembold, H. and Lackner, B. (1985) Convenient method for the determination of picomole amounts of juvenile hormone. *J. Chromatog.* **323:** 355–361.

Rembold, H. and Sehnal, F. (1987) Juvenile hormones and their titer regulation in *Galleria mellonella. Insect Biochem.* **17:** 997–1001.

Retnakaran, A., Granett, J. and Ennis, T. (1985) Insect growth regulators. In: *Comprehensive Insect Physiology, Biochemistry and Pharmacology,* Vol. 12 (eds Kerkut, G.A. and Gilbert, L.I.), pp. 529–601. Pergamon Press, Oxford.

Reynolds, S.E., Taghert, P.H. and Truman, J.W. (1979) Eclosion hormone and bursicon titres and the onset of hormonal responsiveness during the last day of adult development in *Manduca sexta* (L.). *J. Exp. Biol.* **78:** 77–86.

Reznik, S.Y. and Umarova, T.Y. (1990) The influence of host's age on the selectivity of parasitism and fecundity of *Trichogramma. Entomophaga* **35:** 31–37.

Ribeiro, J.M.C. (1987) Role of saliva in blood-feeding by arthropods. *Annu. Rev. Entomol.* **32:** 463–478.

Ribeiro, J.M.C. (1992) Characterization of a vasodilator from the salivary glands of the yellow fever mosquito *Aedes aegypti. J. Exp. Biol.* **165:** 61–71.

Ribeiro, J.M.C. and Nussenzveig, R.H. (1993) Nitric oxide synthase activity from a hematophagous insect salivary gland. *FEBS Letts* **330:** 165–168.

Ribeiro, J.M.C., Vachereau, A., Modi, G.B. and Tesh, R.B. (1989) A novel vasodilatory peptide from salivary glands of the sand fly *Lutzomyia longipalpis. Science* **243:** 212–214.

Ribeiro, J.M.C., Vaughan, J.A. and Azad, A.F. (1990) A salivary vasodilator in the blood-sucking bug, *Rhodnius prolixus. Brit. J. Pharmacol.* **101:** 932–936.

Ribeiro, J.M., Nussenzveig, R.H. and Tortorella, G. (1994) Salivary vasodilators of *Aedes triseriatus* and *Anopheles gambiae* (Diptera: Culicidae). *J. Med. Entomol.* **31:** 747–753.

Ribeiro, J.M., Schneider, M. and Guimaraes, J.A. (1995) Purification and characterization of prolixin S (nitrophorin 2), the salivary anticoagulant of the blood-sucking bug *Rhodnius prolixus. Biochem J.* **308:** 243–249.

Richards, A.G. (1951) *The Integument of Arthropods.* University of Minnesota Press, Minneapolis.

Richards, A.G. (1978) The chemistry of insect cuticle. In: *Biochemistry of Insects* (ed. Rockstein, M.), pp. 205–232. Academic Press, New York.

Richards, A., Matthews, M. and Christian, P. (1998) Ecological considerations for the environmental impact evaluation of recombinant baculovirus insecticides. *Annu. Rev. Entomol.* **43:** 493–517.

Richards, D.S., Applebaum, S.W., Sliter, T.J., Baker, F.C., Schooley, D.A., Reuter, C.C., Henrich, V.C. and Gilbert, L.I. (1989) Juvenile hormone bisepoxide biosynthesis *in vitro* by the ring gland of *Drosophila melanogaster*: A putative juvenile hormone in the higher Diptera. *Proc. Natl Acad. Sci. USA* **86:** 1421–1425.

Richards, E.H. and Edwards, J.P. (1999a) Parasitism of *Lacanobia oleracea* (Lepidoptera, Noctuidae) by the ectoparasite wasp *Eulophus pennicornis*, results in the appearance of a 27 kDa parasitism-specific protein in host plasma. *Insect Biochem. Mol. Biol.* **29:** 557–569.

Richards, E.H. and Edwards, J.P. (1999b) Parasitism of *Lacanobia oleracea* (Lepidoptera) by the ectoparasite wasp *Eulophus pennicornis*: Effects of venom, parasitization and starvation on host haemocytes. *J. Insect Physiol.* **45:** 1073–1083.

Richards, E.H. and Edwards, J.P. (2000) Parasitism of *Lacanobia oleracea* (Lepidoptera) by the ectoparasite wasp *Eulophus pennicornis*, suppresses haemocyte-mediated recognition of non-self and phagocytosis. *J. Insect Physiol.* **46:** 1–11.

Richards, E.H. and Edwards, J.P. (2001) Proteins synthesized and secreted by larvae of the ectoparasitic wasp, *Eulophus pennicornis. Arch. Insect Biochem. Physiol.* **46:** 140–151.

Richman, A.M., Dimopoulos, D.S. and Kafatos, F.C. (1997) *Plasmodium* activates the innate immune response of *Anopheles gambiae* mosquitoes. *EMBO J.* **16:** 6114–6119.

Richter, K. and Gersch, M. (1983) Electrophysiological evidence of nervous involvement in the control of the prothoracic gland in *Periplaneta americana*. *Experientia* **39:** 917–918.

Riddiford, L.M. (1975) Host hormones and insect parasites. In: *Invertebrate Immunity* (eds Maramorosch, K. and Schope, R.E.), pp. 339–353. Academic Press, New York.

Riddiford, L.M. (1980) Interaction of ecdysteroids and juvenile hormone in the regulation of larval growth and metamorphosis of the tobacco hornworm. In: *Progress in Ecdysone Research* (ed. Hoffmann, J.A.), pp. 409–430. Elsevier, Amsterdam.

Riddiford, L.M. (1985) Hormone action at the cellular level. In: *Comprehensive Insect Physiology, Biochemistry and Pharmacology,* Vol. 8 (eds Kerkut, G.A. and Gilbert, L.I.), pp. 37–84. Pergamon Press, Oxford.

Riddiford, L.M. (1994) Cellular and molecular actions of juvenile hormone. I. General considerations and premetamorphic actions. *Adv. Insect Physiol.* **24:** 213–274.

Riddiford, L.M. (1996) Molecular aspects of juvenile hormone action in insect metamorphosis. In: *Metamorphosis – Postembryonic Reprogramming of Gene Expression in Amphibian and Insect Cells* (eds Gilbert, L.I., Tata, J.R. and Atkinson, B.G.), pp. 223–251. Academic Press, San Diego.

Riddiford, L.M. and Hice, R.H. (1985) Developmental profiles of the mRNAs for *Manduca* arylphorin and two other storage proteins during the final larval instar of *Manduca sexta*. *Insect Biochem.* **15:** 489–502.

Riddiford, L.M. and Law, J.H. (1983) Larval serum proteins of Lepidoptera. In: *The Larval Serum Proteins of Insects* (ed. Scheller, K.), pp. 75–85. Georg Thieme Verlag, New York.

Riddiford, L.M. and Truman, J.W. (1978) Biochemistry of insect hormones and insect growth regulators. In: *Biochemistry of Insects* (ed. Rockstein, M.), pp. 308–357. Academic Press, New York.

Riddiford, L.M. and Truman, J.W. (1993) Hormone receptors and the regulation of insect metamorphosis. *Amer. Zool.* **33:** 340–347.

Riddiford, L.M., Hiruma, K., Lan, Q. and Zhou, B. (1999) Regulation and role of nuclear receptors during larval molting and metamorphosis of Lepidoptera. *Amer. Zool.* **39:** 736–746.

Rivers, D.B. and Denlinger, D.L. (1994a) Developmental fate of the flesh fly, *Sarcophaga bullata*, envenomated by the pupal ectoparasitoid, *Nasonia vitripennis. J. Insect Physiol.* **40:** 121–127.

Rivers, D.B. and Denlinger, D.L. (1994b) Redirection of metabolism in the flesh fly, *Sarcophaga bullata*, following envenomation by the ectoparasitoid, *Nasonia vitripennis*, and correlation of metabolic effects with the diapause status of the host. *J. Insect Physiol.* **40:** 207–215.

Rivers, D.B. and Denlinger, D.L. (1995a) Venom-induced alterations in fly lipid metabolism and its impact on larval development of the ectoparasitoid *Nasonia vitripennis* (Walker) (Hymenoptera: Pteromalidae). *J. Invert. Pathol.* **66:** 104–110.

Rivers, D.B. and Denlinger, D.L. (1995b) Fecundity and development of the ectoparasitic wasp *Nasonia vitripennis* are dependent on host quality. *Entomol. Exp. Appl.* **76:** 15–24.

Rivers, D.B. and Yoder, J.A. (1997) Site-specific effects of a parasitism on water balance and lipid content of the parasitic wasp *Nasonia vitripennis* (Hymenoptera: Pteromalidae). *Eur. J. Entomol.* **94:** 75–82.

Rivers, D.B., Hink, W.F. and Denlinger, D.L. (1993) Toxicity of the venom from *Nasonia vitripennis* (Hymenoptera: Pteromalidae) toward fly hosts, nontarget insects, different developmental stages, and cultured insect cells. *Toxicon* **31:** 755–765.

Rivers, D.B., Pagnotta, M.A. and Huntington, E.R. (1998) Reproductive strategies of 3 species of ectoparasitic wasps are modulated by the response of the fly host *Sarcophaga bullata* (Diptera: Sarcophagidae) to parasitism. *Ann. Entomol. Soc. Am.* **91:** 458–465.

Rivers, D.B., Genco, M. and Sanchez, R.A. (1999) *In vitro* analysis of venom from the wasp *Nasonia vitripennis*: susceptibility of different cell lines and venom-induced changes in plasma membrane permeability. *In vitro Cell. Dev. Biol.–Animal* **35:** 102–110.

Rizki, R.M. and Rizki, T.M. (1990) Parasitoid virus-like particles destroy *Drosophila* cellular immunity. *Proc. Natl Acad. Sci. USA* **87:** 8388–8392.

Roberts, D.B. and Brock, H.W. (1981) The major serum proteins of dipteran larvae. *Experientia* **37:** 103–110.

Robertson, P.L. (1968) A morphological and functional study of the venom apparatus in representatives of some major groups of Hymenoptera. *Aust. J. Zool.* **16:** 133–166.

Robinson, G.E., Strambi, C., Strambi, A. and Feldlaufer, M.F. (1991) Comparison of juvenile hormone and ecdysteroid haemolymph titres in adult worker and queen honey bees (*Apis mellifera*). *J. Insect Physiol.* **37:** 929–935.

Robinson, P.D., Morgan, E.D., Wilson, I.D. and Lafont, R. (1987) The metabolism of ingested and injected [^{3}H]ecdysone by final instar larvae of *Heliothis armigera. Physiol. Entomol.* **12:** 321–330.

Rodrigues, J.C.M., Souza, M.L., O'Reilly, D.R., Velloso, L.M., Pinedo, F.J.R., Razuck, F.B., Ribeiro, B. and Ribeiro, B.M. (2001) Characterization of the ecdysteroid UDP-glucosyltransferase (*egt*) gene of *Anticarsia gemmatalis* nucleopolyhedrovirus. *Virus Genes* **22:**103–112.

Rojas, M.G. (1993) Chemical cues that *Bracon compressitarsis* Wharton and *Bracon thurberiphagae* Muesebeck (Hymenoptera: Braconidae) utilize to localize their host *Anthonomus grandis* Boheman (Coleoptera: Curculionidae). Dissertation. Texas A&M University, College Station, Texas.

Rojas, M.G., Morales-Ramos, J.A. and King, E.G. (1996) *In vitro* rearing of the boll weevil (Coleoptera: Curculionidae) ectoparasitoid *Catolaccus grandis* (Hymenoptera: Pteromalidae) on meridic diets. *J. Econ. Entomol.* **89:** 1095–1104.

Rolle, R.S. and Lawrence, P.O. (1994) Characterization of a 24 kDa parasitism-specific hemolymph protein from pharate pupae of the Caribbean fruit fly, *Anastrepha suspensa*. *Arch. Insect Biochem. Physiol.* **25:** 227–244.

Röller, H., Dahm, K.H., Sweeley, C.C. and Trost, B.M. (1967) The structure of the juvenile hormone. *Angew. Chem. Int. Ed.* **6:** 179–180.

Romans, P., Tu, Z., Ke, Z. and Hagedorn, H.H. (1995) Analysis of a vitellogenin gene of the mosoquito, *Aedes aegypti* and comparisons to vitellogenins from other organisms. *Insect Biochem. Mol. Biol.* **25:** 939–958.

Romans, P., Bhattacharyya, R.K. and Colavita, A. (1998) *Ikirara*, a novel transposon family from the malaria vector mosquito, *Anopheles gambiae*. *Insect Mol. Biol.* **7:** 1–10.

Rose, M.R. (1991) *Evolutionary Biology of Aging*. Oxford University Press, Oxford.

Rotheram, S. (1973a) The surface of the egg of a parasitic insect. I. The surface of the egg and first-instar larva of *Nemeritis*. *Proc. R. Soc. Lond. B.* **183**, 179–194.

Rotheram, S. (1973b) The surface of the egg of a parasitic insect. II. The ultrastructure of the particulate coat on the egg of *Nemeritis*. *Proc. R. Soc. Lond. B.* **183:** 195–204.

Rothschild, M. (1975) Recent advances in our knowledge of the order Siphonaptera. *Annu. Rev. Entomol.* **30:** 371–413.

Rountree, D.B. and Bollenbacher, W.E. (1986) The release of prothoracicotropic hormone in the tobacco hornworm, *Manduca sexta*, is controlled intrinsically by juvenile hormone. *J. Exp. Biol.* **120:** 41–58.

Ruberson, J.R. and Kring, T.J. (1993) Parasitism of developing eggs by *Trichogramma pretiosum* (Hymenoptera: Trichogrammatidae): Host age preference and suitability. *Biol. Contr.* **3:** 39–46.

Rupes, V., Moravec, J., Chmela, J., Ledvinka, J. and Zelenkova, J. (1995) A resistance of head lice (*Pediculus capitis*) to permethrin in Czech Republic. *Cent. Eur. J. Public Health* **3:** 30–32.

Ryan, R.O., Schmidt, J.O. and Law, J.H. (1984) Arylphorin from the haemolymph of the larval honeybee, *Apis mellifera*. *Insect Biochem.* **14:** 515–520.

Rybczynski, R. and Gilbert, L.I. (1994) Changes in general and specific protein synthesis that accompany ecdysteroid synthesis in stimulated prothoracic glands of *Manduca sexta*. *Insect Biochem. Mol. Biol.* **24:** 175–189.

Rybczynski, R. and Gilbert, L.I. (1995) Prothoracicotropic hormone elicits a rapid, developmentally specific synthesis of β-tubulin in an insect endocrine gland. *Dev. Biol.* **169:** 15–28.

Safranek, L. and Williams, C.M. (1980) Studies of the prothoracicotropic hormone in the tobacco hornworm, *Manduca sexta*. *Biol. Bull.* **158:** 141–153.

Sahota, T.S. (1969) Hormonal control of ovarian development and metamorphosis in *Malacosoma pluviale*. *Can. J. Zool.* **47:** 917–920.

Sakurai, S. and Williams, C.M. (1989) Short-loop negative and positive feedback on ecdysone secretion by prothoracic gland in the tobacco hornworm, *Manduca sexta*. *Gen. Comp. Endocrinol.* **75:** 204–216.

Sakurai, S., Kaya, M. and Satake, S.I. (1998) Hemolymph ecdysteroid titer and ecdysteroid-dependent developmental events in the last-larval stadium of the silkworm, *Bombyx mori*: Role of low ecdysteroid titer in larval-pupal metamorphosis and a reappraisal of the head critical period. *J. Insect Physiol.* **44:** 867–881.

Salt, G. (1968) The resistance of insect parasitoids to the defense reactions of their host. *Biol. Rev.* **43:** 200–232.

Sappington, T.W., Hays, A.R. and Raikhel, A.S. (1995) Mosquito vitellogenin receptor: purification, developmental and biochemical characterization. *Insect Biochem. Mol. Biol.* **25:** 807–817.

Sappington, T.W., Hebling, P., Koller, C.N. and Raikhel, A.S. (1998) Activation *in vitro* of vitellogenin uptake by the oocytes of the mosquito, *Aedes aegypti*. *Physiol. Entomol.* **23:** 158–164.

Schepers, E.J., Dahlman, D.L. and Zhang, D. (1998) *Microplitis croceipes* teratocytes: *in vitro* culture and biological activity of teratocyte secreted protein. *J. Insect Physiol.* **44:** 767–777.

Schlinger, E.I. and Hall, J.C. (1960) The biology, behavior, and morphology of *Praon palitans* Muesebeck, an internal parasite of the spotted alfalfa aphid, *Therioaphis maculata* (Buckton) (Hymenoptera: Braconidae, Aphidiinae). *Ann. Entomol. Soc. Am.* **53:** 144–160.

Schmidt, J.M. and Smith, J.J.B. (1985) Host volume measurement by the parasitoid wasp *Trichogramma minutum*: The roles of curvature and surface area. *Entomol. Exp. Appl.* **39:** 213–221.

Schmidt, J.M. and Smith, J.J.B. (1987) Measurement of host curvature by the parasitoid wasp *Trichogramma minutum*, and its effect on host examination and progeny allocation. *J. Exp. Biol.* **129:** 151–164.

Schmidt, J.O. (1982) Biochemistry of insect venoms. *Annu. Rev. Entomol.* **27:** 339–368.

Schmidt, J.O. (1986) Chemistry, pharmacology and chemical ecology of ant venoms. In: *Venoms of the Hymenoptera* (ed. Piek, T.), pp. 425–508. Academic Press, London.

Schmidt, J.O., Blum, M.S. and Overal, W.L. (1986) Comparative enzymology of venoms from stinging Hymenoptera. *Toxicon* **24:** 907–921.

Schmutterer, H. (1990) Properties and potential of natural pesticides from the neem tree, *Azadirachta indica*. *Annu. Rev. Entomol.* **35:** 271–297.

Schneider, F. (1950) Die Entwicklung des Syrphidenparasiten *Diplazon fissorius* Grav. in uni-, oligo- und polyvoltinen Wirten und sein Verhalten bei parasitärer Aktivierung der Diapauselarven durch *Diplazon pectoratorius* Grav. *Mitt. Schweiz. Entomol. Gesell.* **23:** 155–194.

Schneider, F. (1951) Einige physiologische Beziehungen zwischen Syrphidenlarven und ihren Parasiten. *Z. Agnew. Entomol.* **33:** 150–162.

Schooley, D.A. and Baker, F.C. (1985) Juvenile hormone biosynthesis. In: *Comprehensive Insect Physiology, Biochemistry and Pharmacology*, Vol. 7 (eds Kerkut, G.A. and Gilbert, L.I.), pp. 363–389. Pergamon Press, Oxford.

Schooley, D.A. and Edwards, J.P. (1996) Anti juvenile hormones: precocenes to peptides. *Proceedings of the BCPC Brighton Conference, Pests and Diseases 1996*, Vol. 3, pp. 1029–1038. British Crop Protection Council, Farnham.

Schooley, D.A., Baker, F.C., Tsai, L.W., Miller, C.A. and Jamieson, G.C. (1984) Juvenile hormones 0, I and II exist only in Lepidoptera. In: *Biosynthesis, Metabolism and Mode of Action of Invertebrate Hormones* (eds Hoffmann, J. and Porchet, M.), pp. 373–383. Springer-Verlag, Berlin.

Schopf, A. and Rembold, H. (1993) Changes in juvenile hormone titer of gypsy moth larvae by parasitism of *Glyptapanteles liparidis*. *Naturwis.* **80:** 527–528.

Schopf, A., Nussbaumer, C., Rembold, H. and Hammock, B.D. (1996) Influence of braconid *Glyptapanteles liparidis* on the juvenile hormone titer of its larval host, the gypsy moth, *Lymantria dispar*. *Arch. Insect Biochem. Physiol.* **31:** 337–351.

Schroder, R.F.W. and Athanas, M.M. (1995) Egg parasitism of Colorado potato beetle, *Leptinotarsa decemlineata* (Say) by *Edovum puttleri* Grissel (Hymenoptera: Eulophidae) infesting tomato and eggplant in Maryland. *Pest Management in Horticultural Ecosystems* **1:** 1–7.

Schwartz, L.M. and Truman, J.W. (1984) Hormonal control of muscle atrophy and degeneration in the moth *Antheraea polyphemus*. *J. Exp. Biol.* **111:** 13–30.

Scowen, P. (1995) Government restricts the use of carbaryl for head lice. *Prof. Care Mother Child* **5:** 163.

Sehnal, F. (1983) Juvenile hormone analogues. In: *Endocrinology of Insects* (eds Downer, R.G.H. and Laufer, H.), pp. 657–672. Alan R. Liss, New York.

Sequeira, R. and Mackauer, M. (1992) Nutritional ecology of an insect host - parasitoid association: the pea aphid – *Aphidius ervi* system. *Ecology* **73:** 183–189.

Sevala, V.L. and Davey, K.G. (1989) Action of juvenile hormone on the follicle cells of *Rhodnius prolixus*: evidence for a novel regulatory mechanism involving protein kinase. *Experientia* **45:** 355–356.

Sevala, V.L., Davey, K.G. and Prestwich, G.D. (1995) Photoaffinity labelling and characterisation of a juvenile hormone binding protein in the membranes of follicle cells of *Locusta migratoria*. *Insect Biochem. Mol. Biol.* **25:** 267–273.

Shapiro, A.B., Wheelock, G.D., Hagedorn, H.H. Baker, F.C. Tsai, L.W. and Schooley, D.A. (1986) Juvenile hormone and juvenile hormone esterase in adult females of the mosquito *Aedes aegypti*. *J. Insect Physiol.* **32:** 867–877.

Shaver, T.N. and Parrott, W.L. (1970) Growth of neonates and 2-day-old larvae of both *H. zea* and *H. virescens* decreased as the concentration of gossypol increased with neonates affected more than older larvae. *J. Econ. Entomol.* **63:** 1802–1806.

Shaw, M.R. (1981a) Delayed inhibition of host development by the non-paralyzing venoms of parasitic wasps. *J. Invert. Pathol.* **37:** 215–221.

Shaw, M.R. (1981b) Possible food plant differences of *Amphipyra pyramidea* (L.) and *A. berbera svenssoni* Fletcher (Lepidoptera: Noctuidae), and a note on their parasites (Hymenoptera). *Entomologists' Gazette* **32:** 165–167.

Shaw, M.R. (1983) On evolution of endoparasitism: the biology of some genera of Rogadinae (Braconidae). *Contrib. Am. Entomol. Inst.* **20:** 307–328.

Shaw, M.R. (1987) Host associations of species of *Eulophus* in Britain (Hymenoptera: Eulophidae). *Entomologists' Gazette*, **38:** 59–63.

Shaw, M.R. and Huddleston, T. (1991) Classification and biology of braconid wasps (Hymenoptera: Braconidae). *Handbook for the Identification of British Insects No. 7*. Royal Entomological Society of London.

Shelby, K.S. and Webb, B.A. (1994) Polydnavirus infection inhibits synthesis of an insect plasma protein, arylphorin. *J. Gen. Virol.* **75:** 2285–2292.

Shelby, K.S. and Webb, B.A. (1997) Polydnavirus infection inhibits translation of specific growth-associated host proteins. *Insect. Biochem. Mol. Biol.* **27:** 263–270.

Shelby, K.S. and Webb, B.A. (1999) Polydnavirus-mediated suppression of insect immunity. *J. Insect Physiol.* **45:** 507–514.

Sherald, A.F. (1980) Sclerotization and coloration of the insect cuticle. *Experientia* **36:** 143–266.

Shi, X., Gomez, S. and Lawrence, P.O. (1999) A 24 kDa parasitism specific protein from the Caribbean fruit fly *Anastrepha suspensa*: cDNA and deduced amino acid sequence. *Insect Biochem. Mol. Biol.* **29:** 749–755.

Shimizu, T., Tatsuki, Y. and Takeda, N. (1993) Aromatic amino acids in the venom of the braconid parasitoid *Apanteles kariyai*. *Z. Naturforsch.* **48C:** 108–109.

Shirai, Y., Aizono, Y., Iwasaki, T., Yanagida, A., Mori, H., Sumida, M. and Matsubara, F. (1993) Prothoracicotropic hormone is released five times in the 5th-larval instar of the silkworm, *Bombyx mori*. *J. Insect Physiol.* **39:** 83–88.

Shu, S., Park, Y.I., Ramaswamy, S.B. and Srinivasan, A. (1997) Hemolymph juvenile hormone titers in pupal and adult stages of Southwestern corn borer [*Diatrea grandiosella* (Pyralidae)] and relationship with egg development. *J. Insect Physiol.* **43:** 719–726.

Silvers, M.J. and Nappi, A.J. (1986) *In vitro* study of physiological suppression of supernumerary parasites by the endoparasitic wasp *Leptopilina heterotoma*. *J. Parasitol.* **72:** 405–409.

Sinden, R.E., Butcher, G.A., Billker, O. and Fleck, S.L. (1996) Regulation of infectivity of *Plasmodium* in the mosquito vector. *Adv. Parasitol.* **38:** 53–117.

Skinner, W.S., Dennis, P.A. and Quistad, G.B. (1990) Partial characterization of toxins from *Goniozus legneri* (Hymenoptera: Bethylidae). *J. Econ. Entomol.* **83:** 733–736.

Skinner, W.S., Dennis, P.A., Li, J.P., Summerfelt, R.M., Carney, R.L. and Quistad, G.B. (1991) Isolation and identification of paralytic peptides from hemolymph of the lepidopteran insects *Manduca sexta, Spodoptera exigua,* and *Heliothis virescens*. *J. Biol. Chem.* **266:** 12873–12877.

Skinner, W.S., Dennis, P.A. and Quistad, G.B. (1993) Paralytic peptides from hemolymph of the lepidopteran insect *Trichoplusia ni*. *Comp. Biochem. Physiol.* **104C:** 133–135.

Slansky, F. Jr. (1986) Nutritional ecology of endoparasitic insects and their hosts: an overview. *J. Insect Physiol.* **32:** 255–261.

Slavicek, J.M., Popham, H.J.R. and Riegel, C.I. (1999) Deletion of the *Lymantria dispar* multicapsid nucleopolyhedrosis ecdysteroid UDP-glucosyl transferase gene enhances viral killing speed in the last instar of the gypsy moth. *Biol. Contr.* **16:** 91–103.

Slavnova, T.I., Antonov, S.M., Magazanik, L.G., Tonkikh, A.K., Kosovskii, A.V., Sadykov, A. A. and Abduvakhabov, A.A. (1987) Effect of toxin from the venom of the ichneumon *Habrobracon hebetor* (Say) on neuromuscular transmission in insects. *Doklady Akademii Nauk SSR* **297:** 492–494.

Smith, I. and Goodale, C. (1998) Sequence and *in vivo* transcription of *Lacanobia oleracea* granulovirus *egt*. *J. Gen. Virol.* **79:** 405–413.

Smith, O.J. (1952) Biology and behavior of *Microctonus vittatae* Muesebeck. *Univ. Calif. Publ. Entomol.* **9:** 315–344.

Smith, R.C. (1927) Observation on *Euplectrus plathypenae* How. (Chalcididae) a parasite of noctuid larvae. *Bull. Brooklyn Ent. Soc.* **22:** 128–134.

Smith, W.A. (1993) Second messengers and the action of prothoracicotropic hormone in *Manduca sexta*. *Amer. Zool.* **33:** 330–339.

Snigirevskaya, E.S., Hays, A.R. and Raikhel, A.S. (1997) Secretory and internalization pathways of mosquito yolk protein precursors. *Cell Tiss. Res.* **290:** 129–142.

Sokoloff, F. (1994) Identification and management of pediculosis. *Nurse Pract.* **19:** 62–64.

Sokolova, M.I. (1994) A redescription of the morphology of mosquito (Diptera: Culicidae) ovarioles during vitellogenesis. *Bull. Soc. Vector Ecol.* **19:** 53–68.

Soldan, T. and Stary, P. (1981) Parasitogenic effects of *Aphidius smithi* (Hymenoptera: Aphidiidae) on the reproductive organs of the pea aphid, *Acyrthosiphon pisum* (Homoptera: Aphididae). *Acta Entomol. Biochem.* **78:** 243–253.

Soldevila, A.I. and Jones, D. (1993) Expression of a parasitism-specific protein in lepidopteran hosts of *Chelonus* sp. *Arch. Insect Biochem. Physiol.* **24:** 149–169.

Soldevila, A.I. and Jones, D. (1994) Characterization of a novel protein associated with the parasitization of lepidopteran hosts by an endoparasitic wasp. *Insect Biochem. Mol. Biol.* **24:** 29–38.

Soller, M. and Lanzrein, B. (1996) Polydnavirus and venom of the egg-larval parasitoid *Chelonus inanitus* (Braconidae) induce developmental arrest in the prepupa of its host *Spodoptera littoralis* (Noctuidae). *J. Insect Physiol.* **42:** 471–481.

Soller, M., Bownes, M. and Kubli, E. (1999) Control of oocyte maturation in sexually mature *Drosophila* females. *Dev. Biol.* **208:** 337–351.

Song, Q. and Gilbert, L.I. (1994) S6 phosphorylation results from prothoracic hormone stimulation of insect prothoracic glands: a role for S6 kinase. *Dev. Genet.* **15:** 332–338.

Song, Q. and Gilbert, L.I. (1998) Alterations in ultraspiracle (USP) content and phosphorylation state accompany feedback regulation of ecdysone synthesis in the insect prothoracic gland. *Insect Biochem. Mol. Biol.* **28:** 849–860.

Soter, N.A., Lewis, R.A., Corey, E.J. and Austen, K.F. (1983) Local effects of synthetic leukotrienes (LTC4, LTD4, LTE4, and LTB4) in human skin. *J. Invest. Dermatol.* **80:** 115–119.

Spanjer, W., Grosu, L. and Piek, T. (1977) Two different paralysing preparations obtained from a homogenate of the wasp *Microbracon hebetor* (Say). *Toxicon* **15:** 413–421.

Sparagana, S.P., Bhaskaran, G. and Barrera, P. (1985) Juvenile hormone acid methyltransferase activity in imaginal discs of *Manduca sexta* prepupae. *Arch Insect Biochem. Physiol.* **2:** 191–202.

Spencer, H. (1926) Biology of the parasites and hyperparasites of aphids. *Ann. Entomol. Soc. Am.* **19:** 119–157.

Speyer, E.R. (1934) Entomological Investigations. In: *Twentieth Annual Report of the Experimental and Research Station, Cheshunt, Herts*, pp. 70–71. Cheshunt Press, Hertfordshire.

Srivastava, U.S. and Chhibber, N. (1980) Effect of a juvenoid on wing development and differentiation in an aphid *Aphis nerii* B. de F. *Indian J. Exp. Biol.* **18:** 814–817.

Sroka, P. and Vinson, S.B. (1978) Phenoloxidase activity in the haemolymph of parasitized and unparasitized *Heliothis virescens*. *Insect Biochem.* **8:** 399–402.

Staal, G.B. (1975) Insect growth regulators with juvenile hormone activity. *Annu. Rev. Entomol.* **20:** 417–460.

Staal, G.B. (1977a) Essential differences between natural juvenile hormones and juvenile hormone analogues elucidated by use of a substitution assay. *Pontif. Acad. Sci. Scripta Varia* **41:** 333–348.

Staal, G.B. (1977b) Insect control with insect growth regulators based on insect hormones. *Pontif. Acad. Sci. Scripta Varia* **41:** 353–383.

Staal, G.B. (1986) Anti juvenile hormone agents. *Annu. Rev. Entomol.* **31:** 391–429.

Stay, B., Tobe, S.S. and Bendena, W.G. (1994) Allatostatins: identification, primary structures, function and distribution. *Adv. Insect Physiol.* **25:** 267–337.

Stearns, S.C. (1989) Trade-offs in life history evolution. *Funct. Ecol.* **3:** 259–268.

Steel, C.G.H. and Davey, K.G. (1985) Integration in the insect endocrine system. In: *Comprehensive Insect Physiology, Biochemistry and Pharmacology,* Vol. 8 (eds Kerkut, G.A. and Gilbert, L.I.), pp. 37–84. Pergamon Press, Oxford.

Steiner, A.L. (1986) Stinging behaviour of solitary wasps. In: *Venoms of the Hymenoptera.* (ed. Piek, T.), pp. 63–160. Academic Press, London.

Steiner, B., Pfister-Wilhelm, R., Grossniklaus-Bürgin, C., Rembold, H., Treiblmayr, K. and Lanzrein, B. (1999) Titers of juvenile hormone I, II and III in *Spodoptera littoralis* (Noctuidae) from the egg to the pupal moult and their modification by the egg-larval parasitoid *Chelonus inanitus* (Braconidae). *J. Insect Physiol.* **45:** 401–413.

Stettler, P., Trenczek, T., Wyler, T., Pfister-Wilhelm, R. and Lanzrein, B. (1998) Overview of parasitism associated effects on host haemocytes in larval parasitoids and comparison with effects of the egg-larval parasitoid *Chelonus inanitus* on its host *Spodoptera littoralis*. *J. Insect Physiol.* **44:** 817–831.

Stoltz, D.B. (1986) Interactions between parasitoid-derived products and host insects: an overview. *J. Insect Physiol.* **32:** 347–350.

Stoltz, D.B. (1993) The polydnavirus life cycle. In: *Parasites and Pathogens of Insects,* Vol. 1. (eds Beckage, N.E., Thompson, S.N. and Federici, B.A.), pp. 167–187. Academic Press, New York.

Stoltz, D.B. and Cook, D.I. (1983) Inhibition of host phenoloxidase activity by parasitoid Hymenoptera. *Experientia* **39:** 1022–1024.

Stoltz, D.B. and Guzo, D. (1986) Apparent haemocytic transformations associated with parasitoid-induced inhibition of immunity in *Malacosoma disstria* larvae. *J. Insect Physiol.* **32:** 377–388.

Stoltz, D.B. and Vinson, S.B. (1979a) Viruses and parasitism in insects. *Adv. Virus Res.* **24:** 125–171.

Stoltz, D.B. and Vinson, S.B. (1979b) Penetration into caterpillar cells of virus-like particles injected during oviposition by parasitoid ichneumonid wasps. *Can. J. Zool.* **25:** 207–216.

Stoltz, D.B., Vinson, S.B. and MacKinnon, E.A. (1976) Baculovirus-like particles in the reproductive tracts of female parasitoid wasps. *Can. J. Microbiol.* **22:** 1013–1023.

Stoltz, D.B., Krell, P., Summers, M.D. and Vinson, S.B. (1984) Polydnaviridae – A proposed family of insect viruses with segmented, double-stranded, circular DNA Genomes. *Intervirology* **21:** 1–4.

Stoltz, D.B., Guzo, D., Belland, E.R., Lucarotti, C.J. and MacKinnon, E.A. (1988) Venom promotes uncoating *in vitro* and persistence *in vivo* of DNA from a braconid polydnavirus. *J. Gen. Virol.* **69:** 903–907.

Strambi, C., Strambi, A., de Reggi, M.L., Hirn, M.H. and Delagge, M.A. (1981) Radioimmunoassay of insect juvenile hormones and of their diol derivatives. *Eur. J. Biochem.* **118:** 401–406.

Strand, M.R. (1986) The physiological interactions of parasitoids with their hosts and their influence in reproductive strategies. In: *Insect Parasitoids* (eds Waage, J. and Greathead, D.), pp. 97–136. Academic Press, New York.

Strand, M.R. and Dover, B.A. (1991) Developmental disruption of *Pseudoplusia includens* and *Heliothis virescens* larvae by the calyx fluid and venom of *Microplitis demolitor*. *Arch. Insect Biochem. Physiol.* **81:** 131–145.

Strand, M.R. and Pech, L.L. (1995) Immunological basis for compatibility in parastioid-host relationships. *Annu. Rev. Entomol.* **40:** 31–56.

Strand, M.R. and Wong, E.A. (1991) The growth and role of *Microplitis demolitor* teratocytes in parasitism of *Pseudoplusia includens*. *J. Insect Physiol.* **37:** 503–515.

Strand, M.R., Ratner, S. and Vinson, S.B. (1980) Maternally induced host regulation by the egg parasitoid *Telenomus heliothidis*. *Physiol. Entomol.* **8:** 469–475.

Strand, M.R., Quarles, J.M., Meola, S.M. and Vinson, S.B. (1985) Cultivation of teratocytes of the egg parasitoid *Telenomus heliothis* (Hymenoptera: Scelionidae) *in vitro*. *Cell. Devel. Biol.* **21:** 361–367.

Strand, M.R., Meola, S.M. and Vinson, S.B. (1986) Correlating pathological symptoms in *Heliothis virescens* eggs with development of the parasitoid *Telenomus heliothidis*. *J. Insect Physiol.* **32:** 389–402.

Strand, M.R., Johnson, J.A. and Culin, J.D. (1988a) Developmental interactions between the parasitoid *Microplitis demolitor* (Hymenoptera: Braconidae) and its host *Heliothis virescens* (Lepidoptera: Noctuidae). *Ann. Entomol. Soc. Amer.* **81:** 822–830.

Strand, M.R., Vinson, S.B., Nettles, W.C. Jr. and Xie, Z.U. (1988b) *In vitro* culture of the egg parasitoid *Telenomus heliothidis*: the role of teratocytes and medium consumption in development. *Entomol. Exp. Appl.* **46:** 71–78.

Strand, M.R., Dover, B.A. and Johnson, J.A. (1990) Alterations in the ecdysteroid and juvenile hormone esterase profiles of *Trichoplusia ni* parasitized by the polyembryonic wasp *Copidosoma floridanum*. *Arch. Insect Biochem. Physiol.* **13:** 41–51.

Strand, M.R., Goodman, W.G. and Baehrecke, E.H. (1991) The juvenile hormone titer of *Trichoplusia ni* and its potential role in the embryogenesis of the polyembryonic wasp *Copidosoma floridanum*. *Insect Biochem.* **21:** 205–214.

Strand, M.R., McKenzie, D.I., Grassl, V., Dover, B.A. and Aiken, J.M. (1992) Persistence and expression of *Microplitis demolitor* polydnavirus in *Pseudoplusia includens*. *J. Gen. Virol.* **73:** 1627–1635.

Strand, M.R., Johnson, J.A., Noda, T. and Dover, B.A. (1994) Characterization of monoclonal antibodies generated against components of the venom of *Microplitis demolitor*. *Arch. Insect Biochem. Physiol.* **23:** 123–136.

Strand, M.R., Hayakawa, Y. and Clark, K.D. (2000) Plasmatocyte spreading peptide (PSP1) and growth blocking peptide (GBP) are multifunctional homologs. *J. Insect Physiol.* **46:** 817–824.

Sullivan, D.J. (1988) Hyperparasites. In: *Aphids. Their Biology, Natural Enemies and Control.* World Crop Pests, Vol. 2B (eds Minks, A.K. and Harrewijn, P.), pp. 189–203. Elsevier, Amsterdam.

Sundaram, M., Palli, S.R., Krell, P.J., Sohi, S.S., Dhadialla, T.S. and Retnakaran, A. (1998) Basis for selective action of a synthetic molting hormone agonist RH-5992 on lepidopteran insects. *Insect Biochem. Mol. Biol.* **28:** 693–704.

Tagashira, E. and Tanaka, T. (1998) Parasitic castration of *Pseudaletia separata* by *Cotesia kariyai* and its association with polydnavirus gene expression. *J. Insect Physiol.* **44:** 733–744.

Takada, H. (1973) Studies on aphid hyperparasites of Japan, I. Aphid hyperparasites of the genus *Dendrocerus* Ratzeburg occurring in Japan (Hymenoptera: Ceraphronidae). *Ins. Matsum.* **2:** 1–37.

Takasu, K. and Overholt, W.A. (1996) Oviposition behaviour of *Cotesia flavipes* and *Goniozus indicus*, larval parasitoids of the lepidopteran stemborer *Chilo partellus*. *Jap. J. Appl. Ent. Zool.* **40:** 308–310.

Tanaka, H.Y., Nakamura, K., Sakakibara, M., Nagata, S. and Kataoka, H. (1999) Identification of a prothoracicostatic peptide in the larval brain of the silkworm, *Bombyx mori*. *J. Biol. Chem.* **274:** 31169–31173.

Tanaka, T. (1987) Calyx and venom fluids of *Apanteles kariyai* (Hymenoptera: Braconidae) as factors that prolong larval period of the host, *Pseudaletia separata* (Lepidoptera: Noctuidae). *Ann. Entomol. Soc. Amer.* **80:** 530–533.

Tanaka, T. and Vinson, S.B. (1991a) Depression of prothoracic gland activity of *Heliothis virescens* by venom and calyx fluids from the parasitoid, *Cardiochiles nigriceps*. *J. Insect Physiol.* **37:** 139–144.

Tanaka, T. and Vinson, S.B. (1991b) Interaction of venoms with the calyx fluids of three parasitoids, *Cardiochiles nigriceps, Microplitis croceipes* (Hymenoptera: Braconidae), and *Campoletis sonorensis* (Hymenoptera: Ichneumonidae) in effecting a delay in the pupation of *Heliothis virescens* (Lepidoptera: Noctuidae). *Ann. Entomol. Soc. Am.* **84:** 87–92.

Tanaka, T. and Wago, H. (1990) Ultrastructural and functional maturation of teratocytes of *Apanteles kariyai*. *Arch. Insect Biochem. Physiol.* **13:** 187–197.

Tanaka, T., Agui, N. and Hiruma, K. (1987) The parasitoid *Apanteles kariyai* inhibits pupation of its host, *Pseudaletia separata*, via disruption of prothoracicotropic hormone release. *Gen. Comp. Endocrinol.* **67:** 364–374.

Tanaka, T., Tagashira, E. and Sakurai, S. (1994) Reduction of testis growth of *Pseudaletia separata* larvae after parasitization by *Cotesia kariyai*. *Arch. Insect Biochem. Physiol.* **26:** 111–122.

Tawfik, A.I., Vedrova, A. and Sehnal, F. (1999) Ecdysteroids during ovarian development and embryogenesis in solitary and gregarious *Schistocerca gregaria*. *Arch. Insect. Biochem. Physiol.* **41:** 134–143.

Taylor, C.W. (1986) Calcium regulation in insects. *Adv. Physiol.* **19:** 155–186.

Taylor, T. and Jones, D. (1990) Isolation and characterization of the 32.5 kDa protein from the venom of and endoparasitic wasp. *Biochim. Biophys. Acta* **1035:** 37–43.

Taylor, T.H.C. (1933) An investigation of the banana-scab moth, *Nacoleia octasema* Meyr., and its parasites, in Java, and the introduction of one of its parasites into Fiji. Part 1. *Fiji Dept. Agric. J.* **6:** 3–13.

Taylor, T.H.C. (1937) *The Biological Control of an Insect in Fiji. An Account of the Coconut-leafmining Beetle and its Parasite Complex.* Commonwealth Institute of Entomology, London.

Teal, P.E., Proveaux, A.T. and Heath, R.R. (2000) Analysis and quantitation of insect juvenile hormones using chemical ionization ion-trap mass spectrometry. *Analyt. Biochem.* **277:** 206–213.

Telfer, W.H. and Kunkel, J.G. (1991) The function and evolution of insect storage hexamers. *Annu. Rev. Entomol.* **36:** 205-228.

Telfer, W.H., Keim, P.S. and Law, J.H. (1983) Arylphorin, a new protein from *Hyalophora cecropia:* Comparisons with calliphorin and manducin. *Insect Biochem.* **13:** 601–613.

Temperley, M.C. (1930) Life history notes on the banana fruit-eating caterpillar (*Tiracola plagiata* Wlk.). *Queensland J. Agric.* **33:** 251–261.

Theilmann, D.A. and Summers, M.D. (1986) Molecular analysis of *Campoletis sonorensis* virus DNA in the lepidopteran host *Heliothis virescens*. *J. Gen. Virol.* **67:** 1961–1969.

Theilmann, D.A. and Summers, M.D. (1987) Physical analysis of the *Campoletis sonorensis* virus multipartite genome and identification of a family of tandemly repeated elements. *J. Virol.* **61:** 2589–2598.

Theilmann, D.A. and Summers, M.D. (1988) Identification and comparison of *Campoletis sonorensis* virus transcripts expressed from four genomic segments in the insect hosts *Campoletis sonorensis* and *Heliothis virescens*. *Virology* **167:** 329–341.

Theopold, U., Krause, E. and Schmidt, O. (1994) Cloning of a VLP-protein coding gene from a parasitoid wasp *Venturia canescens*. *Arch. Insect Biochem. Physiol.* **26:** 137–145.

Thompson, C.G., Scott, D.W., and Wickman, B.E. (1981) Long-term persistence of the nuclear polyhedrosis virus of the Douglas-fir tussock moth, *Orgyia pseudosugata* (Lepidoptera: Lymantriidae) in forest soil. *Environ. Entomol.* **10:** 254–255.

Thompson, M., Feldlaufer, M., Lozano, R., Rees, H., Lusby, W., Svoboda, J. and Wilzer, K. (1987) Metabolism of [^{14}C]hydroxyecdysone 26-phosphate in the tobacco hornworm, *Manduca sexta* L., to a new ecdysteroid conjugate: 26-[^{14}C]hydroxyecdysone 22-glucoside. *Arch. Insect Biochem. Physiol.* **4:** 1–15.

Thompson, M.J., Svoboda, J.A., Lozano, R. and Wilzer, K.R. (1988) Profile of free and conjugated ecdysteroids and ecdysteroid acids during embryonic development of *Manduca sexta* (L.) following maternal incorporation of [^{14}C] cholesterol. *Arch. Insect Biochem. Physiol.* **7:** 157–172.

Thompson, S.N. (1983) Biochemical and physiological effects of metazoan endoparasites on their host species. *Comp. Biochem. Physiol.* **74B:** 183–211.

Thompson, S.N. (1986) Effect of the parasite *Hyposotes exiquae* (Viereck) on the carbohydrate metabolism of its host *Trichoplusia ni* (Hubner). *J. Insect Physiol* **32:** 287–293.

Thompson, S.N. (1993) Redirection of host metabolism and effects on parasite nutrition. In: *Parasites and Pathogens of Insects,* Vol. 1: (eds Beckage N.E., Thompson S.N. and Federici B.A.), pp. 125–144. Academic Press, San Diego.

Thompson, S.N. and Cohen, A.C. (1984) Nitrogen elimination in *Trichoplusia ni* parasitized by the insect parasite *Hyposoter exiguae*. *Comp. Biochem. Physiol.* **78A:** 785–786.

Thompson, S.N. and Dahlman, D.L. (1998) Aberrant nutritional regulation of carboyhydrate synthesis by parasitized *Manduca sexta* L. *J. Insect Physiol.* **44:** 745–753

Thompson, W.R. (1955) Hosts of the Hymenoptera (Calliceratid to Evanid). In: *A Catalogue of the Parasites and Predators of Insect Pests, Section 2: Host parasite catalogue; Part 3.* Commonwealth Agricultural Bureau Ottawa, Canada.

Thomsen, M. (1927) Some observations on the biology and anatomy of a cocoon-making chalcid larva, *Euplectrus bicolor* Swed. *Videnk. Med. Nat. For. Kjob.* **84:** 73–89.

Thyagaraja, B.S., Gelman, D.B., Masler, E.P., Kelly, T.J., Bell, R.A. and Imberski, R.B. (1993) A role for the terminal abdominal segment in the larval and pupal molts of the gypsy moth *Lymantria dispar* L: Discovery of a developmental control center in the anal papilla pouch. *J. Insect Physiol.* **39:** 969–680.

Tipping, P.W., Holko, C.A., Abdul-Baki, A.A. and Aldrich, J.R. (1999) Evaluating *Edovum puttleri* Grissell and *Podisus maculiventris* (Say) for augmentative biological control of Colorado potato beetle in tomatoes. *Biol. Cont.* **16:** 35–42.

Togashi, I. (1963) A comparative morphology of the poison glands in adults of ichneumon-flies (Hymenoptera: Ichneumonidae). *Kontyû* **31:** 297–304.

Toister-Achituv, M. and Faktor, O. (1997) Transcriptional analysis and promoter activity of the *Spodoptera littoralis* multicapsid nucleopolyhedrovirus ecdysteroid UDP-glucosyltransferase gene. *J. Gen. Virol.* **78:** 487–491

Tojo, S., Nagata, J.G. and Kobayashi, M. (1980) Storage proteins in the silk moth, *Bombyx mori. Insect Biochem.* **10:** 289–303.

Tojo, S., Kiguchi, K. and Kimura, S. (1981) Hormonal control of storage protein synthesis and uptake by the fat body in the silkworm, *Bombyx mori*. *J. Insect Physiol.* **27:** 491–497.

Tojo, S., Morita, M., Agui, N. and Hiruma, K. (1985) Hormonal regulation of phase polymorphism and storage-protein fluctuation in the common cutworm *Spodoptera litura*. *J. Insect Physiol.* **31:** 283–292.

Trautmann, K.H., Masner, P., Schuler, A., Suchy, M. and Wipf, H.-K. (1974) Evidence of the juvenile hormone Methyl (2E,6E)-10-11-epoxy-3,7,11-trimethyl-2,6-dodecadienoate (JH-3) in insects of four Orders. *Zeit. Naturforsch.* **29C:** 757–759.

Treacy, M.F., All, J.N. and Ghidui, G.M. (1997) Effect of ecdysteroid UDP-glucosyltransferase gene deletion on efficacy of a baculovirus against *Heliothis virescens* and *Trichoplusia ni* (Lepidoptera: Noctuidae). *J. Econ. Entomol.* **90:** 1207–1214.

Truesdell, P.F., Koladich, P.M., Kataoka, H., Kojima, K., Suzuki, A., McNeil, J.N., Mizoguchi, A., Tobe, S.S. and Bendena, W.G. (2000) Molecular characterisation of a cDNA from the true armyworm *Pseudaletia unipuncta* encoding *Manduca sexta* allatotropin peptide. *Insect Biochem. Mol. Biol.* **30:** 691–702.

Truman, J.W. (1985) Hormonal control of ecdysis. In: *Comprehensive Insect Physiology, Biochemistry and Pharmacology*, Vol. 8 (eds Kerkut, G.A. and Gilbert, L.I.), pp. 413–440. Pergamon Press, Oxford.

Truman, J.W. (1996) Ecdysis control sheds another layer. *Science* **271:** 40–42.

Truman, J.W. and Dominick, O.S. (1983) Endocrine mechanisms organizing invertebrate behaviour. *BioScience* **33:** 546–541.

Truman, J.W. and Riddiford, L.M. (1970) Neuroendocrine control of ecdysis in silkmoths. *Science* **167:** 1624–1626.

Truman, J.W., Riddiford, L.M. and Safranek, L. (1974) Temporal patterns of response to ecdysone and juvenile hormone in the epidermis of the tobacco hornworm, *Manduca sexta*. *Dev. Biol.* **39:** 247–262.

Truman, J.W., Taghert, P.H., Copenhaver, P.F., Tublitz, N.J. and Schwartz, L.M. (1981) Eclosion hormone may control all ecdyses in insects. *Nature* **291:** 70–71.

Truman, J.W., Rountree, D.B., Reiss, S.E. and Schwartz, L.M. (1983) Ecdysteroids regulate the release and action of eclosion hormone in the tobacco hornworm, *Manduca sexta* (L.). *J. Insect Physiol.* **29:** 895–900.

Uematsu, H. (1981a) Bionomics of *Euplectrus kuwanae* Crawford (Hymenoptera: Eulophidae), a parasitoid of *Argyrogramma albostriata* (Bremer et Grey). *Appl. Entomol. Zool.* **16:** 57–59.

Uematsu, H. (1981b) The ovipositional behaviour in *Euplectrus kuwanae* Crawford (Hymenoptera: Eulophidae), a parasitoid of *Argyrogramma albostriata* (Bremer et Grey) (Lepidoptera: Noctuidae). *App. Ent. Zool.* **16:** 443–450.

Uematsu, H. (1986) Preovipositional stinging of an external parasitoid, *Euplectrus kuwanae* Crawford (Hymenoptera: Eulophidae). *Jpn. J. Appl. Ent. Zool.* **30:** 55–57.

Uematsu, H. and Sakanoshita, A. (1987) Effects of venom from an external parasitoid, *Euplectrus kuwanae* (hymenoptera: Eulophidae) on larval ecdysis of *Agyrogramma albostriata* (Lepidoptera: Noctuidae). *Appl. Entomol. Zool.* **22:** 139–144.

Ullmann, S.L. (1973) Oogenesis in *Tenebrio molitor*: histological and autoradiological observations on pupal and adult ovaries. *J. Embryol. Exp. Morph.* **30:** 179–217.

Unni, B., Barrera, P., Muszynska-Pytel, M., Bhaskaran, G. and Dahm, K.H. (1993) Partial characterization of allatinhibin, a neurohormone of *Manduca sexta*. *Arch. Insect Biochem. Physiol.* **24:** 173–185.

van Alphen, J.J. and Vet, L.E. (1986) An evolutionary approach to host finding and selection. In: *Insect Parasitoids* (eds Waage, J. and Greathead, D.), pp. 23–61. Academic Press, London.

van de Veire, M. (1993) First observations in glasshouse sweet peppers in Belgium and laboratory rearing of the parasitic wasp *Eulophus pennicornis* (Hym.: Eulophidae). *Entomophaga* **38:** 61–62.

Van Handel, E. and Lea, A.O. (1984) Vitellogenin synthesis in blood-fed *Aedes aegypti* in the absence of head, thorax and ovaries. *J. Insect Physiol.* **30:** 871–875.

van Marle, J. (1977) Structure and histochemistry of the venom glands of the wasps *Microbracon hebetor* Say and *Philanthus triangulum* F. *Toxicon* **15:** 529–539.

van Marle, J. and Piek, T. (1986) Morphology of the venom apparatus. In: *Venoms of the Hymenoptera* (ed. Piek, T.), pp.17–44. Academic Press, London.

van Veen, J.C. and van Wijk, M.L.E. (1985) The unique structure and functions of the ovipositor of the non-paralyzing ectoparasitoid *Colpoclypeus florus* Walk. (Hym., Eulophidae) with special reference to antennal sensilla and immature stages. *Z. Agnew. Entomol.* **99:** 511–531.

van Veen, J.C. and van Wijk, M.L.E. (1987) Parasitization strategy in the non-paralyzing ectoparasitoid *Colpoclypeus florus* (Hym., Eulophidae) towards its common summer host *Adoxophyes orana* (Lep., Tortricidae). 1. Host finding, acceptance and utilization. *J. Appl. Entomol.* **104:** 402–417.

van Veen, J.C., Dijkstra, L. and Dalla Monta, L. (1985) Parasitization success of geographically different *Colpoclypeus florus* Walk. ectoparasitoids and their hybrids (Hym., Eulophidae). *Redia* **68:** 219–228.

Vargas, L., Paguia, P. and DeWilde, J. (1976) Juvenile hormone titers in penultimate and last instar larvae of *Pieris brassicae* and *Barathra brassicae*, in relation to the effect of juvenoid application. *Experientia* **32:** 249–251.

Vargas, L., Kulcsar, P., Fekete, J., Bihatsi-Karsai, E. and Lelik, L. (1992) JH titres measured by GC-MS, in the hemolymph of *Mamestra oleracea* larvae reared under different photoperiodic conditions. In: *Insect Juvenile Hormone Research* (eds Mauchamp, B., Couillaud, F. and Baehr, J.C.), pp. 45–50. Institut National de la Recherche Agronomique, Paris.

Varricchio, P., Falabella, P., Sordetti, R., Graziani, F., Malva, C. and Pennacchio, F. (1999) *Cardiochiles nigriceps* polydnavirus: molecular characterization and gene expression in parasitized *Heliothis virescens* larvae. *Insect Biochem. Mol. Biol.* **29:** 1087–1096.

Vasconcelos, S.D.V., Cory, J.S., Wilson, K.R., Sait, S.M. and Hails, R.S. (1996) The effect of baculovirus infection on the mobility of *Mamestra brassicae* L. (Lepidoptera: Noctuidae) larvae at different developmental stages. *Biol. Cont.* **7:** 299–306.

Veval, E.J. (1942) On the bionomics of *Aphidius matricariae* Hal., a braconid parasite of *Myzus persicae* Sulz. *Parasitology* **34:** 141–151.

Vinson, G.P. (1987) The stimulation of steroidogenesis by corticotrophin: the role of the intracellular regulatory peptides and proteins. *J. Endocrinol.* **114:** 163–165.

Vinson, S.B. (1970) Development and possible functions of teratocytes in the host-parasite association. *J. Invert. Pathol.* **16:** 93–101.

Vinson, S.B. (1972a) Factors involved in successful attack on *Heliothis virescens* by the parasitoid *Cardiochiles nigriceps*. *J. Invert. Pathol.* **20:** 118–125.

Vinson, S.B. (1972b) Effect of the parasitoid, *Campoletis sonorensis* on the growth of its host *Heliothis virescens*. *J. Insect Physiol.* **18:** 1509–1514.

Vinson, S.B. (1975) Biochemical coevolution between parasitoids and their hosts. In: *Evolutionary Strategies of Parasitic Insects and Mites* (ed. Price, P.W.), pp. 14–48. Plenum Press, New York.

Vinson, S.B. (1976) Host selection by insect parasitoids. *Annu. Rev. Entomol.* **21:** 109–113.

Vinson, S.B. (1988) Physiological studies of parasitoids reveal new approaches to the biological control of insect pests. *ISI Atlas of Science* **1:** 25–32.

Vinson, S.B. (1990a) How parasitoids deal with the immune system of their host: an overview. *Arch. Insect Biochem. Physiol.* **13:** 3–27.

Vinson, S.B. (1990b) Physiological interactions between the host genus *Heliothis* and its guild of parasitoids. *Arch. Insect Biochem. Physiol.* **13:** 63–81.

Vinson, S.B. (1993a) Interactions between the insect endocrine system and the immune system. In: *Insect Immunity,* Vol. 1 (ed. Pathak, J.P.), pp. 103–112. Oxford & IBH Publishing, New Delhi.

Vinson, S.B. (1993b) Suppression of insect immune systems by parasitic Hymenoptera. In: *Insect Immunity,* Vol. 1 (ed. Pathak, J.P.), pp. 171–182. Kluwer, Dordrecht.

Vinson, S.B. (1994) Parasitoid *in vitro* rearing: success and challenges. In: *Techniques of Insect Rearing for the Development of Integrated Pest and Vector Management Strategies,* Vol. 1 (ed. Ochieng-Odero, J.P.), pp. 49104. ICIPE Science Press, Nairobi.

Vinson, S.B. (1997) Invasion of the Red Imported Fire ant. *Amer. Entomolog.* **43:** 23–39.

Vinson, S.B. (1998) The general host selection behavior of parasitoid Hymenoptera and a comparison of initial strategies utilized by larvaphagous and oophagous species. *Biol. Cont.* **11:** 79–96.

Vinson, S. B. and Barbosa, P. (1987) Interrelationships of nutritional ecology of parasitoids. In. *Nutritional Ecology of Insects, Mites, and Spiders* (eds Stansky, F. and Rodriguez, J.G.), pp. 673–695. John Wiley & Sons, Inc., New York.

Vinson, S.B. and Barras, D.J. (1970) Effects of the parasitoid, *Cardiochiles nigriceps*, on the growth, development, and tissues of *Heliothis virescens*. *J. Insect Physiol.* **16:** 1329–1338.

Vinson, S.B. and Hegazi, E.M. (1998) A possible mechanism for the physiological suppression of conspecific eggs and larvae following superparasitism by solitary endoparasitoids. *J. Insect Physiol.* **44:** 703–712.

Vinson, S.B. and Iwantsch, G.F. (1980a) Host Regulation by Insect Parasitoids. *Q. Rev. Biol.* **55:** 143–165.

Vinson, S.B. and Iwantsch, G.F. (1980b) Host suitability for insect parasitoids. *Annu. Rev. Entomol.* **25:** 397–419.

Vinson, S.B. and Mourad, A.K. (2000) The response of *Cardiochiles nigriceps* (Hymenoptera: Braconidae) larvae to conspecific competitors. In. *Hymenoptera: Evolution, Biodiversity and Biological Control* (eds Austin, A.D. and Dowton, M.), pp. 46–49. CSIRO Publishing, Collingwood, Australia.

Vinson, S.B. and Scott, J.R. (1974) Ultrastructure of teratocytes of *Cardiochiles nigriceps* Viereck (Hymenoptera: Braconidae). *Int. J. Insect Morphol. Embryol.* **3:** 293–304.

Vinson, S.B. and Scott, J.R. (1975) Particles containing DNA associated with the oocyte of an insect parasitoid. *J. Invert. Pathol.* **25:** 375–378.

Vinson, S.B., Edson, K.M. and Stoltz, D.B. (1979) Effect of a virus associated with the reproductive tract of the parasitoid wasp, *Campoletis sonorensis*, on host weight gain. *J. Invert. Pathol.* **34:** 133–137.

Vinson, S.B., Mourad, A.K. and Sebesta, D.K. (1994) Sources of possible regulatory factors in *Cardiochiles nigriceps* (Hymenoptera: Braconidae). *Arch. Insect Biochem. Physiol.* **26:** 197–209.

Visser, B.J., Spanjer, W., de Klonia, H., Piek, T., van der Meer, C. and van der Drift, A.C.M. (1976) Isolation and some biochemical properties of a paralysing toxin from the venom of the wasp *Microbracon hebetor* (Say). *Toxicon* **14:** 357–370.

Visser, B.J., Labruyere, W.T., Spanjer, W. and Piek, T. (1983) Characterization of two paralyzing protein toxins (A-MTX and B-MTX), isolated from a homogenate of the wasp, *Microbracon hebetor* (Say). *Comp. Biochem. Physiol.* **75B:** 523–530.

Volkman, L.E. (1997) Nucleopolyhedrovirus interactions with their insect hosts. *Adv. Virus. Res.* **48:** 313–348.

Voukassovitch, P. (1925) Sur la biologie de *Goniozus claripennis* Forst., parasite d'*Oenophthira pilleriana* Schiff. [Toulouse] *Soc. d'Hist. Nat. Bull.* **52:** 225–246.

Wago, H. and Kitano, H. (1985) Effects of venom from *Apanteles glomeratus* on the hemocytes and hemolymph of *Pieris rapae crucivora. Appl. Ent. Zool.* **20:** 103–110.

Wall, R. and Berberet, R.C. (1974) The life-cycle of *Euplectrus plathypenae*, a gregarious external parasitoid of peanut foliage feeders in Oklahoma. *Environ. Entomol.* **3:** 744–746.

Walther, C. and Reinecke, M. (1983) Block of synaptic vesicle exocytosis without block of Ca^{2+}-influx – An ultrastructural analysis of the paralyzing action of *Habrobracon* venom on locust motor-nerve terminals. *Neurosci.* **9:** 213–224.

Wang, X., Ooi, B.G. and Miller, L.K. (1991) Baculovirus vectors for multiple gene expression and for occluded virus production. *Gene* **100:** 131–137.

Wang, Y. and Telfer, W.H. (1997) cAMP-stimulated termination of vitellogenesis in *Hyalophora cecropia*: formation of a diffusion barrier and the loss of patency. *J. Insect Physiol.* **43:** 675–684.

Wang, Y., Jiang, H. and Kanost, M.R. (1999) Biological activity of *Manduca sexta* paralyic and plasmatocyte spreading peptide and primary structure of its hemolymph precursor. *Insect Biochem. Mol. Biol.* **29:** 1075–1086.

Wang, Z. and Haunerland, N.H. (1991) Ultrastructural study of storage protein granules in fat body of the corn earworm, *Heliothis zea. J. Insect Physiol.* **37:** 353–363.

Wang, Z. and Haunerland, N.H. (1992) Fate of differentiated fat body tissues during metamorphosis of *Helicoverpa zea. J. Insect Physiol.* **38:** 199–213.

Wang, Z. and Haunerland, N.H. (1994) Storage protein uptake in *Helicoverpa zea*: arylphorin and VHDL share a single receptor. *Arch. Insect Biochem. Physiol.* **26:** 15–26.

Wani, M., Yagi, S. and Tanaka, T. (1990) Synergistic effect of venom, calyx and teratocytes of *Apanteles kariyai* on the inhibition of larval pupal ecdysis of the host, *Pseudaletia separata. Entomol. Exp. Appl.* **57:** 101–104.

Warren, J., Steiner, B., Dorn, A., Pak, M. and Gilbert, L.I. (1986) Metabolism of ecdysteroids during the embryogenesis of *Manduca sexta. J. Liquid Chromatog.* **9:** 1759-1782.

Warren, J.T., Sakurai, S., Rountree, D.B., Gilbert, L.I., Lee, S.-S. and Nakanishi, K. (1988) Regulation of the ecdysteroid titer of *Manduca sexta:* reappraisal of the role of the prothoracic glands. *Proc. Natl Acad. Sci. USA* **85:** 958–962.

Washburn, J.O., Haas-Stapleton, E.J., Tan, F.F., Beckage, N.E. and Volkman, L.E. (2000) Co-infection of *Manduca sexta* larvae with polydnavirus from *Cotesia congregata* increases susceptibility to fatal infection by *Autographa californica* M nucleopolyhedrovirus. *J. Insect Physiol.* **46:** 179–190.

Watson, R.D., Ciancio, M.J., Gunnar, W.P., Gilbert, L.I. and Bollenbacher, W.E. (1985) Regulation of insect prothoracic glands: Existence of a haemolymph stimulatory factor in *Manduca sexta. J. Insect Physiol.* **31:** 487–494.

Watson, R.D., Spaziani, E. and Bollenbacher, W.E. (1989) Regulation of ecdysone biosynthesis in Insects and Crustaceans: a comparison. In: *Ecdysone: from Chemistry to Mode of Action* (ed. Koolman, J.), pp. 188–203. George Thieme Verlag, Stuttgart-New York.

Weaver, R.J., Marris, G.C., Olieff, S.M., Mosson, H.J. and Edwards, J.P. (1997) The role of ectoparasitoid venom in the regulation of haemolymph ecdysteroid titres in a noctuid moth. *Arch. Insect Biochem. Physiol.* **35:** 169–178.

Weaver, R.J., Edwards, J.P., Bendena, W. and Tobe, S.S. (1998) Structures, functions and occurrence of insect allatostatic peptides. In: *Recent Advances in Arthropod Endocrinology* (eds Coast, G.M. and Webster, S.G.), pp. 3–32. Cambridge University Press, Cambridge.

Webb, B.A. (1998) Polydnavirus biology, genome structure and evolution. In: *The Insect Viruses.* (eds Miller, L.K. and Ball, A.), pp. 105–139. Plenum Publishing Corporation, New York.

Webb, B.A. and Dahlman, D.L. (1985) Developmental pathology of *Heliothis virescens* larvae parasitized by *Microplitis croceipes* parasite-mediated host developmental arrest. *Arch. Insect Biochem. Physiol.* **2:** 131–143.

Webb, B.A. and Dahlman, D.L. (1986) Ecdysteroid influence on the development of the host *Heliothis virescens* and its endoparasite *Microplitis croceipes. J. Insect Physiol.* **32:** 339–345.

Webb, B.A. and Luckhart, S. (1996) Factors mediating short- and long-term immune suppression in a parasitized insect. *J. Insect Physiol.* **42:** 33–40.

Webb, B.A. and Riddiford, L.M. (1988) Synthesis of two storage proteins during larval development of the tobacco hornworm, *Manduca sexta. Dev. Biol.* **130:** 671–681.

Webb, B.A. and Summers, M.D. (1990) Venom and viral expression products of the endoparasitic wasp *Campoletis sonorensis* share epitopes and related sequences. *Proc. Natl Acad. Sci.* USA **87:** 4961–4965.

Webb, T.J and Hurd, H. (1995a) The use of a monoclonal antibody to detect parasite induced reduction in vitellogenin sequestration by ovaries of *Tenebrio molitor*. *J. Insect Physiol.* **41:** 745–751.

Webb, T.J. and Hurd, H. (1995b) *Hymenolepis diminuta*-induced fecundity reduction in *Tenebrio molitor* may be caused by changes in hormone binding to *Tenebrio molitor* ovaries. *Parasitology* **110:** 565–571.

Webb, T.J. and Hurd, H. (1996) *Hymenolepis diminuta*: metacestode-induced reduction in the synthesis of the yolk protein, vitellogenin, in the fat body of *Tenebrio molitor*. *Parasitology* **110:** 565–571.

Webb, T.J. and Hurd, H. (1999) Direct manipulation of insect reproduction by agents of parasite origin. *Proc. R. Soc. Lond. B* **266:** 1537–1541.

Webster, F.M. and Phillips, W.J. (1912) The spring grain-aphis or "Green Bug". *U.S. Dept. Agr. Bureau Ent. Bull.* **110:** 153.

Webster, J.P. and Macdonald, D.W. (1995) Parasites of wild brown rats (*Rattus norvegicus*) on UK farms. *Parasitology* **111:** 247–255.

Weirich, G.F. (1989) Enzymes involved in ecdysone metabolism. In: *Ecdysone: from Chemistry to Mode of Action* (ed. Koolman, J.), pp. 174–180. Georg-Thieme Verlag, Stuttgart.

Weisel-Eichler, A., Haspel, G. and Libersat, F. (1999) Venom of a parasitoid wasp induces prolonged grooming in the cockroach. *J. Exp. Biol.* **202:** 957–964.

Wharton, R.A. (1993) Bionomics of the Braconidae. *Annu. Rev. Entomol.* **38:** 121–143.

White, D. (1968) Postnatal treatment of the cabbage aphid with a synthetic juvenile hormone. *J. Insect Physiol.* **14:** 901–912.

White, D. and Lamb, R.J. (1968) Effect of a synthetic juvenile hormone on adult cabbage aphids and their progeny. *J. Insect Physiol.* **14:** 395–402.

Whitfield, J.B. (1992a) Phylogeny of the non-aculeate Apocrita and the evolution of parasitism in the Hymenoptera. *J. Hym. Res.* **1:** 3–14.

Whitfield, J.B. (1992b) The polyphyletic evolution of endoparasitism in the cyclostome lineages of Braconidae (Hymenoptera). *Syst. Entomol.* **17:** 273–286.

Whitfield, J.B. (1998) Phylogeny and evolution of host-parasitoid interactions in Hymenoptera. *Annu. Rev. Entomol.* **43:** 129–151.

Whiting, A.R. (1967) The biology of the parasitic wasp *Mormoniella vitripennis* (= *Nasonia vitripennis*) (Walker). *Q. Rev. Biol.* **42:** 333–406.

Wilson, D.D. and Ridgway, R.L. (1975) Morphology, development, and behavior of the immature stages of the parasitoid, *Campoletis sonorensis* (Hymenoptera: Ichneumonidae). *Ann. Entomol. Soc. Amer.* **68:** 191–196.

Wilson, J.W. (1933) The biology of parasites and predators of *Laphygmae exigua* reared during the season of 1932. *Florida Entomol.* **17:** 7–11.

Wilson, K.R., O'Reilly, D.R., Hails, R.S. and Cory, J.S. (2000) Age-related effects of the *Autographa californica* multiple nucleopolyhedrovirus *egt* gene in the cabbage looper (*Trichoplusia ni*). *Biol. Contr.* **19:** 57–63.

Windass, J.D., Duncan, R.E., Baule, V.J. and Christian, P.D. (1996) International Patent – WO 96/16171.

Wing, K.D. (1988) RH-5849, a nonsteroidal ecdysone agonist: effects on a *Drosophila* cell line. *Science* **241:** 467–469.

Wing, K.D., Slawecki, R.A. and Carlson, G.R. (1988) RH-5849, a nonsteroidal ecdysone agonist: effects on larval Lepidoptera. *Science* **241:** 470–472.

Woodhead, A.P., Stay, B., Seidel, S.L., Khan, M.A. and Tobe, S.S. (1989) Primary structure of four allatostatins: neuropeptide inhibitors of juvenile hormone synthesis. *Proc. Natl Acad. Sci. USA* **86:** 5997–6001.

Woodward, D.F., Gary, R.K. Jr., Spada, C.S., Hawley, S.B. and Nieves, A.L (1989) Time-dependent differences in the vasopermeability response to intradermal peptidoleukotrienes. *J. Lipid Mediat.* **1:** 63–73.

Worden, B.D., Parker, P.G. and Pappas, P. (2000) Parasites reduce attractiveness and reproductive success in male grain beetles. *Anim. Behav.* **59:** 543–550.

Wyatt, G.R. (1980) The fat body as a protein factory. In: *Insect Biology in the Future* (eds Locke, M. and Smith, D.S.), pp. 201–225. Academic Press, London.

Wyatt, G.R. and Davey, K.G. (1996) Cellular and molecular actions of juvenile hormone. II Roles of juvenile hormone in adult insects. *Adv. Insect Physiol.* **23:** 1–155.

Wyatt, G.R. and Pan, M.L. (1978) Insect plasma proteins. *Annu. Rev. Biochem.* **47:** 779–817.

Yagi, S. and Tanaka, T. (1992) Retardation of testis development in the armyworm, *Pseudaletia separata,* parasitized by the braconid wasp, *Cotesia kariyai. Invertebr. Reprod. Devel.* **22:** 151–157.

Yamamoto, A.C., Doetzer, A.K. and Foerster, L.A. (1998) Effect of temperature on the development of *Euplectrus ronnai* (Brethes) (Hymenoptera: Eulophidae) parasitizing *Pseudaletia sequax* Franclemont (Lepidoptera: Noctuidae) and impact on food consumption of the host larvae. *Acta Biologica Paranaense* **27:** 85–95.

Yang, B.H., Yang, B.L., Savani, R.C. and Turley, E.A. (1994) Identification of a common hyaluron-binding motif in the hyaluron-binding proteins RHAMM, CD44 and Link protein. *EMBO J.* **13:** 286–296.

Yao, T.P., Forman, B.M., Jiang, Z., Cherbas, L., Chen, J.D., McKeown, M., Cherbas, P. and Evans, R.M. (1993) Functional ecdysone receptor is the product of EcR and *Ultraspiracle* genes. *Nature* **366:** 476–479.

Yoder, J.A., Theriot, G.C. and Rivers, D.B. (1996) Venom from *Nasonia vitripennis* alters water loss from the flesh fly, *Sarcophaga bullata. Entomol. Exp. Appl.* **81:** 235–238.

Zhang, D. and Dahlman, D.L. (1989) *Microplitis croceipes* teratocytes cause developmental arrest of *Heliothis virescens* larvae. *Arch. Insect Biochem. Physiol.* **12:** 51–61.

Zhang, D., Dahlman, D.L. and Gelman, D.B. (1992) Juvenile hormone esterase activity and ecdysteroid titer in *Heliothis virescens* larvae injected with *Microplitis croceipes* teratocytes. *Arch. Insect Biochem. Physiol.* **20:** 231–242.

Zaviezo, T. and Mills, N. (1999) Aspects of the biology of *Hyssopus pallidus* (Hymenoptera: Eulophidae), a parasitoid of the codling moth (Lepidoptera: Olethreutidae). *Environ. Entomol.* **28:** 748–754.

Zimin, Y.S. (1930) On the biology of *Euplectrus bicolor* Swed. as a parasite of noctuid larvae. *Bull. N. Caucas. Pl. Prot. Stn*. **vi-vii:** 99–106.

Zimowska, G., Rembold, H. and Bayer, G. (1989) Juvenile hormone identification, titer, and degradation during the last larval stadium of *Spodoptera littoralis. Arch. Insect Biochem. Physiol.* **12:** 1–14.

Zitnan, D., Kingan, T.G., Kramer, S.J. and Beckage, N.E. (1995a) Accumulation of neuropeptides in the cerebral neurosecretory system of *Manduca sexta* larvae parasitized by the braconid wasp *Cotesia congregata. J. Comp. Neurol.* **356:** 83–100.

Zitnan, D., Kingan, T.G. and Beckage, N.E. (1995b) Parasitism-induced accumulation of FMRFamide-like peptides in the gut innervation and endocrine cells of *Manduca sexta. Insect Biochem. Mol. Biol.* **25:** 669–678.

Zitnan, D., Kingan, T.G., Hermesman, J.L. and Adams, M.E. (1996) Identification of ecdysis-triggering hormone from an epitracheal endocrine system. *Science* **271:** 88–91.

Zitnan, D., Ross, L.S., Zitnanova, I., Hermesman, J.L., Gill, S.S., and Adams, M.E. (1999) Steroid induction of a peptide hormone gene leads to orchestration of a defined behavioral sequence. *Neuron* **23:** 523–535.

Index